T0135113

Lecture Notes in Computer Science 14662

Founding Editors

Gerhard Goos
Juris Hartmanis

The series Lecture Notes in Computer Science (LNCS), including its subseries Lecture Notes in Artificial Intelligence (LNAI) and Lecture Notes in Bioinformatics (LNBI), has established itself as a medium for the publication of new developments in computer science and information technology research, teaching, and education.

LNCS enjoys close cooperation with the computer science R & D community, the series counts many renowned academics among its volume editors and paper authors, and collaborates with prestigious societies. Its mission is to serve this international community by providing an invaluable service, mainly focused on the publication of conference and workshop proceedings and postproceedings. LNCS commenced publication in 1973.

Yuval Emek

Editor

Structural Information and Communication Complexity

31st International Colloquium, SIROCCO 2024
Vietri sul Mare, Italy, May 27–29, 2024
Proceedings

 Springer

Editor
Yuval Emek
Faculty of Data and Decision Sciences
Technion – Israel Institute of Technology
Haifa, Israel

ISSN 0302-9743   ISSN 1611-3349 (electronic)
Lecture Notes in Computer Science
ISBN 978-3-031-60602-1   ISBN 978-3-031-60603-8 (eBook)
https://doi.org/10.1007/978-3-031-60603-8

This Springer imprint is published by the registered company Springer Nature Switzerland AG
The registered company address is: Gewerbestrasse 11, 6330 Cham, Switzerland

If disposing of this product, please recycle the paper.

Preface

This volume contains the papers presented at SIROCCO 2024: 31st International Colloquium On Structural Information and Communication Complexity held on May 27–29, 2024, in the enchanting setting of Vietri sul Mare, Salerno, Italy.

SIROCCO is devoted to the study of the interplay between structural knowledge, communication, and computing in decentralized systems of multiple communicating entities. Special emphasis is given to innovative approaches leading to better understanding of the relationship between computing and communication.

This year, SIROCCO received 49 submissions. Each submission was reviewed by at least 3 program committee members with the help of expert subreviewers. The committee decided to accept 24 submissions as regular papers and 6 as brief announcements. In addition to the contributed regular papers and brief announcements, this volume also includes an invited paper from Giovanni Viglietta, one of the keynote speakers. Of the regular papers, "Stability of P2P Networks Under Greedy Peering" by Lucianna Kiffer and Rajmohan Rajaraman received the Best Paper Award and "Near-Optimal Fault Tolerance for Efficient Batch Matrix Multiplication via an Additive Combinatorics Lens" by Keren Censor-Hillel, Yuka Machino, and Pedro Soto received the Best Student Paper Award.

I would like to thank the keynote speakers Shiri Chechik (Tel Aviv University, Israel), Fabian Kuhn (University of Freiburg, Germany), and Giovanni Viglietta (University of Aizu, Japan) for their insightful talks, as well as Shay Kutten (Technion - Israel Institute of Technology, Israel) for his featured talk as the recipient of the 2024 SIROCCO Innovation in Distributed Computing Prize.

I would also like to thank the authors who submitted their work to SIROCCO this year and the program committee members and subreviewers for their valuable reviews and discussions. The SIROCCO steering committee, chaired by Keren Censor-Hillel, provided help and guidance throughout the process. The EasyChair system was used to handle the submission of papers and to manage the review process. Last but not least, I am grateful to the local arrangements chair Gianluca De Marco who made the conference possible together with his local organization team.

May 2024   Yuval Emek

Organization

Program Committee Chair

Yuval Emek   Technion - Israel Institute of Technology, Israel

Program Committee

Karine Altisen	Université Grenoble Alpes, France
Shantanu Das	Aix-Marseille University, France
Peter Davies	Durham University, UK
Michal Dory	University of Haifa, Israel
Fabien Dufoulon	Lancaster University, UK
Orr Fischer	Weizmann Institute of Science, Israel
Seth Gilbert	National University of Singapore, Singapore
William K. Moses Jr.	Durham University, UK
Amos Korman	CNRS - FILOFOCS and University of Haifa, Israel
Irina Kostitsyna	KBR at NASA Ames Research Center, USA
Frederik Mallmann-Trenn	King's College London, UK
Yannic Maus	TU Graz, Austria
Darya Melnyk	TU Berlin, Germany
Alexandre Nolin	CISPA Helmholtz Center for Information Security, Germany
Debasish Pattanayak	Carleton University, Canada
Ami Paz	LISN - CNRS & Paris Saclay University, France
Mor Perry	The Academic College of Tel-Aviv-Yaffo, Israel
Mikaël Rabie	IRIF - Université Paris Cité, France
Peter Robinson	Augusta University, USA
Will Rosenbaum	Amherst College, USA
Joel Rybicki	Humboldt University of Berlin, Germany
Jared Saia	University of New Mexico, USA
Volker Turau	Hamburg University of Technology, Germany
Jennifer Welch	Texas A&M University, USA
Prudence Wong	University of Liverpool, UK
Yukiko Yamauchi	Kyushu University, Japan

Organizing Committee

Gianluca De Marco	Università di Salerno, Italy
Rocco Zaccagnino	Università di Salerno, Italy
Rosalba Zizza	Università di Salerno, Italy

Steering Committee

Keren Censor-Hillel	Technion - Israel Institute of Technology, Israel
Paola Flocchini	University of Ottawa, Canada
Tomasz Jurdziński	University of Wroclaw, Poland
Merav Parter	Weizmann Institute of Science, Israel
Sergio Rajsbaum	Universidad Nacional Autonoma de Mexico, Mexico
Stefan Schmid	TU Berlin, Germany

Additional Reviewers

Adhikari, Ramesh
Akhoondian Amiri, Saeed
Aspnes, James
Attiya, Hagit
Binsky, Hadar
Bourreau, Yann
Bramas, Quentin
Chalopin, Jérémie
Cournier, Alain
Daymude, Joshua
Defalque, Geoffrey
Devismes, Stéphane
Fritsch, Robin
Gloyd, James
Jakob, Manuel

Khoury, Seri
Kim, Yonghwan
Kokkou, Maria
Meir, Uri
Narayanan, Ananth
Nguyen, Minh Hang
Pai, Shreyas
Petit, Franck
Saia, Jared
Scheideler, Christian
Shibata, Masahiro
Tan, Ming Ming
Vacus, Robin
Zamaraev, Viktor

SIROCCO Prize for Innovation in Distributed Computing (Invited Talk)

Laudatio for Shay Kutten

We are pleased to announce that the 2024 Prize for Innovation in Distributed Computing is awarded to Prof. Shay Kutten from the Technion - Israel Institute of Technology. Shay is a distinguished figure in the realm of distributed computing, and his significant contributions to the field, particularly in the area of proof labeling schemes, underscore his expertise and influence within the SIROCCO community.

Inspired by the P versus NP paradigm, the concept of proof labeling schemes represents an abstract proof system designed to verify graph properties in a distributed manner. Shay's seminal work which appeared in PODC 2005 [3] (see [4] for the journal version), in collaboration with Amos Korman and David Peleg, introduced and formalized this concept, laying the foundation for further exploration and research. Notably, Shay's earlier collaborations with Yehuda Afek and Moti Yung [1, 2] implicitly paved the way for the development of proof labeling schemes, highlighting his longstanding dedication to advancing the field. Since its formalization in 2005, the concept of proof labeling schemes has attracted considerable attention from a large number of researchers, leading to the publication of hundreds of papers. The study of proof labeling schemes has also influenced many other concepts in the domain of distributed proof systems to the extent that it can be seen as the main driving force of this research domain.

Throughout his career, Shay has continued to delve into the study of proof labeling schemes, publishing numerous papers dedicated to this topic and closely related areas. In particular, his contributions have been instrumental in applying proof labeling schemes and local verification techniques to the design of self-stabilizing algorithms, thereby propelling the development of self-stabilizing systems.

For his pioneering advancements in proof labeling schemes and their application to self-stabilizing systems, we are proud to present the 2024 Prize for Innovation in Distributed Computing to Prof. Shay Kutten.

The 2024 Award committee:

Pierre Fraigniaud (IRIF and CNRS, France)
Magnús M. Halldórsson (Reykjavik University, Iceland)
Tomasz Jurdziński (University of Wroclaw, Poland)
Merav Parter (Weizmann Institute, Israel)
Andréa W. Richa (Arizona State University, USA)
Christian Scheideler (University of Paderborn, Germany)
Stefan Schmid (Technical University of Berlin, Germany)

References

1. Afek, Y., Kutten, S., Yung, M.: Memory-efficient self stabilizing protocols for general networks. In: van Leeuwen, J., Santoro, N. (eds.) Distributed Algorithms. WDAG 1990. LNCS, vol. 486, pp. 15–28. Springer, Heidelberg (1991). https://doi.org/10.1007/3-540-54099-7_2
2. Afek, Y., Kutten, S., Yung, M.: The local detection paradigm and its application to self-stabilization. Theor. Comput. Sci. **186**(1-2), 199–229 (1997)
3. Korman, A., Kutten, S., Peleg, D.: Proof labeling schemes. In: Aguilera, M.K., Aspnes, J. (eds.) Proceedings of the Twenty-Fourth Annual ACM Symposium on Principles of Distributed Computing, PODC 2005, Las Vegas, NV, USA, 17–20 July 2005, pp. 9–18. ACM (2005)
4. Korman, A., Kutten, S., Peleg, D.: Proof labeling schemes. Distrib. Comput. **22**(4), 215–233 (2010)

Contents

Brief Announcements

Invited Paper

History Trees and Their Applications

Giovanni Viglietta[✉]

University of Aizu, Aizuwakamatsu, Japan
viglietta@gmail.com
https://giovanniviglietta.com

Abstract. In the theoretical study of distributed communication networks, *history trees* are a discrete structure that naturally models the concept that anonymous agents become distinguishable upon receiving different sets of messages from neighboring agents. By conveniently organizing temporal information in a systematic manner, history trees have been instrumental in the development of optimal deterministic algorithms for networks that are both anonymous and dynamically evolving.

This note provides an accessible introduction to history trees, drawing comparisons with more traditional structures found in existing literature and reviewing the latest advancements in the applications of history trees, especially within dynamic networks. Furthermore, it expands the theoretical framework of history trees in new directions, also highlighting several open problems for further investigation.

Keywords: history tree · anonymous network · dynamic network

1 Introduction

A *distributed communication network* consists of a finite number of independent computational units known as *agents*, which can send each other messages and modify their internal states based on the messages they receive. These networks generally operate in synchronous or asynchronous *communication steps*, where agents send messages through the links of a (multi)graph, which may be static or dynamic, directed or undirected, connected or disconnected, etc.[1]

A network is *anonymous* if its agents are initially indistinguishable, i.e., they lack unique identifiers and can only be told apart by external input or due to the network's layout. In such networks, it is typically assumed that all agents run the same local *deterministic* algorithm. It is important to note that allowing for randomness would enable agents to generate unique identifiers with high probability, thereby compromising the study's focus on anonymity.

A common algorithmic approach in this setting involves assigning a "weight" to each agent, which is then distributed among neighbors at each step according to specific rules. This process is analyzed using standard stochastic methods to understand how these weights eventually converge to a common value [6,13].

---

[1] Although this note does not focus on *population protocols*, they can also be modeled within this framework as dynamic communication networks where finite-state agents interact in pairs consisting of an initiator and a responder having different roles.

Y. Emek (Ed.): SIROCCO 2024, LNCS 14662, pp. 3–23, 2024.
https://doi.org/10.1007/978-3-031-60603-8_1

Although this "averaging technique" enabled the development of general algorithms for anonymous networks, one of its major limitations is its disregard for the network's structural and topological characteristics, providing minimal insight into these aspects. Furthermore, analyzing these algorithms tends to be technically cumbersome, yielding only asymptotic estimates for how quickly the network reaches convergence or stability.

Another line of research, initiated by Angluin [1], explores discrete and algebraic structures, such as the *views* of Yamashita–Kameda [17] and the *graph fibrations* and *minimum bases* of Boldi–Vigna [4]. This approach allows algorithms to fully leverage the network's structure and usually permits a more precise analysis of running times. However, the theoretical frameworks of views and graph fibrations were developed for networks with unchanging topologies, and their successful application has been limited to such static networks.

History trees are a newer data structure that inherently includes a temporal dimension and was specifically designed for networks with dynamic topologies. The introduction of history trees has recently led to the development of optimal linear-time algorithms for anonymous dynamic networks [9,11] and state-of-the-art general algorithms for congested anonymous dynamic networks [10].

In Sect. 2, we outline the basic architecture of history trees, drawing comparisons with Yamashita–Kameda's views and Boldi–Vigna's minimum bases. Additionally, we showcase a straightforward linear-time algorithm for dynamic networks that illustrates the practicality of history trees. In Sect. 3, we discuss more advanced applications of history trees in challenging network situations, while also pointing out a number of areas that remain open for investigation.

2   Basic Structure and Algorithms

In this section we focus on anonymous networks operating in synchronous steps, modeled as undirected dynamic multigraphs with no port awareness. Other network models will be discussed in Sect. 3.

Throughout this section, the reader may find it beneficial to examine the software available at https://github.com/viglietta/Dynamic-Networks. The repository includes a simulator that supports the creation and visualization of history trees of user-defined dynamic networks and the testing of fundamental algorithms.

2.1   History Trees and Construction of a View

Consider a dynamic network of six anonymous agents whose first four communications steps are as in Fig. 1. Before the first step (i.e., at time $t = 0$), each agent only knows its own input, which in our example is represented by a color: either cyan or yellow. In the first communication step, both cyan agents (which are indistinguishable) receive three messages from yellow agents (also indistinguishable). Thus, the cyan agents remain indistinguishable at time $t = 1$.

However, two yellow agents receive two messages from cyan agents, but the other two receive only one message. Thus, at time $t = 1$, the two yellow agents

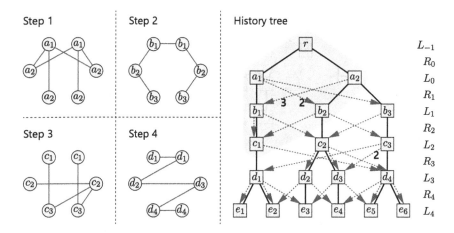

Fig. 1. The first communication steps of a dynamic network and the corresponding levels of its history tree. Initial inputs are represented by agents' colors (cyan and yellow). Labels on agents and nodes have been added for the reader's convenience and indicate classes of indistinguishable agents. The portion of the history tree with a green background is the view of the two agents labeled c_1 after two communication steps. (Color figure online)

labeled b_2 are still indistinguishable from each other, but are distinguishable from the ones labeled b_3, because they have a different "history".

In general, since the network is synchronous, at step $t > 0$ (i.e., between time $t - 1$ and time t) all agents simultaneously send messages and simultaneously receive them (as unordered multisets). Two messages are assumed to be identical if and only if they are sent by indistinguishable agents (possibly by the same agent). By definition, at time $t = -1$ no two agents are distinguishable; at time $t = 0$ they are distinguishable if and only if they have different inputs. At time $t > 0$, two agents are distinguishable if and only if they are distinguishable at time $t - 1$ or receive different multisets of messages at step t.

History trees were introduced in [9] to study how and when anonymous agents in a (dynamic) network become distinguishable. The history tree $\mathcal{H}_G$ of a network G is an infinite tree subdivided into *levels*, as in Fig. 1, where the nodes in level L_t represent the equivalence classes of agents that are indistinguishable at time t. The number of agents in the class represented by a node v is called the *anonymity* of v and is denoted as $\mathbf{a}(v)$. In our example, $\mathbf{a}(a_1) = 2$ and $\mathbf{a}(a_2) = 4$.

The root r represents all agents in the network. In general, the *children* of a node $v \in L_{t-1}$, i.e., the nodes $v_1, v_2, \ldots, v_k \in L_t$ that are connected to v by a *black edge*, represent a partition of the set of agents represented by v. Thus,

$$\mathbf{a}(v) = \sum_{i=1}^{k} \mathbf{a}(v_i). \tag{1}$$

On the other hand, the *red edges* in R_t represent messages sent and received at step t. That is, a directed red edge (v, u) with *multiplicity* m, where $v \in L_{t-1}$

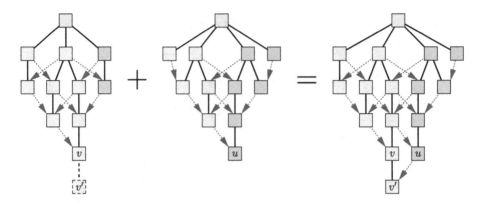

Fig. 2. Updating the view of an agent represented by node v after it receives the view of an agent represented by u as a message. The two views are matched and merged starting from the roots, v gets a child v', and a new red edge from u to v' is added.

and $u \in L_t$, indicates that, at step t, each agent represented by node u receives exactly m (identical) messages from agents represented by v.

Assuming that their internal memory and message sizes are unbounded, the agents in a network G can locally construct portions of the history tree called *views*. More precisely, if an agent p at time t is represented by a node $v \in L_t$, then the view $\mathcal{V}_G^t(p)$ is defined as the portion of $\mathcal{H}_G$ spanned by all directed paths from the root r to v, using black and red edges indifferently (black edges are assumed to be directed away from the root r). The node v is called the *bottom* of the view. Figure 1 shows an example of a view within a history tree.

The distributed construction of views can be achieved via an iterative process, assuming that all agents send their current view to all their neighbors at every step. Upon receiving a multiset M of messages, an agent p integrates its current view $\mathcal{V}_G^t(p)$ with all views in M. This is done by a straightforward match-and-merge algorithm, starting from the root and proceeding level by level, as exemplified in Fig. 2. The result is the smallest tree containing $\mathcal{V}_G^t(p)$ and all views in M as induced subtrees. To obtain $\mathcal{V}_G^{t+1}(p)$, the agent p creates a child v' for the bottom node v of $\mathcal{V}_G^t(p)$ and connects the bottom nodes of all views in M to v' by red edges with the appropriate multiplicities. The node v' is now the bottom of $\mathcal{V}_G^{t+1}(p)$ and represents p (and possibly other agents) at time $t + 1$.

It is a simple observation that the view $\mathcal{V}_G^t(p)$ contains all the information that p can possibly extract from the network G after t communication steps [9].

2.2   Related Structures for Static Networks

We will now focus on *static* anonymous networks, which are well understood thanks to the works of Yamashita–Kameda, Boldi–Vigna, and other authors.

If G is a static network of n agents, once the classes of indistinguishable agents remain unchanged for a single communication step, they must remain unchanged forever. Thus, the number of nodes in the levels of $\mathcal{H}_G$ strictly increases at every level until it stabilizes, say, at level L_s. Therefore, if $|L_s| = \widehat{n}$, we have $0 \leq s < \widehat{n} \leq n$, because $|L_0| \geq 1$ and each node represents at least one agent.

We come to the conclusion that in any network of size n, if two agents are indistinguishable at time $t = n - 1$, they will remain indistinguishable thereafter. This observation was first made by Norris (though formulated differently) in [14], where she also raised the question of whether the same applies to agents in two distinct networks of the same size n.[2] We restate Norris' question below.

Open Problem 1. Let G and G' be two disjoint static networks of n agents each. If an agent in G and an agent in G' have isomorphic views at time $t = n - 1$, do they have isomorphic views at all times?

Let us consider the subgraph of $\mathcal{H}_G$ induced by the nodes in levels L_s and L_{s+1}, as highlighted in Fig. 3 with a green background. Contracting the black edges in this subgraph, we obtain a directed multigraph named $\widehat{G}$.

This is an important structure that has been given many names and equivalent definitions in the literature. Essentially, $\widehat{G}$ could be defined as the smallest directed multigraph that gives rise to a history tree isomorphic to $\mathcal{H}_G$. Yamashita and Kameda call $\widehat{G}$ the *quotient graph* of G and define it in graph-theoretical terms using their own notion of *view* of a static network [16,17], whereas Boldi and Vigna call $\widehat{G}$ the *minimum base* of G and give it a topological definition in terms of *graph fibrations* [2–5].[3]

Notably, since the results concerning static networks from these authors are presented with reference to $\widehat{G}$, they can be directly rephrased in the language of history trees. The advantage of using history trees lies in their ability to readily offer timing information on when agents become distinguishable. This feature makes history trees particularly suitable for dynamic networks, where $\widehat{G}$ is not well defined and the lack of topological regularity prevents the straightforward deduction of temporal data.

Before delving into applications of history trees to dynamic networks in Sect. 2.3, let us draw a closer comparison between views of history trees and views in the sense of Yamashita–Kameda [17]. While it is important not to confuse these two concepts, they bear a deep relationship illustrated in Fig. 3.

For Yamashita and Kameda, the view $\mathcal{T}_G(p)$ of an agent p in a static network G is an infinite tree rooted at p, where each node of depth k represents a distinct

---

[2] By treating the disjoint union of the two networks as a single network of size $2n$, it becomes evident that agents that are indistinguishable at time $t = 2n - 1$ are indistinguishable forever. This point was already noted by Conway in [8, Chapter 1, Theorem 7], albeit phrased in terms of Moore machines. Although $t = 2n - 1$ might not be an optimal bound if both networks have size n (hence Norris' question), Fig. 4 implies that $t = 2n$ is optimal if they have sizes n and $n + 1$, respectively. Indeed, if the two cyan agents are colored yellow, they become distinguishable after $2n$ steps, matching the upper bound given by the combined number of agents, $2n + 1$.

[3] More specifically, G is fibered onto the fibration-prime graph $\widehat{G}$, and the fibers are precisely the classes of agents represented by the nodes of L_s in the history tree $\mathcal{H}_G$.

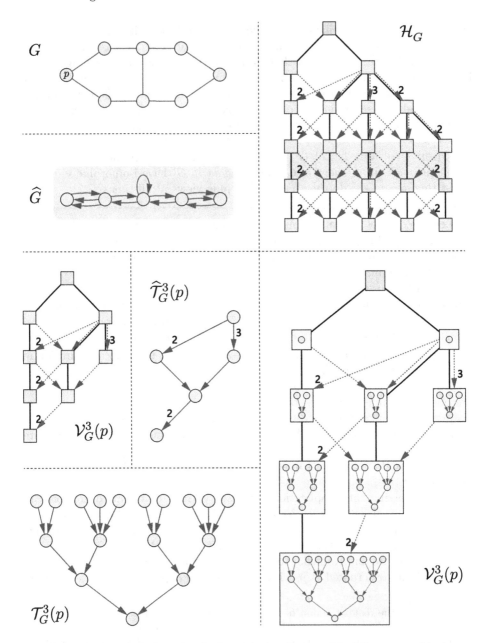

Fig. 3. A static network G with a distinguished agent p and its history tree $\mathcal{H}_G$ (truncated at level L_4). At stabilization, the directed graph of the red edges between non-branching levels (highlighted in green in $\mathcal{H}_G$) is isomorphic to the Boldi–Vigna minimum base $\widehat{G}$. The view $\mathcal{V}_G^3(p)$ after step 3 is related to the Yamashita–Kameda view $\mathcal{T}_G^3(p)$ truncated at depth 3, as well as to its Tani folded view $\widehat{\mathcal{T}}_G^3(p)$, as illustrated. The precise correspondence between these structures is described in Sect. 2.2. (Color figure online)

walk of length k in G terminating in p. As it turns out, $\mathcal{T}_G(p)$ truncated at depth k, denoted as $\mathcal{T}_G^k(p)$, contains precisely the information encoded in $\mathcal{V}_G^k(p)$.

Indeed, starting from $\mathcal{V}_G^k(p)$, we can construct $\mathcal{T}_G^k(p)$ as follows. Let us consider the subgraph spanned by the directed paths in $\mathcal{V}_G^k(p)$ terminating in the bottom node using only red edges. Such a subgraph is a directed acyclic graph isomorphic to the so-called *folded view* $\widehat{\mathcal{T}}_G^k(p)$. This is a structure devised by Tani in [15] to compress $\mathcal{T}_G^k(p)$ from exponential size to polynomial size by conflating equivalent nodes. Finally, $\widehat{\mathcal{T}}_G^k(p)$ can easily be unraveled to reconstruct $\mathcal{T}_G^k(p)$.

Conversely, starting from $\mathcal{T}_G^k(p)$, one can construct an inventory of "fragments" by listing all subtrees hanging from different nodes and truncating them at all possible depths. The isomorphism classes of these fragments, when ordered by their height, correspond to the levels of $\mathcal{V}_G^k(p)$, with the "empty fragment" being the root. Then, the parent of a fragment f in $\mathcal{V}_G^k(p)$ is determined by deleting all the leaves of f. Similarly, the fragments of $\mathcal{V}_G^k(p)$ that are connected to f via red edges (and the multiplicities of such red edges) match the subtrees that dangle from the root's children within f. Thus, we can construct $\mathcal{V}_G^k(p)$ from $\mathcal{T}_G^k(p)$.

2.3 Basic Applications to Dynamic Networks

We will now describe a general algorithmic technique that can be used to solve a wide range of fundamental problems in networks operating in synchronous steps, modeled as *dynamic* undirected multigraphs that are connected at every step. Albeit being extremely straightforward, this technique achieves optimal running times and matches in efficiency the best algorithms for static networks.

As argued in Sect. 2.1, the agents in a network have a distributed algorithm for constructing their view of the history tree and update it at every communication step. Since an agent's view encodes all the information that the agent can infer from the network, in principle we can reduce any problem about anonymous networks to a problem about views of history trees.

Of course, this is true assuming that all agents have enough internal memory and can send large-enough messages to store their views. After t steps, the size of a view is $O(tn^2 \log M)$ bits, where n is the total number of agents and M is the maximum number of messages sent by any agent in a single step. Indeed, such a view has t levels, each of which contains at most n nodes and n^2 incoming red edges of multiplicity at most M. In particular, if the view construction algorithm runs for a polynomial number of steps, it requires internal memory and messages of polynomial size. For the time being, we will assume that this is not an issue.

The basic technique relies on the following observation. If two nodes u and v in the same level L_i of the history tree are *non-branching*, i.e., they have a unique child u' and v' respectively, and there are red edges (u, v') with multiplicity $m_{u,v'} > 0$ and (v, u') with multiplicity $m_{v,u'} > 0$, then we can count the number of messages exchanged by the agents represented by u and by v as

$$m_{v,u'}\, \mathbf{a}(u) = m_{u,v'}\, \mathbf{a}(v) \tag{2}$$

because communication links are bidirectional. Thus, since we know the multiplicities of the red edges, we can infer the ratio of the anonymities involved.

Now, if L_i is a level where all nodes are non-branching, and since the network is connected at every step by assumption, we can repeatedly apply Eq. (2) to compute the ratios of the anonymities of all nodes in L_i. For example, since level L_1 in Fig. 1 is non-branching, we can easily deduce that $\mathbf{a}(b_1) = \mathbf{a}(b_2) = \mathbf{a}(b_3)$. Then, using Eq. (1), we can extend the computation to all previous levels, and obtain for instance that $2\mathbf{a}(a_1) = \mathbf{a}(a_2)$. The reader may apply the same method to the history tree in Fig. 3 to deduce that the ratio between the number of yellow agents and the number of cyan agents is 7.

Once we have this information, we can solve problems such as *Average Consensus*: if the agents are given input numbers, we can compute the fraction of agents that have each input and use it as a weight to compute the mean input.

If we are given additional information, such as the total number of agents or the number of agents that have a certain input, we can also compute how many agents have each input. For instance, in Fig. 1, if we know that there are two cyan agents, we deduce that there are four yellow ones, because $2\mathbf{a}(a_1) = \mathbf{a}(a_2)$.

In particular, if we know that the network contains a unique distinguished agent, typically referred to as a *leader*, then we can solve the *Counting problem*, which asks to compute the total number of agents, n. Thus, in Fig. 3, if we know that p is the unique leader, we can deduce from the history tree that $n = 8$.

Let us discuss the efficiency of this method. Since $|L_{-1}| = 1$ and the number of nodes per level is at most n, it follows that it is sufficient to inspect the history tree up to level L_{n-1} to find a non-branching level and the relative red edges.

A subtler issue is how long it takes before the views of all agents acquire all nodes (and hence all edges) up to L_{n-1}, so that all agents can carry out correct computations locally. In Fig. 1, for instance, since level L_0 is non-branching in the highlighted view (because node b_3 is not in the view), at time $t = 2$ the cyan agents may be deceived and incorrectly deduce that $3\mathbf{a}(a_1) = 2\mathbf{a}(a_2)$.

Recall that the network is assumed to be connected at all steps. So, it takes fewer than n steps for information to travel from any agent to any other agent; in other words, the *dynamic diameter* d is at most $n - 1$. Indeed, if $k < n$ agents have some information at time t, then at least $k + 1$ agents have it at time $t + 1$.

Hence, it takes at most d steps for a node in the history tree to appear in the views of all agents. In particular, at time $n + d - 1 \leq 2n - 2$, all agents have the entirety of L_{n-1} in their views, and thus have enough information to do correct computations. For example, in Fig. 3, knowing all levels of $\mathcal{H}_G$ up to L_3 is enough, but these levels entirely appear in the view of p only at step 7. In fact, the diameter of the network G is $d = 4$.

This technique yields an algorithm for Average Consensus that *stabilizes* in $2n - 2$ steps. That is, if all agents attempt to compute the average of the input values after every step, they may guess it incorrectly for some time, but will be all correct after $2n - 2$ steps at the latest. The same upper bound holds for the Counting problem, assuming the presence of a unique leader in the network.

A downside of this method is that it provides no certificate of correctness, and therefore the algorithm is supposed to run indefinitely. However, if n is known

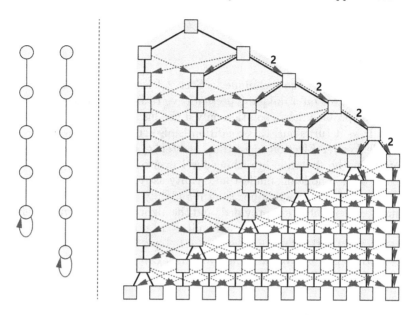

Fig. 4. A static network with two connected components of sizes n and $n + 1$. The history tree shows that the two agents in cyan become distinguishable after $2n - 1$ communication steps. Therefore, in static networks of n agents with a unique leader, no Counting algorithm can terminate in $2n - 2$ steps or stabilize in $2n - 4$ steps. (Color figure online)

to the agents, they can simply count $2n - 2$ steps, do all computations on the first $n - 1$ levels, and then terminate returning the correct output. Similarly, if an agent knows the dynamic diameter d, it can terminate at time t if its view contains a non-branching level L_i with $i < t - d$, because all levels up to L_{i+1} are guaranteed to be entirely in the view. This occurs by the time $t = n + d - 1$.

Finally, let us discuss lower bounds. Figure 4 shows two networks of sizes n and $n + 1$ whose leaders have the same view up to step $2n - 2$. Since agents with equal views must return equal outputs (assuming they execute the same algorithm), it easily follows that there is no Counting algorithm that terminates in $2n - 2$ steps or stabilizes in $2n - 4$ steps in both networks. For if there were such an algorithm, then both leaders would return the same output, and at least one of them would be incorrect. Note that this almost matches the running time of the stabilizing Counting algorithm given above, which is $2n - 2$ steps. This leaves only a small gap: namely, whether $2n - 3$ steps are actually sufficient.

> **Open Problem 2.** Can a Counting algorithm stabilize in $2n - 3$ steps in all connected undirected dynamic networks with a unique leader?

As for Average Consensus, if the two cyan agents in Fig. 4 are colored yellow, then they have the same view up to step $2n - 1$. This implies that no Average Consensus algorithm can stabilize in $2n - 3$ steps in all networks, and therefore the one described above, which stabilizes in $2n - 2$ steps, is optimal.

3   Variations and Extensions

3.1   Leader Election

Another application of the technique in Sect. 2.3 is *Leader Election*, where all agents have to agree on a unique representative to be identified as the "leader". Of course, this problem has a solution only if the history tree contains a node of anonymity 1; as it turns out, this condition is also sufficient.

A simple Leader Election algorithm, at step t, computes the ratios of all anonymities in the first non-branching level occurring after $L_{\lfloor t/2 \rfloor}$ and deterministically picks a node of smallest anonymity as representing the leader. This strategy eventually elects a unique leader if and only if it is possible to do so, but no certificate of correctness is provided, and the algorithm has to run indefinitely.

However, if n is known, then anonymities in non-branching levels can be computed exactly with a known delay of $n-1$ steps. In this case, any agent that becomes distinguishable from all others at time t can be identified with certainty by time $t + 2n - 2$, allowing all agents to terminate.

3.2   Terminating Counting

In Sect. 2.3 we gave a Counting algorithm for connected undirected dynamic networks with a unique leader that stabilizes in $2n - 2$ steps. At the cost of n additional communication steps, we can implement a correctness certificate and make all agents return n and terminate. The algorithm's details are somewhat intricate and can be found in [9]; here we will merely sketch the main ideas.

If we know the anonymities of a node u and of all its children, then u is called a *guesser*. If there is a red edge directed from u to a node v of unknown anonymity, we can write an equation similar to Eq. (2) to count messages exchanged by the corresponding agents. Solving this equation yields a *guess* $\mathbf{g}(v)$ on $\mathbf{a}(v)$ in terms of known anonymities and multiplicities of red edges. As it turns out, $\mathbf{g}(v) \geq \mathbf{a}(v)$ always holds; moreover, if v has no siblings, $\mathbf{g}(v) = \mathbf{a}(v)$.

Clearly, the nodes representing the leader are always guessers, because their anonymity is 1. Using these nodes, we can make initial guesses on at least one node per level. The question is how we can determine which guesses are correct without knowing the whole history tree, but only a view of it.

Let us define the *weight* $\mathbf{w}(v)$ of a node v as the number of guesses that have been made on nodes in the subtree hanging from v (including on v itself). Then, v is said to be *heavy* if $\mathbf{w}(v) \geq \mathbf{g}(v)$. The key observation is that, if guesses are *well spread*, i.e., no two sibling nodes are assigned a guess, then the deepest heavy node necessarily has a correct guess.

Hence, in Fig. 5, the nodes in (a)–(d) with a blue number have a correct guess, but the one in (e) may not. For (a), this is obvious: since $\mathbf{g}(v) = 1$ and $\mathbf{g}(v) \geq \mathbf{a}(v)$, we must conclude that $\mathbf{a}(v) = 1$ and the guess is correct. As for (b), if the lower node u has a correct guess, then the upper node v has a correct guess too, because $2 = \mathbf{g}(v) \geq \mathbf{a}(v) \geq \mathbf{a}(u) = \mathbf{g}(u) = 2$, hence $\mathbf{g}(v) = \mathbf{a}(v)$. Otherwise, the guess on u is incorrect, and so u must have a sibling (perhaps not in the

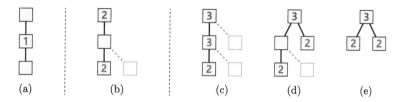

Fig. 5. The numbers in the nodes are guesses made according to the algorithm in Sect. 3.2. Blue numbers indicate heavy nodes; their guesses are necessarily correct except for the one in (e), where guesses are not well spread. (Color figure online)

view). Hence, $2 = \mathbf{g}(v) \geq \mathbf{a}(v) \geq 2$, and the guess on v is correct. The reader may verify that a similar line of reasoning applies to (c) and (d) but not to (e).

It is not difficult to see that, if there are at least n well-spread nodes with guesses, then some of them must be heavy, and thus the deepest of them is necessarily correct. In this way, we can steadily determine the anonymities of new nodes. Eventually, some of these nodes become guessers themselves and can be used to make new guesses, and so on. Since the network is connected at all steps, it can be shown that only $2n - 2$ levels of the history tree are required for this algorithm to eventually identify nodes with correct guesses on all branches.

Adding up the anonymities of nodes on all branches produces an estimate n' on the actual size of the network, n. In general we have $n' \leq n$, because the current view may not include all branches of the history tree. To confirm this estimate, it is sufficient to wait an additional n' steps, which is the longest time it may take for at least one missing branch to appear in the view, if one exists.

Overall, this Counting algorithm terminates in $3n - 2$ steps in the worst case, leaving a small gap with the lower bound of $2n - 2$ steps given in Sect. 2.3.

> **Open Problem 3.** Can a Counting algorithm terminate in $2n + O(1)$ steps in all connected undirected dynamic networks with a unique leader?

The previous Counting algorithm was generalized in [11] to networks where the leader may not be unique, but there is a known number $\ell \geq 1$ of indistinguishable leaders, or *supervisors*. Note that the history tree may contain several branches corresponding to leaders; the challenge posed by this scenario is that the individual anonymities of such branches are unknown (although their sum is ℓ).

The state-of-the-art solution given in [11] involves subdividing the history tree into ℓ intervals, each of at most $(\ell + 1)n$ levels. The total running time is roughly $(\ell^2 + \ell + 1)n$ steps, making this algorithm impractical unless ℓ is small.

In fact, it is a major open question whether a larger ℓ may actually be helpful.

> **Open Problem 4.** In connected undirected dynamic networks with a known number ℓ of leaders, how does the optimal running time of terminating Counting algorithms scale with ℓ?

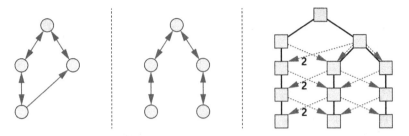

Fig. 6. Two directed static networks with a unique leader and no outdegree awareness that have a different number of agents but the same history tree.

3.3 Directed Networks

Generalizing the scenario of Sect. 2, we may consider *directed networks*, i.e., networks with unidirectional links. It is important to recognize that, in these networks, even solving basic problems such as Average Consensus or Counting with a unique leader requires some additional assumptions, as Fig. 6 demonstrates.

A reasonable approach is to assume that, in each step, agents are aware of their outdegree either before (*early outdegree awareness*) or after (*late outdegree awareness*) they transmit their messages. The key distinction is that in the early awareness model, an agent can incorporate its current outdegree into the messages it sends. Here we will focus on the late outdegree awareness model, which is less advantageous especially in dynamic networks.

The basic history tree architecture can be extended by attaching outdegrees to black edges, as in Fig. 7. We can then identify a non-branching level L_i in the history tree, as in Sect. 2.3, and use outdegrees and multiplicities of red edges to write linear equations generalizing Eq. (2), as shown in Fig. 7.

The resulting homogeneous system is represented by an *irreducible* matrix A, assuming the network is strongly connected at every step. Note that $A = \lambda I - P$, where $\lambda > 0$ and $P \geq 0$ is also an irreducible matrix. In our example,

$$A = \begin{bmatrix} 3 & -1 & 0 & 0 \\ 0 & 2 & -1 & -2 \\ -2 & 0 & 1 & 0 \\ 0 & 0 & -1 & 1 \end{bmatrix} = \begin{bmatrix} 3 & 0 & 0 & 0 \\ 0 & 3 & 0 & 0 \\ 0 & 0 & 3 & 0 \\ 0 & 0 & 0 & 3 \end{bmatrix} - \begin{bmatrix} 0 & 1 & 0 & 0 \\ 0 & 1 & 1 & 2 \\ 2 & 0 & 2 & 0 \\ 0 & 0 & 1 & 2 \end{bmatrix} = 3I - P.$$

We have $Ax = 0$, where $x > 0$ is the vector of anonymities of the nodes in L_i. Hence, $Px = \lambda x$. By the Perron-Frobenius theorem, λ is a *simple* eigenvalue of P, and thus 0 is a simple eigenvalue of A. In other words, the *nullity* of A is 1, which means we can solve the linear system in terms of a single free variable.

Thus, we can find the ratios between the anonymities of all nodes in L_i, or compute them exactly if there is a known number $\ell \geq 1$ of leaders in the network.

This technique yields stabilizing Average Consensus and Counting algorithms for strongly connected directed dynamic networks with late outdegree awareness, generalizing the ones in Sect. 2.3. The stabilization time of these algorithms is again $2n - 2$ steps, which is optimal.

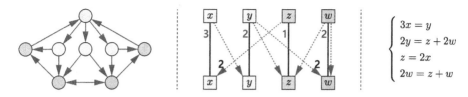

Fig. 7. A communication step of a directed network with late outdegree awareness and a non-branching level of its history tree. Blue numbers indicate outdegrees. (Color figure online)

We now present a *terminating* Counting algorithms for directed networks with late outdegree awareness and a known number $\ell \geq 1$ of leaders. Every agent waits until its view contains a long-enough interval $\mathcal{I}$ of non-branching levels. The goal is to find an upper bound U_i on the anonymity of a node v_i in each branch B_i within $\mathcal{I}$. To start, we take $U_1 = \ell$, where B_1 is any leader branch.

The algorithm repeatedly uses branches with a known upper bound to determine new upper bounds. Namely, let B_i have a known upper bound $U_i \geq \mathbf{a}(v_i)$ and consider all red edges (v, v'), where v is a descendant of v_i within $\mathcal{I}$. Each of these yields an *estimate* δU_i on $\mathbf{a}(v')$, where δ is the outdegree corresponding to the child of v in B_i. Note that, if v has a unique child in the history tree, at most δU_i messages are sent from agents represented by v, and so $\delta U_i \geq \mathbf{a}(v')$. Nonetheless, even if v is not branching in a view, it may still branch in the history tree, and in this case δU_i may not be a correct upper bound. However, this undesirable event may occur at most $U_i - 1$ times. Thus, as soon as a branch B_j receives estimates from B_i on U_i nodes, we take the maximum estimate as a correct upper bound U_j on the anonymity of the *deepest* such node v_j in B_j.

When we finally have an upper bound for all branches in $\mathcal{I}$, we wait $\sum_i U_i$ additional steps to confirm that there are no branches missing from the view. Then we run the previous stabilizing Counting algorithm and output the result.

Assuming that the network is simple and strongly connected at all steps, this algorithm terminates in $2^{O(n \log n)}$ steps, which is likely far from optimal.

> **Open Problem 5.** Can a Counting algorithm terminate in a *polynomial* number of steps in all strongly connected directed dynamic simple networks with (early or late) outdegree awareness and a unique leader?

3.4 Disconnected Networks

For networks that are not necessarily connected at all steps, we define a *communication round* as a minimal sequence of consecutive steps whose communication multigraphs have a (strongly) connected *sum* (constructed by adding together their adjacency matrices). A network is *τ-union-connected* if every block of τ consecutive steps contains a communication round. It was observed in [11] that τ and the dynamic diameter d are related by the tight inequalities $\tau \leq d \leq \tau(n-1)$.

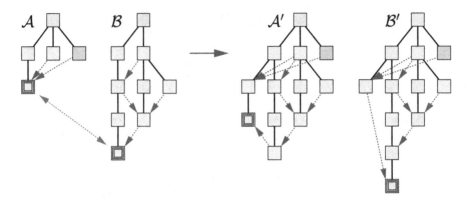

Fig. 8. Two agents in a semi-synchronous network share their views $\mathcal{A}$ and $\mathcal{B}$, which are updated as $\mathcal{A}'$ and $\mathcal{B}'$. Note that red edges may go upward or skip multiple levels. The highlighted nodes are the "bottom nodes" of their respective views, which can be identified as the only sinks (assuming black edges are directed away from the root).

Clearly, any non-trivial *terminating* computation is impossible if the agents know nothing about τ. On the other hand, with knowledge of τ, all of the previous algorithms can be straightforwardly adapted to τ-union-connected networks. It is sufficient for each agent to accumulate incoming messages at every step, updating its view only once every τ steps. This adaptation slows down all running times by a factor of τ. However, this is worst-case optimal, since a τ-union-connected network may be devoid of links except at steps that are multiples of τ.

As for *stabilizing* computations, they can be done even with no knowledge of τ, again with a worst-case optimal slowdown by a factor of τ, as detailed in [11].

3.5 Semi-Synchronous Networks

In a network operating *semi-synchronously*, any agent may unpredictably be *inactive* at any step. An inactive agent does not communicate with other agents and does not even update its state. This model is called "asynchronous" by Boldi–Vigna [3,5] and generalizes the *asynchronous starts* of Charron-Bost–Moran [7], but we will give the term "asynchronous" a more extensive meaning in Sect. 3.6.

The concept of a round and the parameter τ are defined as in Sect. 3.4. A notable distinction from the synchronous models discussed so far is that agents in a semi-synchronous network cannot reliably count communication steps, because they do not know for how many steps they have been inactive. Hence, two views shared by two agents may not have the same height, as the ones in Fig. 8.

Consequently, a red edge no longer has to connect a level to the next, but can span any number of levels. However, it is always possible to restore this property by adding *dummy nodes* to represent inactive agents, resulting in an *equalized* view (Fig. 9). Thus, we can reduce any semi-synchronous network to an equivalent synchronous one with the same parameter τ and apply the methods outlined in Sect. 3.4 to obtain algorithms with the same running times.

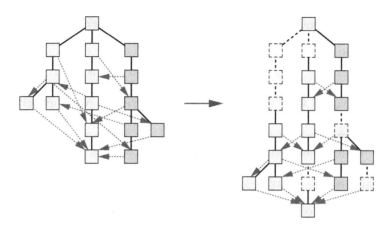

Fig. 9. Equalizing the view of an agent in a semi-synchronous network by adding dummy nodes. In the resulting view, every red edge connects a level to the next.

3.6 Asynchronous Networks

A network is *asynchronous* if messages can take an arbitrary, independent, and unpredictable amount of time to reach their destinations. Such a network is necessarily directed, and therefore some form of outdegree awareness is needed, as argued in Sect. 3.3. A *round* is now any minimal interval of time such that the messages that are sent and received form a strongly connected multigraph.

Observe that, even in asynchronous networks, time can always be discretized, assuming that no agent can send an infinite number of messages in a finite time.

Also, doing non-trivial computations with termination in asynchronous networks with no knowledge about the duration of a round is impossible.

However, there is a simple stabilization algorithm, as follows. When an agent sends some messages, it expands its view by adding a child to the bottom node and attaches its outdegree to the corresponding black edge, as in Sect. 3.3. When it receives a multiset of messages, it updates its view as in Sect. 3.5.

Then, the agent seeks the first interval of levels constituting a round where all nodes are non-branching. The outdegrees and red edges in this interval yield a system of equations that is solved as in the stabilizing algorithm of Sect. 3.3.

Such a non-branching interval occurs in the history tree after at most $n - 1$ rounds and appears in the agent's view in at most $n - 1$ additional rounds. The total stabilization time is therefore $2n - 2$ rounds, which is worst-case optimal.

3.7 Port Awareness

A widely studied scenario, especially within static networks, is when each agent has a local numbering for its incoming links (*input port awareness*) or outgoing links (*output port awareness*). It is evident that, although input port awareness enables each agent in a static network to identify all messages sent by the same neighbor, this feature has no clear meaning or effect in dynamic networks.

In contrast, output port awareness has a significant impact on both static and dynamic networks, as it not only implies outdegree awareness, but allows agents to assign a different tag to each message they send within a communication step. This is helpful in breaking network symmetry, thus facilitating certain computations. The history tree architecture for outdegree awareness of Sect. 3.3 can also be adapted to this model by simply attaching a port number to each red edge.

As an example, output port awareness combined with a unique leader makes the Counting problem particularly simple even in strongly connected directed dynamic networks. Indeed, all messages sent by the leader in a communication step have different tags, and so all agents that receive them become distinguishable. Generalizing, every node on a red path starting at a leader node must have an anonymity of 1. When a whole level L_i in a view consists of such nodes, we can check if the outdegree of each node in L_i matches the number of its outgoing red edges. If so, all agents have been accounted for, and we can return $n = |L_i|$.

This Counting algorithm terminates in $2n - 1$ steps, greatly improving upon the one in Sect. 3.2. Observe that the lower bound in Fig. 4 does not hold for networks with output port awareness, which may allow for even faster solutions.

Open Problem 6. Can a Counting algorithm terminate in fewer than $2n - 2$ steps in all strongly connected directed dynamic networks with output port awareness and a unique leader?

A less obvious fact is that, in any non-branching level of the history tree of a strongly connected directed dynamic network with output port awareness, all nodes must have the same anonymity.[4] Indeed, if A and B are two classes of indistinguishable agents represented by nodes in a non-branching level, and an agent in A receives a message tagged k from an agent in B, then so do all agents in A (or else their node would branch). Since no agent in B can send more than one message tagged k, we have $|A| \leq |B|$. Repeating this reasoning for all pairs of communicating classes, and recalling that the network is strongly connected, we conclude that all classes must have the same size.

A remarkable consequence is that, with output port awareness, a unique leader can be elected if and only if it is possible to assign unique identifiers to all agents (also known as the *Naming problem*). A terminating algorithm for both problems, if n is known, is easily obtained by combining the Leader Election algorithm in Sect. 3.1 (which was designed for undirected networks only) with the stabilizing technique for directed networks in Sect. 3.3.

3.8   Varying Inputs

Suppose that agents have inputs that may change at every step. Note that the history tree architecture already supports attaching inputs to nodes, and so the basic stabilizing algorithms previously discussed can become *streaming algorithms* that adaptively return the correct output with an amortized delay of $n - 1$ steps.

---

[4] This observation, limited to static networks, is also found in [2, Theorems 3 and 10].

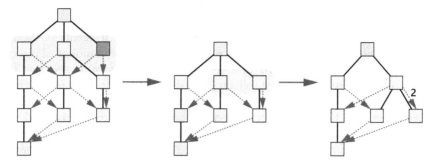

Fig. 10. A routine in the universal self-stabilizing protocol: delete level L_0 in the view, and then repeatedly merge nodes whose corresponding sub-views are isomorphic.

3.9 Self-stabilization

A network algorithm is *self-stabilizing* if it returns the correct output regardless of the initial state of the agents. This implies tolerance to memory corruption, transient faults, and agents dynamically joining and leaving the network.

Let us assume the network to be synchronous, directed, and dynamic. We remark that most existing self-stabilizing protocols, such as Boldi–Vigna's [5], are specifically designed for static networks, and cause agents to perpetually and incorrectly reset their states when executed in highly dynamic networks.

We will give a *universal* self-stabilizing protocol that constructs coherent views, allowing to convert any stabilizing algorithm into a self-stabilizing one (note that non-trivial terminating computations are impossible in this scenario).

The core idea is that an agent can deliberately "forget" old information by deleting level L_0 of its current view, connect level L_1 to the root, and merge equivalent nodes from top to bottom to restore a well-formed view, as in Fig. 10.

If the number of agents n is known, they can simply update their views as usual (resetting their state if it does not encode a well-formed view) and delete old levels when their number exceeds some fixed threshold, e.g., $2n - 2$. Eventually, this protocol produces views that correctly describe the last $2n - 2$ communication steps, which are enough for the stabilizing algorithms of Sects. 2.3 and 3.3.

On the other hand, if n is unknown, every agent updates its view as usual, but deletes level L_0 every *two* steps. This is controlled by a local binary flag that is toggled at every step. However, before merging its view with a neighbor's view of different height, it deletes the top levels of the taller view to match the other. Also, if the taller view was its own, the agent does not toggle its flag for a step.

In $O(n)$ steps, all agents have views of equal height, as well as equal flags. Moreover, their views eventually describe an increasingly large number of previous communication steps, enabling any stabilizing algorithm to work correctly.

Open Problem 7. Is there a universal self-stabilizing protocol for *semi-synchronous* strongly connected directed dynamic networks?

3.10   Finite-State Stabilization

The stabilizing algorithms described so far assume that all agents constantly update their views at every step, which requires an unlimited amount of internal memory. To mitigate this, we will now give a *universal finite-state* protocol that enables the conversion of any stabilizing algorithm into one that uses a finite amount of memory (as a function of n), albeit with an extra delay.

Let us assume the network to be synchronous, undirected, connected, and dynamic. A level L_i in a view is said to be *suitable* if every node in L_{i-1} has a unique child in L_i and the red edges between these two levels are compatible with a connected network. Recall that the basic stabilizing algorithms of Sect. 2.3 do all computations using only the shallowest suitable level of a view.

Now, when two neighboring agents share views whose shallowest suitable levels are isomorphic, they do not merge their views. Essentially, this removes a link from the communication network for that step. Also, when an agent receives no views that it can merge, it skips updating its view altogether, i.e., it remains inactive for that step. Note that this makes the network semi-synchronous, so views must be *equalized* prior to being used by the protocol (see Sect. 3.5).

According to this protocol, if all agents remain inactive forever, they all share the same suitable level, which is enough to perform correct computations. Otherwise, some inactive agent acquires relevant information within $n-1$ steps and reactivates. Moreover, since there are at most $n-1$ branching nodes in the history tree, the agents can incorrectly guess a suitable level at most $n-1$ times. Thus, the protocol is finite-state, but introduces an overhead of $O(n^2)$ steps.

Open Problem 8. Is there a universal finite-state stabilizing protocol for connected undirected dynamic networks with an overhead of $O(n)$ steps?

Another open problem is to design a protocol that is both self-stabilizing and finite-state. In Sect. 3.9 we already gave one, but it assumes n to be known.

Open Problem 9. Is there a universal finite-state self-stabilizing protocol for connected undirected dynamic networks of unknown size?

3.11   Memoryless Computation

An agent is *memoryless* if its state is reset at every communication step, meaning its entire memory is erased after it sends messages and before receiving messages from neighbors. The goal is to design a universal protocol that enables memoryless agents to construct coherent views of *some* history tree related to the network. This would allow basic algorithms to run correctly within the memoryless model.

Open Problem 10. Under what assumptions is there a universal memoryless protocol for anonymous networks?

3.12   Congested Networks

In a *congested network* of n agents, communication links have a logarithmic bandwidth, and therefore the size of each message is limited to $O(\log n)$ bits.[5]

It is evident that the basic technique of encoding a view as a single message is not applicable in a congested network, because the size of a view grows polynomially at every step (see Sect. 2.3). Furthermore, any naive approach that simply subdivides a view into smaller pieces to be transmitted over multiple steps is clearly ineffective in anonymous dynamic networks.

Correct protocols for constructing views in congested dynamic networks are found in [10]. If the dynamic diameter d is known, then $O(\log n)$ bits of information can be reliably *broadcast* in phases of d steps. This makes it possible for agents to share enough information to construct a history tree that is compatible with their network, one level at a time. Every node in this history tree has a unique label of logarithmic size, which also identifies all agents represented by that node. This protocol can be used to solve the Counting problem in $O(dn^2)$ steps.

If d is unknown but there is a unique leader, there is a more complex protocol that attempts to estimate d by implementing a *reset module* that repeatedly doubles the estimate every time a broadcasting error is detected. The leader coordinates the process, ensuring that all agents agree on the same version of the history tree. This yields a Counting algorithm that terminates in $O(n^3)$ steps.

Open Problem 11. Is there a terminating Counting algorithm for congested dynamic networks *with more than one leader*?

It was proved in [12] that solving the Counting problem in congested networks by broadcasting "tokens" requires at least $\Omega(n^2/\log n)$ steps. On the other hand, there is a solution in $O(n^2)$ steps if messages have size $O(n \log n)$ bits [10].

Open Problem 12. Can a Counting algorithm terminate in $O(n^2)$ steps in all congested dynamic networks with a unique leader?

We find it fitting to conclude this note with an open problem that is unlikely to have a solution using history trees.

Open Problem 13. Is there a terminating Counting algorithm for congested dynamic networks where agents have *logarithmic-sized memory*?

---

[5] Agents do not necessarily know this limit, since they generally do not know n. Nonetheless, if they attempt to send larger messages, the network's behavior is undefined. Ideally, algorithms should be designed to automatically avoid this situation.

Acknowledgments. The author expresses profound gratitude to collaborators Jérémie Chalopin, Giuseppe A. Di Luna, and Federico Poloni for numerous inspiring discussions. This work was supported by the JSPS KAKENHI grant 23K10985, the University of Aizu Competitive Research Funding for FY 2023, and the University of Salerno.

Disclosure of Interests. The author has no competing interests to declare that are relevant to the content of this article.

References

1. Angluin, D.: Local and global properties in networks of processors. In: Proceedings of the 12th ACM Symposium on Theory of Computing (STOC '80), pp. 82–93 (1980)
2. Boldi, P., Codenotti, B., Gemmell, P., Shammah, S., Simon, J., Vigna, S.: Symmetry breaking in anonymous networks: characterizations. In: Proceedings of the 4th Israeli Symposium on Theory of Computing and Systems (ISTCS '96), pp. 16–26 (1996)
3. Boldi, P., Vigna, S.: Computing anonymously with arbitrary knowledge. In: Proceedings of the 18th Annual ACM Symposium on Principles of Distributed Computing (PODC '99), pp. 181–188 (1999)
4. Boldi, P., Vigna, S.: Fibrations of graphs. Discret. Math. **243**(1), 21–66 (2002)
5. Boldi, P., Vigna, S.: Universal dynamic synchronous self-stabilization. Distrib. Comput. **15**, 137–153 (2002)
6. Charron-Bost, B., Lambein-Monette, P.: Computing outside the box: average consensus over dynamic networks. In: Proceedings of the 1st Symposium on Algorithmic Foundations of Dynamic Networks (SAND '22), pp. 10:1–10:16 (2022)
7. Charron-Bost, B., Moran, S.: The firing squad problem revisited. Theor. Comput. Sci. **793**, 100–112 (2019)
8. Conway, J.H.: Regular Algebra and Finite Machines. Chapman & Hall, Boca Raton (1971)
9. Di Luna, G.A., Viglietta, G.: Computing in anonymous dynamic networks is linear. In: Proceedings of the 63rd IEEE Symposium on Foundations of Computer Science (FOCS '22), pp. 1122–1133 (2022)
10. Di Luna, G.A., Viglietta, G.: Brief announcement: efficient computation in congested anonymous dynamic networks. In: Proceedings of the 42nd ACM Symposium on Principles of Distributed Computing (PODC '23), pp. 176–179 (2023)
11. Di Luna, G.A., Viglietta, G.: Optimal computation in leaderless and multi-leader disconnected anonymous dynamic networks. In: Proceedings of the 37th International Symposium on Distributed Computing (DISC '23), pp. 18:1–18:20 (2023)
12. Dutta, C., Pandurangan, G., Rajaraman, R., Sun, Z., Viola, E.: On the complexity of information spreading in dynamic networks. In: Proceedings of the 24th Annual ACM-SIAM Symposium on Discrete Algorithms (SODA '13), pp. 717–736 (2013)
13. Kowalski, D.R., Mosteiro, M.A.: Polynomial counting in anonymous dynamic networks with applications to anonymous dynamic algebraic computations. J. ACM **67**(2), 11:1–11:17 (2020)
14. Norris, N.: Universal covers of graphs: isomorphism to depth $n - 1$ implies isomorphism to all depths. Discret. Appl. Math. **56**(1), 61–74 (1995)

15. Tani, S.: Compression of view on anonymous networks-folded view–. IEEE Trans. Parallel Distrib. Syst. **23**(2), 255–262 (2012)
16. Yamashita, M., Kameda, T.: Computing on an anonymous network. In: Proceedings of the 7th ACM Symposium on Principles of Distributed Computing (PODC '88), pp. 117–130 (1988)
17. Yamashita, M., Kameda, T.: Computing on anonymous networks. I. Characterizing the solvable cases. IEEE Trans. Parallel Distrib. Syst. **7**(1), 69–89 (1996)

Regular Papers

An Analysis of Avalanche Consensus

Ignacio Amores-Sesar(ID), Christian Cachin(ID), and Philipp Schneider(✉)(ID)

University of Bern, Bern, Switzerland
{ignacio.amores,christian.cachin,philipp.schneider2}@unibe.ch

Abstract. A family of leaderless, decentralized consensus protocols, called Snow consensus was introduced in a recent whitepaper by Yin et al. These protocols address limitations of existing consensus methods, such as those using proof-of-work or quorums, by utilizing randomization and maintaining some level of resilience against Byzantine participants. Crucially, Snow consensus underpins the Avalanche blockchain, which provides a popular cryptocurrency and a platform for running smart contracts.

Snow consensus algorithms are built on a natural, randomized routine, whereby participants continuously sample subsets of others and adopt an observed majority value until consensus is achieved. Additionally, Snow consensus defines conditions based on participants' local views and security parameters. These conditions indicate when a party can confidently finalize its local value, knowing it will be adopted by honest participants.

Although Snow consensus algorithms can be formulated concisely, there is a complex interaction between randomization, adversarial influence, and security parameters, which requires a formal analysis of their security and liveness. Snow protocols form the foundation for Avalanche-type blockchains, and this work aims to increase our understanding of such protocols by providing insights into their liveness and safety characteristics. First, we analyze these Snow protocols in terms of latency and security. Second, we expose a design issue where the trade-off between these two is unfavorable. Third, we propose a modification of the original protocol where this trade-off is much more favorable.

1 Introduction

Establishing consensus is one of the most fundamental tasks in distributed computing, for instance to implement atomic broadcast, to synchronize processes, or to elect leaders. Distributed blockchains and in particular cryptocurrencies rely on consensus to ensure proper operation, and therefore trust in these systems, which has put increased focus on new kinds of consensus algorithms. In the consensus problem we consider n parties of which some are potentially faulty. Every party has some input value, which we often refer to as opinion in this article. We say that a protocol that coordinates communication and local computations

This work has been funded by the Swiss National Science Foundation (SNSF) under grant agreement Nr. 200021_188443 (Advanced Consensus Protocols) and a grant from Avalanche, Inc. to the University of Bern.

Y. Emek (Ed.): SIROCCO 2024, LNCS 14662, pp. 27–44, 2024.
https://doi.org/10.1007/978-3-031-60603-8_2

of all parties solves consensus when everyone agrees on a single opinion, which was also the input of at least one party.

The consensus problem becomes challenging in the presence of Byzantine faults, i.e., parties that can deviate from the protocol. In particular, reaching consensus is impossible deterministically in the asynchronous setting with even one fault [8] and in the synchronous setting with one third or more Byzantine faults [12], except if one would use cryptographic signatures. Real protocols that solve consensus in the context of blockchains have to navigate around these impossibility results, while also optimizing other criteria; chief among them are the latency until a transaction is finalized, the throughput of transactions, resource consumption, scalability, and resiliency against adversarial parties, which often necessitates trade-offs [5, 10].

One particularly simple consensus design sacrifices determinism and works along the following principle. Each party continually samples random subsets of other parties and adjusts its opinion based on the observed sample according to certain rules. There has been extensive research that explores such mechanisms, see the recent survey [2]. It has been shown for a subset of protocols of this type that they can be expected to converge very rapidly to a state of stable consensus (with a limited adversary, a bounded number of opinions, and in a synchronous network) [3,4,6,7,9]. Additionally, such mechanisms have the advantage that parties need only few such samplings to be relatively certain what the consensus opinion will be, resulting in near-linear message complexity.

A whitepaper [13] released in 2019 exploits this design and introduces the Avalanche protocol, which forms the basis of the Avalanche blockchain infrastructure and its services. Avalanche gained popularity and reach due to competitive characteristics in the performance spectrum of latency, throughput, scalability, and resource consumption [11].[1] In particular, [13] introduces the *Snow family* of binary consensus protocols that build on this principle of random samplings, which can be adapted to maintain consistency of the corresponding Avalanche blockchain network.

The simplest protocol of this family, known as *Slush*, works as follows. Each party continuously samples the opinion of $k \geq 2$ others; if such a sampling contains an opinion different from its own at least α times (for some $\alpha > k/2$) then the party adopts this opinion as its own. Slush can be considered as a self-organizing mechanism that it is likely to converge to a stable consensus relatively quickly and remain there, even in the presence of a limited number of parties that deviate from the protocol. The whitepaper also introduces the *Snowflake* and *Snowball protocols*, which add mechanisms to finalize an opinion of a node based on past queries which reflects how stable the observed majority is. The level of confidence in a majority can be controlled with a security parameter β.

The complex interaction between performance characteristics, security level, and the involved parameters k, α, and β makes the analysis of Snow-type consensus protocols challenging. The whitepaper [13] relies primarily on empirical observations and informal explanations to motivate its design choices. Currently,

---

[1] For instance, through its AVAX token, Avalanche ranks in the top 10 among the "Layer-1" blockchains by market capitalization (as of December 2023).

a formal understanding of the performance and security characteristics of Snow protocols is lacking.

Overview and Contributions. We focus on bridging the gap in understanding of Snow consensus protocols, which we consider as a necessary first step for an encompassing analysis of the complete Avalanche blockchain protocol, which builds upon Snow consensus (although the Avalanche protocol itself is beyond the scope of this work). First we explore the performance of Snow protocols, beginning with the self-organizing, binary consensus mechanism of Slush. In Sect. 3, we express the progress toward a stable consensus per round depending on the distribution of opinions and parameters $k \geq 2$ and $\alpha > \frac{k}{2}$, which gives insights into the evolution of the system (cf. Fig. 1 for a visualization).

In Sect. 4, we show that coming close to a consensus already requires a minimum of $\Omega\left(\frac{\log n}{\log k}\right)$ rounds, even in the absence of adversarial influence (see Theorem 1, a simpler, weaker form is given in Corollary 1). Furthermore, we generalize upper bounds from the so-called Median and 3-Majority consensus protocols [3,6] in the Gossip model (discussed in Sect. 4) and establish that Slush reaches a stable consensus in $O(\log n)$ rounds (see Theorem 2), which holds even when an adversary can influence up to $O(\sqrt{n})$ parties.

We interpret these results in the following way. Even assuming that the performance of Slush matches the lower bound, increasing the parameter k yields only a limited speed-up of $O(\log k)$ (usually $k \ll n$). Furthermore, since the message complexity per round is $\Theta(kn)$ Slush has the advantage of near-linear message complexity for small k, which is negated if k increases significantly (e.g., sampling sizes close to n). We conclude that values higher than $k = 20$ as suggested originally [13] have diminishing benefit and an unfavorable trade-off in terms of message complexity. In Sect. 5 we show that the lower bound $\Omega\left(\frac{\log n}{\log k}\right)$ rounds extends to Snowflake and Snowball.

In Sect. 6 we analyze the security mechanisms of the Snow protocols, that deal with the possibility of failing to achieve consensus due to the randomized nature of the algorithm or adversarial influence. The protocol provides a security parameter β to control the probability of such a failure. This has an unfavorable trade-off as we show in Sect. 6, specifically, a negligible probability of failure (w.r.t. β) and a latency to finalize a value that is at most polynomial in β are mutually exclusive (see Theorem 3). In Sect. 7 we propose a solution for this issue by replacing the security mechanism of Snowball with a simple mechanism that achieves security with all but negligible probability (w.r.t. β) in $O(\beta + \log n)$ rounds (Theorem 4 and Corollary 4).

Due to space restrictions, this paper takes the form of an extended abstract with all proofs, further discussion on related work and supplementary material deferred to the full paper.

2 Preliminaries

Before moving to the technical parts of the analysis, we introduce the definitions and modeling assumptions that we use throughout the paper. In part, this work aims to be a supplementary of the whitepaper [13] in the style of other theoretical works that study randomized, self-stabilizing consensus protocols [3,4,6,7,9] and our nomenclature, definitions and modeling assumptions are a composition of those.

2.1   Model

Communication. We consider a fully connected network of n parties with identifiers $\mathcal{N} = \{1, \ldots, n\}$. Parties communicate by sending point to point messages. For the message transfer we assume the synchronous message passing model, where there is a fixed period of time until any given message is delivered. In fact, the synchronous setting allows to assume that time is slotted into discrete rounds and all messages sent in the previous round have arrived by the next round. This also allows us to use the number of rounds as a proxy for algorithm running time.

Consensus. In the general problem setup there are m opinions in the network and each party has an initial opinion, however note that in the context of Snow protocols we typically have $m = 2$. Snow protocols can be seen as self-stabilizing protocols and we define a stable state where almost all parties have the same opinion and the likelihood to revert from this state is low (see Sect. 7 for some properties of this stable state).

Definition 1 (State of Stable Consensus). *The system is in a state of stable consensus if at least $n-o(n)$ parties have the same opinion.*

Randomization or the presence of an adversary implies that at any point in time there is a non-zero chance that a stable consensus is reverted. Therefore, in the context of blockchain applications, parties need to eventually finalize or *decide* on an opinion. In that sense we define the consensus problem as follows.

Definition 2 (Consensus Problem). *A protocol solves this problem if the following conditions are satisfied.*

Termination: *Every party eventually* decides *on some opinion.*
Validity: *If all parties propose the same value, then all parties decide on that value.*
Integrity: *No party decides* twice.
Agreement: *No two parties decide* differently.

We consider this consensus problem under the influence of the following adversary.

Definition 3 (F-Bounded Adversary). *An F-bounded adversary can set the opinion of up to F (undecided) parties at the beginning of each round to one of the m opinions.*

Randomization, Security and Latency. The consensus protocols we consider in this work are randomized, and we work with some standard definitions (more details in the full paper). We give a few basic definitions and principles pertaining to the probabilistic security properties of some protocol (used in Definition 7).

Definition 4 (Negligible Function). *A function f is negligible if for any polynomial π there is a constant $\lambda_0 \geq 0$, s.t., for any $\lambda \geq \lambda_0$ it is $f(\lambda) \leq \pi(\lambda)$.*

We often use that for any constant $c > 0$, the function $f(\lambda) = e^{-c \cdot \lambda}$ is a negligible w.r.t. λ. We usually aim for a certain security threshold given by a variable γ.

Definition 5 (All But Negligible Probability). *An event is said to occur with all but negligible probability with respect to some parameter λ if the probability of the event not happening is a function in λ that is negligible w.r.t. λ.*

In the literature, randomized consensus protocols are often shown to be successful *with high probability*, which expresses the probability of failure as a function that decreasing inversely with the input size n of the problem (here n is the number of parties). This is often quite convenient as it eliminates any other variable from the analysis, compared to defining some fixed failure threshold.

Definition 6 (With High Probability). *An event is said to hold with high probability (w.h.p.), if there exists a constant $c \geq 1$ such that the event occurs with probability at least $1 - n^{-c}$ for sufficiently large n.*

The disadvantage of the notion w.h.p. is that it usually looks at the asymptotic behavior of a system (i.e., for large n), which does not provide a fixed security level for small n, which can be a requirement in practice. We provide a link between the two notions in the full paper.

We also consider an F-bounded adversary, which introduces a non-zero chance to delay a consensus for any fixed period of time. Hence, we can only hope to make the probability of failure of such a protocol negligibly small, while at the same time maintaining a reasonable latency (i.e., number of rounds until consensus). To connect the notions of failure probability and latency we make the following provisions.

Let E be an event or condition during the execution of some protocol $\mathcal{P}$, e.g., E describes the event that $\mathcal{P}$ successfully establishes consensus, see Definition 2. The protocols we are investigating typically depend on so called security parameters. Increasing the security parameter increases the likelihood of success but typically has detrimental effects on the running time. To quantify this, we formally define further below what it means that E holds with all but negligible probability with respect to protocol $\mathcal{P}$. Roughly speaking, as we increase the running time of a protocol measured in the security parameter λ, the probability that some condition E (e.g., consensus) has not been established yet, decreases super-polynomially in λ.

Definition 7 (Negligible Probability, Security Parameter). *An event E holds with all but negligible probability or equivalently its complement $\overline{E}$ has negligible probability in a protocol $\mathcal{P}$, if the following holds. There exists a polynomial ρ such that for any polynomial π there exists a value $\lambda_0 \geq 0$, such that for all*

$\lambda \geq \lambda_0$ _the following holds. If_ $\mathcal{P}$ _is executed for at least_ $\rho(\lambda)$ _rounds it holds that_ $\mathbb{P}(\overline{E}) \leq 1/\pi(\lambda)$, _i.e.,_ $\mathbb{P}(E) > 1 - 1/\pi(\lambda)$. _We call the value_ $\lambda \geq \lambda_0$ _a security parameter._

In this article, we often talk about randomly sampling a set of k parties, which means that we take a uniform random sample of size k from the set of all n parties _with repetition_, which is modeled by the binomial distribution. Note that sampling _without repetition_, represented by the hyper-geometric distribution, approaches the binomial distribution as the ratio n/k becomes larger. Therefore, our results approximate the case without repetition well for $n \gg k$ (the usual case). The assumption of uniform sampling with repetition makes our analysis much more feasible, in particular it removes undesirable marginal cases (e.g. k-samples for $n < k$) allowing use to use continuous functions to describe certain aspects of the system.

2.2   The Snow Family

The Snow family consists of three consensus protocols based on random sampling instead of the traditional quorum intersection. This approach allows the snow family to reduce the message complexity by sending messages to a constant number of parties each round. Here, we provide an informal description of each protocol, whereas a more detailed pseudocode can be found in the full paper.

Slush. The first protocol in the Snow family is Slush. Each party j runs the Slush protocol in local rounds which we assume are synchronous, however in the general protocol every party may have a different round value. Party j starts the round by randomly sampling k parties for their opinion, with k a constant value. If at least α parties respond with the same opinion b, party j adopts it as its own opinion and starts a new round. The value α must be strictly larger than $\frac{k}{2}$. If there is no α-majority, party j keeps its opinion and starts a new round. The Slush algorithm has a hard-coded number of rounds defining the protocol's end. When party j reaches the maximum round, it _decides_(b) its candidate value b. Note that in our analysis in Sect. 3 and 4 we analyze the time until Slush reaches a state of stable consensus (Definition 1) and assume that this hard coded maximum round does not exist or is sufficiently large to not play a role.

Snowflake. The main limitation of the Slush algorithm is the hard-coded number of rounds. The number of rounds needs to be relatively high to guarantee consensus even in the worst case (which the case when the network starts in a balanced state: half of the parties _proposes_(0) and the other half _proposes_(1)). Snowflake aims to address this issue by modifying the termination condition of Slush.

In the Snowflake protocol, party j counts the number of consecutive queries with an α-majority for opinion b. If j observes β consecutive rounds with α-majority for b, party j _decides_(b). The intuition behind this termination rule is the following: the probability of obtaining β consecutive α-majorities for opinion b is small when expressed as a function of β, unless almost every party has b

as candidate value in the network. This termination rule allows for an adaptive running time based on the state of the network. Looking ahead, we show how this termination rule forces the Snowflake protocol to choose between a high confidence in the agreement property and polynomial running time (Theorem 3).

Snowball. The Snowball protocol introduces another modification how parties change their opinion. In Snowflake, the change of opinion is only based on the outcome of the last query, i.e., it is stateless. By contrast, in Snowball, party j considers the past queries in order to decide whether to change its opinion or not. Party j changes its opinion value from b to b' when the number of α-majorities for b' surpasses the number of α-majorities for value b since the beginning of the execution. The idea behind considering the whole history of the protocol is to make it less likely for a party with opinion b to switch to b' when the prevalent opinion in the network is b, thus possibly reducing the number of rounds until termination. Looking ahead, we show that this routine does not reduce the number of rounds until termination in expectation (Lemma 11).

Avalanche. The Snow consensus protocols serve as foundation for the Avalanche consensus [1,13]. Avalanche employs a classification system to group transactions into conflicting sets and subsequently applies a tailored adaptation of the Snowball algorithm to each of these conflict sets. To optimize communication efficiency, Avalanche establishes connections between distinct instances of the Snowball consensus, enabling the reuse of messages and, consequently, reducing message complexity. However, due to these interdependencies, Avalanche is unable to inherit the liveness properties from the Snow family [1]. Nevertheless, it is noteworthy that the security of Avalanche remains equivalent to that of the Snowball protocol [13].

3 Dynamics of Slush

The whitepaper [13] observes that the Slush consensus protocol converges to a stable consensus very fast in practice. Concrete claims are made pertaining to the time to consensus, but no conclusive proof is given. In this section we analyze the rate of convergence of Slush towards a consensus, which will later also inform the rate of convergence of Snowflake and Snowflake (Sect. 5).

3.1 Expected Rate of Progress of Slush

We start by investigating the expected rate of progress of Slush, which characterizes the dynamics of Slush and how it depends on the parameters α and k. We will later show that other Snow protocols (Snowflake, Snowball) behave similar in terms of the required number of rounds to consensus.

We make the following definitions and assumptions. First we assume that all parties have an initial opinion 0 or 1, so no party has initially the opinion $\perp$ (the case where there exist parties with opinion $\perp$ can be disregarded for the lower bound and is handled separately for the upper bound). Recall that the set of parties N is numbered from 1 to n and assume rounds are numbered $0, 1, 2, \ldots$.

- Let $X_{ij} \in \{0,1\}$ be the current opinion of party j after round i of the Slush protocol was executed ($X_{0,j}$ describes the initial opinion of party j).
- Let Y_{ij} be the number of replies with opinion 1 that party j obtains in its sample of k parties in round i. Note that $Y_{ij} \sim \mathrm{Bin}(k,p_i)$.
- Let the state of the network be $S_i := \sum_{j=1}^{n} X_{ij}$, which describes the total number of parties whose current opinion is 1 in round i.
- Let $p_i := S_i/n$ be the relative share of parties with opinion 1 in round i, which corresponds to the probability that a sampled party has state 1.
- For $i \geq 1$, we define as $\Delta_i := S_i - S_{i-1}$, i.e., the absolute progress to 1-consensus (or 0-consensus for negative values).
- Let $\delta_i := \mathbb{E}(\Delta_i)/n$ be the expected relative progress in round i. We will later show that δ_i can also be expressed as a function $\delta : [0,1] \to \mathbb{R}$ that only depends on p_i (when viewing k, α as fixed values), such that $\delta_i = \delta(p_i)$. Subsequently, we establish a relation between the $\delta(p_i)$ for varying parameters k, α, in which case we denote it as $\delta^{k,\alpha}(p_i)$ (however, for conciseness we will refrain from using this superscript if only single, fixed values for k and α are involved).

Note that for $i \geq 1$, the quantities $X_{ij}, Y_{ij}, S_i, \Delta_i$, are random variables. This is not the case for the *expected* relative progress δ_i, which, for fixed α, k, can be expressed only in terms of p_i, i.e., δ_i can be expressed as a function $\delta(p_i)$ that depends only on p_i.

Lemma 1. *Let $k/2 < \alpha \leq k$. Then*

$$\delta_i = \delta(p_i) := \sum_{\ell=\alpha}^{k} \binom{k}{\ell} \left[p_i^\ell (1-p_i)^{k-\ell+1} - (1-p_i)^\ell p_i^{k-\ell+1} \right].$$

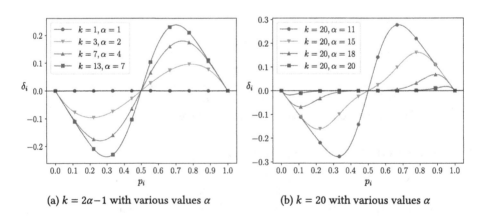

(a) $k = 2\alpha - 1$ with various values α (b) $k = 20$ with various values α

Fig. 1. Plots of $\delta(p_i)$ for different parameters k and α. For $k = 2\alpha - 1$ the expected progress for larger α dominates those for smaller (note that in the extreme case $k = \alpha = 1$ there is no expected progress). For fixed k the opposite is true. The combination $k = 20, \alpha = 15$ was suggested by the whitepaper [13]. Note that $\delta(p_i)$ is point-symmetric with respect to the point $(\frac{1}{2}, 0)$.

3.2   Mapping out the Dependency of δ_i on k and α

Recall that we define $\delta^{k,\alpha}(p_i)$ as the function defining the rate of progress for parameters α, k depending on p_i. It is interesting that for $k = 2\alpha - 1$ we have $\delta^{k,\alpha} = \delta^{k,\alpha+1}$ as the first summand in the expression $\delta^{k,\alpha}(p_i)$ from Lemma 1 is zero. Another interesting observation is that in the marginal case $k = \alpha = 1$, we have $\delta^{k,\alpha} = 0$ (see Fig. 1a), i.e., there is no expected progress and Slush essentially degenerates into a random walk, and it is not too hard to show that in this case Slush takes $\Omega(n^2)$ rounds in expectation.

In this section we establish non-trivial relations and claims for $\delta^{k,\alpha}(p_i)$. First, we show that $\delta^{k,\alpha}(p_i)$ is always larger (in absolute value) than $\delta^{k,\alpha'}(p_i)$ for $\alpha' > \alpha$ (see Fig. 1b for a visualized example of this claim). This means that choosing the smallest α with $\alpha > k/2$ (i.e., $\alpha = \lceil \frac{k+1}{2} \rceil$) is best in terms of the expected rate of progress $\delta^{k,\alpha}(p_i)$. While this is useful on its own regarding the choice of α in practice, we utilize this later to essentially eliminate α from the of analysis for the lower bound of the rate of convergence to consensus.

Lemma 2. *For fixed k, let $k/2 < \alpha \le k$ and consider $\alpha' > \alpha$. Then for any p_i we have $|\delta^{k,\alpha}(p_i)| \ge |\delta^{k,\alpha'}(p_i)|$.*

In particular, for $\alpha = \lceil \frac{k+1}{2} \rceil$ we can express $\delta^{k,\alpha}(p_i)$ in a different form, which we will use frequently in subsequent lemmas to simplify (or enable) subsequent proofs.

Lemma 3. *Let $\alpha = \lceil \frac{k+1}{2} \rceil$. Then*

$$\begin{cases} \delta^{k,\alpha}(p_i) = \left[\sum_{\ell=\alpha}^{k} \binom{k}{\ell} p_i^{\ell}(1-p_i)^{k-\ell} \right] - p_i, & k \ odd \\ \delta^{k,\alpha}(p_i) = \left[\sum_{\ell=\alpha}^{k} \binom{k}{\ell} p_i^{\ell}(1-p_i)^{k-\ell} \right] - p_i + \binom{k}{\alpha-1} p_i^{\alpha}(1-p_i)^{\alpha-1}, & k \ even. \end{cases}$$

The next two lemmas show, perhaps surprisingly, that for odd $k = 2\alpha - 1$ the expected progress functions $\delta^{k,\alpha}(p_i)$ and $\delta^{k-1,\alpha}(p_i)$ coincide. When combined with Lemma 2, this will later, when we cover lower bounds, allow us to focus our analysis solely on odd values of the form $k = 2\alpha - 1$ as the corresponding claim for even $k = 2\alpha - 2$ is implied.

Lemma 4. *For $k > \alpha \ge 1$ the following equation holds*

$$\sum_{\ell=\alpha}^{k} \binom{k}{\ell} p_i^{\ell}(1-p_i)^{k-\ell} = \left[\sum_{\ell=\alpha}^{k-1} \binom{k-1}{\ell} p_i^{\ell}(1-p_i)^{k-\ell-1} \right] + \binom{k-1}{\alpha-1} p_i^{\alpha}(1-p_i)^{k-\alpha}.$$

Lemma 5. *Let $k = 2\alpha-1$ and $\alpha \ge 2$. Then $\delta^{k,\alpha} = \delta^{k-1,\alpha}$.*

As our final structural claim in this subsection, we show that for $k = 2\alpha-1$ or $k = 2\alpha-2$, the expected progress towards consensus $|\delta^{k,\alpha}(p_i)|$ does not decrease as α (and thereby k) increases, an example is given in Fig. 1a. We will use the this lemma later to extend an upper bound for the number of rounds to a consensus from small values of α to large (where $k = 2\alpha-2$ or $k = 2\alpha-1$).

Lemma 6. *Let $k = 2\alpha-1$ and $k' = 2\alpha'-1$ for $\alpha > \alpha' > 1$. Then for any p_i we have $|\delta^{k,\alpha}(p_i)| \geq |\delta^{k',\alpha'}(p_i)|$. The same holds for even $k = 2(\alpha-1)$ and $k' = 2(\alpha'-1)'$.*

Note that while increasing α with $k = 2\alpha-1$ or $k = 2\alpha$ does in fact strictly increase $|\delta^{\alpha,k}(p_i)|$ thereby speeding up the expected time to consensus, this effect is rather limited, as we shall see in the next section.

4   Bounding the Time to Consensus for Slush

We will now show how the expected progress $\delta(p_i)$ can be used to obtain bounds for the number of rounds required to obtain consensus.

4.1   Lower Bound

As the system converges to consensus, arguably the most critical phase is when the network is roughly in balanced state, i.e., where fractions of parties with opinion 0 and 1 are roughly equal ($p_i \approx 1/2$) and where progress $\delta(p_i)$ towards consensus is close to 0, see Fig. 1.

To lead us out of a potential perfect balance, the system can only rely on pure randomness to gain some small initial imbalance, as the expected progress is 0. (The best one can hope for is a deviation of $p_i \approx 1/2 + c/\sqrt{n}$ for some constant c within reasonable time bounds, due to the central limit theorem). After that initial perturbation the convergence to consensus crucially depends on how fast the expected progress for the next round grows in parameter p_i.

Indicative for the change in progress is the derivative of $\delta(p_i)$, whose upper bound is useful to analyze the case where the system moves to a 1-consensus (w.l.o.g., due to a symmetry argument). Intuitively, this limits how fast the expected progress increases from an almost balanced state. We will first restrict ourselves to the case $k = 2\alpha-1$, as the previous section gives us all tools to extend this result to general k and α, as will be shown formally afterwards.

Lemma 7. *Let $k = 2\alpha - 1$. For $p_i \geq 1/2$ it holds that $\frac{\partial \delta(p_i)}{\partial p_i} \leq k - 1$.*

Next, we show that the progress towards consensus in a single round is limited, in particular around the balanced state. Here we utilize two tools. First of all, we employ Lemma 7 that bounds the progress around an almost balanced state but only for the "well behaved" case $k = 2\alpha-1$. Second, we use Lemma 5 to extend this to even Lemma 2 to extend it to any k and α for which $\frac{k}{2} < \alpha \leq k$.

Lemma 8. *Let $k \geq 2$ and $\frac{k}{2} < \alpha \leq k$ and $S_i \geq \frac{n}{2}$ (w.l.o.g.). Then $\Delta_{i+1} > (k-1)\left(S_i - \frac{n}{2}\right) + t\sqrt{n}$ with probability at most $\frac{1}{t^2}$, for any $t \geq 1$.*

Building on the previous lemma, we can give the following probabilistic bound for the number of parties that have opinion 1 after i rounds.

Lemma 9. *Let $k \geq 2$, $\frac{k}{2} < \alpha \leq k$. Assume the system is in a roughly balanced state with $S_0 \leq \frac{n}{2} + f(n)$ for $f(n) = \sqrt{n \log n}$. Then for any $i \leq \frac{\log n}{c}$ it holds $S_i > \frac{n}{2} + (k+1)^i f(n)$ with probability at most $1/c$ for any $c \geq 1$.*

We have all tools to deduce the lower bound for the required number of rounds of Slush to get moderately close to a consensus state with some moderate probability.

Theorem 1. *For $k \geq 2$ and any $\frac{k}{2} < \alpha \leq k$ and sufficiently large n, running Slush for at most $\frac{\log n}{3 \log(k+1) \log \gamma}$ rounds there is a majority opinion with at least $\frac{n}{2} + \frac{n}{\gamma}$ parties with probability at most $\frac{1}{\log(k+1) \log \gamma}$ for any constant $\gamma \geq 2$.*

We express the theorem above in a simpler, albeit weaker form.

Corollary 1. *For $k \geq 2$ and any $\frac{k}{2} < \alpha \leq k$, Slush takes $\Omega\left(\frac{\log n}{\log k}\right)$ rounds in expectation to reach a stable consensus (as defined in Definition 1).*

4.2   Upper Bound

We show how to use the structural insights about Slush with respect to parameters k and α to extend known upper bounds for the so called Median protocol, the 3-Majority protocol and the 2-Choices protocol. These are usually conceptualized for the case of multiple (> 2) opinions and are defined as follows.

Definition 8 (cf. [2]).   *The* Median protocol *assumes some globally known total order among opinions. In each round, each party samples the opinion of two others and adopts the median among those two and its own.*

In the 3-Majority *protocol, in each round, each party samples the opinion of three others and adopts the majority opinion, or picks a random opinion among the three in case of a tie.*

In the 2-Choices *protocol, in each round, each party samples the opinion of two others and then applies the 3-Majority rule, defaulting to its own opinion in case of a tie.*

We make the following observation.

Remark 1. Assume that all parties have initially only one of two opinions (i.e., the binary case, in particular, there are no parties with opinion $\perp$). Then the Median protocol, 2-choices protocol and Slush for $k = 2$ and $\alpha = 2$ are all equivalent. This is because in the *binary case*, in all three protocols a given party will switch its own opinion if and only if it samples two parties that both have a different opinion from its own. Under the same circumstances and for the same reason, the 3-Majority protocol is equivalent to Slush for $k = 3$ and $\alpha = 2$ (exploiting that there can never be a tie in the binary case).

There has been extensive research on the dynamics of the Median, 2-Choices and 3-Majority protocol (Definition 8) and the techniques are for the most part analogous or at least quite similar if the number of opinions is kept constant. We have already established the lower bound of $\Omega\left(\frac{\log n}{\log k}\right)$, i.e., the *additional* speedup one can gain by increasing the query size k diminishes very fast. Therefore, we do not deem it particularly worthwhile to show an upper bound that strictly improves on the $O(\log n)$ bound for the aforementioned cases.

Furthermore, it is *not* the scope of this paper to give detailed proofs of slight generalizations of those for the protocols from Definition 8. To keep this paper reasonably self-contained we showcase how these proofs generalize to Slush with arbitrary $k \geq 2$ and $\alpha = \lceil \frac{k+1}{2} \rceil$. We will give an extended proof sketch that shows how the existing proof techniques generalize to obtain the following theorem. For more details we refer to the according sources (in particular [3,6]).

Theorem 2. *Let $k \geq 2$ and $\alpha = \lceil \frac{k+1}{2} \rceil$. Then Slush reaches a state where all but $n - O(\sqrt{n})$ parties have the same opinion in $O(\log n)$ rounds with high probability, even in the presence of a $\sqrt{n}$-bounded adversary.*

We can translate the above result into the notion of concentration with all but negligible probability (Definition 5) by adding a factor of β to the running time that gives more control over the level of security in particular for small n (see full paper for the details). Specifically, the corollary conforms to Definition 7, as the runtime is polynomial in β.

Corollary 2. *Let $k \geq 2$ and $\alpha = \lceil \frac{k+1}{2} \rceil$. Then Slush reaches a stable consensus in $O(\log n + \beta)$ rounds with all but negligible probability (with respect to β), even in the presence of a $\sqrt{n}$-bounded adversary.*

5   Dynamics of Snowflake and Snowball

In this section we are going to extend the lower bound for Slush derived in Sect. 4 to the Snowflake and Snowball protocols. Note that the quantities $S_i, p_i, \Delta_i, \delta_i$ can be defined the same as in Slush, see Sect. 3, since the variables only depend on the opinion attribute of parties, which is present in all three protocols. However, the actual (expected) changes of these quantities in this section can and will differ from those in Slush. We will denote these quantities with superscripts (slush, flake, ball) in case we compare them over protocols (but avoid this whenever possible). Moreover, we condition the results of this section on the assumption that no node decides (finalizes) their opinion, before the system reaches a stable majority and consider the repercussions at the end.

Snowball, as explained in Sect. 2.2, augments the consensus mechanism from Slush with the concept of *confidence* associated to the current value, which influences the decision of a party to change its opinion. In a nutshell, in the Snowball protocol a node changes its opinion when the cumulative number of queries with a majority for the new opinion exceeds that for the old opinion. In the Slush protocol, the variable S_i was sufficient to describe the expected progress required to predict the evolution of the system, which is not the case in Snowball anymore, since aforementioned confidence levels play a crucial role.

Definition 9. *Define the set L_i^c to be the set of parties in round i such that $\text{cnt}(1) - \text{cnt}(0) = c$ (where $\text{cnt}(b)$ is the number of queries of a given party that had a majority of opinion b). We further divide the set L_i^0 in two subsets $L_i^{0,0}$ and $L_i^{0,1}$. Parties in L_i^0 ($L_i^{0,1}$) that have opinion 0 (1) belong to $L_i^{0,0}$ ($L_i^{0,1}$).*

The variable S_i can be reconstructed as follows: $S_i = |L_i^{0,1}| + \sum_{c>0} |L_i^c|$. Given a round $i > 0$, the set of parties $\mathcal{N}$ is contained in $\bigcup_{c=-i}^{i} L_i^c$ as the end of round i, since every party performed i queries by the end of round i, thus the difference in counts c is bounded between $-i$ and i.

Remark 2. Consider the collection $\mathcal{L}_i := \{L_i^c\}_{c=-i}^{i} \cup \{L_i^{0,0}, L_i^{0,1}\}$ of disjoint sets. The evolution of the system in the next round $i+1$ can now be described using this set $\mathcal{L}_i$. After a query is performed a party in L_i^c moves to set L_i^{c+1} if the query had a majority for 1, to L_i^{c-1} if the query had a majority for 0, or L_i^c if the query had no majority. The only parties that can change value after round i are the parties contained in L_i^0.

Definition 10. *For $i \geq 1$, we define the absolute progress as $\Delta_i := S_i - S_{i-1} = |L_i^{0,1}| - |L_{i-1}^{0,1}| + \sum_{c=1}^{i}(|L_i^c| - |L_{i-1}^c|)$, i.e., the number of parties with 1 in their view in round i minus the number of parties with 1 in their view in round $i-1$. As before, we define the expected relative progress as $\delta_i := \mathbb{E}(\Delta_i)/n$.*

The following Lemma shows that in Snowball, Δ_i is only affected by parties migrating from $L_{i-1}^{0,1}$ or $L_{i-1}^{0,0}$ in round $i-1$ to L_i^{-1} or L_i^1 in round i, respectively.

Lemma 10. *The absolute progress can be expressed as $\Delta_i = |\Lambda_i^1| - |\Lambda_i^0|$, where $\Lambda_i^0 := L_i^{-1} \cap L_{i-1}^{0,1}$ and $\Lambda_i^1 := L_i^1 \cap L_{i-1}^{0,0}$.*

An interesting interpretation of Lemma 10 if the following. Since the parties contained in the set L_i^0 are the only parties that can change their opinion, the expected progress of Snowball in a given round is the same as the expected progress of Slush restricted to the parties in L_i^0. We formalize this intuition in the following lemma.

Lemma 11. *The expected absolute progress of Snowball is at most as high as in Slush, i.e., $\delta_i^{ball} \leq \delta_i^{slush}$.*

Lemma 11 states that the expected progress of the Snowball protocol is upper-bounded by the expected progress of the Slush protocol. We conclude that the expected number of rounds even to reach majority of a constant fraction of nodes of one opinion of the Snowball protocol is lower-bounded by the Slush protocol, if no node decides prematurely. Note that the same is clearly true for as Snowflake which is essentially equal to Slush if no node decides prematurely.

Corollary 3 (cf. Corollary 1). *For $k \geq 2$ and any $\frac{k}{2} < \alpha \leq k$, Snowflake and Snowball take $\Omega\left(\frac{\log n}{\log k}\right)$ rounds in expectation to reach a state state of stable consensus, assuming that nodes do not decide before such a state is reached.*

Note that since the decision mechanism in Snowflake and Snowball implies that no node can decide before β rounds have passed the corollary implies a lower bound of $\Omega\left(\min\left(\frac{\log n}{\log k}, \beta\right)\right)$ rounds. Furthermore, we will see in Sect. 6 that the dependence of the runtime on β behaves much worse than $\Omega(\beta)$ as the adversary can exploit the decision mechanism to delay a decision *super-polynomially* in β.

6   Security of Snowflake and Snowball

We show that Snowflake and Snowball has a vulnerability towards an adversary that intends to delay consensus (as defined in Definition 2). In particular, there is an unfavorable trade-off between confidence of success and latency. In particular, the mechanism that Snowflake and Snowball protocols use introduces a security parameter β to control the probability of failure of obtaining a consensus (according to Definition 2). We show that this mechanism to make decision allows an adversary to delay the decision of any given party when using the consensus mechanisms of Snowflake and Snowball for a super-polynomial number of rounds in β. This is independent of the current state of the system, i.e., the claim is true even if the system is in a state of a stable consensus (see Definition 1) and is true for a weaker notion of the F-bounded adversary from Definition 3 for a small F.

Definition 11. *A weak F-bounded adversary controls up to F undecided parties whose state (opinion) it can set once each round. We call these influenced parties and in particular we assume that the adversary can reset any decision on some opinion.*

We start by giving a lower bound for the probability that some party samples a majority of influenced parties.

Lemma 12. *The probability that a random sample of k parties contains at least α that are influenced by a weak F-bounded adversary is at least $\left(\frac{F}{n}\right)^k$.*

Note that even though the probability above decreases with k, the parameter k is considered a small, constant sized tuning parameter [13] and is not a proper security parameter, particularly since the message complexity scales in $\Omega(kn)$.

Interestingly, the lemma shows that even in a stable consensus (according to Definition 1), i.e., where almost all nodes share the same opinion, even a weak adversary can create a small but inherent "background noise", i.e., an expected fraction $\left(\frac{F}{n}\right)^k$ of all parties can be reverted by the adversary to the minority opinion each round, because it gains a majority in a sample.

This situation is what the security mechanism of Snowflake and Snowball protocols is intended for as it makes parties decide and finalize an opinion by introducing an according mechanism with a security parameter β. One condition for some party to decide an opinion, is that it must have at least β consecutive queries with an α-majority of the same opinion, consider the full paper for detailed pseudocode. (Note that Snowball imposes an additional condition for parties to decide based on the history of queries, which, however, only delays the decision of a party down even further).

The idea behind this mechanism is to reduce the probability that an adversary can make some party accept the *minority* opinion, since sampling the minority opinion β times in a row has a probability that is negligible with respect to β. This mechanism is flawed in the sense that even an adversary that influences just a single party can abuse it to introduce a delay to the decision of a party that scales badly in β.

Lemma 13. *In the Snowflake and Snowball protocols, there exists a value $c > 1$ which is constant in β, such that a weak F-bounded adversary (Definition 11, for any $F \geq 2$) can ensure that the probability that a given party decides within $c^{\beta-1}/2$ queries (rounds) is at most $4/c^{2(\beta-1)}$. To enforce this, the adversary needs no information on the current state of the network.*

Using the lemma above, we show that Snowflake and Snowball cannot satisfy Definition 7, which states the conditions for a mechanism that provides a decent trade-off between security and performance. Specifically, the following theorem shows that having consensus with all but negligible probability w.r.t., β and a polynomial runtime in β are mutually exclusive.

Theorem 3. *In the Snowflake and Snowball protocol with a weak F-bounded adversary ($F \geq 2$) the following properties are mutually exclusive*

- *The protocol ensures consensus with all but negligible probability with respect to β (cf. Definition 2).*
- *Parties decide with less than $\pi(\beta)$ queries with all but negligible probability with respect to β, for any fixed polynomial π.*

Note that this holds even when the definition of consensus is restricted to those parties which are not influenced by the adversary.

7   Reconciling Security and Fast Consensus

Theorem 3 shows that the Snowflake and Snowball protocols cannot achieve consensus within a polynomial number of rounds with all but negligible probability with respect to security parameter β, due to the termination condition. We propose a modification of the Slush protocol that we call *Blizzard* which incorporates confidence levels as a termination criterion (a description is given, pseudocode is available in the full paper). Importantly, Blizzard leaves the basic dynamics of the Slush protocol intact, in particular, it neglects the mechanism of Snowball, where parties change their current opinion depending not only on the current but also on past queries, which is not helpful for the time to converge to a stable consensus as shown in Sect. 5.

This modification is arguably simpler than the corresponding mechanisms in Snowball and works as follows. Each party maintains two counters, which track the total number of queries that contained at least α of opinion 0 or 1, respectively (an α-majority). A decision in favor of one opinion is made if the corresponding counter has a decisive lead over the other. The lead that is required is of the order $O(\log n + \beta)$ and we show that this also corresponds to the number of rounds until consensus is established with all but negligible probability (w.r.t. β).

The idea is that within the given time frame, the network will reach a state of a stable consensus, and it will remain close to this state for a sufficiently long time such that each party can establish a lead in the counter for the opinion which is in the majority. Furthermore, unanimity is ensured because the required lead is large enough such that no party can accidentally make a "premature" decision, i.e., reaching the threshold even when no stable majority has been established yet.

Note that as time progresses and given an adversary that controls at least α parties, there is always a small but non-zero chance that a system reverts from a state of stable consensus and even switches majorities. However, we show that once the system is in a stable consensus, it will not "slide back" too far within a given time frame, i.e., one opinion retains an overwhelming majority for a sufficient amount of time, even in the presence of an adversary. We start with a lemma about the probabilities to sample an α-majority of an opinion given that one opinion has a majority.

Lemma 14. *Let $S_i \geq \frac{15n}{16}$ and $\alpha = \lceil \frac{k+1}{2} \rceil$. Then $\mathbb{P}(Y_{ij} \geq \alpha) \geq p_i$ and $\mathbb{P}(Y_{ij} \leq k-\alpha) \leq 4(1-p_i)^2$.*

The next lemma shows that once the network is in a stable consensus state, it will very likely conserve a majority for a certain time frame.

Lemma 15. *Let $s, t \geq 1$ with $s \leq \frac{t}{2}$ and $t \leq \sqrt{n}/16$. Assume the number of parties with opinion 1 is currently $S_0 \geq n - s\sqrt{n}$. Then $S_i \geq n - t \cdot \sqrt{n}$ for at least $i \leq T := \min\left(\frac{\sqrt{n}}{32t}, \frac{t}{4}\right)$ rounds with probability at least e^{-t^2}. This holds even with a $\sqrt{n}$-bounded adversary.*

While Lemma 15 captures the stability of a state of almost consensus in a more general way, we will use it in the following form, by specifying some parameters.

Lemma 16. *Let $s \geq 1$ be a constant and assume the number of parties with opinion 1 is $S_0 \geq n - s\sqrt{n}$. For an arbitrary constant $\beta \geq s/2$, let $T \geq \beta$ and $T \in o(n^{1/4})$. Then $S_i \geq \frac{15n}{16}$ for at least T rounds with probability at least $1 - e^{-\beta^2}$. This holds for sufficiently large n even with a $\sqrt{n}$-bounded adversary.*

We will now use the stability properties of a network that is in a state of stable consensus (Definition 1) to show that Blizzard ensures consensus (Definition 2) with all but negligible probability after $O(\beta + \log n)$ rounds. Recall that Blizzard essentially corresponds to running an instance of Slush, where each node maintains counters that track how often an α-majority was observed for opinion 0 and 1, respectively. If the difference in counters reaches a (sufficiently large) threshold τ, then a decision for the opinion with the larger counter is made. In the following theorem we use a variable running time for Slush (due to the fact that as speedups up to $\log k$ over our upper bound can not be excluded, see Theorems 1 and 2).

Theorem 4. *Algorithm Blizzard with a threshold $\tau := 2T_{Slush}$ ensures consensus with all but negligible probability (w.r.t. β) after at most $7T_{Slush}$ rounds, assuming that T_{Slush} is the number of rounds until Slush reaches a state where at least $n - O(\sqrt{n})$ parties have the same opinion and T_{Slush} is a sufficiently large multiple of β. This holds even with a $\sqrt{n}$-bounded adversary.*

The theorem above is given in a more general way, depending on the number of rounds until Slush reaches a stable consensus. Given that we have already shown an upper bound of $O(\log n + \beta)$ for this (Corollary 2), we rephrase the theorem as follows.

Corollary 4. *Blizzard ensures consensus with all but negligible probability after at most $O(\log n + \beta)$ rounds. This holds even with a $\sqrt{n}$-bounded adversary.*

8   Conclusion

With the goal of improving latency in mind, we deduce two main recommendations for changes to Snow-style consensus protocols as they are deployed in the Avalanche network today. First, for a given k we recommend to choose $\alpha > k/2$ as small as possible, i.e., $\alpha = \lceil \frac{k+1}{2} \rceil$, as this promises a better performance compared to α closer to k (see Lemma 2 or Fig. 1b). Second, we propose to change the termination condition of the Snowflake and Snowball protocols where we observe an unfavorable trade-off between security and latency (see Corollary 3) to the simpler one of the Blizzard protocol. This modification will resolve the observed issue (see Theorem 4 and Corollary 4). We caveat our recommendations by noting that there might be other considerations than the asymptotic performance aspects analyzed in this paper.

From a theoretical point of view, we see our results on the performance of Slush (Theorem 1 and 2) as a natural continuation of the corresponding analysis of the randomized, self-stabilizing consensus protocols in the GOSSIP model where the sample size is at most 3 (such as the Median Protocol, the 2-Choices Protocol and the 3-Majority Protocol, see Definition 8). These protocols have been analyzed with respect to performance as a function of the initial number of opinions and it was shown that in "most" conditions they converge quite fast (a discussion on this is provided in the full paper). An interesting avenue of future research is the adaptation of Slush to multiple opinions, a party adopts a new opinion if it has a simple majority in a sampling of size k, i.e., the k-Majority protocol. We believe that our technical work on Slush gives insights how such a k-Majority protocol would perform on multiple opinions.

References

1. Amores-Sesar, I., Cachin, C., Tedeschi, E.: When is spring coming? A security analysis of avalanche consensus. In: 26th International Conference on Principles of Distributed Systems, OPODIS 2022, 13–15 December 2022, Brussels, Belgium. Ed. by Eshcar Hillel, Roberto Palmieri, and Etienne Rivière, vol. 253. LIPIcs. Schloss Dagstuhl - Leibniz-Zentrum für Informatik, pp. 10:1–10:22 (2022). https://doi.org/10.4230/LIPIcs.OPODIS.2022.10
2. Becchetti, L., Clementi, A., Natale, E.: Consensus dynamics: an overview. SIGACT News **51**(1), 58–104 (2020). https://doi.org/10.1145/3388392.3388403
3. Becchetti, L., et al.: Simple dynamics for plurality consensus. Distrib. Comput. **30**(4), 293–306 (2017). https://doi.org/10.1007/s00446-016-0289-4

4. Becchetti, L., et al.: Stabilizing consensus with many opinions. In: Proceedings of the Twenty-Seventh AnnualACM-SIAM Symposium on Discrete Algorithms, SODA 2016, Arlington, VA, USA, 10–12 January 2016. Ed. by Robert Krauthgamer. SIAM, 2016, pp. 620–635 (2016). https://doi.org/10.1137/1.9781611974331.ch46

5. Buterin, V.: The Scalability Trilemma—Why Sharding is Great: Demystifying the Technical Properties. https://vitalik.ca/general/2021/04/07/sharding.html.2017

6. Doerr, B., et al.: Stabilizing consensus with the power of two choices. In: SPAA 2011: Proceedings of the 23rd Annual ACM Symposium on Parallelism in Algorithms and Architectures, San Jose, CA, USA, 4–6 June 2011 (Co-located with FCRC 2011). Ed. by Rajmohan Rajaraman and Friedhelm Meyer auf der Heide. ACM, 2011, pp. 149–158 (2011). https://doi.org/10.1145/1989493.1989516

7. Elsässer, R., et al.: Brief announcement: rapid asynchronous plurality consensus. In: Proceedings of the ACM Symposium on Principles of Distributed Computing, PODC 2017, Washington, DC, USA, 25–27 July 2017. Ed. by Elad Michael Schiller and Alexander A. Schwarzmann. ACM, 2017, pp. 363–365 (2017). https://doi.org/10.1145/3087801.3087860

8. Fischer, M.J., Lynch, N.A., Paterson, M.S.: Impossibility of distributed consensus with one faulty process. J. ACM $32(2)$, 374–382 (1985). https://doi.org/10.1145/3149.214121

9. Ghaffari, M., Lengler, J.: Nearly-tight analysis for 2-choice and 3-majority consensus dynamics. In: Proceedings of the 2018 ACM Symposium on Principles of Distributed Computing, PODC 2018, Egham, United Kingdom, 23–27 July 2018. Ed. by Calvin Newport and Idit Keidar. ACM, 2018, pp. 305–313 (2018). https://dl.acm.org/citation.cfm?id=3212738

10. Gilbert, S., Lynch, N.: Brewer's conjecture and the feasibility of consistent, available, partition-tolerant web services. SIGACT News $33(2)$, 51–59 (2002)

11. Gramoli, V., et al.: Diablo: a benchmark suite for blockchains. In: Proceedings of the Eighteenth European Conference on Computer Systems, EuroSys 2023, Rome, Italy, 8–12 May 2023. Ed. by Giuseppe Antonio Di Luna et al. ACM, 2023, pp. 540–556 (2023). https://doi.org/10.1145/3552326.3567482

12. Pease, M., Shostak, R., Lamport, L.: Reaching agreement in the presence of faults. J. ACM $27(2)$, 228–234 (1980). https://doi.org/10.1145/322186.322188

13. Rocket, T., et al.: Scalable and Probabilistic Leaderless BFT Consensus through Metastability. CoRR abs/1906.08936 (2019). arXiv: 1906.08936, http://arxiv.org/abs/1906.08936

Awake Complexity of Distributed Minimum Spanning Tree

John Augustine[1] , William K. Moses Jr.[2](✉) , and Gopal Pandurangan[3]

[1] Indian Institute of Technology Madras, Chennai, India
augustine@cse.iitm.ac.in
[2] Durham University, Durham, UK
wkmjr3@gmail.com
[3] University of Houston, Houston, TX, USA
gopal@cs.uh.edu

Abstract. The *awake complexity* of a distributed algorithm measures the number of rounds in which a node is awake. When a node is not awake, it is *sleeping* and does not do any computation or communication and spends very little resources. Reducing the awake complexity of a distributed algorithm can be relevant in resource-constrained networks such as sensor networks, where saving energy of nodes is crucial. Awake complexity of many fundamental problems such as maximal independent set, maximal matching, coloring, and spanning trees have been studied recently.

In this work, we study the awake complexity of the fundamental distributed minimum spanning tree (MST) problem and present the following results.

- **Lower Bounds.**
  1. We show a lower bound of $\Omega(\log n)$ (where n is the number of nodes in the network) on the awake complexity for computing an MST that holds even for randomized algorithms.
  2. To better understand the relationship between the awake complexity and the round complexity (which counts both awake and sleeping rounds), we also prove a *trade-off* lower bound of $\tilde{\Omega}(n)$ (throughout, the $\tilde{O}$ notation hides a polylog n factor and $\tilde{\Omega}$ hides a $1/(\text{polylog } n)$ factor) on the product of round complexity and awake complexity for any distributed algorithm (even random-

Part of the work was done while the William K. Moses Jr. was a Post Doctoral Fellow at the University of Houston.
J. Augustine was supported, in part, by DST/SERB MATRICS Grant MTR/2018/001198 and the Centre of Excellence in Cryptography Cybersecurity and Distributed Trust under the IIT Madras Institute of Eminence Scheme and by the VAJRA visiting faculty program of the Government of India.
W. K. Moses Jr. was supported, in part, by NSF grants CCF-1540512, IIS-1633720, and CCF-1717075 and BSF grant 2016419.
G. Pandurangan was supported, in part, by NSF grants CCF-1540512, IIS-1633720, and CCF-1717075 and BSF grant 2016419 and by the VAJRA visiting faculty program of the Government of India.

Y. Emek (Ed.): SIROCCO 2024, LNCS 14662, pp. 45–63, 2024.
https://doi.org/10.1007/978-3-031-60603-8_3

ized) that outputs an MST. Our lower bound is shown for graphs having diameter ranging from $\tilde{\Omega}(\sqrt{n})$ to $\tilde{\Omega}(n)$.

– **Awake-Optimal Algorithms.**

1. We present a distributed randomized algorithm to find an MST that achieves the optimal awake complexity of $O(\log n)$ (with high probability). Its round complexity is $O(n \log n)$. We note that by our trade-off lower bound, in general (in terms of n), this is the best round complexity (up to logarithmic factors) for an awake-optimal algorithm.

2. We also show that the $O(\log n)$ awake complexity bound can be achieved deterministically as well, by presenting a distributed *deterministic* algorithm that has $O(\log n)$ awake complexity and $O(n \log^5 n)$ round complexity. We also show how to reduce the round complexity to $O(n \log n \log^* n)$ at the expense of a slightly increased awake complexity of $O(\log n \log^* n)$.

– **Trade-Off Algorithms.** To complement our trade-off lower bound, we present a parameterized family of distributed algorithms that gives an essentially optimal trade-off (up to polylog n factors) between the awake complexity and the round complexity. Specifically we show a family of distributed algorithms that find an MST of the given graph with high probability in $\tilde{O}(D + 2^k + n/2^k)$ round complexity and $\tilde{O}(n/2^k)$ awake complexity, where D is the network diameter and integer k is an input parameter to the algorithm. When $k \in [\max\{\lceil 0.5 \log n \rceil, \lceil \log D \rceil\}, \lceil \log n \rceil]$, we can obtain useful trade-offs.

Our work is a step towards understanding resource-efficient distributed algorithms for fundamental global problems such as MST. It shows that MST can be computed with any node being awake (and hence spending resources) for only $O(\log n)$ rounds which is significantly better than the fundamental lower bound of $\tilde{\Omega}(\text{Diameter}(G) + \sqrt{n})$ rounds for MST in the traditional CONGEST model, where nodes can be active for at least so many rounds.

Keywords: Minimum Spanning Tree · Sleeping model · energy-efficient · awake complexity · round complexity · trade-offs

1   Introduction

We study the distributed minimum spanning tree (MST) problem, a central problem in distributed computing. This problem has been studied extensively for several decades starting with the seminal work of Gallagher, Humblet, and Spira (GHS) in the early 1980s [16]; for example, we refer to the survey of [31] that traces the history of the problem till the state of the art. The round (time) complexity of the GHS algorithm is $O(n \log n)$ rounds, where n is the number of nodes in the network. The round complexity of the problem has been continuously improved since then and now tight optimal bounds are known.[1] It is now

---

[1] Message complexity has also been well-studied, see e.g., [31], but this is not the focus of this paper, although our algorithms are also (essentially) message optimal—see Sect. 1.2.

well-established that $\Theta(D + \sqrt{n})$ is essentially (up to logarithmic factors) a tight bound for the round complexity of distributed MST in the standard CONGEST model [11, 14, 26, 33]. The lower bound applies even to randomized Monte Carlo algorithms [11], while deterministic algorithms that match this bound (up to logarithmic factor) are now well-known (see e.g., [15, 26, 30, 32]). Thus, the round complexity of the problem in the traditional CONGEST distributed model is settled (see also the recent works of [18, 20]). In the CONGEST model, any node can send, receive, or do local computation in any round and only $O(\log n)$-sized messages can be sent through any edge per round.

MST serves as a basic primitive in many network applications including efficient broadcast, leader election, approximate Steiner tree construction etc. [31, 32]. For example, an important application of MST is for energy-efficient broadcast in wireless networks which has been extensively studied, see e.g., [1, 23].[2]

In resource-constrained networks such as sensor networks, where nodes spend a lot of energy or other resources over the course of an algorithm, a round complexity of $\tilde{O}(D + \sqrt{n})$ (which is essentially optimal) to construct an MST can be large. In particular, in such an algorithm, a node can be active over the entire course of the algorithm. It is worth studying whether MST can be constructed in such a way that each node is active only in a small number of rounds—much smaller than that taken over the (worst-case) number of rounds—and hence could use much less resources such as energy.

Motivated by the above considerations, in recent years several works have studied energy-distributed algorithms (see e.g., [3–7, 24]). This paper studies the distributed MST problem in the *sleeping model* [7]. In this model (see Sect. 1.1), nodes can operate in two modes: *awake* and *sleeping*. Each node can choose to enter the awake or asleep state at the start of any specified round. In the sleeping mode, a node cannot send, receive, or do any local computation; messages sent to it are also lost. The resources utilized in sleeping rounds are negligible and hence only awake rounds are counted. The goal in the sleeping model is to design distributed algorithms that solve problems in a small number of awake rounds, i.e., have small *awake complexity* (also called *awake time*), which is the (worst-case) number of awake rounds needed by any node until it terminates. This is motivated by the fact that, if the awake complexity is small, then every node takes only a small number of rounds during which it uses a significant amount of resources. For example, in ad hoc wireless or sensor networks, a node's energy consumption depends on the amount of time it is actively communicating with nodes. In fact, significant amount of energy is spent by a node even when it is just waiting to hear from a neighbor [7]. On the other hand, when a node is sleeping—when all of its radio devices are switched off—it spends little or no energy. While the main goal is to minimize awake complexity, we would also like to minimize the (traditional) *round complexity* (also called *time complexity*

---

[2] Note that for energy-efficient broadcast, the underlying graph is *weighted* and it is known that using an MST to broadcast minimizes the total cost [1, 23].

or run time) of the algorithm, which counts the (worst-case) total number of rounds taken by any node, including both awake and sleeping rounds.

The work of Barenboim and Maimon [3] shows that global problems such as broadcast and constructing a (arbitrary) spanning tree (but not an MST on weighted graphs, which is required for applications mentioned earlier) can be accomplished in $O(\log n)$ awake complexity (and $\tilde{\Theta}(n)$ round complexity) in the sleeping (CONGEST) model (see also the related result of [4]—Sect. 1.3.). The above work is significant because it shows that even such global problems can be accomplished in a very small number of awake rounds, bypassing the $\Omega(D)$ lower bound on the round complexity (in the traditional model). In this work, we focus on another fundamental global problem, namely MST, and study its awake complexity. We show that MST can be solved in $O(\log n)$ rounds awake complexity which we show is also *optimal*, by showing a matching *lower bound*. The upper bound is in contrast to the classical $\tilde{\Omega}(D + \sqrt{n})$ lower bound on the round complexity (in the traditional CONGEST model) where nodes can be awake for at least so many rounds. Another key issue we study is the relationship between awake complexity and round complexity of MST. It is intriguing whether one can obtain an algorithm that has both optimal awake and round complexity. We show that, in general (i.e., for graphs with diameter ranging from $\tilde{\Omega}(\sqrt{n})$ to $\tilde{\Theta}(n)$), this is not possible by showing a *trade-off* lower bound of $\tilde{\Omega}(n)$ on the product of awake and round complexity (see Sect. 1.2).

1.1 Distributed Computing Model and Complexity Measures

Distributed Network Model. We are given a distributed network modeled as an arbitrary, undirected, connected, weighted graph $G(V, E, w)$, where the node set V ($|V| = n$) represent the processors, the edge set E ($|E| = m$) represents the communication links between them, and $w(e)$ is the weight of edge $e \in E$.

The network diameter is denoted by D, also called the hop-diameter (that is, the unweighted diameter) of G, and in this paper by diameter we always mean hop-diameter. We assume that the weights of the edges of the graph are all distinct. This implies that the MST of the graph is unique. (The definitions and the results generalize readily to the case where the weights are not necessarily distinct.)

Each node hosts a processor with limited initial knowledge. We assume that nodes have unique IDs, and at the beginning of the computation each node is provided its ID as input and the weights of the edges incident to it. We assume node IDs are of size $O(\log n)$ bits. We assume that each node has ports (each port having a unique port number); each incident edge is connected to one distinct port. We also assume that nodes know n, the number of nodes in the network. For the deterministic algorithm, we make the additional assumption that nodes know the value of N, an upper bound on the largest ID of any node. Nodes initially do not have any other global knowledge and have knowledge of only themselves.

Nodes are allowed to communicate through the edges of the graph G and it is assumed that communication is *synchronous* and occurs in rounds. In particular, we assume that each node knows the current round number, starting from round 1. In each round, each node can perform some local computation (which happens instantaneously) including accessing a private source of randomness, and can exchange (possibly distinct) $O(\log n)$-bit messages with each of its neighboring nodes. This is the traditional *synchronous CONGEST* model.

As is standard, the goal, at the end of the distributed MST computation, is for every node to know which of its incident edges belong to the MST.

Sleeping Model and Complexity Measures. The *sleeping model* [7] is a generalization of the traditional model, where a node can be in either of the two states—sleeping or awake—before it finishes executing the algorithm (locally). (In the traditional model, each node is always awake until it finished the algorithm). Initially, we assume that all nodes are awake. That is, any node v, can decide to *sleep* starting at any (specified) round of its choice; we assume all nodes know the correct round number whenever they are awake. It can *wake up* again later at any specified round and enter the *awake* state. In the sleeping state, a node does not send or receive messages, nor does it do any local computation. Messages sent to it by other nodes when it was sleeping are lost. This aspect makes it especially challenging to design algorithms that have a small number of awake rounds, since one has to carefully coordinate the transmission of messages.

Let A_v denote the number of awake rounds for a node v before termination. We define the *(worst-case) awake complexity* as $\max_{v \in V} A_v$. For a randomized algorithm, A_v will be a random variable and our goal is to obtain high probability bounds on the awake complexity. Apart from minimizing the awake complexity, we also strive to minimize the overall (traditional) *round complexity* (also called *run time or time complexity*), where both, sleeping and awake rounds, are counted.

1.2 Our Contributions and Techniques

We study the awake complexity of distributed MST and present the following results (see Table 1 for a summary).

Table 1. Summary of our Results.

Algorithm	Type	Awake Time (AT)	Run Time (RT)	AT Lower Bound	AT × RT Lower Bound
*RANDOMIZED-MST	Randomized	$O(\log n)$	$O(n \log n)$	$\Omega(\log n)$	$\tilde{\Omega}(n)$†
DETERMINISTIC-MST	Deterministic	$O(\log n)$	$O(n \log^5 n)$	$\Omega(\log n)$	$\tilde{\Omega}(n)$†
Modified DETERMINISTIC-MST	Deterministic	$O(\log n \log^* n)$	$O(n \log n \log^* n)$	$\Omega(\log n)$	$\tilde{\Omega}(n)$†
*♣TRADE-OFF-MST	Randomized	$\tilde{O}(n/2^k)$	$\tilde{O}(D + 2^k + n/2^k)$	-	-

* The algorithm outputs an MST with high probability.
♣ This algorithm takes integer k as an input parameter.
† Holds for (some) graphs with diameter $\tilde{\Omega}(\sqrt{n})$.
Our lower bounds also apply to Monte Carlo randomized algorithms with constant success probability.
n is the number of nodes in the network and N is an upper bound on the largest ID of a node.

1. Lower Bound on the Awake Complexity of MST. We show that $\Omega(\log n)$ is a lower bound on the awake complexity of constructing an MST, even for randomized Monte Carlo algorithms with constant success probability (Sect. 2.1). We note that showing lower bounds on the awake complexity is different from showing lower bounds for round complexity in the traditional LOCAL model. In the traditional LOCAL model, *locality* plays an important role in showing lower bounds. In particular, to obtain information from a r-hop neighborhood, one needs at least r rounds and lower bounds use indistinguishability arguments of identical r-hop neighborhoods to show a lower bound of r to solve a problem. For example, this approach is used to show a lower bound of $\Omega(D)$ for leader election or broadcast even for randomized algorithms [25]. Lower bounds in the CONGEST model are more involved and exploit bandwidth restriction to show stronger lower bounds for certain problems, e.g., MST has a round complexity lower bound of $\tilde{\Omega}(D + \sqrt{n})$ rounds. In contrast, as we show in this paper, MST can be solved using only $O(\log n)$ awake rounds and requires a different type of lower bound argument for awake complexity.

We first show a randomized lower bound of $\Omega(\log n)$ awake complexity for *broadcast* on a line of n nodes. Our broadcast lower bound is an adaptation of a similar lower bound shown for the energy complexity model [4]. Our randomized lower bound is more general and subsumes a deterministic lower bound of $\Omega(\log n)$ shown in [3] for computing a spanning tree called the Distributed Layered Tree (by a reduction this lower bound applies to broadcast and leader election as well). The deterministic lower bound follows from the *deterministic message complexity* lower bound of $\Omega(n \log n)$ on leader election on rings (which also requires an additional condition that node IDs should be from a large range) to show that some node should be awake at least $\Omega(\log n)$ rounds. Note that this argument does not immediately apply for randomized algorithms, since such a message complexity lower bound does not hold for randomized (Monte Carlo) leader election [25]. On the other hand, our lower bound uses probabilistic arguments to show that some node has to be awake for at least $\Omega(\log n)$ rounds for accomplishing broadcast in a line. We then use a reduction to argue the same lower bound for MST problem on a *weighted ring*. We believe that the broadcast lower bound is fundamental to awake complexity and will have implications for several other problems as well.

2. Lower Bounds Relating Awake and Round Complexities. An important question is whether one can optimize awake complexity and round complexity *simultaneously* or whether one can only optimize one at the cost of the other. Our next result shows the latter is true by showing a (existential) lower bound on the *product* of awake and round complexities. Specifically, we construct a family of graphs with diameters ranging between $\tilde{\Omega}(\sqrt{n})$ to $\tilde{O}(n)$. Time-optimal MST algorithms for these graphs will have round complexity within polylog n factors of their diameters. We show that any distributed algorithm on this graph family that requires only $\tilde{O}(\text{Diameter}(G))$ rounds must have an awake complexity of at least $\tilde{\Omega}(n/\text{Diameter}(G))$ for any distributed algorithm. This holds even for Monte-Carlo randomized algorithms with constant success probability. The precise result is stated in Theorem 2 (see Sect. 2.2). In other words, the product of the round complexity and awake complexity is $\tilde{\Omega}(n)$. We note that this product lower bound is shown for graphs with diameter $\tilde{\Omega}(\sqrt{n})$, and we leave open whether a similar bound holds for graphs with much smaller diameters. In particular, our trade-off lower bound does not rule out a $\tilde{O}(D + \sqrt{n})$ round complexity and $O(\log n)$ awake complexity algorithm for graphs with diameters much smaller than (essentially) $\sqrt{n}$.

Our lower bound technique for showing a conditional lower bound on the awake complexity (conditional on upper bounding the round complexity) can be of independent interest. We use a lower bound graph family that is similar to that used in prior work (e.g., [11,33]), but our lower bound technique uses communication complexity to lower bound awake complexity by lower bounding the *congestion caused in some node*. This is different from the *Simulation Theorem* technique ([11]) used to show unconditional lower bounds for *round complexity* in the traditional setting. The main idea is showing that to solve the distributed set disjointness problem in less than c rounds, at least $\tilde{\Omega}(n/c)$ bits have to send through some node that has small (constant) degree. This means that the node has to be awake for essentially $\tilde{\Omega}(n/c)$ rounds. By using standard reductions, the same lower bound holds for MST. The technique is quite general and could be adapted to show similar conditional lower bounds on the awake complexity for other fundamental problems such as shortest paths, minimum cut etc.

3. Awake-Optimal Algorithms and Techniques. We present a distributed randomized algorithm (Sect. 3.2) that has $O(\log n)$ awake complexity which is optimal since it matches the lower bound shown. The round complexity of our algorithm is $O(n \log n)$ and by our trade-off lower bound, this is the best round complexity (up to logarithmic factors) for an awake-optimal algorithm. We then show that the awake-optimal bound of $O(\log n)$ can be obtained deterministically by presenting a deterministic algorithm. However, the deterministic algorithm has a slightly higher round complexity of $O(n \log^5 n)$, assuming that the node IDs are in the range $[1, N]$ and $N = O(\text{poly } n)$ is known to all nodes. We also show that one can reduce the deterministic round complexity to $O(n \log n \log^* n)$ at the cost of slightly increasing the awake complexity to $O(\log n \log^* n)$ rounds.

Our algorithms use several techniques for constructing an MST in an awake-efficient manner. Some of these can be of independent interest in designing such algorithms for other fundamental problems.[3] Our main idea is to construct a spanning tree called the *Labeled Distance Tree (LDT)*. An LDT is a rooted oriented spanning tree such that each node is labeled by its distance from the root and every node knows the labels of its parent and children (if any) and the label of the root of the tree. We show that an LDT can be constructed in $O(\log n)$ awake complexity and in a weighted graph, the LDT can be constructed so that it is an MST. While LDT construction is akin to the classic GHS/Boruvka algorithm [16], it is technically challenging to construct an LDT that is also an MST in $O(\log n)$ awake rounds (Sect. 3). As in GHS algorithm, starting with n singleton fragments we merge fragments via minimum outgoing edge (MOE) in each phase. The LDT distance property allows for finding an MOE in $O(1)$ awake

---

[3] As an example, the $O(n \log n \log^* n)$ deterministic algorithm is useful in designing an MIS algorithm with small awake and round complexities [12]—see Sect. 1.3.

rounds, since broadcast and convergecast can be accomplished in $O(1)$ rounds. On the other hand, maintaining the property of LDT when fragments are merged is non-trivial. A key idea is that we show that merging can be accomplished in $O(1)$ awake rounds by merging only fragment chains of *constant* length (details under "Technical challenges" in Sect. 3.2).[4] We develop a technical lemma that shows that this merging restriction still reduces the number of fragments by a constant factor in every phase and hence the algorithm takes overall $O(\log n)$ awake rounds.

Our tree construction is different compared to the construction of trees in [3,4]. In particular, a tree structure called as Distributed Layered Tree (DLT) is used in Barenboim and Maimon [3]. A DLT is a rooted oriented spanning tree where the vertices are labeled, such that each vertex has a greater label than that of its parent, according to a given order and each vertex knows its own label and the label of its parent. Another similar tree structure is used in [4]. A crucial difference between the LDT construction and the others is that it allows fragments to be merged via desired edges (MOEs), unlike the construction of DLT, for example, where one has to merge along edges that connect a higher label to a lower label. This is not useful for MST construction. Another important difference is that the labels used in LDTs scale with the number of nodes, whereas the labels in DLTs scale with the maximum ID assigned to any node. As the running time for both constructions are proportional to the maximum possible labels and the ID range is usually polynomially larger than the number of nodes, the running time to construct a DLT is much larger than the running time to construct an LDT.

4. Trade-Off Algorithms. We present a parameterized family of distributed algorithms that show a trade-off between the awake complexity and the round complexity and essentially (up to a polylog n factor) matches our product lower bound of $\tilde{\Omega}(n)$.[5] Specifically we show a family of distributed algorithms that find an MST of the given graph with high probability in $\tilde{O}(D + 2^k + n/2^k)$ running time and $\tilde{O}(n/2^k)$ awake time, where D is the network diameter and integer $k \in [\max\{\lceil 0.5 \log n \rceil, \lceil \log D \rceil\}, \lceil \log n \rceil]$ is an input parameter to the algorithm. Notice that when $D = O(\sqrt{n})$, the round complexity can vary from $\tilde{O}(\sqrt{n})$ to $\tilde{O}(n)$, and we can choose (integer) $k \in [\lceil 0.5 \log n \rceil, \lceil \log n \rceil]$ from to get the $\tilde{O}(n)$ product bound for this entire range. On the other hand, when $D = \omega(\sqrt{n})$, the round complexity can vary from $\tilde{O}(D)$ to $\tilde{O}(n)$, and we can choose $k \in [\lceil \log D \rceil, \lceil \log n \rceil]$ and get a family of algorithms with (essentially) optimal round complexities from $\tilde{O}(D)$ to $\tilde{O}(n)$.

Due to lack of space, full details of all algorithms (including omitted ones) and omitted proofs are given in the full paper [2].

---

[4] Consider the supergraph where the fragments are nodes and the MOEs are edges. A fragment chain is one such supergraph that forms a path. The exact details of the supergraphs formed are slightly different and explained in the relevant section, but this idea is beneficial to understanding.

[5] The product lower bound of $\tilde{\Omega}(n)$ is shown for graphs with diameter at least $\tilde{\Omega}(\sqrt{n})$. Hence, the near tightness claim holds for graphs in this diameter range.

1.3 Related Work and Comparison

The sleeping model and the awake complexity measure was introduced in a paper by Chatterjee, Gmyr and Pandurangan [7] who showed that MIS in general graphs can be solved in $O(1)$ rounds *node-averaged* awake complexity. Node-averaged awake complexity is measured by the *average* number of rounds a node is awake. The (worst-case) awake complexity of their MIS algorithm is $O(\log n)$, while the worst-case complexity (that includes all rounds, sleeping and awake) is $O(\log^{3.41} n)$ rounds. Subsequently, Ghaffari and Portmann [19] developed a randomized MIS algorithm that has worst-case complexity of $O(\log n)$, while having $O(1)$ node-averaged awake complexity (both bounds hold with high probability). They studied approximate maximum matching and vertex cover and presented algorithms that have similar node-averaged and worst-case awake complexities. These results show that the above fundamental local symmetry breaking problems have $O(\log n)$ (worst-case) awake complexity as is shown for global problems such as spanning tree [3] and MST (this paper). In a recent result, Dufoulon, Moses Jr., and Pandurangan [12] show that MIS can be solved in $O(\log \log n)$ (worst-case) awake complexity which is exponentially better than previous results. But the round complexity is $O(poly(n))$. It then uses the *deterministic* LDT construction algorithm of this paper to obtain an MIS algorithm that has a slightly larger awake complexity of $O(\log \log n \log^* n)$, but significantly better round complexity of $O(\text{polylog } n)$. The existence of a deterministic LDT algorithm is crucial to obtaining their result.

Barenboim and Maimon [3] showed that many problems, including broadcast, construction of a spanning tree, and leader election can be solved deterministically in $O(\log n)$ awake complexity. They also showed that fundamental symmetry breaking problems such as MIS and $(\Delta + 1)$-coloring can be solved deterministically in $O(\log \Delta + \log^* n)$ awake rounds in the LOCAL model, where Δ is the maximum degree. More generally, they also define the class of *O-LOCAL* problems (that includes MIS and coloring) and showed that problems in this class admit a deterministic algorithm that runs in $O(\log \Delta + \log^* n)$ awake time and $O(\Delta^2)$ round complexity. Maimon [27] presents trade-offs between awake and round complexity for O-LOCAL problems.

While there is significant amount of work on energy-efficient distributed algorithms over the years we discuss those that are most relevant to this paper. A recent line of relevant work is that Chang, Kopelowitz, Pettie, Wang, and Zhan and their follow ups [4–6,8,9] (see also the references therein and its follow up papers mentioned below and also much earlier work on energy complexity in radio networks e.g., [21,22,28]). This work defines the measure of *energy complexity* which is the same as (worst-case) awake complexity (i.e., both measures count only the rounds that a node is awake). While the awake complexity used here and several other papers [3,7,19] assumes the usual CONGEST(or LOCAL) communication model (and hence the model can be called *SLEEPING-CONGEST* (or *SLEEPING-LOCAL*)), the energy complexity measure used in [6] (and also papers mentioned above) has some additional communication restrictions that pertain to radio networks (and can be called *SLEEPING-RADIO* model). The most important being that nodes can only broadcast messages (hence the same message is sent to all neighbors) and when a node transmits, no other neighboring node can. (Also a node cannot transmit and listen in the same round.) The energy model has a few variants depending on how collisions are handled. There is a version of the SLEEPING-RADIO model called "Local" where collisions are ignored and nodes can transmit messages at the same time; this is essentially same as SLEEPING-LOCAL model, apart from the notion that in a given round a node can transmit only the same message to its neighbors. In particular, upper bounds in the radio model apply directly to the sleeping model. In particular, we use a recent result due to Dani and Hayes [9] that computes breadth-first search (BFS) tree with $O(\text{polylog}(n))$ energy complexity in the radio model as a subroutine in our MST tradeoff algorithm. Also, algorithms in the SLEEPING-CONGEST model can be made to work in the SLEEPING-RADIO model yielding similar bounds (with possibly a $O(\text{polylog}(n))$ multiplicative factor) to the energy/awake complexity.

Lower bounds shown in the local version of the SLEEPING-RADIO model apply to other models including SLEEPING-LOCAL (and SLEEPING-CONGEST). For example, Chang, Dani, Hayes, He, Li, and Pettie [4] show a lower bound $\Omega(\log n)$ on the energy complexity of broadcast which applies also to randomized algorithms. This lower bound is shown for the local version of their model, and this result holds also for the awake complexity in the sleeping model. We adapt this lower bound result to show a $\Omega(\log n)$ lower bound on the awake complexity of MST even for randomized algorithms.

2   Lower Bounds

We begin by showing that $\Omega(\log n)$ is an *unconditional* lower bound on the awake time for MST. This shows that our algorithms presented in Sect. 3 achieve optimal awake complexity. We then show a lower bound of $\tilde{\Omega}(n)$ on the product of the awake and round complexities. This can be considered as a conditional lower bound on awake complexity, conditioned on an upper bound on the round complexity.

2.1   Unconditional Lower Bound on Awake Complexity of MST

We consider a ring of $\Theta(n)$ nodes with random weights on edges. The two largest weighted edges will be apart by a hop distance of $\Omega(n)$ with constant probability and any MST algorithm must detect which one has the lower weight. Clearly, this will require communication over either one of the $\Omega(n)$ length paths between the two edges. Under this setting, we get the following theorem.

Theorem 1. *Any algorithm to solve MST with probability exceeding 1/8 on a ring network comprising $\Theta(n)$ nodes requires $\Omega(\log n)$ awake time even when message sizes are unbounded.*

2.2   Lower Bound on the Product of Awake and Round Complexity

We adapt the lower bound technique from [10] to the sleeping model and show a lower bound on the product of the awake and round complexities, thereby exposing an inherent trade off between them.

The high level intuition is as follows. Note that both endpoints of an edge must be awake in a round for $O(\log n)$ bits to be transmitted across in that round. When only one of the endpoints is awake, the awake node can sense the other node is asleep, which is $O(1)$ bits of information. There is no transfer of information when both endpoints are asleep. Thus, if an edge $e = (u, v)$ must transmit B bits when executing an algorithm in the CONGEST model, then, either u or v must be awake for at least $\Omega(B/\log n)$ rounds. Thus congestion increases awake time. Our goal is to exploit this intuition to prove the lower bound.

We use a communication complexity based reduction to reduce *set disjointness* (SD) in the classical communication complexity model to the *distributed set disjointness* (DSD) problem in the sleeping model and then extend DSD to the minimum spanning tree problem via an intermediate *connected spanning subgraph* (CSS), both to be solved in the sleeping model.

In the SD problem, two players, Alice and Bob, possess two k-bit strings $a = (a_i)_{1 \leqslant i \leqslant k}$ and $b = (b_i)_{1 \leqslant i \leqslant k}$, respectively. They are required to compute an output bit $\mathsf{Disj}(a, b)$ that is 1 iff there is no $i \in [k]$ such that $a_i = b_i = 1$ (i.e., the inner product $\langle a, b \rangle = 0$), and 0 otherwise. Alice and Bob must compute $\mathsf{Disj}(a, b)$ while exchanging the least number of bits. It is well-known that any protocol that solves SD requires $\Omega(k)$ bits (on expectation) to be exchanged between Alice and Bob even if they employ a randomized protocol [34] that can fail with a small fixed probability $\varepsilon > 0$.

Note that the $\Omega(k)$ lower bound (on the number of bits exchanged) for SD problem applies to the asynchronous setting which is the default model in classical 2-party communication complexity. On the other hand, we assume the synchronous setting for the sleeping model. However, it can be shown using the Synchronous Simulation Theorem [29] (see also Lemma 2.2 in [13]) that essentially the same lower bound (up to a $(\log k)$ factor) holds in the synchronous setting as well if one considers algorithms that run in (at most) polynomial in k number of rounds. Thus for the rest of the proof, we assume a $\Omega(k/\log k)$ lower bound for the SD problem for the *synchronous* 2-party communication complexity if we consider only polynomial (in k) number of rounds.

The DSD problem is defined on a graph G_{rc} that is schematically shown in Fig. 1. Let r and c be two positive integers such that $rc + \Theta(\log n) = n$ (the network size). We focus on the regime where $c \in \omega(\sqrt{n}\log^3 n)$ and $r \in o(\sqrt{n}/\log^3 n)$. The graph comprises r rows (or parallel paths) p_ℓ, $1 \leqslant \ell \leqslant r$, with p_1 referring to the parallel path at the bottom. Each parallel path comprises c nodes arranged from left to right with the first node referring to the leftmost node and the last node referring to the rightmost node. The first and last nodes in p_1 are designated Alice and Bob because they are controlled by the players Alice and Bob in our reduction. Alice (resp., Bob) is connected to the first node (resp., last node) of each p_ℓ, $2 \leqslant \ell \leqslant r$. Additionally, we pick $\Theta(\log n)$ equally

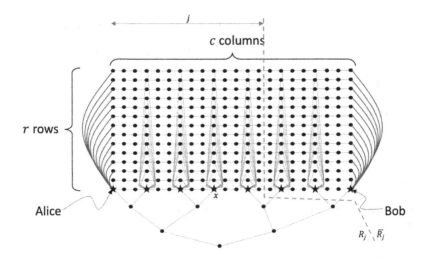

Fig. 1. Construction of network graph G_{rc} for proving lower bound. The vertices in X are shown as stars (there is a binary tree at the bottom having the nodes in X as its leaves). One such $x \in X$ is labeled. The cut induced by an R_j is shown in dotted lines.

spaced nodes X (of cardinality that is a power of two) from p_1 such that the first and last nodes in p_1 are included in X. For each $x \in X$, say at position j in p_1, we add edges from x to the jth node in each p_ℓ, $2 \leqslant \ell \leqslant r$. Using X as leaves, we construct a balanced binary tree. We will use I to denote the internal nodes of this tree. Alice is in possession of bit string a and Bob is in possession of b and, to solve DSD, they must compute $\mathsf{Disj}(a, b)$ in the sleeping model over the network G_{rc}. The outputs of all other nodes don't matter; for concreteness, we specify that their outputs should be empty.

In the CSS problem defined again on G_{rc}, some edges in G_{rc} are marked and at least one node in the network must determine whether the marked edges form a connected spanning subgraph of G_{rc}. For the MST problem, we require edges in G_{rc} to be weighted and the goal is to construct a minimum spanning tree of G_{rc} such that the endpoints of each MST edge e are aware that e is an MST edge. Both CSS and MST must be solved in the sleeping model.

G_{rc} is constructed such that any node can reach some $x \in X$ within $O(c/\log n)$ steps and any pair of nodes in X are within $O(\log \log n)$ steps (through the tree). Recall that $c \in \omega(\sqrt{n} \log^3 n)$. Thus, we have the following observation.

Observation 1. *The network graph G_{rc} has diameter $D \in \Theta(c/\log n)$. Moreover, $D \in \omega(\sqrt{n} \log^* n)$. Therefore, DSD, CSS, and MST (if edges in G_{rc} are assigned weights) can be computed in $O(D) = O(c/\log n)$ rounds [17].*

Reduction from SD → DSD.

Lemma 1. *Consider an algorithm P in the sleeping model that solves DSD on G_{rc} with $c \in \omega(\sqrt{n} \log^3 n)$ and $r \in o(\sqrt{n}/\log^3 n)$ in T (worst-case) rounds such* *that $T \in o(c)$ (and we know such an algorithm exists from Observation 1, in particular because $D \in O(c/\log n)$). Then, the awake time of P must be at least $\Omega(r/\log^2 n)$. This holds even if P is randomized and has an error probability that is bounded by a small constant $\varepsilon > 0$.*

Proof. Suppose for the sake of contradiction P runs in time T and has an awake complexity of $o(r/\log^2 n)$. Then, we can show that Alice and Bob can simulate P in the classical communication complexity model and solve SD on r bits by exchanging only $o(r/\log n)$ bits which will yield a contradiction to the SD lower bound. We establish this by showing that Alice and Bob can simulate P to solve SD in the classical communication complexity model.

We show this simulation from Alice's perspective. Bob's perspective will be symmetric. Recall that p_ℓ is the ℓth parallel path. Let p_ℓ^j, $1 \leqslant j \leqslant c$, denote the first j vertices of path p_ℓ. We define R_j to be the union of all p_ℓ^j and I (recall that I is the set of the internal nodes of the binary tree), i.e., $R_j = (\bigcup_{\ell=1}^r p_\ell^j) \cup I$. Note that R_j induces a cut $(R_j, \bar{R}_j)$ that is shown in Fig. 1. Alice begins by simulating R_{c-1} in round 1 as she knows the state of all nodes in R_{c-1}. Initially, all the information needed for the simulation (including which nodes are awake and which are asleep) is available for Alice because the structure of G_{rc} is fully known (except for Bob's input). At each subsequent round t, Alice simulates R_{c-t}.

As the simulation progresses, in each round $t > 1, t \leqslant T \in o(c)$, all inputs will be available except for the new bits that may enter I through nodes in $\bar{R}_{c-t}$. Note that information about whether a node should be awake/asleep will be available at that node based on the execution in the previous round. Alice will not need to ask for the bits needed by p_ℓ^{c-t} because she simulated all nodes in p_ℓ^{c-t+1}, $1 \leqslant \ell \leqslant r$, in the previous round. Note that the portion simulated by Bob will encompass the portion from which Alice may need bits from Bob, so Bob will indeed have the bits requested by Alice. In order to continue the simulation, Alice borrows bits that P transmitted from $\bar{R}_{c-t}$ to $I \cap R_{c-t}$ from Bob. Suppose during the course of the simulation in the communication complexity model, B bits are borrowed from Bob. Then nodes in I must have been awake for a collective total of at least $\Omega(B/\log n)$ rounds (because each message of $O(\log n)$ bits must be received by a node that is awake in P).[6] This implies that at least one node in I must have been awake for $\Omega(B/\log^2 n)$ rounds because $|I| \in O(\log n)$ and the number of edges incident to nodes in I is also $O(\log n)$ (since nodes in I are of constant degree).

Since the node awake time is $o(r/\log^2 n)$ for P, B must be $o(r/\log n)$, accounting for the fact that each node awake time can potentially transmit $O(\log n)$ bits. But this contradicts the fact that SD requires $\Omega(r/\log n)$ bits in the synchronous communication complexity model.

---

[6] Note that all the B bits cannot solely come through a row path of length c, since we are restricting $T \in o(c)$. In other words, each of the bits has to go through at least one node in I.

Reduction from DSD → CSS. We show a reduction (in the full version) from DSD → CSS by encoding a given DSD problem instance as a CSS instance in the following manner. Recall that in DSD, Alice and Bob have bit strings a and b, respectively, of length r each. Furthermore, recall that Alice (resp., Bob) is connected to first node (resp., last node) of each p_ℓ, $2 \leqslant \ell \leqslant r$.

Lemma 2. *Suppose there is a protocol Q in the sleeping model that solves CSS on G_{rc} with $c \in \omega(\sqrt{n}\log^3 n)$ and $r \in o(\sqrt{n}/\log^3 n)$ in T rounds such that $T \in o(c)$. Then, the node awake time of Q must be at least $\Omega(r/\log^2 n)$. This holds even if Q is randomized and has an error probability that is bounded by a small constant $\varepsilon > 0$.*

Reduction from CSS → MST. Recall that CSS is a decision problem that requires a subset of the edges in the network graph G_{rc} to be marked; we are to report whether the marked edges form a spanning subgraph of G_{rc}. MST on the other hand is a construction problem. It takes a weighted network graph and computes the minimum spanning tree. A reduction from CSS to MST can be constructed by assigning a weight of 1 for marked edges in the CSS instance and n for all other edges and asking if any edge of weight n is included in the MST. This leads us to the following lemma.

Lemma 3. *Suppose there is a protocol M in the sleeping model that solves MST on G_{rc} with $c \in \omega(\sqrt{n}\log^3 n)$ and $r \in o(\sqrt{n}/\log^2 n)$ in T rounds such that $T \in o(c)$. Then, the node awake time of M must be at least $\Omega(r/\log^2 n)$. This holds even if M is randomized and has an error probability that is bounded by a small constant $\varepsilon > 0$.*

Thus, we can conclude with the following theorem.

Theorem 2. *Consider positive integers r and c such that $rc+\Theta(\log n) = n$ (the network size) and $c \in \omega(\sqrt{n}\log^3 n)$. (Thus, c can range between $\tilde{\Omega}(\sqrt{n})$ to $\tilde{O}(n)$ with $r \in \Theta(n/c)$.) Then there exists graphs with diameter at least $\omega(\sqrt{n}\log^3 n)$ such that any randomized algorithm for MST in the sleeping model that runs in time $T \in o(c)$ rounds and guaranteed to compute the MST with probability at least $1 - \varepsilon$ for any small fixed $\varepsilon > 0$ has worst case awake complexity at least $\Omega(r/\log^2 n)$.*

As an illustrative example of this trade off, let $c = n^{3/4}$ and $r \in \Theta(n^{1/4})$. Consider an MST algorithm that takes $o(n^{3/4})$ rounds; Observation 1 guarantees the existence of such an algorithm. Then, Theorem 2 implies that its awake complexity is at least $\tilde{\Omega}(n^{1/4})$.

3   MST Algorithms with Optimal Awake Complexity

In this section, we present our algorithm to construct an MST that takes optimal awake complexity. We first present a toolbox of procedures which are used repeatedly in the subsequent algorithm. We then develop a randomized algorithm that creates an MST in optimal awake time.

3.1   Main Ideas and Useful Procedures

The algorithm we develop in this section can be seen as variations of the classic GHS algorithm to find the MST, adapted to optimize the awake time of the algorithm. Recall that each phase of the GHS algorithm consists of two steps. Step (i) corresponds to finding the minimum outgoing edges (MOEs) for the current fragments and step (ii) involves merging these fragments.

Our algorithm works in phases where at the end of each phase, we ensure that the original graph has been partitioned into a forest of node-disjoint trees that satisfy the following property. For each such tree, all nodes within the tree know the ID of the root of the tree (called *fragment ID*), the IDs of their parents and children in the tree, if any, and their *distance from the root* (note that it is the hop distance, ignoring the weights) of that tree. We call each such tree a *Labeled Distance Tree (LDT)* and a forest of such trees a *Forest of Labeled Distance Trees (FLDT)*. By the end of the algorithm we design, our goal is to have the FLDT reduced to just one LDT which corresponds to the MST of the original graph. The challenge is to construct an LDT (which will also be an MST) in an awake-optimal manner.

The purpose of maintaining such a structure is that we know how to design fast awake procedures to propagate information within an LDT, which we describe below. Specifically, we know how to set up a schedule of rounds for nodes to wake up in such that (i) information can be passed from the root to all nodes in an LDT in $O(1)$ awake rounds, (ii) information can be passed from a child to the root in $O(1)$ awake rounds, and (iii) information can be spread from one LDT to its neighbors in $O(1)$ awake rounds. There are several well-known procedures typically associated with GHS such as broadcast, upcast-min, etc. that make use of such information propagation. Once we have an LDT, it is easy to implement these procedures in constant awake time. For the processes described below, it is assumed that the initial graph has already been divided into an FLDT where each node u knows the ID of the root, $root$, of the tree it belongs to (i.e., the fragment ID of the tree), the IDs of u's parent and children in that tree, and u's distance to $root$. First of all, we define a transmission schedule that is used in each of the procedures and will be directly utilized in the algorithms. Then we briefly mention the procedures typically associated with GHS.

Transmission Schedule of Nodes in a Tree. We describe the function $Transmission\text{-}Schedule(root, u, n)$, which takes a given node u and a tree rooted at $root$ (to which u belongs) and maps them to a set of rounds in a block of $2n + 1$ rounds. $Transmission\text{-}Schedule(root, u, n)$ may be used by node u to determine which of the $2n + 1$ rounds to be awake in. Consider a tree rooted at $root$ and a node u in that tree at distance i from the root. For ease of explanation, we assign names to each of these rounds as well. For all non-root nodes u, the set of rounds that $Transmission\text{-}Schedule(root, u, n)$ maps to includes rounds $i, i + 1, n + 1, 2n - i + 1$, and $2n - i + 2$ with corresponding names Down-Receive, Down-Send, Side-Send-Receive, Up-Receive, and Up-Send, respectively. $Transmission\text{-}Schedule(root, root, n)$ only maps to the set containing rounds 1, $n + 1$, and $2n + 1$ with names Down-Send, Side-Send-Receive, and Up-Receive, respectively.[7]

Observation 2. *Procedure* FRAGMENT-BROADCAST(n), *run by all nodes in a tree, allows the root of that tree to transmit a message to each node in its tree in $O(n)$ running time and $O(1)$ awake time.*

Observation 3. *Procedure* UPCAST-MIN(n), *run by all nodes in a tree, allows the smallest value among all values held by the nodes, if any, to be propagated to the root of the tree in $O(n)$ running time and $O(1)$ awake time.*

Observation 4. *Procedure* TRANSMIT-NEIGHBOR(n), *run by all nodes in a tree, allows each node in the tree to transmit a message, if any, to its parent and children in $O(n)$ running time and $O(1)$ awake time.*

Observation 5. *Procedure* TRANSMIT-ADJACENT(n), *run by all nodes in a tree, allows each node in a tree to transfer a message, if any, to neighboring nodes belonging to other trees in $O(n)$ running time and $O(1)$ awake time.*

In the course of our algorithms, we ensure that nodes stay synchronized, i.e., time can be viewed in blocks of $2n + 1$ rounds such that all nodes start their first schedule at the same time (and end them at the same time) and continue to start (and end) future schedules at the same time.

3.2   Awake-Optimal Randomized Algorithm

Brief Overview. We describe Algorithm RANDOMIZED-MST. We use a synchronous variant of the classic GHS algorithm (see e.g., [31,32]) to find the MST of the graph, where we modify certain parts of the algorithm in order to ensure that the awake time is optimal. Each phase of the classic GHS algorithm consists of two steps. Step (i) corresponds to finding the minimum outgoing edges (MOEs) for the current fragments and step (ii) involves merging these fragments. In the current algorithm, we introduce a new step between those two steps. Specifically, we utilize randomness to restrict which MOEs are considered "valid" for the current phase by having each fragment leader flip a coin and only considering MOEs from fragments whose leaders flip tails to those that flip heads. This restriction ensures that the diameter of any subgraph formed by fragments and valid MOEs is *constant*. This restriction, coupled with careful analysis that includes showing that the number of fragments decreases by a constant factor in each phase on expectation, guarantees that the MST of the original graph is obtained after $O(\log n)$ phases with high probability.

Technical Challenges. As mentioned above, one of the key changes we make is to restrict the MOEs to a subset of "valid" ones. This is to address a key technical

---

[7] In this description, we assumed that *Transmission-Schedule*($\cdot, \cdot, n$) was started in round 1. However, if *Transmission-Schedule*($\cdot, \cdot, n$) is started in round r, then just add $r - 1$ to the values mentioned here and in the previous sentence.

challenge. When we merge two fragments together, one of those fragments must internally re-orient itself and update its internal values (including distance to the root). This re-alignment and updation takes $O(1)$ awake time. If we have a chain of fragments, say of diameter d, we may have to perform this re-alignment procedure $d-1$ times since the re-alignment of fragments is sequential in nature.[8] As a result, if we do not control the length of chains of connected components formed by the fragments and their MOEs, we risk blowing up the awake time of the algorithm. We use randomness to ensure the diameter of any such connected component is a constant.

As a result of the above change, we have a second technical challenge. Because we reduce the number of valid MOEs, we have to be careful to argue that a sufficient number of fragments are merged together in each phase so that after $O(\log n)$ phases, we end up with exactly one fragment with high probability. We provide such a careful argument, which is somewhat different from the usual argument used in GHS style algorithms.

Theorem 3. *Algorithm* RANDOMIZED-MST *is a randomized algorithm to find the MST of a graph with high probability in $O(n \log n)$ running time and $O(\log n)$ awake time.*

Acknowledgments. We thank Orr Fischer for the useful idea that helped to reduce the run time of the deterministic awake-optimal algorithm, in particular, the coloring procedure. We thank Fabien Dufoulon for helpful discussions.

References

1. Ambühl, C.: An optimal bound for the MST algorithm to compute energy efficient broadcast trees in wireless networks. In: Caires, L., Italiano, G.F., Monteiro, L., Palamidessi, C., Yung, M. (eds.) ICALP 2005. LNCS, vol. 3580, pp. 1139–1150. Springer, Heidelberg (2005). https://doi.org/10.1007/11523468_92
2. Augustine, J., Moses Jr., W.K., Pandurangan, G.: Awake complexity of distributed minimum spanning tree. arXiv preprint arXiv:2204.08385 (2022)
3. Barenboim, L., Maimon, T.: Deterministic logarithmic completeness in the distributed sleeping model. DISC 2021. LIPIcs, vol. 209, pp. 10:1–10:19. https://doi.org/10.4230/LIPIcs.DISC.2021.10
4. Chang, Y.J., Dani, V., Hayes, T.P., He, Q., Li, W., Pettie, S.: The energy complexity of broadcast. In: PODC 2018, pp. 95–104 (2018)
5. Chang, Y.J., Dani, V., Hayes, T.P., Pettie, S.: The energy complexity of BFS in radio networks. In: PODC 2020, pp. 273–282 (2020)
6. Chang, Y., Kopelowitz, T., Pettie, S., Wang, R., Zhan, W.: Exponential separations in the energy complexity of leader election. ACM Trans. Algorithms **15**(4), 49:1–49:31 (2019). conference version: STOC 2017

---

[8] To observe the sequential nature of the re-alignment, consider a chain with three fragments, say $A \leftarrow B \leftarrow C$. Suppose A maintains its orientation. The nodes in B must be processed first and must update their distance to A. Only then can the nodes of C accurately update their distance to A (after the node u in C connected to the node v in B learns v's updated distance to A).

7. Chatterjee, S., Gmyr, R., Pandurangan, G.: Sleeping is efficient: MIS in $O(1)$-rounds node-averaged awake complexity. In: PODC 2020, pp. 99–108 (2020)
8. Dani, V., Gupta, A., Hayes, T.P., Pettie, S.: Wake up and join me! An energy-efficient algorithm for maximal matching in radio networks. In: DISC 2021, pp. 19:1–19:14 (2021)
9. Dani, V., Hayes, T.P.: How to wake up your neighbors: safe and nearly optimal generic energy conservation in radio networks. In: DISC 2022. LIPIcs, vol. 246, pp. 16:1–16:22 (2021). https://doi.org/10.4230/LIPIcs.DISC.2022.16
10. Das Sarma, A., et al.: Distributed verification and hardness of distributed approximation. In: STOC 2011, pp. 363–372 (2011)
11. Das Sarma, A., et al.: Distributed verification and hardness of distributed approximation. SIAM J. Comput. **41**(5), 1235–1265 (2012)
12. Dufoulon, F., Moses Jr., W.K., Pandurangan, G.: Distributed MIS in o(log log n) awake complexity. In: PODC 2023 (2023)
13. Dufoulon, F., Pai, S., Pandurangan, G., Pemmaraju, S.V., Robinson, P.: The message complexity of distributed graph optimization. ITCS 2024
14. Elkin, M.: Unconditional lower bounds on the time-approximation tradeoffs for the distributed minimum spanning tree problem. In: STOC 2004, pp. 331–340 (2004)
15. Elkin, M.: A simple deterministic distributed MST algorithm, with near-optimal time and message complexities. PODC 2017, pp. 157–163
16. Gallager, R., Humblet, P., Spira, P.: A distributed algorithm for minimum-weight spanning trees. TOPLAS **5**(1), 66–77 (1983)
17. Garay, J.A., Kutten, S., Peleg, D.: A sublinear time distributed algorithm for minimum-weight spanning trees. SICOMP **27**(1), 302–316 (1998)
18. Ghaffari, M., Haeupler, B.: Distributed algorithms for planar networks II: low-congestion shortcuts, MST, and min-cut. In: SODA 2016, pp. 202–219 (2016)
19. Ghaffari, M., Portmann, J.: Average awake complexity of MIS and matching. In: SPAA 2022, pp. 45–55 (2022)
20. Haeupler, B., Wajc, D., Zuzic, G.: Universally-optimal distributed algorithms for known topologies. In: STOC 2021, pp. 1166–1179 (2021)
21. Jurdzinski, T., Kutylowski, M., Zatopianski, J.: Efficient algorithms for leader election in radio networks. In: PODC 2002, pp. 51–57 (2002)
22. Kardas, M., Klonowski, M., Pajak, D.: Energy-efficient leader election protocols for single-hop radio networks. In: ICPP 2013, pp. 399–408 (2013)
23. Khan, M., Pandurangan, G., Kumar, V.S.A.: Distributed algorithms for constructing approximate minimum spanning trees in wireless sensor networks. IEEE Trans. Parallel Distrib. Syst. **20**(1), 124–139 (2009)
24. King, V., Phillips, C.A., Saia, J., Young, M.: Sleeping on the job: energy-efficient and robust broadcast for radio networks. Algorithmica **61**(3), 518–554 (2011). https://doi.org/10.1007/s00453-010-9422-0
25. Kutten, S., Pandurangan, G., Peleg, D., Robinson, P., Trehan, A.: On the complexity of universal leader election. J. ACM **62**(1) (2015)
26. Kutten, S., Peleg, D.: Fast distributed construction of small k-dominating sets and applications. J. Algorithms **28**(1), 40–66 (1998)
27. Maimon, T.: Sleeping model: local and dynamic algorithms (2021)
28. Nakano, K., Olariu, S.: Randomized leader election protocols in radio networks with no collision detection. In: Goos, G., Hartmanis, J., van Leeuwen, J., Lee, D.T., Teng, S.-H. (eds.) ISAAC 2000. LNCS, vol. 1969, pp. 362–373. Springer, Heidelberg (2000). https://doi.org/10.1007/3-540-40996-3_31

29. Pandurangan, G., Peleg, D., Scquizzato, M.: Message lower bounds via efficient network synchronization. Theor. Comput. Sci. **810**, 82–95 (2020). https://doi.org/10.1016/J.TCS.2018.11.017

30. Pandurangan, G., Robinson, P., Scquizzato, M.: A time- and message-optimal distributed algorithm for minimum spanning trees. In: STOC 2017, pp. 743–756 (2017)

31. Pandurangan, G., Robinson, P., Scquizzato, M.: The distributed minimum spanning tree problem. Bull. EATCS **125** (2018)

32. Peleg, D.: Distributed Computing: A Locality Sensitive Approach. SIAM, USA (2000)

33. Peleg, D., Rubinovich, V.: A near-tight lower bound on the time complexity of distributed minimum-weight spanning tree construction. SIAM J. Comput. **30**(5), 1427–1442 (2000)

34. Razborov, A.A.: On the distributional complexity of disjointness. Theor. Comput. Sci. **106**(2), 385–390 (1992)

All You Need are Random Walks: Fast and Simple Distributed Conductance Testing

Tuğkan Batu[1(✉)] [ID], Amitabh Trehan[2(✉)] [ID], and Chhaya Trehan[1,3(✉)] [ID]

[1] London School of Economics and Political Science, London, UK
t.batu@lse.ac.uk
[2] Durham University, Durham, UK
amitabh.trehan@durham.ac.uk
[3] University of Sheffield, Sheffield, UK
c.trehan@sheffield.ac.uk

Abstract. We propose a simple and time-optimal algorithm for property testing a graph for its conductance in the CONGEST model. Our algorithm takes only $O(\log n)$ rounds of communication (which is known to be optimal), and consists of simply running multiple random walks of $O(\log n)$ length from a certain number of random sources, at the end of which nodes can decide if the underlying network is a good conductor or far from it. Unlike previous algorithms, no aggregation is required even with a smaller number of walks. Our main technical contribution involves a tight analysis of this process for which we use spectral graph theory. We introduce and leverage the concept of sticky vertices which are vertices in a graph with low conductance such that short random walks originating from these vertices end in a region around them.

The present state-of-the-art distributed CONGEST algorithm for the problem by Fichtenberger and Vasudev [MFCS 2018], runs in $O(\log n)$ rounds using three distinct phases: building a rooted spanning tree (*preprocessing*), running $O(n^{100})$ random walks to generate statistics (*Phase 1*), and then convergecasting to the root to make the decision (*Phase 2*). The whole of our algorithm is, however, similar to their Phase 1 running only $O(m^2) = O(n^4)$ walks. Note that aggregation (using spanning trees) is a popular technique but spanning tree(s) are sensitive to node/edge/root failures, hence, we hope our work points to other more distributed, efficient and robust solutions for suitable problems.

Keywords: Graph Conductance · Distributed Property Testing · Random Walks

1 Introduction

Checking whether a distributed network satisfies a certain property is an important problem. For example, this knowledge may be used to choose appropriate

© The Author(s), under exclusive license to Springer Nature Switzerland AG 2024
Y. Emek (Ed.): SIROCCO 2024, LNCS 14662, pp. 64–82, 2024.
https://doi.org/10.1007/978-3-031-60603-8_4

algorithms to be run on the network for certain tasks. For instance, the randomised leader election algorithm of [20] works in sublinear time if the underlying graph is a good expander but not otherwise. However, it may be hard to efficiently *verify* certain global graph properties in the CONGEST model of distributed computing. In this model, each vertex of the input graph acts as a processing unit and works in conjunction with other vertices to solve a computational problem. The computation proceeds in synchronous rounds, in each of which every vertex can send an $O(\log n)$-bits message to each of its neighbours, do some local computations and receive messages from its neighbours.

Distributed decision problems are tasks in which the vertices of the underlying network have to collectively decide whether the network satisfies a global property $\mathcal{P}$ or not. If the network indeed satisfies the property, then all vertices must accept and, if not, then at least one vertex in the network must reject. For many global properties, lower bounds on the number of rounds of computation of the form $\tilde{\Omega}(\sqrt{n} + D)$ are known for distributed decision, where n is the number of vertices and D is the diameter of the network. (See [8]). It makes sense to relax the decision question and settle for an approximate answer in these scenarios as is done in the field of property testing (see [14,17]) in the sequential setting. A property testing algorithm in the sequential setting arrives at an approximate decision about a certain property of the input by *querying* only a small portion of it. Specifically, an ϵ-tester for a graph property $\mathcal{P}$ is a randomised algorithm that can distinguish between graphs that satisfy $\mathcal{P}$ and the graphs that are ϵ-far from satisfying $\mathcal{P}$ with high constant probability. An m-edge graph G is considered ϵ-far from satisfying $\mathcal{P}$ if one has to modify (add or delete) more than $\epsilon \cdot m$ edges of G for it to satisfy $\mathcal{P}$. Two-sided error testers may err on all graphs, while one-sided error testers have to present a witness when rejecting a graph. The cost of the algorithm is measured in the number of queries made. (See [14–17] for a detailed exposition of the subject.)

Distributed Property Testing: A *distributed property testing* problem is a relaxed variant of the corresponding decision problem: if the input network satisfies a property, then, with sufficiently high probability, all the vertices accept but if the input network is ϵ-far from satisfying the property, then at least one vertex rejects. The definition of "farness" in it remains the same as in the classical setting. The complexity measures are the number of communication rounds and the number of messages exchanged during the execution of the tester. Distributed property testing has been an active area of research recently. The work of [4] was the first to present a distributed algorithm (for finding near-cliques) with a property testing flavour. Later, in [5], the authors did a more detailed study of distributed property testing. There has been further study on the topic (see [10] and [12]) in the specific context of subgraph freeness.

Conductance Testing: We address the problem of testing the conductance of an unweighted, undirected graph $G = (V, E)$ in the CONGEST model. Throughout, we denote $|V|$ by n and $|E|$ by m. A distributed conductance tester can be a useful pre-processing step for some distributed algorithms (such as [20]) which perform better on graphs with high conductance. The test can help determine whether to proceed with the algorithm or not.

Given a graph $G = (V, E)$, and a set $A \subseteq V$, the *volume* of A (denoted $\mathrm{vol}(A)$) is the sum of degrees of vertices in A. We say that a graph $G = (V, E)$ is an α-*conductor* if every $U \subseteq V$ such that $\mathrm{vol}(U) \leq \mathrm{vol}(V)/2$ has conductance at least α. Here, the conductance of a set A is defined as $E(A, V \setminus A)/\mathrm{vol}(A)$, where $E(A, V \setminus A)$ is the number of edges crossing between A and $V \setminus A$. A closely related property of graphs is their expansion. We call a graph $G = (V, E)$ an α-*vertex expander* if every $U \subseteq V$ such that $|U| \leq |V|/2$ has at least $\alpha|U|$ neighbours. Here, a vertex $v \in V \setminus U$ is a neighbour of U if it has at least one edge incident to some $u \in U$. Similarly, G is called an α-*edge expander*, if every $U \subseteq V$ such that $|U| \leq |V|/2$ has at least $\alpha|U|$ edges crossing between U and $V \setminus U$. For a constant d, a graph $G = (V, E)$ is called a *bounded-degree graph* with degree bound d if every $v \in V$ has degree at most d. In this case, both the vertex and edge expansions are bounded by a constant (depending on d) times the conductance. Testing expansion (essentially testing conductance) in the bounded degree model has been studied for a long time in the classic centralised property testing model. In this setting, the problem of testing expansion was first studied by Goldreich and Ron [18] and later by [7]. Specifically, Czumaj and Sohler showed that given parameters $\alpha, \epsilon > 0$, the tester proposed by [18] accepts all graphs with vertex expansion larger than α, and rejects all graphs that are ϵ-far from having vertex expansion at least $\alpha' = \Theta(\alpha^2/\log n)$. Their work was followed by the state of the art results by [19] and [25]. Both these papers present $\tilde{O}(n^{1/2+\mu}/\alpha^2)$-query testers (for a small constant $\mu > 0$) for distinguishing between graphs that have expansion at least α and graphs that are ϵ-far from having expansion at least $\Omega(\alpha^2)$. In the general graph model (with no bound on degrees), Li, Pan and Peng [21] presented a conductance tester. Their tester essentially pre-processes the graph and turns it into a bounded degree graph while preserving (roughly) its expansion and size and then uses a tester for bounded degree graphs.

In the last few years, the same problem has also been addressed in non-sequential models of computing such as MPC [23] and distributed CON-GEST [11]. There are earlier papers studying distributed random walks whose results can be adapted towards conductance testing e.g. [24, 26] . However, these results yield large gaps in the conductance of the graphs that are accepted ($\Omega(\alpha)$) and that of the graphs that are rejected ($O(\alpha^2/\mathrm{polylog}(n))$). The first distributed algorithm for the specific task of testing the conductance of an input graph that we are aware of is by Fichtenberger and Vasudev [11]. This can test the conductance of the input network in the unbounded-degree graph model (like ours).

A typical algorithm for the problem in the sequential, as well as non-sequential, models can be thought of as running in two *phases* (after possibly a *pre-processing phase*). In the first phase, the algorithm performs a certain number of short ($O(\log n)$-length) random walks from a randomly chosen starting vertex. The walks should mix well on a graph with high conductance and should take longer to mix on a graph which is far from having high conductance (at least from some fraction of starting vertices). In the second phase, the algorithm then checks whether those walks mixed well or not. For that, the algorithm gleans some information from every vertex in the graph and computes some aggregate function. Specifically in the classical and MPC settings, the algorithms count the total number of pairwise collisions between the endpoints of the walks.

The distributed algorithm for the problem by Fichtenberger and Vasudev [11] precedes the first phase by building a rooted BFS spanning tree of the input graph.[1] This spanning tree is used for collecting information from the endpoints of the random walks in the second phase. Specifically, their algorithm estimates the discrepancy of the endpoint probability distribution from the stationary distribution by collecting the estimate of discrepancy on each endpoint at a central point. If the overall discrepancy is above a certain threshold, the algorithm rejects the graph. This process of building a spanning tree and collecting information at the root to decide if the property holds or not takes a global and centralised view of the testing process.

The following natural question arises in the context of the second phase:

Question 1. Is it possible to execute the second phase without computing a global aggregate function?

In the classic setting, one strives for testers that make a sublinear (in n or m) number of queries which translates to running a sublinear number of walks. With only a sublinear $(O(\sqrt{n}))$ number of walks, one hardly expects to see any useful information by itself on any individual vertex or in a small constant neighbourhood around it to know if the walks mixed well or not. Therefore, one has to rely on an aggregate function such as the total number of pairwise collisions between the endpoints of the walks. In the non-sequential settings such as distributed CONGEST, one can utilise the parallelism to run a superlinear number of short walks while keeping the run time proportional to the length of the walks. This inspires us to stick to Question 1 in distributed setting and investigate what information one should store at each vertex during phase 1 and how it should be processed *locally* to allow each node to decide locally whether it is part of a good conductor or not. This leads us to our main result provided all the nodes know n and m beforehand. Note that one can overcome the requirement of knowing m by performing a rooted spanning tree construction as in [11] and using this tree to count the number of edges. Note that we will not use this tree for collecting information about the random walks.

Our Results: Our main result is the algorithm presented in Sect. 3 (Pseudocodes 1 and 2). The main theorem is restated below:

Theorem (*Theorem 3*). *For an input graph $G = (V, E)$, and parameters $0 < \alpha < 1$ and $\epsilon > 0$, the distributed algorithm described in Sect. 3*

---

[1] If the construction of BFS tree takes longer than $O(\log n)$ rounds the algorithm rejects without proceeding to the first phase since all good conductors have small diameter. However, a bad conductor such as a dumbbell graph may also have small diameter, so their algorithm still needs to proceed with the test after the successful construction of the spanning tree.

– *outputs Accept, with probability at least 2/3, on every vertex of G if G is an α-conductor.*
– *outputs Reject, with probability at least 2/3, on at least one vertex of G if G is ϵ-far from any $(\alpha^2/2880)$-conductor.*

The algorithm uses $O_{\epsilon,\alpha}(\log n)$ communication rounds.

To be precise, the algorithm runs $2m^2$ walks of length $\frac{32}{\alpha^2} \log n$ from each of $\theta(1/\epsilon)$ starting vertices with the number of communication rounds equal to length of each random walk i.e. $\frac{32}{\alpha^2} \log n$ and messages at most $O(m \log n)$. A lower bound theorem in [11] (Theorem 2) states that any distributed tester with this gap requires $\Omega(\log(n+m))$ rounds of communication even in LOCAL model, hence, our running time is optimal.

Testing in a Single Phase: The advantage of not having to collect global information is that it lets us do away with the wasteful construction of a spanning tree and information accumulation at the root. Since we do not need to construct a spanning tree, we do not need a pre-processing phase unlike [11].

Note that setting up a spanning tree creates multiple points of failures for the aggregation phase. One could attempt to handle failure of the root of a single tree by setting up multiple spanning trees simultaneously. However, note that a single node failure (of a node internal to all these trees) could disconnect all these trees and if this happens early in the phase 2, we may not get enough information for the root(s) to make their decisions. Since our phase 2 is 'instant' i.e. involves no communication, we do not have any failure issues. This opens up the possibility of a fully-fault-tolerant tester for dynamic networks if a fault-tolerant phase 1 (i.e. fault-tolerant random walks) could be designed.

1.1   Technical Overview

In this section, we give a general overview of the concepts used in our algorithm. Like all the previous algorithms for conductance testing, we perform a certain number of random walks from a randomly selected starting vertex. To boost the success probability of the process, we repeat this process in parallel from a constant number of randomly selected starting vertices. The main technical challenge in running random walks in parallel from different starting vertices is the congestion on the edges. As done by [11], we overcome this problem by not sending the entire trace of the walk from its current endpoint to the next. For each starting point q and for all the walks going from u to v, we simply send the ID of q and the number of walks destined for v to v. At the end of this process, for each starting point q, we simply store at each vertex v, the number of walks that ended at v. Finally, each vertex $v \in V$ looks at the information stored at v to check if the number of walks received from any starting vertex is more than a certain threshold. If so, it outputs *Reject* and, otherwise, it outputs *Accept*.

Stickyness Helps: To show that the number of walks received by a vertex v is sufficient to decide whether v is part of a good conductor or not, we proceed as follows. A technical lemma from [22] implies that if a graph G is ϵ-far from being an α-conductor, then there exists a set $S \in V$ of sufficiently low conductance (of cut $(S, V \setminus S)$, see Definition 1)) and sufficiently high volume. It follows intuitively that it is likely that a short random walk starting from a randomly selected starting vertex in S should not go very far and end in S. In particular, we show that there exist a subset $P \subseteq S$ such that short walks starting from any $v \in P$ end in a large enough region T (subset of S) around v. We make this notion precise by using spectral graph theory to show that a large portion of the volume of low-conductance set S (as described above) belongs to *sticky* vertices. We call a vertex $v \in S$ sticky if there exists a set $T \subseteq S$ such that $v \in T$ and short random walks starting from v end in T with a *sufficiently* high probability. We define trap(v, T, ℓ) as the probability that an ℓ length walk starting from $v \in T \subseteq S$ ends in T.

Trap Probability: Observe that our definition of trap probability is slightly different from the one generally used in the analysis of similar problems. The notion of trap probability is generally used to bound the probability of an ℓ-step random walk staying in a specific set for its entire duration (in each of the ℓ steps). See for example the definitions of *remain* and *escape* probabilities in [13]. Similarly, Czumaj and Sohler also implicitly use the concept of trap probability in their expansion tester [7] and they also bound the probability of a walk staying inside a set of low conductance for its entire duration. We relax the definition a bit and only care about the walk *ending* in a subset of a low conductance set. This allows us to also use the walks that may have briefly escaped the low conductance region when counting the number of trapped walks. Thus, if we run sufficiently many walks from one of the sticky vertices, then a lot of them will end in a subset T of S and some vertex in T will see a lot more walks than any vertex in a good conductor should. To ensure that we pick one of the sticky vertices as a starting vertex, we sample each vertex to be a starting vertex with appropriate probability.

Spectral Approach: In the analysis of the convergence behaviour of random walks (to the stationary distribution) using the eigenvectors and eigenvalues of the random walk matrix M (first introduced by Kale And Seshadhri [19] and later refined by [11] for unbounded degree graphs), they divide the set of eigenvectors of M into *heavy* and *light* sets. All the eigenvectors with eigenvalues above a certain threshold (appropriately chosen) are considered heavy and the rest and considered light. This lets one drop all the light eigenvectors from the analysis since their contribution to the convergence behaviour is minimal. In our analysis, we use a similar technique where we focus on the heavy eigenvectors of the random walk matrix M to lower bound the trap probability of a random walk from an appropriately chosen starting vertex. To further tighten our analysis, we further divide the set H of heavy eigenvectors into the *heaviest* eigenvector $\vec{e}_1$ (with the maximum eigenvalue 1) and the set $H \setminus \{\vec{e}_1\}$. We use both (but separately and not as one bundle H) in our analysis. This also makes intuitive sense since the heaviest eigenvector makes the maximum contribution to the trap probability and treating it separately tightens our bound. We note that [19] and [11] analyse a different measure - the discrepancy between their final endpoint probability distribution and the stationary distribution; and the contribution of $\vec{e}_1$ to this measure is zero, so their analysis does not benefit from segregating the heaviest eigenvector from the set H of the heavy eigenvectors.

Organisation: The rest of the paper is organised as follows. In Sect. 2, we provide necessary definitions and state some basic lemmas that are used in rest of the paper. In Sect. 3, we provide a detailed description of our testing algorithm. Section 3.1 is dedicated to the proof of our main theorem. For any missing proofs not provided here, we refer the reader to the full version of the paper [3].

2   Preliminaries

Let $G = (V, E)$ be an unweighted, undirected graph on n vertices and m edges. We assume that the vertices of G have unique identifiers. For a given vertex $v \in V$, $\deg(v)$ denotes the *degree* of v. For sets $A, B \subseteq V$, we denote by $E(A, B)$ the number of edges that have one endpoint in A and the other in B. A cut is a partition of the vertices into two disjoint subsets. Given a graph $G = (V, E)$ any subset $S \subseteq V$ defines a cut denoted by $(S, \overline{S})$, where $\overline{S} = V \setminus S$.

Definition 1. *Given a cut $(S, \overline{S})$ in G, the conductance of $(S, \overline{S})$ is defined as* $\frac{E(S,\overline{S})}{\min\{\mathrm{vol}(S),\mathrm{vol}(\overline{S})\}}$, *where* $\mathrm{vol}(A) = \sum_{v \in A} \deg(v)$. *Alternatively, we also refer to the conductance of a cut $(S, \overline{S})$ as the* conductance of set S. *The conductance of a graph is the minimum conductance of any cut in the graph.*

Throughout the paper, all vectors $\vec{x} \in \mathbb{R}^n$ are column vectors. For a vector $\vec{x} \in \mathbb{R}^n$, we denote by $\vec{x}^{\mathsf{T}}$ the transpose of $\vec{x}$. For two vectors $\vec{x}$ and $\vec{y}$ in $\mathbb{R}^n$, $\langle \vec{x}, \vec{y} \rangle$ denotes their inner product. We denote the $n \times n$ adjacency matrix of the input graph G by A, where $A_{ij} = 1$, if $(i, j) \in E$ and 0 otherwise. Let D denote the $n \times n$ diagonal degree matrix of G, where $D_{ij} = \deg(i)$ if $i = j$ and 0 otherwise. The main technical tool in our analysis will be random walks on the input graph G. We denote a random walk by its transition matrix M. For a pair of vertices $u, v \in V$, let M_{uv} be the probability of visiting u from v in one step of M. In the standard definition of a random walk, M_{uv} is defined as $1/\deg(v)$ if there is an edge from u to v, and 0 otherwise.

We use a slightly modified version of the standard random walk called a *lazy random walk*. A *lazy random walk* currently stationed at $v \in V$, stays at v with probability $1/2$ and, with the remaining probability $1/2$, it visits a neighbour of v uniformly at random. Let M be a lazy random walk on G, the transition probabilities for M are defined as follows: for a pair $v, w \in V$ such that $v \neq w$, $M_{wv} = \frac{1}{2\deg(v)}$, if $(v, w) \in E$ and 0, otherwise. Further, for $v \in V$, we define $M_{vv} = 1/2$. Algebraically, M can be expressed as $M = \frac{1}{2}(I + AD^{-1})$, where I is the $n \times n$ identity matrix. Let π be the stationary distribution of M. In the stationary distribution of a lazy random walk, each vertex is visited with probability proportional to its degree. More formally, $\pi = \frac{\vec{d}}{2m}$, where $\vec{d}$ is an n-dimensional vector of vertex degrees and m is the number of edges in G.

In the following, we provide a brief exposition of relevant concepts from spectral graph theory. We refer the reader to the textbook [6] by Fan Chung for a detailed treatment of the subject. Note that for irregular graphs, M is an asymmetric matrix and may not have an orthogonal set of eigenvectors. For analyzing random walks on G in terms of the eigenvalues and eigenvectors of its associated matrices, we rely on a related symmetric matrix called the *normalized Laplacian* of G denoted by N. The normalized Laplacian N is defined as

$$N = I - D^{-1/2}AD^{-1/2}.$$

We show below a way to express M in terms of N.

$$M = 1/2(I + AD^{-1}) = I - 1/2(I - AD^{-1})$$
$$= I - 1/2D^{1/2}(I - D^{-1/2}AD^{-1/2})D^{-1/2} = D^{1/2}(I - N/2)D^{-1/2}.$$

Since N is a symmetric matrix, it has a set of n real eigenvalues and a corresponding set of mutually orthogonal eigenvectors. Throughout we let $\omega_1 \leq \omega_2 \leq \ldots \leq \omega_n$ denote the set of eigenvalues and $\vec{\zeta}_1, \vec{\zeta}_2, \ldots, \vec{\zeta}_n$ denote the corresponding set of eigenvectors. It is well known that eigenvalues of N are $0 = \omega_1 \leq \omega_2 \leq \ldots \leq \omega_n \leq 2$. It is easy to verify that $\sqrt{d} = (\sqrt{d_1}, \sqrt{d_2}, \ldots, \sqrt{d_n})$ is the first eigenvector $\vec{\zeta}_1$ of N with corresponding eigenvalue $\omega_1 = 0$. Each of the orthogonal eigenvectors $\vec{\zeta}_i$ can be normalized to be a unit vector as $\vec{e}_i = \vec{\zeta}_i/\|\vec{\zeta}_i\|_2$ Together, these orthogonal unit eigenvectors define an orthonormal basis for $\mathbb{R}^n$. Observe that the first unit eigenvector $\vec{e}_1$ of this orthonormal basis is $\sqrt{d}/\sqrt{2m}$. Also observe that the stationary distribution π of M is equal to $D^{1/2}\vec{e}_1/\sqrt{2m}$.

It is easy to verify that for every unit eigenvector $\vec{e}_i$ of N, we have a corresponding right eigenvector $D^{1/2}\vec{e}_i$ of M with eigenvalue $1 - \omega_i/2$.

On a connected, graph, a lazy random walk M can be viewed as a reversible, aperiodic Markov chain with state space V and transition matrix M.

Definition 2. *Let M be a reversible, aperiodic Markov chain on a finite state space V with stationary distribution π. Furthermore, let $\pi(S) = \sum_{v \in S} \pi(v)$. The Cheeger constant or conductance $\phi(M)$ of the chain is defined as*

$$\phi(M) = \min_{S \subset V : \pi(S) \leq 1/2} \frac{\sum_{x \in S, y \in V \setminus S} \pi(x)M(y, x)}{\pi(S)}.$$

Here $M(y, x)$ is the probability of moving to state y from state x in one step.

The definition of the lazy random walk matrix M and the fact that the stationary distribution π of our lazy random walk is $\vec{d}/2m$ together imply that the Cheeger constant $\phi(M)$ (henceforth, ϕ_*) of the walk M is

$$\phi_* = \min_{U \in V, \text{vol}(U) \leq \text{vol}(V)/2} \frac{E(U, V \setminus U)}{2 \cdot \text{vol}(U)}.$$

For an α-conductor, we get that

$$\phi_* = \frac{\alpha}{2}. \tag{1}$$

3   A Distributed Algorithm for Testing Conductance

Given an input graph $G = (V, E)$, a conductance parameter α, and a distance parameter ϵ, our distributed conductance tester distinguishes, with probability at least $2/3$, between the case where G is an α-conductor and the case where G is ϵ-far from being an $\Omega(\alpha^2)$-conductor. A key technical lemma from [22] implies that, if G is ϵ-far from being an $\Omega(\alpha^2)$-conductor, then there exists a low-conductance cut $(S, \overline{S})$ such that $\mathrm{vol}(S) \geq \epsilon m/10$. We build on this lemma to show, using spectral methods (see Lemma 3 and Corollary 1), that there exist a set of *sticky* vertices with high enough volume in S. Recall that a vertex x in a low-conductance set S is sticky if there exists a large enough subset $T \subset S$ such that $x \in T$ and a short random walk starting from x *ends* in T with a sufficiently high probability. Intuitively, random walks starting from sticky vertices tend to stick to a small region around them. This leads to some vertex in the graph receive more than their *fair share* of number of walks. On the other, hand if a graph is a good conductor, then the random walks from anywhere in the graph mix very quickly. This ensures that the fraction of the total number of walks received by any vertex in the graph is proportional to its degree.

We randomly sample a set Q of $\Omega(1/\epsilon)$ *source* vertices (each vertex $v \in V$ is sampled with probability proportional to its degree). Then we run a certain number of short random walks from each source in Q. Since a large part of the volume of our high-volume, low-conductance set S consists of only sticky vertices in a bad conductor, some vertex in Q will be sticky with sufficiently large probability. We use the number of walks received by each vertex from a specific source as a test criteria. We implement sampling of the set Q by having each vertex v sample itself by flipping a biased coin with probability $5000 \cdot \deg(v)/(2\epsilon m)$. Therefore, we get that $|Q| = 5000/\epsilon$ in expectation. It follows from Chernoff bound that the probability of Q having more than $5500/\epsilon$ vertices is at most $e^{-23/\epsilon} \leq e^{-23}$. Then, we perform K random walks of length ℓ starting from each of the chosen vertices in Q. The exact values of these parameters are specified later in the sequel. The pseudocode of the algorithm is presented in Algorithm 1. At any point before the last step of random walks, each vertex $v \in V$ contains a set W of tuples $(q, count, i)$, where $count$ is the number of walks of length i originating from source q currently stationed at v. All these walks are advanced by one step (for ℓ times) by invoking Algorithm 2. At the end of the last step of the walks, Algorithm 2 outputs a set of tuples C_v. Each tuple in C_v is of the form $(q, count)$, where $count$ is the total number of ℓ-step walks starting at q that ended at v. Then, in Algorithm 1, processor at vertex v goes over every tuple $(q, count)$ in C_v (see Lines 11 to 15 of Algorithm 1), and if the $count$ value of any of them is above a pre-defined threshold τ, it outputs *Reject*. If none of the tuples have their $count$ value above threshold, it outputs *Accept*. The exact value of τ is specified later in the sequel.

When advancing the walks originating at a source $q \in Q$ by one step, we do not send the full trace of every random walk. Instead, for every source $q \in Q$, every vertex $v \in V$ only sends a tuple (q, k, i) to its neighbour w indicating that k random walks originating at q have chosen w as their destination in their ith step.

Since the size of Q is constant with high enough probability, we will not have to send more than a constant number of such tuples on each edge. Moreover, each tuple can be encoded using $O(\log n)$ bits (given the values of parameters ℓ and K specified in the sequel). Hence, we only communicate $O(\log n)$ bits per edge in any round with high probability. To ensure no congestion, we check the length of every message (Algorithm 2: lines 13–15). If a message appears too large to send, we simply output *Reject* on the host vertex and abort the algorithm. Note that the number of tuples we ever have to send along any edge is upper bounded by $|Q|$ and $|Q| \leq 5500\epsilon$, with probability at least $1 - e^{-23}$. Therefore, we may rarely abort the algorithm before completing the execution of the random walks. If that happens, then the probability of accepting an α-conductor is slightly reduced. Hence the following observation follows:

Observation 1. *Algorithm 1 rejects an α-conductor due to congestion with probability at most e^{-23}.*

We set the required parameters of Algorithm 1 as follows:

- the number of walks $K = 2m^2$,
- the length of each walk $\ell = \frac{32}{\alpha^2} \log n$
- the rejection threshold for vertex $v \in V$, $\tau_v = m \cdot \deg(v) \cdot (1 + 2n^{-1/4})$.

Algorithm 1. Distributed algorithm running at vertex v for testing conductance.

1: **Algorithm** DISTRIBUTED-GRAPH-CONDUCTANCE-TEST$(G, \epsilon, \alpha, \ell, K)$
2:     ▷ The algorithm performs K random walks of length ℓ from a set Q of $\Theta(1/\epsilon)$ starting vertices, where every starting vertex is sampled randomly from V.
3:     ℓ : The length of each random walk
4:     K : The number of walks
5:     W_v : Set of tuples $(q, count, i)$ ▷ where *count* is the number of walks originating at source q currently stationed at v
6:     C_v : Set of tuples $(q, count)$ ▷ where *count* is the total number of ℓ step walks starting at q that ended at v
7:     τ_v : maximum number of ℓ-length walks v should see from a given source on an α-conductor.
8:     Flip a biased coin with probability $p = 5000 \deg(v)/(\epsilon 2m)$ to decide whether to start K lazy random walks.
9:     If chosen, initialise W_v as $W_v \leftarrow \{(v, K, 0)\}$.
10:     Call Algorithm 2 for ℓ synchronous rounds.
11:     **while** there is some tuple $(q, count)$ in C_v **do**
12:         **if** $count > \tau_v$ **then**   ▷ Received too many walks from q.
13:             Output *Reject* and stop all operations.
14:         **else**
15:             Remove $(q, count)$ from C_v
16:     Output *Accept*

Algorithm 2. Algorithm for moving random walks stationed at v by one step.

1: **Algorithm** MOVE-WALKS-AT-v
2:     W_v : Set of tuples $(q, count, i)$ ▷ where $count$ is the number of walks
   originating at source q currently stationed at v just after their ith step..
3:     D_v : Set of tuples $(q, count, dest)$ ▷ where $count$ is the number of walks
   starting at q that are to be forwarded to $dest$.
4:     C_v : Set of tuples $(q, count)$ ▷ where $count$ is the total number of walks
   starting at q that have v as their final destination or endpoint.
5:     $D_v \leftarrow \emptyset$.
6:     **while** there is some tuple (q, k, i) in W_v **do**
7:         **if** $i \neq L$ **then**   ▷ If not the last step, process the next set of destinations.
8:             Draw the next set of destinations for the k walks and update the set D_v.
9:             Remove (q, k, i) from W_v
   ▷ If last step of the walks, update how many ended at v.
10:             Update C_v to reflect the k walks that ended in v.
   ▷ Prepare the messages to be sent
11:     **while** there is some tuple $(q, count, dest)$ in D_v **do**
12:             Add tuple $(q, count, i + 1)$ to the message to be sent to $dest$
   ▷ Check each message for length
13:     For each message M to be sent
14:     **if** the number of tuples in $M > 5500/\epsilon$ **then**
15:             Output $Reject$ and stop all operations.
16:     Send all the messages to their respective destinations.
   ▷ Process the messages received
17:     For each source s from which a total of s_{count} walks are received,
18: add tuple $(s, s_{count}, i + 1)$ to W_v

3.1   Analysis of the Algorithm

The main idea behind our algorithm is that, in a bad conductor, a random walk would converge to the stationary distribution more slowly and would initially get trapped within sets of vertices with small conductance. We provide a lower bound on the probability of an ℓ-step random walk starting from a vertex chosen at random (with probability proportional to its degree) from a subset T of a low-conductance set S finishing at some vertex in T.

Definition 3. *For a set $T \subseteq V$, and a vertex $u \in T$, let* $\text{trap}(u, T, \ell)$ *(henceforth trap probability) denote the probability of an ℓ-step random walk starting from $u \in T$ finishing at some vertex in T. When the starting vertex is chosen at random from T with probability proportional to its degree, we denote by* $\text{trap}(T, \ell)$ *the average trap probability (weighted by vertex degrees) over set T:* $\text{trap}(T, \ell) = \frac{1}{\text{vol}(T)} \sum_{u \in T} \deg(u) \cdot \text{trap}(u, T, \ell)$.

Given a set S of conductance at most δ and $T \subseteq S$, we establish a relationship between the average trap probability $\text{trap}(T, \ell)$ and conductance δ of S in the next two lemmas. We first consider the case $T = S$ in Lemma 1, and then obtain a bound when T is a subset (of sufficiently large volume) of S in Lemma 3.

Lemma 1. *Consider a set $S \subseteq V$ such that the cut $(S, \overline{S})$ has conductance at most δ. Then, for any integer $\ell > 0$, the following holds*

$$\mathrm{trap}(S, \ell) \geq \frac{\mathrm{vol}(S)}{2m} + \left(\frac{5}{6} - \frac{\mathrm{vol}(S)}{2m}\right)(1 - 3\delta)^\ell.$$

Proof. Let $\mathbb{1}_S$ denote the n-dimensional indicator vector of set S. We pick a source vertex $v \in S$ with probability $\deg(v)/\mathrm{vol}(S)$, where $\deg(v)$ is degree of v and $\mathrm{vol}(S)$ is the sum of degrees of vertices in set S. This defines an initial probability distribution denoted by $\vec{p}_S^0$ on the vertex set V, where, $\vec{p}_S^0(v) = d(v)/\mathrm{vol}(S)$ for $v \in S$ and $\vec{p}_S^0 = 0$ for $v \notin S$. Note that $\vec{p}_S^0 = \frac{1}{\mathrm{vol}(S)}D\mathbb{1}_S$, where, D is the diagonal degree matrix of G. Denote by M the transition matrix of a lazy random walk on G. The endpoint probability distribution $\vec{p}_S^\ell$ of an ℓ-step lazy random walk on G starting from a vertex chosen from S according to $\vec{p}_S^0$ is

$$\vec{p}_S^\ell = M^\ell \vec{p}_S^0 = (1/\mathrm{vol}(S))M^\ell D\mathbb{1}_S.$$

Recall that M can be expressed in terms of the normalized Laplacian matrix $N = I - D^{-1/2}AD^{-1/2}$ of G as $M = D^{1/2}(I - N/2)D^{-1/2}$. (See Sect. 2.)

It follows therefore that $\vec{p}_S^\ell = (1/\mathrm{vol}(S))\left(D^{1/2}(I - N/2)D^{-1/2}\right)^\ell D\mathbb{1}_S$.

The trap probability $\mathrm{trap}(S, \ell)$ of an ℓ-step lazy random walk starting from a random vertex in S picked according to $\vec{p}_S^0$ can be expressed as the inner product of vectors $\vec{p}_S^\ell$ and $\mathbb{1}_S$:

$$\mathrm{trap}(S, \ell) = \frac{1}{\mathrm{vol}(S)}\mathbb{1}_S^\mathsf{T}M^\ell D\mathbb{1}_S = \frac{1}{\mathrm{vol}(S)}\mathbb{1}_S^\mathsf{T}\left(D^{1/2}(I - N/2)D^{-1/2}\right)^\ell D\mathbb{1}_S$$

$$= \frac{1}{\mathrm{vol}(S)}(D^{1/2}\mathbb{1}_S)^\mathsf{T}(I - N/2)^\ell(D^{1/2}\mathbb{1}_S).$$

Recall that $0 = \omega_1 \leq \omega_2 \leq \cdots \leq \omega_n < 2$ are the eigenvalues of N and $\vec{e}_1, \vec{e}_2, \ldots, \vec{e}_n$ denote the corresponding unit eigenvectors. We can express $D^{1/2}\mathbb{1}_S$ in the orthonormal basis defined by the eigenvectors of N as $D^{1/2}\mathbb{1}_S = \sum_i \alpha_i \vec{e}_i$. It follows that

$$\sum_i \alpha_i^2 = \langle D^{1/2}\mathbb{1}_S, D^{1/2}\mathbb{1}_S \rangle = \mathrm{vol}(S). \tag{2}$$

Taking the quadratic form of N for vector $D^{1/2}\mathbb{1}_S$, we get

$$(D^{1/2}\mathbb{1}_S)^\mathsf{T}N(D^{1/2}\mathbb{1}_S) = (D^{1/2}\mathbb{1}_S)^\mathsf{T}I(D^{1/2}\mathbb{1}_S) - (D^{1/2}\mathbb{1}_S)^\mathsf{T}(D^{-1/2}AD^{-1/2})(D^{1/2}\mathbb{1}_S)$$

$$= \mathrm{vol}(S) - (D^{1/2}\mathbb{1}_S)^\mathsf{T}(D^{-1/2}AD^{-1/2})(D^{1/2}\mathbb{1}_S) = \mathrm{vol}(S) - \mathbb{1}_S^\mathsf{T}A\mathbb{1}_S.$$

Note that the term $\mathbb{1}_S^\mathsf{T}A\mathbb{1}_S$ corresponds to the number of edges in $S \times S$. Therefore. it follows that

$$(D^{1/2}\mathbb{1}_S)^\mathsf{T}N(D^{1/2}\mathbb{1}_S) = E(S, \overline{S}).$$

Since the conductance of the cut $(S, \overline{S})$ is at most δ, we have that

$$(D^{1/2}\mathbb{1}_S)^\mathsf{T}N(D^{1/2}\mathbb{1}_S) = E(S, \overline{S}) \leq \delta \cdot \mathrm{vol}(S). \tag{3}$$

Expressing $D^{1/2}\mathbb{1}_S$ as $\sum_i \alpha_i \vec{e}_i$, the quadratic form of N for $D^{1/2}\mathbb{1}_S$ can also be written as

$$(D^{1/2}\mathbb{1}_S)^\mathsf{T} N (D^{1/2}\mathbb{1}_S) = \left(\sum_i \alpha_i \vec{e}_i\right)^\mathsf{T} N \left(\sum_i \alpha_i \vec{e}_i\right) = \sum_i \alpha_i^2 \omega_i. \qquad (4)$$

Combining from Eqs. (2), (3) and (4), we get that

$$(D^{1/2}\mathbb{1}_S)^\mathsf{T} (I - N/2)(D^{1/2}\mathbb{1}_S) = \sum_i \alpha_i^2 - \frac{1}{2}\sum_i \alpha_i^2 \omega_i \geq \mathrm{vol}(S) - \frac{\delta}{2}\mathrm{vol}(S). \qquad (5)$$

Recall that $0 = \omega_1 \leq \omega_2 \leq \ldots \leq \omega_n < 2$ are the eigenvalues of N and let $\vec{e}_1, \vec{e}_2, \ldots, \vec{e}_n$ be the corresponding unit eigenvectors. Correspondingly, we can define a set of eigenvalues $1 = \lambda_1 \geq \lambda_2 \geq \ldots \geq \lambda_n > 0$ and the same set of eigenvectors $\vec{e}_1, \vec{e}_2, \ldots, \vec{e}_n$ for $I - N/2$. Notice that for each i, $\lambda_i = 1 - \omega_i/2$. With this translation of eigenspace, we get that

$$(D^{1/2}\mathbb{1}_S)^\mathsf{T} (I - N/2)(D^{1/2}\mathbb{1}_S) = \sum_i \alpha_i^2 \lambda_i.$$

We call the quantity $\sum_i \alpha_i^2$ the *coefficient sum of the eigenvalue set*. We also call an eigenvalue λ_i of $I - N/2$ (and the corresponding eigenvector $\vec{e}_i$) *heavy* if $\lambda_i \geq 1 - 3\delta$. We denote by H the index set of the heavy eigenvalues and let $\overline{H}$ be the index set of the rest. Since $\sum_i \alpha_i^2 \lambda_i \geq (1 - \delta/2)\mathrm{vol}(S)$ is large for a set with small conductance, we expect many of the coefficients α_i^2 corresponding to heavy eigenvalues to be large. This would slow down the convergence of the random walk and make the trap probability for our low-conductance set S large. The following lemma establishes a lower bound on the contribution of the index set H to the coefficient sum.

Lemma 2. *For $\{\alpha_i\}_i, H$, and s as defined above, $\sum_{i \in H} \alpha_i^2 \geq \frac{5}{6}\mathrm{vol}(S)$.*

See the full version [3] for the proof. Lemma 2 gives us a lower bound on the average trap probability of set S in terms of the conductance of the cut $(S, \overline{S})$.

$$\begin{aligned}
\mathrm{trap}(S, \ell) &= \frac{1}{\mathrm{vol}(S)}(D^{1/2}\mathbb{1}_S)^\mathsf{T} (I - N/2)^\ell (D^{1/2}\mathbb{1}_S) \\
&= \frac{1}{\mathrm{vol}(S)}\left(\sum_i \alpha_i \vec{e}_i\right)^\mathsf{T} (I - N/2)^\ell \left(\sum_i \alpha_i \vec{e}_i\right) \\
&= \frac{1}{\mathrm{vol}(S)}\left(\sum_i \alpha_i \vec{e}_i\right)^\mathsf{T} \left(\sum_i \alpha_i \lambda_i^\ell \vec{e}_i\right) = \frac{1}{\mathrm{vol}(S)}\sum_i \alpha_i^2 \lambda_i^\ell.
\end{aligned}$$

Further, focusing on the contribution of the index set H to the trap probability,

$$\begin{aligned}
\mathrm{trap}(S, \ell) &= \frac{1}{\mathrm{vol}(S)}\sum_i \alpha_i^2 \lambda_i^\ell \geq \frac{1}{\mathrm{vol}(S)}\sum_{i \in H} \alpha_i^2 \lambda_i^\ell \\
&= \frac{1}{\mathrm{vol}(S)}\left(\alpha_1^2 \lambda_1 + \sum_{i \in H \setminus \{1\}} \alpha_i^2 \lambda_i^\ell\right) \\
&\geq \frac{1}{\mathrm{vol}(S)}\left(\alpha_1^2 + (5\,\mathrm{vol}(S)/6 - \alpha_1^2)(1 - 3\delta)^\ell\right). \qquad (6)
\end{aligned}$$

The last inequality follows by the definition of a heavy eigenvalue and by Lemma 2 (we have that $\sum_{i \in H} \alpha_i^2 \geq 5 \operatorname{vol}(S)/6$). By definition, $\lambda_1 = 1$ and $\vec{e}_1 = \sqrt{\vec{d}}/\sqrt{2m}$, where, $\vec{d}$ is the vector of vertex degrees. It follows that $\alpha_1 = \langle D^{1/2} \mathbb{1}_S, \vec{e}_1 \rangle = \operatorname{vol}(S)/\sqrt{2m}$. Plugging in the values of α_1 in (6), we get

$$
\begin{aligned}
\operatorname{trap}(S, \ell) &\geq \frac{1}{\operatorname{vol}(S)} \left(\frac{(\operatorname{vol}(S))^2}{2m} + \left(\frac{5 \operatorname{vol}(S)}{6} - \frac{(\operatorname{vol}(S))^2}{2m} \right) (1 - 3\delta)^\ell \right) \\
&= \frac{\operatorname{vol}(S)}{2m} + \left(\frac{5}{6} - \frac{\operatorname{vol}(S)}{2m} \right) (1 - 3\delta)^\ell.
\end{aligned}
$$

Next lemma states that every subset $T \subset S$ of large enough volume has high trap probability.

Lemma 3. *Consider sets $T \subseteq S \subseteq V$, such that the cut $(S, \overline{S})$ has conductance at most δ and that $\operatorname{vol}(T) = (1 - \eta) \operatorname{vol}(S)$ for some $0 < \eta < 5/6$, then for any integer $\ell > 0$, there exists a vertex $v \in T$ such that*

$$
\operatorname{trap}(v, T, \ell) \geq \frac{\operatorname{vol}(T)}{2m} + \left(\frac{5}{6}(1 - \sqrt{(6\eta)/5})^2 - \frac{\operatorname{vol}(T)}{2m} \right) (1 - 3\delta)^\ell. \tag{7}
$$

Refer to the full version [3] for the proof.

Lemma 3 implies the following corollary.

Corollary 1. *Consider a set $S \subset V$, such that the cut $(S, \overline{S})$ has conductance at most δ. Given any $0 < \eta < 5/6$ and integer $\ell > 0$, there exist a set of volume at least $\eta \cdot \operatorname{vol}(S)$, such that every vertex in this set is sticky. In other words, for every vertex v is this set, there exists $T \subseteq S$ of volume $\operatorname{vol}(T) = (1 - \eta) \cdot \operatorname{vol}(S)$ such that $\operatorname{trap}(v, T, \ell)$ is given by (7).*

Refer to the full version [3] for the proof.

We build on the following combinatorial lemma from [22]:

Lemma 4 (Lemma 9 of [22]). *Let $G = (V, E)$ be an m-edge graph. If there exists a set $P \subseteq V$ such that $\operatorname{vol}(P) \leq \epsilon m/10$ and the subgraph $G[V \setminus P]$ that is induced by the vertex set $V \setminus P$ has conductance at least ϕ', then there exists an algorithm that modifies at most ϵm edges of G to get a graph $G' = (V, E')$ with conductance at least $\phi'/3$.*

In the following lemma, we show the existence of a high volume set A with low enough conductance in a graph that is far from being a good conductor.

Lemma 5. *Let $G = (V, E)$ be an n-vertex, m-edge graph such that G is ϵ-far from having conductance at least $\alpha^2/2880$, then there exists a set $A \subseteq V$ such that $\operatorname{vol}(A) \geq \epsilon m/10$ and conductance of cut $(A, \overline{A})$ is at most $\alpha^2/960$.*

Refer to the full version [3] for the proof.

Finally, we need the following classical relation between the conductance or Cheeger constant of a Markov chain and its second largest eigenvalue.

Theorem 2 ([1,2,9]). *Let P be a reversible lazy chain (i.e., for all x, $P(x,x) \geq 1/2$) with Cheeger constant ϕ_*. Let λ_2 be the second largest eigenvalue of P. Then, $\frac{\phi_*^2}{2} \leq 1 - \lambda_2 \leq 2\phi_*$.*

We can now state our main theorem.

Theorem 3. *For an input graph $G = (V, E)$, parameters $0 < \alpha < 1$ and $\epsilon > 0$, the distributed algorithm described in Sect. 3*

- *outputs Accept, with probability at least 2/3, on every vertex of G if G is an α-conductor.*
- *outputs Reject, with probability at least 2/3, on at least one vertex of G if G is ϵ-far from any $(\alpha^2/2880)$-conductor.*

The algorithm uses $O(\log n/\alpha^2)$ communication rounds.

Proof. Let us start by showing that, with high enough probability, the algorithm outputs *Accept* on every vertex if G is an α-conductor. By Observation 1, we may reject G and abort the algorithm due to congestion with probability at most e^{-23}. For now, let us assume that this event did not occur. Denote by λ_2 the second largest eigenvalue of the lazy random walk matrix M on G. It is well known (see, e.g., [27]) that, for a pair $u, v \in V$, $\left|M^\ell(v,u) - \deg(v)/(2m)\right| \leq \lambda_2^\ell \leq e^{-\ell(1-\lambda_2)}$. It follows from Theorem 2 that $\left|M^\ell(v,u) - \frac{\deg(v)}{2m}\right| \leq e^{-\ell\phi_*^2/2} \leq e^{-\frac{\ell\alpha^2}{8}}$, where the second inequality above follows from the fact that, for a random walk on an α-conductor, $\phi_* = \alpha/2$ (see (1)). Thus, in an α-conductor, for $\ell = (32/\alpha^2)\log n$, any starting vertex $u \in V$ and a fixed vertex $v \in V$, we have that

$$\deg(v)/(2m) - 1/n^4 \leq M^\ell(v,u) \leq \deg(v)/(2m) + 1/n^4.$$

Recall that the number K of random walks and rejection threshold τ_v for vertex v are set as $K = 2m^2$ and $\tau_v = m \cdot \deg(v) \cdot (1 + 2n^{-1/4})$. Let $X_{u,v}$ denote the number of random walks starting from u that ended in v. It follows that

$$\mathbb{E}X_{u,v} = K \cdot M^\ell(v,u) \leq 2m^2 \cdot \frac{\deg(v)}{2m} + \frac{2m^2}{n^4} \leq m \cdot \deg(v) + 1.$$

The random variable $X_{u,v}$ is the sum of K independent Bernoulli trials with success probability $M^\ell(v,u)$. Applying multiplicative Chernoff bounds, we get the following for large enough n.

$$\Pr[X_{u,v} > (1 + n^{-1/4}) \cdot \mathbb{E}[X_{u,v}]] < \exp(-n^{-1/2} \cdot (m \cdot \deg(v) + 1)/3) \leq \exp(\frac{-n^{1/2}}{3}).$$

The second inequality above follows from the fact that $m \cdot \deg(v) > n - 1$ for a connected graph. Thus, each vertex y receives at most

$$(1+n^{-1/4}) \cdot \mathbb{E}[X_{u,v}] \leq (1+n^{-1/4}) \cdot (m \cdot \deg(v)+1) < m \cdot \deg(v) + 2m/n^{1/4} \cdot \deg(v)$$

walks from u, with probability at least $1-n^{-2}$. Taking union bound over all $y \in V$ and all starting vertices u, we get that, with high probability, our algorithm outputs *Accept* on every vertex of G for every starting point if G is an α-conductor. Finally, taking the union bound over the events that we rejected due to congestion or due to receiving too many walks at some vertex, the claim follows.

Next, we analyse the probability of rejecting if G is far from having the desired conductance. By Lemma 5, there exists a set $S \subset V$, with $\mathrm{vol}(S) \geq \epsilon m/10$, such that the conductance of S is at most $\alpha^2/960$. Further applying Corollary 1 with $\eta = 5/486$ on S as above, we get that there exists a set P of sticky vertices such that $\mathrm{vol}(P) \geq (5\,\mathrm{vol}(S))/486$ and for every $v \in P$, there exists a set $T \subseteq S$, $\mathrm{vol}(T) = (481\,\mathrm{vol}(S))/486$, such that

$$
\begin{aligned}
\mathrm{trap}(v, T, \ell) &\geq \frac{\mathrm{vol}(T)}{2m} + \left(\frac{160}{243} - \frac{\mathrm{vol}(T)}{2m} \right) \left(1 - \frac{\alpha^2}{320} \right)^{\ell} \\
&\geq \frac{\mathrm{vol}(T)}{2m} + \left(\frac{160}{243} - \frac{\mathrm{vol}(T)}{2m} \right) \cdot e^{-\alpha^2 \ell/160}
\end{aligned}
$$

where the last inequality follows from that $1 - x > e^{-2x}$, for $0 \leq x < 1/2$, provided that $\alpha^2/320 \leq 1/2$. For $\ell = (32/\alpha^2) \log n$ and $\mathrm{vol}(T) \leq \mathrm{vol}(S) \leq \mathrm{vol}(V)/2 = m$, we get that, for every $v \in P$, $\mathrm{trap}(v, T, \ell) \geq \frac{\mathrm{vol}(T)}{2m} + \frac{77}{486} \cdot \frac{1}{n^{1/5}}$.

Let us assume that a vertex $u \in P \subset A$ is picked as the starting vertex of $K = 2m^2$ random walks in G. By Corollary 1, a set T with $\mathrm{vol}(T) = (481\,\mathrm{vol}(S))/486$ with $\mathrm{trap}(v, T, \ell) \geq \mathrm{vol}(T)/2m + 77/486 \cdot n^{-1/5}$ will exist, for every $v \in P$. Also note that $\mathrm{vol}(T) \leq \mathrm{vol}(S) \leq \mathrm{vol}(V) = 2m$. For some appropriate constant c_1, $\mathrm{vol}(T) = c_1 \cdot m = \Theta(m)$. Further, let $Y_{u,T}$ be the number of walks that ended in the set T (corresponding to u as in Corollary 1) after ℓ steps. It follows that

$$
\mathbb{E} Y_{u,T} \geq K \cdot \left(\frac{\mathrm{vol}(T)}{2m} + \frac{77}{486} n^{-1/5} \right) \geq m \cdot \mathrm{vol}(T) + \frac{154}{486} \cdot \frac{m^2}{n^{1/5}}.
$$

Let $c_2 = \frac{77}{486}$. By an application of Chernoff bound, we get

$$
\begin{aligned}
\Pr &\left[Y_{u,T} < \left(1 - 3(m \cdot \mathrm{vol}(T))^{-1/2} \right) \mathbb{E}[Y_{u,T}] \right] \\
&< \exp \left(-4(m \cdot \mathrm{vol}(T))^{-1} \cdot \mathbb{E}[Y_{u,T}] \right) \\
&\leq \exp \left(-4(m \cdot \mathrm{vol}(T))^{-1} \cdot \left(\mathrm{vol}(T) \cdot m + 2c_2 m^2 n^{-1/5} \right) \right) \\
&\leq \exp \left(-4 \cdot (1 + \Theta(n^{-1/5})) \right) < 1/10.
\end{aligned}
$$

With probability at least $9/10$, the total number of walks received by set T is

$$
\begin{aligned}
&\geq (1 - 3(m \cdot \mathrm{vol}(T))^{-1/2}) \left(m \cdot \mathrm{vol}(T) + 2c_2 m^2/n^{1/5} \right) \\
&\geq m \cdot \mathrm{vol}(T) + 2c_2 m^2/n^{1/5} - 3\sqrt{m}\sqrt{\mathrm{vol}(T)} - 6c_2 m^{3/2} (\mathrm{vol}(T))^{-1/2} n^{-1/5}.
\end{aligned}
$$

The number of walks received by any vertex $v \in T$ is minimum when the walks within T have mixed well reaching their stationary distribution with respect to T. It follows that the number of walks received by a vertex $v \in T$ is at least

$$\deg(v) \cdot \left(m + 2c_2 \frac{m^2}{n^{1/5} \cdot \mathrm{vol}(T)} - 3 \frac{\sqrt{m} \cdot \sqrt{\mathrm{vol}(T)}}{\mathrm{vol}(T)} - 6c_2 \frac{m^{3/2}}{(\mathrm{vol}(T))^{3/2} n^{1/5}} \right).$$

Recalling that $\mathrm{vol}(T) = c_1 m$, the expected number of walks received by a vertex $v \in T$ is at least

$$\deg(v) \cdot \left(m + 2 \frac{c_2}{c_1} \frac{m}{n^{1/5}} - \frac{3}{\sqrt{c_1}} - 6 \frac{c_2}{(c_1)^{3/2} n^{-1/5}} \right) = \deg(v) \cdot \left(m + 2 \frac{c_2}{c_1} \frac{m}{n^{1/5}} - O(1) \right).$$

Therefore, on average, vertex $v \in T$ of degree $\deg(v)$ receives more than the threshold $\tau_v = m \cdot \deg(v) + 2m \cdot \deg(v)/n^{1/4}$ number of walks for large enough n. Thus, some vertex in T will receive more than τ_v walks and output *Reject*.

Let $\mathcal{E}$ be the event that none of the vertices in P is sampled to be one of the starting points in Q. Since each vertex $u \in V$ is sampled with probability $5000 \cdot \deg(v)/(2\epsilon \cdot m)$ and $\mathrm{vol}(P) \geq (5\epsilon \cdot m)/4860$, it follows that

$$Pr[\mathcal{E}] \leq (1 - 5000 \cdot \deg(v)/(2\epsilon \cdot m))^{\frac{5\epsilon \cdot m}{4860}} \leq e^{-2.5} = 0.08.$$

Taking a union bound over the probability of the event $\mathcal{E}$ and the probability of set T around a starting vertex $u \in P$ not receiving enough walks, we get that with probability at most $0.1 + 0.08 = 0.18$, no vertex will output *Reject*. Thus, our distributed algorithm will output *Reject* with probability at least $2/3$, on at least one vertex of G. Finally, the upper bound on the number of communication rounds follows from the length $\ell = \frac{32}{\alpha^2} \log n$ of each random walk.

References

1. Alon, N.: Eigenvalues and expanders (1986). https://doi.org/10.1007/BF02579166
2. Alon, N., Milman, V.D.: Eigenvalues, expanders and superconcentrators (extended abstract). In: 25th Annual Symposium on Foundations of Computer Science. IEEE Computer Society (1984). https://doi.org/10.1109/SFCS.1984.715931
3. Batu, T., Trehan, A., Trehan, C.: A distributed conductance tester without global information collection (2023). https://doi.org/10.48550/arXiv.2305.14178
4. Brakerski, Z., Patt-Shamir, B.: Distributed discovery of large near-cliques. Distrib. Comput. **24**, 79–89 (2009). https://doi.org/10.1007/s00446-011-0132-x
5. Censor-Hillel, K., Fischer, E., Schwartzman, G., Vasudev, Y.: Fast distributed algorithms for testing graph properties. Distrib. Comput. **32**, 41–57 (2019). https://doi.org/10.1007/s00446-018-0324-8
6. Chung, F.R.K.: Spectral Graph Theory. AMS, Providence (1997). https://bookstore.ams.org/cbms-92
7. Czumaj, A., Sohler, C.: Testing Expansion in Bounded-Degree Graphs. Combinatorics, Probability & Computing (2010). https://doi.org/10.1017/S096354831000012X

8. Das Sarma, A., et al.: Distributed verification and hardness of distributed approximation. In: Proceedings of the Forty-Third Annual ACM Symposium on Theory of Computing, STOC 2011 (2011). https://doi.org/10.1145/1993636.1993686
9. Dodziuk, J.: Difference equations, isoperimetric inequality and transience of certain random walks. Trans. Am. Math. Soc. **284**(2) (1984). http://www.jstor.org/stable/1999107
10. Even, G., et al.: Three notes on distributed property testing. In: 31st International Symposium on Distributed Computing (DISC) (2017). http://drops.dagstuhl.de/opus/volltexte/2017/7984
11. Fichtenberger, H., Vasudev, Y.: A two-sided error distributed property tester for conductance. In: 43rd International Symposium on Mathematical Foundations of Computer Science, MFCS 2018 (2018). https://doi.org/10.4230/LIPIcs.MFCS.2018.19
12. Fischer, O., Gonen, T., Oshman, R.: Distributed property testing for subgraph-freeness revisited. CoRR (2017). http://arxiv.org/abs/1705.04033
13. Gharan, S.O., Trevisan, L.: Approximating the expansion profile and almost optimal local graph clustering. In: 53rd Annual IEEE Symposium on Foundations of Computer Science, FOCS 2012 (2012). https://doi.org/10.1109/FOCS.2012.85
14. Goldreich, O., Goldwasser, S., Ron, D.: Property testing and its connection to learning and approximation. In: Proceedings of 37th Conference on Foundations of Computer Science (1996). https://doi.org/10.1109/SFCS.1996.548493
15. Goldreich, O.: Introduction to Testing Graph Properties. Springer, Heidelberg (2010). https://doi.org/10.1007/978-3-642-16367-8_7
16. Goldreich, O.: Introduction to Property Testing. Cambridge University Press, Cambridge (2017). https://doi.org/10.1017/9781108135252
17. Goldreich, O., Ron, D.: Property testing in bounded degree graphs. In: Proceedings of the Twenty-Ninth Annual ACM Symposium on Theory of Computing, STOC 1997 (1997). https://doi.org/10.1145/258533.258627
18. Goldreich, O., Ron, D.: On testing expansion in bounded-degree graphs. Electron. Colloquium Comput. Complex. (2000). https://eccc.weizmann.ac.il/eccc-reports/2000/TR00-020/index.html
19. Kale, S., Seshadhri, C.: An expansion tester for bounded degree graphs. SIAM J. Comput. **40** (2011). https://doi.org/10.1137/100802980
20. Kutten, S., Pandurangan, G., Peleg, D., Robinson, P., Trehan, A.: Sublinear bounds for randomized leader election. In: Distributed Computing and Networking (2013). https://doi.org/10.1007/978-3-642-35668-1_24
21. Li, A., Pan, Y., Peng, P.: Testing conductance in general GR PHS. Electron. Colloquium Comput. Complex. **TR11** (2011). https://api.semanticscholar.org/CorpusID:15721089
22. Li, A., Peng, P.: Testing small set expansion in general graphs. In: 32nd International Symposium on Theoretical Aspects of Computer Science, STACS 2015 (2015). https://doi.org/10.4230/LIPIcs.STACS.2015.622
23. Łącki, J., Mitrović, S., Onak, K., Sankowski, P.: Walking randomly, massively, and efficiently. In: Proceedings of the 52nd Annual ACM SIGACT Symposium on Theory of Computing, STOC 2020 (2020). https://doi.org/10.1145/3357713.3384303
24. Molla, A.R., Pandurangan, G.: Distributed computation of mixing time. In: Proceedings of the 18th International Conference on Distributed Computing and Networking, ICDCN 2017. Association for Computing Machinery (2017). https://doi.org/10.1145/3007748.3007784

25. Nachmias, A., Shapira, A.: Testing the expansion of a graph. Inf. Comput. **208**, 309–314 (2010). https://doi.org/10.1016/j.ic.2009.09.002
26. Sarma, A.D., Molla, A.R., Pandurangan, G.: Distributed computation of sparse cuts via random walks. In: Proceedings of the 16th International Conference on Distributed Computing and Networking, ICDCN 2015. Association for Computing Machinery (2015). https://doi.org/10.1145/2684464.2684474
27. Sinclair, A.: Algorithms for Random Generation and Counting: A Markov Chain Approach. Birkhauser Verlag (1993). https://dl.acm.org/doi/10.5555/140552

k-Center Clustering in Distributed Models

Leyla Biabani[1]($\boxtimes$) and Ami Paz[2] (iD)

[1] Eindhoven University of Technology, Eindhoven, The Netherlands
`l.biabani@tue.nl`
[2] LISN—CNRS & Paris-Saclay University, Paris, France
`ami.paz@lisn.fr`

Abstract. The k-center problem is a central optimization problem with numerous applications for machine learning, data mining, and communication networks. Despite extensive study in various scenarios, it surprisingly has not been thoroughly explored in the traditional distributed setting, where the communication graph of a network also defines the distance metric.

We initiate the study of the k-center problem in a setting where the underlying metric is the graph's shortest path metric in three canonical distributed settings: the LOCAL, CONGEST, and CLIQUE models. Our results encompass constant-factor approximation algorithms and lower bounds in these models, as well as hardness results for the bi-criteria approximation setting.

Keywords: k-Center clustering · Distributed graph algorithms · Shortest path metric

1 Introduction

1.1 Distributed k-Center

The k-center problem is a key optimization problem, seeks to locate a small set of points in space ("centers") that minimize the maximal distance from them to any other point. The problem was extensively studied in the centralized setting, where the points are taken from a metric space, with or without guarantees on the metric. Over the years, there was also some work on the k-center problem in parallel and distributed models, looking to solve the problem faster using multiple computational units. As in the centralized setting, the points and metric under consideration in these works are taken from some metric space, and it is usually assumed that distances are given to the computational units by an oracle.

A natural setting for the k-center problem is that of a network. We consider a graph representing a communication network in a natural way, and the goal is to make k of the nodes into centers while minimizing the maximal distance between these nodes and all others. Here, the metric space is not arbitrary, but is the *graph metric* on the graph itself. This is significant, for example, when

Y. Emek (Ed.): SIROCCO 2024, LNCS 14662, pp. 83–100, 2024.
https://doi.org/10.1007/978-3-031-60603-8_5

determining server placement within a network to minimize the maximum delay, as seen in scenarios like content distribution over the Internet.

Two related, yet different problems are the metric *facility location* problem, where the goal is to minimize the average distance from the servers (or equivalently, the sum of distances), and the *online k-server* problem, where the points and servers can move in space.

In this work, we initiate the study of k-center in the distributed setting, where an undirected graph represents both the communication network and the problem's metric. We address the problem in the popular distributed models of LOCAL, CONGEST, and CLIQUE, and derive upper and lower bounds. While the problem was studied earlier in some of these distributed models, it is important to note that all the prior work considered points in an arbitrary metric space that, independent of the communication graph. As far as we know, our work is the first to consider the natural setting where the input graph represents both the network used for computation and the metric on which k-center is solved.

1.2   Our Results and Techniques

We present upper and lower bounds for the k-center problem in different distributed models. The metric considered is the shortest-path metric, and the graphs are unweighted unless otherwise specified.

The LOCAL Model. A simple and natural model for communication networks is the LOCAL model [22], where the network's nodes communicate by synchronously exchanging messages of unbounded size. The local computation time is neglected, and the complexity measure in this case is the number of communication rounds. In this model, any problem is solvable in D rounds, where D is the diameter of the graph, i.e., the largest distance between two nodes in it. Hence, our upper bounds should be read as the minimum of the given value and D, and the lower bounds only apply for graphs with a diameter larger than the lower bound value.

We start Sect. 3 by giving a relatively simple algorithm that finds a $(2k + \epsilon)$-approximate solution to the k-center problem in $O(k/\epsilon)$ rounds of the LOCAL model. We then show that reducing the approximation ratio to be below $k - 1$ requires $\Omega(n)$ time. We stress that $D \leq n$, so in $\Omega(n)$ time the nodes can aggregate the full structure of the network and find an exact solution.

Put differently, our results present two extremes. On the one hand, if a large approximation ratio of at least $2k + \epsilon$ is allowed, the problem is simple—for ϵ, k constants, it is solvable in constant time. On the other hand, if a lower approximation ratio is required, e.g., a constant that does not depend on k, then the running time is so large that the nodes can also compute an optimal solution.

The CONGEST Model. The CONGEST model is a restrictive variant of the LOCAL model, where the messages are limited to $O(\log n)$ bits. Specifically, problems are no longer trivially solvable in $O(D)$ time, and getting a constant approximation is solvable in non-trivial time, but is challenging.

In Sect. 4 we present 2-approximation CONGEST algorithm for the k-center problem running in $O(kD)$ rounds. It constructs different BFS trees, sometimes from multiple sources simultaneously, and simulates a centralized greedy approximation algorithm for the k-center problem. On the other hand, in Sect. 5, we prove that improving the approximation ratio to below $4/3$, and hence also finding an exact solution, requires a much longer time: any algorithm for this problem, even randomized, must take $\tilde{\Omega}(n/k)$ rounds, and this is true even for graphs of diameter as small as 12.

This lower-bound proof is rather involved: it uses a reduction to communication complexity with a twist. Proving CONGEST lower bounds by reducing them to communication complexity is common in the literature [1,10,28], where usually the solution for the CONGEST problem directly implies an answer communication complexity problem. In our reduction, a new post-processing phase is added, where the players must do extra computation and also communicate more after getting the solution to the (approximate) k-center problem and before finding an answer to the communication problem.

The CLIQUE Model. Finally, in Sect. 6 we consider the more recent congested clique model, denoted CLIQUE. It resembles the CONGEST, but with an all-to-all communication—the communication graph is a clique, and the input graph is a subgraph of it on the same set of nodes. One might think that the $O(kD)$-round CONGEST algorithm will directly translate to an $O(k)$-round CLIQUE algorithm, as the communication network has a diameter 1. However, this it not the case: the CONGEST algorithm builds BFS trees in $O(D)$ time, and this step cannot be translated to an $O(1)$-round subroutine in the CLIQUE model.

At a high level, we show that finding a 2-approximate solution for the k-center problem in this model can be done deterministically in the same time as computing all-pairs-shortest paths, which is $O(n^{1/3})$ for weighted graphs, and $O(n^{0.158})$ for unweighted graphs [5]. Interestingly, if ω, the exponent of the (centralized) matrix multiplication problem, will be discovered to be 2, these algorithms will also have an improved running time of $O(n^{\epsilon'})$ for any $\epsilon' > 0$, or even a poly-logarithmic time. Previous work implies that a $(2 + \epsilon)$-approximate solution can be found in only $O(\text{poly} \log n)$ rounds, even in weighted graphs [2]. By allowing higher approximation ratios we can get even faster algorithms, such as a $(4+\epsilon)$-approximation in $O(\text{poly} \log \log n + k)$ rounds, and up to an $O(\log n)$-approximation in $O(k)$ rounds.

En route, we prove a new result regarding approximate k-center. A simple greedy algorithm [15] finds a 2-approximation of k-center, assuming the distance between every two nodes is known. We extend this claim in Lemma 4, to show that if only a one-way, multiplicative α-approximation of the distances is known, then a similar greedy algorithm gives a (2α)-approximate solution to the k-center problem. Hence, by studying distributed k-center, we also provide new insights to the centralized case, which could be of independent interest.

Unfortunately, current techniques cannot establish lower bounds in the CLIQUE model. Any non-trivial lower bound in this model will imply circuit complexity lower bounds, solving a long-standing and notoriously hard open problem [13].

2   Preliminaries

2.1   The k-Center Problem

Consider a graph $G = (V, E)$ with nodes V and edges E where $|V| = n$. The edges in set E may have weights or be unweighted, and we mainly focus on the former case, and state it explicitly when this is not the case. The length of a path is determined by the number of edges in the path when E is unweighted, and by the total weight of the edges in the path when E is weighted. The diameter of the graph G, denoted D, is the maximum distance between any two nodes in V, i.e., $D = \max_{u,v \in V} d(u, v)$. For any node $u \in V$, the eccentricity is $\mathrm{ecc}(u) = \max_{v \in V} d(u, v)$, so we also have $D = \max_{u \in V} \mathrm{ecc}(u)$.

Let $k \in \mathbb{N}$ be a given parameter. In the k-center problem, we aim to find a set $S^* \subseteq V$ with a size of at most k that minimizes the maximum distance of any node of V to its nearest center in S. More formally, we seek a set S^* with $|S^*| \leq k$ that minimizes the value $\max_{v \in V} \min_{s \in S^*} d(v, s)$. This value is denoted as $\mathrm{OPT}_k(G)$ or simply OPT_k when the context is clear. For $\alpha \geq 1$, an algorithm is considered α-approximation if it can compute a solution S such that the distance of any node $v \in V$ to its nearest node in S is at most $\alpha \cdot \mathrm{OPT}_k$.

2.2   Computational Models

We consider three common computational models for studying distributed graph algorithms, namely the LOCAL, CONGEST and CLIQUE models. We model a communication network using its graph, with nodes representing computational units and edges representing communication links. We use n for the number of nodes (computational units) and assume they have unique ids in $\{1, \ldots, n\}$ (specifically, there is always a node with id 1). The computation proceeds in synchronous rounds, where in each round each node sends messages to its neighbors, receives messages form them, and updates its local state accordingly. The input is local, in the sense that each node initially knows only its id, list of neighbors, and if there are inputs such as edge weights, then also the weights of its incident edges. The outputs are similarly local, e.g., at the end of the algorithm's execution each node should know if it is a center or not.

In the LOCAL model [27], the message sizes are unbounded, and an r-round algorithm is equivalent to having each node deciding by its distance-r neighborhood [22]. The CONGEST model [10,28] is similarly defined, but each node is limited to sending $O(\log n)$-bit messages to each of its neighbors in each round, where messages to different neighbors might be different from one another. This model allows each node to send, e.g., its id or the ids of some of its neighbors in a single round, but not its full list of neighbors. A common primitive in this model is that of construction a BFS tree from a node. This is sometimes extended to multiple BFS trees, or trees of bounded depth. See, e.g., [17,21].

Finally, we model a network with a congested all-to-all overly network by the CLIQUE model [13,23]. In this model, the input is a network as before, but the communication is less limited: in each round, each node can send $O(\log n)$-bit messages to each other node in the graph, and not only to its neighbors. This allows the nodes, e.g., to re-distribute the inputs in constant time [11,20], and compute all-pairs-shortest-paths in sub-linear time [5].

2.3   Communication Complexity

We prove lower bounds for the *k*-center problem in the CONGEST model using a reduction to *communication complexity*, a well-studied topic in theoretical computer science [19,30].

In the two-party *set disjointness* (henceforth: disjointness) communication complexity problem, two players referred to as Alice and Bob get two ℓ-bit strings x (Alice's string) and y (Bob's string), and communicate by exchanging messages on a reliable asynchronous channel. Their goal is to decide if the sets represented by the indicator vectors x and y are disjoint or not, i.e., if there is an index i such that $x[i] = y[i] = 1$, in which case we say they are not disjoint and the players must output 0, or otherwise, the sets are disjoint and they should output 1.

Alice and Bob follow some protocol indicating who should send messages at each step and what message to send. The communication complexity of a protocol (as a function of ℓ is the maximal number of bits they exchange when executing it, and the deterministic communication complexity of a problem the minimal communication complexity of a protocol solving the problem. The randomized communication complexity is similarly defined, but the players may also use random bits when executing the protocol, and the success probability is at least 2/3, when taken on the choice of random bits.

For the disjointness problem, there is an $\Omega(\ell)$ lower bound, which holds for deterministic and randomized algorithms alike [19, Example 3.22].

2.4   Related Work

To our surprise, the *k*-center problem was not studied in traditional distributed computing models. We survey below works in related computational models, and works on the related problem of metric facility location in distributed settings.

k-Center in Related Computational Models. Bandyapadhyay, Inamdar, Pai, and Pemmaraju [2] studied the *k*-center problem, along with the related uncapacitated facility location and *k*-median problems. They consider the *k*-machine model (the parameter *k* here need not be the same as in *k*-center) which is closely related to the CLIQUE model. As mentioned, their work implies a randomized $(2 + \epsilon)$-approximation $O(\text{poly} \log n)$-time *k*-center algorithm in the CLIQUE model. Chiplunkar, Kale, and Ramamoorthy [8] studied approximate *k*-center in *streaming* models, and also in a parallel model where multiple processors perform local computations and then send the results to a central processor. Surprisingly, these seem to be the only works on the *k*-center problem in distributed settings. Cruciani, Forster, Goranci, Nazari, and Skarlatos [9] recently studied the *k*-center problem in a centralized *dynamic graphs* setting.

Metric Facility Location. The metric facility location problem has attracted more attention in distributed settings than the k-center problem. Works on this problem consider models that share similarities with the ones we consider here, although typically the models are not precisely identical.

The CONGEST *Model.* In an unpublished manuscript, Briest, Degener, Kempkes, Kling, and Pietrzyk [4] studied metric facility location in the CONGEST model. They focused on a bipartite graph, with one side representing facilities and the other representing clients. Some of their results appeared in the thesis of Pietrzyk [29]. This work improved upon previous work that also consider metric facility location in bipartite setting in the CONGEST model [25,26].

The CLIQUE *Model.* Gehweiler, Lammersen, and Sohler [14] studied metric facility location in a model resembling the CLIQUE. They present a 3-round randomized algorithm that gives a constant approximation factor, based on the method of Mettu and Plaxton [24].

Distributed Large-Scale Computational Models. Inamdar, Pai, and Pemmaraju [18] studied metric facility location in the CLIQUE model, MPC and k-machine, and gave an $O(1)$-approximation algorithm using a Mettu-Plaxton-style algorithm. When considering the CLIQUE model with what they call "implicit metric", their model coincides with ours. They also consider another input regime ("explicit metric"), but not in the LOCAL and CONGEST models, and not for the k-center problem. Their work improves upon earlier works on metric facility location in the CLIQUE model [3,16].

3 The k-Center Problem in the LOCAL Model

In this section, we consider the k-center problem in the LOCAL model. We start with a fast and simple $((2+\epsilon)k)$-approximation algorithm, Algorithm 1.

Lemma 1. *For any $\epsilon > 0$ there is a deterministic $O(k/\epsilon)$-round algorithm in the LOCAL model that gives a $((2+\epsilon)k)$-approximate solution for the k-center problem.*

Proof. Let $t = 2 + \frac{4}{\epsilon}$. The algorithm starts by having the node with id 1 (or any other arbitrary node) initiating the construction of a BFS tree of depth tk. After tk rounds, the nodes report back up the tree if the tree construction algorithm terminated in all its branches (i.e., reached all the nodes) or not. If it terminated, we have $D \leq tk$, and in another $O(tk)$ rounds node 1 can aggregate all the graph structure and find an optimal solution. Otherwise, node 1 becomes the only center.

We claim that this simple algorithm gives a $(2 + \epsilon)k$ approximate solution; clearly, we only need to prove it for the case $D > tk$. The eccentricity of node 1 is at most D, which gives an upper bound on ALG, the quality of the solution given by the algorithm.

Algorithm 1. A $((2 + \epsilon)k)$-approximation in the LOCAL model

---

1: $t \leftarrow 2 + \frac{4}{\epsilon}$
2: Perform a BFS from v_1 for tk rounds
3: Report back on the tree if terminated on all branches under you
4: **if** All branches terminated **then**
5:     Aggregate all the graph to v_1
6:     Locally compute an optimal solution
7:     Disseminate the solution on the BFS tree
8: **else** Only v_1 marks itself as a center

Recall that OPT_k is the largest distance from a node to its center in an optimal solution. Let π be the shortest path between two nodes of the largest distance, i.e., two nodes of distance D. In an optimal solution, there are at least $D + 1 - k$ non-center nodes in π. Therefore, there is a center that covers at least $(D + 1 - k)/k$ nodes of π. Hence, OPT_k is at least $(D + 1 - k)/(2k) = D/2k + 1/2k - 1/2 \geq D/2k - 1$.

Note that $D > tk$ implies $D/(tk) > 1$. The approximation ratio is thus

$$\frac{\text{ALG}}{\text{OPT}_k} \leq \frac{D}{D/2k - 1} \leq \frac{D}{D/2k - D/(tk)} = \frac{2tk}{t - 2} = (2 + \epsilon)k$$

by the choice of t, as required.

This algorithm utilizes the fact that local computation is not limited in the LOCAL model, so v_1 can find an optimal solution. However, it is not strictly necessary to use all this computational power: in case v_1 has to locally solve k-center, it can instead compute a 2-approximation in a greedy manner. The approximation ratio is still as required, and the local computation now takes polynomial time.

Next, we move to the main result of the section: a lower bound for the k-center problem in the LOCAL model. Our lower bound also works for bi-criteria algorithms, which return at most βk centers instead of at most k centers, for some $\beta \geq 1$; if the algorithm returns at most k centers, we can simply set $\beta = 1$. Our lower bound states that any t-round algorithm cannot achieve an approximation ratio better than $k - \frac{k^2 + k(\beta k - 1)(2t + 1)}{n + k}$. If $k^2 + k(\beta k - 1)(2t + 1) < n + k$ holds, then our lower bound states that it is not possible to get an approximation ratio better than $k - 1$. The typical case is the non-bi-criteria one, with k being a constant, and there our lower bound translates to stating that getting an approximation ratio better than $k - 1$ requires linear time.

Theorem 1. *Let* $t, k \in \mathbb{N}$, $\beta \geq 1$. *Any t-round bi-criteria deterministic LOCAL algorithm that solves the k-center problem and reports at most βk centers as the solution, cannot have an approximation ratio better than* $k - \frac{k^2 + k(\beta k - 1)(2t + 1)}{n + k}$, *where $n > 2\beta kt$ is the size of the graph and $\beta \geq 1$.*

At a high-level, the proof considers a communication graph which is a cycle, an algorithm in the LOCAL model that is faster than the lower bound, and the centers chosen by it. In t rounds, each node (and specifically, each chosen center) can gather information only from a segment of $2t+1$ nodes. We thus create a new cycle by concatenating all the $(2t+1)$-node segments around the chosen centers. By the choice of the parameters, this leaves some "leftover nodes" that are not in any segment, which are concatenated after all the segments. When executing the same algorithm on the new cycle, the same nodes as before will become centers, as they gather exactly the same information. The "leftover nodes" are now far from all the chosen centers, rendering a bad approximation ratio, as claimed.

Proof. Let $\mathcal{A}$ be an algorithm that finds an (approximate, bi-criteria) solution for the k-center problem. Let $k' \leq \beta k$ be the maximum number of centers that $\mathcal{A}$ reports for any cycle of length n, and let C be such a cycle. For simplicity, we label the nodes of C with numbers $1, \ldots, n$ in a clockwise order. Assume that $c_1 < \ldots < c_{k'}$ are the centers returned by $\mathcal{A}$ in clockwise order. Since $n > 2kt$, there exists at least two consecutive centers such that the distance between them is more than $2t$. We introduce a new cycle C' by re-arranging the nodes and show that the solution that $\mathcal{A}$ finds for C' has an approximation ratio of at least $k - \frac{k^2 + k(\beta k - 1)(2t+1)}{n+k}$.

To build the cycle C', we first define segments $S_1, \ldots, S_{k'}$ of C as follows. Each segment S_i is of the form $S_i = [b_i, e_i]$, which refers to the nodes b_i to e_i in the clockwise order. Roughly speaking, each S_i is a segment of $2t + 1$ nodes centered around c_i, but since such segments might overlap, the exact definition is a bit more subtle.

We define $e_i := \min(c_i + t, c_{i+1} - 1)$ for any $1 \leq i < k'$ and $e_{k'} := \min(c_{k'} + t, n + c_1 - 1)$. We also define $b_1 := \max(c_1 - t, n - c_{k'} + 1)$ and $b_i := \max(c_i - t, e_{i-1} + 1)$ for any $1 < i \leq k'$. Note that e_k may be larger than n, in which case we consider $S_{k'}$ as the segment starting at b_i and ending at $e_{k'} - n$ in the clockwise order, i.e. $S_{k'} = [b_{k'}, n] \cup [1, e_{k'} - n]$. Similarly, b_1 might be smaller than 1, in which case we consider nodes $b_1 + n$ to e_1 in the clockwise order, i.e. $S_1 = [b_1 + n, n] \cup [1, e_1]$.

Since we assumed $n > 2\beta kt$, then there is an i^* such that the distance between c_{i^*} and c_{i^*+1} is more than $2t$ (we consider c_1 as c_{i^*+1} if $i^* = k'$). To build C', we concatenate the segments $S_{i^*+1}, S_{i^*+2}, \ldots, S_k$ followed by $S_1, S_2, \ldots, S_{i^*}$, and finally all the remaining nodes of C, in a clockwise order (see Fig. 1).

Now, we claim that there is an execution of $\mathcal{A}$ on C' that reports $c_1, \ldots, c_{k'}$ as the centers. Since the distance between c_{i^*} and c_{i^*+1} is more than $2t$, then the neighbourhood of length t for the nodes c_{i^*} and c_{i^*+1} is disjoint. Along with the definition of the segments, we can conclude that the distance-t neighborhood for any node c_i is the same in both C and C'. Hence, any node c_i receives the same information for both graphs C and C', and therefore, each node c_i makes the same decisions. Since $c_1, \ldots, c_k$ chose to be the centers in C, they also chose to be the centers in C'. Moreover, we assumed that k' is the maximum number of centers that $\mathcal{A}$ reports for a cycle of length n, and hence, no other node chooses to be a center.

We next discuss the approximation ratio. According to the definition of S_i, each S_i is of size at most $2t + 1$. Therefore, there are at most $(k' - 2)(2t + 1)$

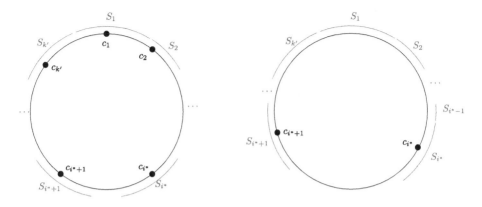

Fig. 1. Illustration for the proof of Theorem 1. Left: cycle C. Right: cycle C'

nodes in $S_{i^*+2}, S_{i^*+3} \ldots, S_k, S_1, S_2, \ldots, S_{i^*-1}$. Then, in the clockwise path from c_{i^*+1} to c_{i^*} in C' there are at most $(t+1) + (k'-2)(2t+1) + (t+1) = (k'-2)(2t+1) + 2t+2$ nodes. This means that the distance from c_{i^*+1} to c_{i^*} in C' in the anti-clockwise order is at least $n - ((k'-2)(2t+1) + 2t+2) + 1 = n - (k'-1)(2t+1)$, and note that there is no center in this path. Hence, there is a node in the anti-clockwise path from c_{i^*+1} to c_{i^*} in C' such that its distance to the nearest center is at least $(n - (k'-1)(2t+1))/2$.

On the other hand, the optimal solution has $\mathrm{OPT}_k \leq \lceil (n-k)/(2k) \rceil$. To see this, we can choose the k centers such that the distance between any two consecutive centers in the clockwise order is $\lceil n/k \rceil$ or $\lfloor n/k \rfloor$. Thus, $\mathrm{OPT}_k \leq \lceil (n-k)/(2k) \rceil$ holds and the approximation ratio of Algorithm $\mathcal{A}$ is at least

$$\frac{(n - (k'-1)(2t+1))/2}{\lceil (n-k)/(2k) \rceil} \geq \frac{(n - (k'-1)(2t+1))/2}{(n-k)/(2k) + 1} = k\frac{n - (k'-1)(2t+1)}{n - k + 2k}$$

$$= k - \frac{k^2 + k(k'-1)(2t+1)}{n+k} \geq k - \frac{k^2 + k(\beta k - 1)(2t+1)}{n+k},$$

which finishes the proof.

4 A 2-Approximation in the CONGEST Model

In this section, we show how to achieve a 2-approximate k-center clustering in the CONGEST model in $O(kD)$ rounds, where D is the diameter of the underlying graph. Algorithm 2 presents an overview of our technique. In the following, we explain each part of this algorithm in detail.

As opposed to the LOCAL algorithm, here we do not need to use the assumption that the nodes ids are in $\{1, \ldots, n\}$, and assuming they are taken from $\{1, \ldots, \mathrm{poly}\, n\}$ suffices. The first step of Algorithm 2 is finding the node $v_{\min}$ with the minimum id, which can be done in $O(D)$ rounds (and can be skipped if the ids are in $\{1, \ldots, n\}$). To do this, we start a BFS from all nodes. In each

---

Algorithm 2. A 2-approximation in the CONGEST model

1: Find the node $v_{\min}$ with minimum id
2: $S \leftarrow \{v_{\min}\}$
3: **for** $k - 1$ times **do**
4:     Perform a BFS from all the nodes in S
5:     Let v^* be the furthest node from S, breaking ties by id
6:     $S \leftarrow S \cup \{v^*\}$
7: Each node in S marks itself as a center

---

round, each node may receive messages of BFS trees from multiple sources. If this happens, such a node only continue the BFS from the source with the minimum id it has seen so far, and ignores the BFS's for all other sources. Therefore, the only source that its BFS is not paused after D rounds is $v_{\min}$. Each leaf of a BFS tree reports the termination of the construction up the tree, and back to the tree's parent. Hence, $v_{\min}$ becomes aware that it is the node with the lowest id in $O(D)$ rounds. In another $O(D)$ time, $v_{\min}$ disseminates along its BFS tree the depth of this tree, and all nodes learn this value, to which we refer as D'. Observe that $D' \leq D < 2D'$, and $2D'$ will be used when an upper bound on D is needed (e.g., for the time of each iteration described below).

We next set $S = \{v_{\min}\}$, and preform $k - 1$ iterations. In each iteration, a BFS is performed, where the sources are all nodes in S. Each node chooses to join the first BFS tree which reaches to it, breaking ties arbitrarily. After $O(D)$ rounds, the BFS is done and each node knows its distance to S. We next need to find the node with minimum id among the furthest nodes to S, which we refer to it as v^*. This can be done in $O(D)$ rounds. In each round, every node informs its neighbors which node has the maximum distance among those it is currently aware of, and if it knows more than one node with maximum distance, it only reports the one with the smallest id. After D' such rounds, node v^* knows that it is the furthest node to S with minimum id. Finally, v^* can be added to S. At the end of the algorithm, the nodes in S mark themselves as centers.

The proof for the approximation ratio of our algorithm comes from the approximation ratio of the known greedy approach by Gonzalez [15], which we applied in our algorithm. In Lemma 2 we formally state this approximation.

Lemma 2 ([15]). *Let G be a graph, and $k \geq 1$ be an integer. Assume $S_1 = v_1$, where $v_1 \in G$ is an arbitrary node. For each $1 < i \leq k$, we have $S_i = S_{i-1} \cup \{v_i\}$, where v_i is a node of G with maximum distance to S_{i-1}. Then, S_k is a 2-approximate solution for k-center of G.*

We can now summarize this section in Theorem 2.

Theorem 2. *There exists a deterministic 2-approximation algorithm for the k-center problem in the CONGEST model that needs $O(kD)$ communication rounds.*

5   A Lower Bound in the CONGEST Model

In this section, we show a lower bound on the number of communication rounds for any algorithm for the k-center problem in the CONGEST model with an approximation ratio better than $4/3$. We start with a lower bound for the 1-center problem and then extend it for the k-center problem for a general k.

5.1   A Lower Bound for the 1-Center Problem

To prove the lower bound, we use a graph that was introduced by Abboud, Censor-Hillel, Khoury, and Paz [1] in order to prove a lower bound for computing the radius of a graph. Let x and y be two binary strings of length ℓ. The graph $G_{x,y}$ is built as follows, on $n = \Theta(\ell)$ nodes (see Fig. 2). On a high level, the graph consists of two main sets of ℓ nodes each, A and B, and a path on 4 nodes: $c_A, \bar{c}_A, c_B, \bar{c}_B$. The input strings x and y for a set-disjointness problem are used to set which edges from A to $\bar{c}_A$ exist (representing x), and which edges form B to $\bar{c}_B$ exist (representing y). The rest of the graph is built in order to guarantee that the optimal solution for the 1-center problem, OPT_1 satisfies $\mathrm{OPT}_1 = 4$ if and only if x and y are disjoint. Hence, by simulating a 1-center algorithm and finding the distance from the chosen center to all other nodes, Alice and Bob can decide if x and y are disjoint. Finally, if the algorithm is too fast, Alice and Bob can simulate it with too little communication, contradicting the communication-complexity lower bound for disjointness.

We now present the graph and the proof in detail. The graph $G_{x,y}$ consists of node sets A, B, F_A, T_A, F_B, and T_B, as well as nodes c_A, $\bar{c}_A$, c_B, and $\bar{c}_B$. The set A is the set of ℓ nodes $a^0, a^1, \ldots, a^{\ell-1}$ and the set B is the set of ℓ nodes $b^0, b^1, \ldots, b^{\ell-1}$. For $S \in \{A, B\}$, the set F_S consists of $\lfloor \log_2 \ell \rfloor$ nodes $f_S^0, f_S^1, \ldots, f_S^{\lfloor \log_2 \ell - 1 \rfloor}$. Similarly, the set T_S contains node $t_S^0, t_S^1, \ldots, t_S^{\lfloor \log_2 \ell - 1 \rfloor}$. The edges of the graph $G_{x,y}$ are described in the following.

- **Edges from A to F_A and T_A and edges from B to F_B and T_B.** Let $0 \leq i < \ell$ and $0 \leq h < \ell$, and let bin_h^i be the h-th bit in the binary representation of i. If $\mathrm{bin}_h^i = 0$, we connect a^i to f_A^h and we connect b^i to f_B^h. Otherwise, if $\mathrm{bin}_h^i = 1$, we connect a^i to t_A^h and we connect b^i to t_B^h. That is to say, we connect a^i to the binary representation of i in the sets F_A and T_A, where F and T represent "false" and "true"; b^i is similarly connected.
- **Edges from $\bar{c}_A$ to A and edge from $\bar{c}_B$ to B.** For each $0 \leq i < \ell$, we connect a^i to $\bar{c}_A$ if and only if $x[i] = 1$, and we connect b^i to $\bar{c}_B$ if and only if $y[i] = 1$.
- **Other edges from c_A, $\bar{c}_A$, c_B and $\bar{c}_B$.** We connect c_A to all nodes in A and we connect c_B to all nodes in B. We also connect $\bar{c}_A$ to all nodes in F_A and T_A, and similarly, we connect $\bar{c}_B$ to all nodes in F_B and T_B.
- **Edges between F_A, T_A, F_B, and T_B.** For every $0 \leq h < \ell$, we connect f_A^h to t_A^h and t_B^h, and we connect t_A^h to f_B^h.
- **Edges from w^0, w^1 and w^2.** We connect w^0 to all nodes in A, and also w^0 to w^1 and w^1 to w^2.

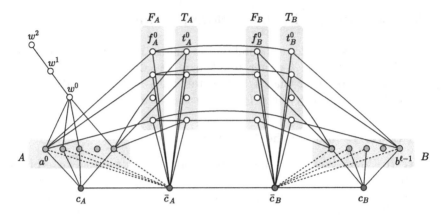

Fig. 2. Illustration of the graph $G_{x,y}$ [1]. The dotted edges depend on the inputs for the disjointness problem.

Now we can have the following claim for the graph $G_{x,y}$ that we defined.

Claim ([1, Claim 4]). Let $G_{x,y}$ be the graph defined above, and V be the set of its nodes. Then, the following holds

1. For every node $u \in V \setminus A$, we have $\text{ecc}(u) \geq 4$.
2. For every node $a^i \in A$ and every $u \in V \setminus \{b^i, c_B\}$ we have $\text{d}(a^i, u) \leq 3$.

The first part of the claim holds since the distance of w^2 to any node in $V \setminus A \cup \{w^0, w^1, w^2\}$ is at least 4, and the distance of c_B to any node in $\{w^0, w^1, w^2\}$ is at least 4. To show the second part of the claim, observe that the distance between a^i and any node in $V \setminus B \cup \{c_B\}$ is clearly at most 3. Besides, if $j \neq i$, there exist $0 \leq h < \ell$ such that $\text{bin}_h^i \neq \text{bin}_h^j$, where bin_h^i and bin_h^j are the h-th bit in the binary representation of i and j, respectively. If $\text{bin}_h^i = 0$ and $\text{bin}_h^j = 1$, the path (a^i, f_A^h, t_B^h, b^j) exists. Otherwise, if $\text{bin}_h^i = 1$ and $\text{bin}_h^j = 0$, the path (a^i, t_A^h, t_B^h, b^j) exists. It means that $\text{d}(a^i, b^j) \leq 3$, and therefore, the second part of the claim holds. The rigorous proof of this claim can be found in [1, Claim 4].

Recall that x and y are not disjoint if there is a $0 \leq i < \ell$ such $x[i] = y[i] = 1$, and otherwise, they are disjoint. In Lemma 3, we show that $\text{OPT}_1(G_{x,y}) = 4$ if and only if x and y are disjoint.

Lemma 3. *If the strings x and y are disjoint then $\text{OPT}_1(G_{x,y}) = 4$. Otherwise, $\text{OPT}_1(G_{x,y}) = 3$.*

Proof. To prove the lemma, we first show that if $\text{OPT}_1 < 4$, then an optimal center of $G_{x,y}$ for the 1-center problem is a node $a^i \in A$, such that $x[i] = y[i] = 1$. Assume that $\text{OPT}_1 < 4$. Then Claim 5.1(1) implies that the optimal center is in A. Let $a^i \in A$ be an optimal center. Then, $\text{d}(a^i, b^i) \leq 3$ should holds since we assume $\text{OPT}_1 < 4$. But $\text{d}(a^i, b^i) \leq 3$ holds only if a^i is connected to $\bar{c}_A$ and b^i is connected to $\bar{c}_B$, which means $x[i] = y[i] = 1$.

We next consider the case that x and y are disjoint. In this case, we have $\mathrm{OPT}_1 \geq 4$ since otherwise, our claim implies that there exists $0 \leq i < \ell$ such that $x[i] = y[i] = 1$ and contradicts the disjointness of x and y. Hence, $\mathrm{OPT}_1 \geq 4$ holds. On the other hand, all nodes are within distance 4 of a^0, and therefore, $\mathrm{OPT}_1 = 4$.

Now, we consider the case that x and y are not disjoint. This means that there exist $0 \leq i < \ell$ such that $x[i] = y[i] = 1$. We show that $\mathrm{ecc}(a^i) = 3$ and then $\mathrm{OPT}_1 = 3$. Claim 5.1(2) states that $\mathrm{d}(a^i, u) \leq 3$ holds for any $u \in V \setminus \{b^i, c_B\}$. Besides, a^i is connected to $\bar{c}_A$ as $x[i] = 1$ and b^i is connected to $\bar{c}_B$ since $y[i] = 1$. Thus, $\mathrm{d}(a^i, b^i) = \mathrm{d}(a^i, c_B) = 3$. Putting everything together we have $\mathrm{OPT}_1 = 3$.

Theorem 3. *Any* CONGEST *algorithm that returns an α-approximation for the 1-center problem with $\alpha < 4/3$, even randomized, must take $\Omega(n/\log^2 n)$ rounds to complete.*

Proof. Assume for contradiction a CONGEST algorithm faster than in the theorem's statement. Let $x, y \in \{0, 1\}^\ell$ be two inputs of Alice and Bob for the disjointness problem. Alice and Bob simulate the algorithm on the graph $G_{x,y}$ described above.

To this end, we split the graph node by $V_A = A \cup F_A \cup T_A \cup \{c_A, \bar{C}_a, w^0, w^1, w^2\}$ and $V_B = V \setminus V_A$ (the nodes on the left and right sides of Fig. 2, respectively). Alice is in charge of the nodes of V_A and Bob on V_B. To simulate a round, they locally simulate the exchange of messages in each of their node sets, and exchange bits to simulate the messages between nodes of V_A and V_B. Since there are $O(\log n)$ edges between the sets, a simulation of a round requires $O(\log^2 n)$ bits of communication.

If the output of the algorithm is a node not in A, Alice and Bob return 1 for the disjointness problem. If it is some node $a^i \in A$, the exchange the bits $x[i]$ and $y[i]$ (for the same index i) and return 0 for the disjointness problem if and only if $x[i] = y[i] = 1$.

For correctness, first note that Alice and Bob return 0 for the disjointness problem only if they find an index such that $x[i] = y[i] = 1$, so this answer must be correct. When they return 1, on the other hand, it might be since the algorithm returned a center $u \notin A$, or a center a^i such that $x[i] = 0$ or $y[i] = 0$ (or both).

If Alice and Bob return 1 because of a center $u \notin A$, then Sect. 5.1(1) guarantees that $\mathrm{ecc}(u) \geq 4$, and by $\alpha < 4/3$ we get that $\mathrm{OPT}_1 > 3$. Have the sets not been disjoint, Lemma 3 guarantees that $\mathrm{OPT}_1 = 3$, a contradiction.

If Alice and Bob return 1 because of a center $a^i \in A$, then we also know that $x[i] = 0$ or $y[i] = 0$. Note that $\mathrm{d}(a^i, b^i) \geq 4$: the nodes of $\mathrm{bin}(a^i)$ and $\mathrm{bin}(b^i)$ are never neighbors, so any 3-path connecting a^i and b^i must go through the edge $(\bar{c}_A, \bar{c}_B)$; but since $x[i] = 0$ or $y[i] = 0$, no such path can exists. Hence, $\mathrm{ecc}(a^i) \geq 4$. As before, if the sets were not disjoint, Lemma 3 guarantees that $\mathrm{OPT}_1 = 3$, and we would have got a contradiction.

For the complexity, let T be the number of rounds used by the algorithm. To simulate these rounds, Alice and Bob exchange $O(T \log^2 n)$ bits. As shown above, they solve the disjointness problem on ℓ bits, which requires them to communicate $\Omega(\ell)$ bits, even when using a randomized algorithm. As $\ell = \Theta(n)$, we get $T = \Omega(n/\log^2 n)$. In fact, Alice and Bob must also exchange the id of the node chosen as the center, but these $\log n$ bits of communication do not affect the asymptotic complexity.

Finally, note that Alice and Bob cannot solve disjointness solely by the output of the algorithm, and need extra communication after it. Hence, we cannot utilize standard reductions such as [1, Theorem 6] and use a non-standard one.

5.2   Extending the Lower Bound to the k-Center Problem

To generalize the lower bound for the 1-center problem to the k-center problem, we consider k copies of the graph $G_{x,y}$ introduced in Sect. 5.1, such that all copies share the node w^2. We refer to this graph as $G_{x,y}^k$.

Claim. In any α-approximate solution for k-center of $G_{x,y}^k$, $\alpha < 3/2$, exactly one center is chosen from each copy.

Proof. Note that Lemma 3 guarantees that in each copy we can choose a center and get a solution with distance at most 4 from each node to a center, and $\text{OPT}_k \leq 4$. If some copy contains no center, then b^0 in this copy has distance at least 6 to w^2, and the solution is at least $(6/4)$-approximation, a contradiction.

Theorem 4. *Any CONGEST algorithm that returns an α-approximation for the k-center problem with $\alpha < 4/3$, even randomized, must take $\Omega(n/(k^2 \log^2 n))$ rounds to complete.*

The proof extends the proof of Theorem 3 by using the graph $G_{x,y}^k$ described above. Note that all copies are for the same input pair (x, y).

Proof. Assume a faster algorithm than in the theorem statement. Alice and Bob simulate this algorithm for ℓ input bits, where $\ell = \Theta(n/k)$, and get an output at the form of k center nodes. If none of the output nodes is of the form $a^i \in A$ for some copy, Alice and Bob answer 1 for the disjointness problems. Otherwise, for every node $a^i \in A$ in the output (perhaps from different copies), they exchange $x[i], y[i]$, and answer 0 for the disjointness problem only if $x[i] = y[i] = 1$ for at least one such i.

For correctness, recall first that Sect. 5.2 guarantees that each copy contains exactly one center.

If Alice and Bob answer 0, this is because they found an index i of intersection and the algorithm is correct. If they answer 1, the proof follows the same lines of the proof of Theorem 3: if the sets are not disjoint then Lemma 3 guarantees each copy has a center a^i with $\text{ecc}(a^i) \leq 3$, and we get $\text{OPT}_k \leq 3$. On the other hand, the algorithm returns at each copy a node $u \notin A$ or a node a^i with $x[i] = 0$ or $y[i] = 0$, which has $d(a^i, b^i) \geq 4$; in both cases, the solution is not an α-approximation for $\alpha < 4/3$.

For the complexity, assume the algorithm takes T rounds. Alice and Bob communicate, for each copy and each round, $O(\log(n/k)\log n)$ bits, which are $O(\log^2 n)$ bits. At the end of the algorithm, they may exchange the indices of at most k centers of the form a^i and their inputs $x[i]$ and $y[i]$ in these locations, with takes requires another $O(k \log n)$ bits. In total, they exchange $O(Tk\log^2 n + k \log n)$ bits, which is $O(Tk\log^2 n)$.

On the other hand, they solve disjointness on $\ell = \Theta(n/k)$ input bits, so they must communicate $\Omega(n/k)$ bits. Hence, $Tk\log^2 n \geq cn/k$ for some constant c, and we get $T = \Omega\left(\frac{n}{k^2 \log^2 n}\right)$ as claimed.

6 A 2-Approximation in the CLIQUE Model

We now present Algorithm 3, which provides an approximate solution to the k-center clustering problem in the CLIQUE model. It is applicable both when the distance between the nodes are weighted and when they are unweighted.

The algorithm consists of two phases. The first phase is computing all-pairs shortest distances or approximating them, using known algorithms.

The second phase is to greedily find the centers, which is done in additional k rounds. In this phase, we first find $v_{\min}$, the node with the minimum id. This is trivial when the ids are in $\{1, \ldots, n\}$, but easy also without this assumption, by having each node send its id to all other nodes. We then set $S \leftarrow \{v_{\min}\}$. Next, we have $k - 1$ iteration, and in each iteration we find v^*, which is the furthest node from S (if there is more than one furthest node, we define v^* as the furthest node with minimum id). To accomplish this, it is enough that each node sends its distance to S to all nodes. Hence, all nodes can know who v^* is in one communication round. At the end of each iteration, we set S to $S \cup \{v^*\}$, and report S as the set of centers at the end of the algorithm. This process is formalized in the following algorithm.

If we calculate the exact all-pairs shortest paths in the first phase of Algorithm 3, Lemma 2 indicates that the set S computed by the algorithm is a 2-approximate solution for k-center.

Approximate distance can be computed much faster than exact ones, and we next show that Algorithm 3

Algorithm 3. Varying approximation ratios in the CLIQUE model

---
1: Compute all-pairs shortest paths
2: Let $v_{\min}$ be the node with minimum id
3: $S \leftarrow \{v_{\min}\}$
4: **for** $k - 1$ times **do**
5:     Let v^* be the furthest node to S
6:     $S \leftarrow S \cup \{v^*\}$
7: Each node in S marks itself as a center

---

can also work with approximate distances, in which case it computes an approximate k-center solution. To prove this, we now prove Lemma 4, which extends Lemma 2 to the scenario where only a multiplicative α-approximation with one-sided error of all-pairs shortest paths is computed in the first phase.

Lemma 4. *For any $\alpha \geq 1$, if an α-approximation of all-pairs shortest paths is computed in the first phase of Algorithm 3, then the set S obtained by the algorithm is a (2α)-approximate solution for the k-center problem.*

Table 1. Approximation times in the CLIQUE model, for constant $0 < \epsilon < 1$

Approx. ratio	Time	Weighted?	Deter.?	Ref
2	$O(n^{1/3} + k)$	No	Yes	[5] & Theorem 5
2	$O(n^{0.158} + k)$	Yes	Yes	[5] & Theorem 5
$2 + \epsilon$	$O(n^{0.158} + k)$	No	Yes	[5] & Theorem 5
$2 + \epsilon$	$O(\text{poly} \log n)$	Yes	No	[2]
$4 + \epsilon$	$O(\log^4 \log n + k)$	No	Yes	[12] & Theorem 5
$6 + \epsilon$	$O(\log^2 n + k)$	Yes	Yes	[6] & Theorem 5
$O(\log n)$	$O(k)$	Yes	No	[7] & Theorem 5

Proof. If $n \leq k$, the algorithm trivially returns all the nodes as the centers, so we assume $n > k$ in the following. Let S be the set returned by Algorithm 3, $\hat{v} \in V \setminus S$ the furthest node from S, and $\hat{r}$ its distance.

First, we claim that $\text{OPT}_k \geq \hat{r}/(2\alpha)$. Since $\hat{v}$ is not added to S, we can conclude that the distance between any two nodes in S is at least $\hat{r}/\alpha$. Thus, the distance between any two nodes in $S \cup \{\hat{v}\}$ is at least $\hat{r}/\alpha$ as well. In addition, in any optimal solution there exist two nodes in $S \cup \{\hat{v}\}$ that have the same nearest center in that optimal solution since $|S \cup \{\hat{v}\}| = k + 1$. Since the distance between these two nodes is at least $\hat{r}/\alpha$, at least one of them has distance at least $\hat{r}/(2\alpha)$ from their common center, implying $\text{OPT}_k \geq \hat{r}/(2\alpha)$. On the other hand, all nodes in V are within distance $\hat{r}$ from S by the definition of $\hat{r}$. Hence, S is a solution with the approximation ratio of at most $\frac{\hat{r}}{\hat{r}/(2\alpha)} = 2\alpha$.

All is left now is to plug fast CLIQUE distance computation algorithms [5–7,12] in the lemma, and we get fast algorithms for exact and approximate k-center. This yields the results detailed in Theorem 5 and Table 1, showing that k-center can be approximated in the same times as all pairs shortest paths computation, up to an additive $O(k)$ time. Note that for the specific case of $(2 + \epsilon)$-approximation, a much faster (randomized) algorithm exists [2], running in $O(\text{poly} \log n)$ rounds.

Theorem 5. *There exists a deterministic CLIQUE algorithm for the k-center problem which gives a 2-approximation in $O(n^{1/3} + k)$ rounds on weighted graphs and $O(n^{0.158} + k)$ rounds on unweighted graphs.*

For every constant $0 < \epsilon < 1$, there exist deterministic algorithms that give a $(2 + \epsilon)$-approximation in $O(n^{0.158} + k)$ rounds on unweighted graphs, a $(4 + \epsilon)$-approximation in $O(\log^4 \log n + k)$ rounds on unweighted graphs, and a $(6 + \epsilon)$-approximation in $O(\log^2 n + k)$ rounds on weighted graphs.

There exists a randomized algorithm which gives a $O(\log n)$-approximation in $O(k)$ rounds on both weighted and unweighted graphs.

Acknowledgement. We thanks Morteza Monemizadeh for many valuable conversations on this project, Michal Dory for discussions regarding distance computation in the CLIQUE model, and the anonymous reviewers of SIROCCO'24 for their helpful comments.

References

1. Abboud, A., Censor-Hillel, K., Khoury, S., Paz, A.: Smaller cuts, higher lower bounds. ACM Trans. Algorithms **17**(4), 30:1–30:40 (2021). https://doi.org/10.1145/3469834
2. Bandyapadhyay, S., Inamdar, T., Pai, S., Pemmaraju, S.V.: Near-optimal clustering in the k-machine model. Theor. Comput. Sci. **899**, 80–97 (2022)
3. Berns, A., Hegeman, J., Pemmaraju, S.V.: Super-fast distributed algorithms for metric facility location. In: Czumaj, A., Mehlhorn, K., Pitts, A., Wattenhofer, R. (eds.) ICALP 2012. LNCS, vol. 7392, pp. 428–439. Springer, Heidelberg (2012). https://doi.org/10.1007/978-3-642-31585-5_39
4. Briest, P., Degener, B., Kempkes, B., Kling, P., Pietrzyk, P.: A distributed approximation algorithm for the metric uncapacitated facility location problem in the congest model. CoRR, abs/1105.1248 (2011)
5. Censor-Hillel, K., Kaski, P., Korhonen, J.H., Lenzen, C., Paz, A., Suomela, J.: Algebraic methods in the congested clique. Distrib. Comput. **32**(6), 461–478 (2019)
6. Censor-Hillel, K., Dory, M., Korhonen, J.H., Leitersdorf, D.: Fast approximate shortest paths in the congested clique. Distrib. Comput. **34**(6), 463–487 (2021)
7. Chechik, S., Zhang, T.: Constant-round near-optimal spanners in congested clique. In: PODC, pp. 325–334. ACM (2022)
8. Chiplunkar, A., Kale, S.S., Ramamoorthy, S.N.: How to solve fair k-center in massive data models. In: ICML. Proceedings of Machine Learning Research, vol. 119, pp. 1877–1886. PMLR (2020)
9. Cruciani, E., Forster, S., Goranci, G., Nazari, Y., Skarlatos, A.: Dynamic algorithms for k-center on graphs. In: Proceedings of the 2024 Annual ACM-SIAM Symposium on Discrete Algorithms (SODA), pp. 3441–3462. SIAM (2024)
10. Das Sarma, A., Holzer, S., Kor, L., Korman, A., Nanongkai, D., Pandurangan, G. Peleg, D. Wattenhofer R.: Distributed verification and hardness of distributed approximation. SIAM J. Comput. **41**(5), 1235–1265 (2012)
11. Dolev, D., Lenzen, C., Peled, S.: "Tri, Tri again": finding triangles and small subgraphs in a distributed setting. In: Aguilera, M.K. (ed.) DISC 2012. LNCS, vol. 7611, pp. 195–209. Springer, Heidelberg (2012). https://doi.org/10.1007/978-3-642-33651-5_14
12. Dory, M., Parter, M.: Exponentially faster shortest paths in the congested clique. J. ACM, **69**(4), 29:1–29:42 (2022)
13. Drucker, A., Kuhn, F., Oshman, R.: On the power of the congested clique model. In: PODC, pp. 367–376. ACM (2014)
14. Gehweiler, J., Lammersen, C., Sohler, C.: A distributed o(1)-approximation algorithm for the uniform facility location problem. Algorithmica **68**(3), 643–670 (2014)
15. Gonzalez, T.F.: Clustering to minimize the maximum intercluster distance. Theor. Comput. Sci. **38**, 293–306 (1985). https://doi.org/10.1016/0304-3975(85)90224-5
16. Hegeman, J.W., Pemmaraju, S.V.: Sub-logarithmic distributed algorithms for metric facility location. Distrib. Comput. **28**(5), 351–374 (2015)
17. Holzer, S., Wattenhofer, R.: Optimal distributed all pairs shortest paths and applications. In: PODC, pp. 355–364. ACM (2012)

18. Inamdar, T., Pai, S., Pemmaraju, S.V.: Large-scale distributed algorithms for facility location with outliers. In: OPODIS. LIPIcs, vol. 125, pp. 5:1–5:16. Schloss Dagstuhl - Leibniz-Zentrum für Informatik (2018)

19. Kushilevitz, E., Nisan, N.: Communication Complexity. Cambridge University Press, New York (1997). ISBN 0-521-56067-5

20. Lenzen, C.: Optimal deterministic routing and sorting on the congested clique. In: Proceedings of the 32nd ACM Symposium on Principles of Distributed Computing (PODC 2013), pp. 42–50 (2013). https://doi.org/10.1145/2484239.2501983

21. Lenzen, C., Patt-Shamir, B., Peleg, D.: Distributed distance computation and routing with small messages. Distrib. Comput. 32(2), 133–157 (2019)

22. Linial, N.: Locality in distributed graph algorithms. SIAM J. Comput. 21(1), 193–201 (1992)

23. Lotker, Z., Pavlov, E., Patt-Shamir, B., Peleg, D.: MST construction in o(log log n) communication rounds. In: SPAA, pp. 94–100. ACM (2003)

24. Mettu, R.R., Plaxton, C.G.: The online median problem. SIAM J. Comput. 32(3), 816–832 (2003)

25. Moscibroda, T., Wattenhofer, R.: Facility location: distributed approximation. In: PODC, pp. 108–117. ACM (2005)

26. Pandit, S., Pemmaraju, S.V.: Return of the primal-dual: distributed metric facility location. In: PODC, pp. 180–189. ACM (2009)

27. Peleg, D.: Distributed Computing: A Locality-sensitive Approach. Society for Industrial and Applied Mathematics (2000). ISBN 0-89871-464-8

28. Peleg, D., Rubinovich, V.: Near-tight lower bound on the time complexity of distributed MST construction. SIAM J. Comput. 30(5), 1427–1442 (2000). https://doi.org/10.1137/S0097539700369740

29. Pietrzyk, P.: Local and online algorithms for facility location. Ph.D. thesis, University of Paderborn (2013)

30. Rao, A., Yehudayoff, A.: Communication Complexity: and Applications. Cambridge University Press (2020). ISBN 9781108497985

Optimal Memory Requirement
for Self-stabilizing Token Circulation

Lélia Blin[1], Gabriel Le Bouder[2], and Franck Petit[3]([⊠])

[1] IRIF, CNRS, Université Paris Cité, Paris, France
[2] LMF, CNRS, Université Paris-Saclay, Gif-sur-Yvette, France
[3] LIP6 CNRS, Sorbonne Université, Paris, France
`franck.petit@sorbonne-universite.fr`

Abstract. In this paper, we consider networks where every transmitted message is received by all of the transmitter's neighbors. Typical such networks are wireless networks, in which to dedicate a message to a specific neighbor, the sender must specify who the recipient is by specifying the recipient's ID. Adding an identifier has a non-negligible cost, more precisely $O(\log n)$ bits in an n-node graph. Token Circulation (TC) is a fundamental problem that consists in guaranteeing that a single token circulates from one node to another, the token fairly visiting every node infinitely often. TC inherently requires that the token holder selects a unique neighboring node to which to pass the token. This paper proposes a solution that overcomes the communication of identifiers. It achieves optimal space complexity for the token circulation problem. The contribution of this paper is fourfold. First, we present $\mathtt{Tok}^{\circ}$, the first deterministic depth-first token circulation algorithm for rooted wireless networks that uses only $O(\log \log n)$ bits of memory per node. This is an exponential improvement compared to the classical addressing. Algorithm $\mathtt{Tok}^{\circ}$ assumes a Destination-Oriented Directed Acyclic Graph (DODAG) spanning the network. Less popular than spanning trees, DODAGs are nonetheless more general and adaptable spanning structures, since they do not need to distinguish one neighboring node as unique parent. Second, $\mathtt{Tok}^{\circ}$ has the very desirable property of being self-stabilizing. This means that starting from a configuration where zero or more than one token circulate in the network, the system is guaranteed to eventually work correctly, *i.e.*, a single token eventually fairly visits every node infinitely often, following a depth-first order. Third, $\mathtt{Tok}^{\circ}$ works under the unfair scheduler, that is the most challenging scheduling assumption of the model. Finally, we show that $\mathtt{Tok}^{\circ}$ is optimal in terms of space complexity. Meaning that, for every n-node network, no TC algorithm can use less than $\Omega(\log \log n)$ bits of memory per node, even given a rooted spanning tree and even without considering self-stabilization.

Keywords: Distributed Algorithms · Self-Stabilization · Token Circulation · Memory Optimization

This work has been partially supported by the ANR projects SKYDATA (ANR-22-CE25-0008) and DREAMY (ANR-21-CE48-0003).

Y. Emek (Ed.): SIROCCO 2024, LNCS 14662, pp. 101–118, 2024.
https://doi.org/10.1007/978-3-031-60603-8_6

1   Introduction

In recent decades, wireless network technology has gained tremendous importance. It not only more and more replaces so far "wired" network installations, but also gives rise to new applications and with them, new tech and soft stacks enabling specific communications. Many distributed algorithms written for wired networks assume that every node v of the system is equipped with a local set of *labels*, sometimes also referred to as *port labels*. This set maps the set of v's neighbors, that is the set of nodes directly linked to v by bidirectional communication links. By using their respective port sets, two neighboring nodes u and v can directly communicate by exchanging messages with each other without interfering with other nodes, and without any further specific information such as IDs. Furthermore, port labels allow to maintain global distributed spanning data structures such as various types of trees, rings, and so on, which makes solving many problems much easier.

By contrast, in wireless networks, nodes have *a priori* no knowledge of their neighborhood. In particular, without further information (particularly, IDs), nodes are unable to distinguish from which node a message is received, neither to pick out a particular node to whom send a message. Thus, in wireless networks, to differentiate the recipient or the sender of a particular message, classically, the message includes the identifier of the recipient. This identifier will be stored (temporarily or permanently) in the receiver's memory. In n-node networks, the inclusion of the identifier requires $\Omega(\log n)$ bits. The dark side effect of the ID's inclusion is that, depending on the task considered, the message size is only determined by the size of the identifiers. Several problems can be specified by a constant number of bits, like the leader election or the token circulation, or by the number of bits depending of the degree of the network, like the coloration of the nodes. In this type of problem, the space complexity is entirely consumed by the inclusion of the identifier in the messages. Moreover, this is not appropriate to many settings, including sensor networks or swarms of robots, in which the memory of each process must be kept as small as possible, ideally constant.

Token Circulation, also referred to as *token passing*, is a fundamental problem to guarantee that a single token circulates from one node to another, and the token fairly visits every node infinitely often. The most popular way to implement such a mechanism in an arbitrary topology deterministically is the *Depth-First Token Circulation* (DFTC, for short). Briefly, this well-known mechanism takes place by executing successive traversals, each of them initiated by a specific node, called the root. In each traversal, the token is passed among the nodes following a depth-first order. A traversal ends when the root holds the token again after every node has been visited by the token at least once. The overall DFTC mechanism is implemented by restarting traversals one after another by the root. By construction, the DFTC mechanism is fair, since the unvisited nodes become visited only once during each traversal cycle.

We are interested in *self-stabilizing* [13] token circulation in distributed systems. A self-stabilizing algorithm must return the system to a correct behavior in finite time, whatever its initial configuration. In the context of token circulation, typical examples of initial arbitrary system configurations are scenarios in which either no token or several tokens exist in the network. To satisfy the requirement of self-stabilization, a distributed token circulation algorithm must return in finite time the system in a configuration which contains exactly one token. Actually, it must do more than that: the algorithm must also guarantee that, eventually, this single token fairly circulates among the nodes. Following the seminal work of Dijkstra [13], token circulation has been widely investigated deterministically in the context of self-stabilization, on rings [9,17], on (spanning) tree networks [8,16,21], and on arbitrary shaped networks [10–12,14,18,19,21], to quote only a few.

All the above algorithms were written in the well known *(atomic-)state* model [13], in the sequel denoted $\mathcal{S}$, a high-level model where communication between nodes is abstracted by the ability for a node to atomically read the content of its neighbor registers, and update its own registers. More precisely, all of them were written assuming the ability for each node to locally select at least one of its neighbors: the one to which to pass the token. Such ability is commonly made with variables called *pointers* that maps $N(v)$, the set of neighbors of v. Given two neighbors u and v, by reading p_v, one of v's pointers, u is able to know whether p_v maps the edge vu, *i.e.*, the link leading to itself or not. Pointers are typically implemented in two ways: (i) the nodes use their identifiers, *i.e.*, in the example, $p_v = u$'s ID, (ii) the model assumes *port labels* that maps $N(v)$, the set of neighbors of v. So, if pointers are implemented with node IDs, then they require $\Omega(\log n)$ bits of memory. If pointers are implemented with port labels, then they need $\Omega(\log \Delta)$ bits (Δ refers to the maximum degree of the network), but nodes must be able to recognize which of its neighbor port labels links it back to itself. Note that the use of pointers is particularly helpful for the problem of Token Circulation, where sending the token to exactly one neighbor is crucial to maintain the unicity of the token.

In this paper, we assume a model where no node is able to locally select one of its neighbors, except by its identifier. In $\mathcal{S}$, this assumption makes the design of algorithms much more challenging, which often require the use of node identifiers or the construction of an underlying vertex coloring. Token circulation algorithms for $\mathcal{S}$ are proposed in [1,2,5]. All these algorithms work over tree topologies and use $\Theta(\log n)$ bits of memory per node.

Aside from focusing on the algorithmic aspects of the self-stabilizing DFTC problem in $\mathcal{S}$, we also aim to reduce the memory required to solve this problem. Note that memory optimization for the self-stabilizing token circulation has been widely addressed in the literature. As mentioned earlier, all of them assume pointers implemented over local ports and achieving $O(\log \Delta)$ bits of memory per node, namely [8,9,13,16,17] on rings or chains, [21] on trees, and [12] on arbitrary networks. Except [9] which assumes a uniform prime size ring, all the above algorithms require a root, *i.e.*, a node with a particular local algorithm with respect to the other nodes.

In $\mathcal{S}$, it has been shown that many standard "one-shot" tasks like leader election and spanning tree construction can be constructed in a self-stabilizing manner with memory $O(\log \log n + \log \Delta)$ bits in any n-node network with maximum degree Δ [6,7]. This indicates that some tasks can be achieved in $\mathcal{S}$ with only sublogarithmic memory. As suggested by the results established in [3], it is much harder to reach $o(\log \log n)$ memory. No previous work addresses the self-stabilizing DFTC in $\mathcal{S}$ without using pointers.

1.1   Contributions

We present $\mathtt{Tok}^\circlearrowleft$, the first deterministic self-stabilizing depth-first token circulation algorithm in $\mathcal{S}$ that uses only $O(\log \log n)$ bits of memory per node. This is an exponential improvement compared to the classical addressing. $\mathtt{Tok}^\circlearrowleft$ does not use any pointer variable. However, like many works in the fields of self-stabilization, it assumes a specific underlying spanning distributed structure called *Destination-Oriented Directed Acyclic Graph*, DODAG for short [20]. DODAGs are bi-directional spanning data structure in which each edge is given an orientation, so that the resulting graph contains no loop, and has exactly one root: a node with only incoming edges from its neighbors. Less popular than other spanning data structures like virtual rings or spanning trees, DODAGs are nonetheless more general and adaptable than trees, since they do not need to distinguish one neighboring node as unique parent. Furthermore, they are particularly desirable in wireless networks because they allow multiple communication paths from every node to one single sink, thus limiting congestion and supporting link failures, as opposed to rooted spanning trees. As a matter of fact, DODAGs are the basic structures used in practice by RPL (Routing Protocol for Low power and lossy networks [23]), which is the standard routing protocol for IPv6-based multi-hop wireless sensor networks [22]. By contrast with rooted spanning trees, DODAGs allow the nodes to have more than one single parent, which drastically increases the difficulty to implement the DFTC.

Another quality of $\mathtt{Tok}^\circlearrowleft$ is that is works under the *unfair* scheduler, the most powerful distributed scheduler. Roughly speaking, a scheduler is considered as an adversary which tries to prevent the protocol to behave as expected. The more powerful the scheduler is, the more it can prevent the algorithm to work correctly. The nuisance power of the scheduler is modeled by the notion of *fairness*. The scheduler is considered to be *fair* if it cannot prevent forever a node to execute an action (if it has one to execute). By contrast, the unfair scheduler can prevent forever a node to execute an action unless if it is the only node able to execute an action.

Finally, we show that $\mathtt{Tok}^\circlearrowleft$ is *optimal* in terms of space complexity. Meaning that, for every n-node network, no token passing algorithm can use less than $\Omega(\log \log n)$ bits of memory per node, even given a rooted spanning tree, less challenging scheduler, and even without considering self-stabilization. Formally:

Theorem 1. $\mathtt{Tok}^\circlearrowleft$ *is a self-stabilizing algorithm for DFTC on DODAGS in $\mathcal{S}$ under the unfair scheduler with $\Theta(\log \log n)$ bits of memory per node.*

Theorem 2. *In $\mathcal{S}$, DFTC requires $\Omega(\log \log n)$ bits of memory per node on tree networks, and under the central scheduler and the synchronous scheduler.*

1.2   Outline of the Paper

In Sect. 2, we describe the model in which $\mathtt{Tok}^\circlearrowleft$ is written and the algorithm itself in Sect. 3. Due to the lack of space, technical parts of the model and the formal proofs are provided in [4]. However, both are sketched in Sect. 4. Finally, we make concluding remarks in Sect. 5.

2   Model and Definitions

2.1   Algorithm Syntax and Semantics

A distributed system is formalized by a n-nodes graph. When two nodes u and v are able to communicate with each other, this bi-directional communication is represented by an edge between u and v. The nodes u and v are said to be *neighbors*. A distributed system is *semi-uniform* if only one node is designated to have a particular role, in this paper we call this node *root*. In a distributed system, all the nodes have the same code, in semi-uniform system the root can have a different code. Let us formalise the semi-uniform distributed system, by an n-node graph $G = (V, E, r)$, where V is the set of nodes, E is the set of edges, and r is the root. We consider an *id-based* system, where every node has a distinct identifier. Note that, like in a classical distributed system, the identifiers are not consecutive between 1 and n, but are taken in $[1, n^c]$ where $c > 1$ is a constant. A system is said anonymous if no node holds an identifier.

Space-complexity in self-stabilization considers only *mutable memory*, the memory whose content changes during the execution of the algorithm. Mutable memory includes the space allocated for algorithm variables. On the other hand, the content of the *non-mutable memory* does *not* change during the execution. It typically stores the code, the constants (including the port label set, if any), and the node identifier. Therefore, IDs size is not a part of the space-complexity.

Each node contains variables and rules. Variable ranges over a domain of values. The variable var_v denotes the variable var located at node v. A rule is of the form [13]:

$$\langle label \rangle : \langle guard \rangle \longrightarrow \langle command \rangle$$

A *guard* is a boolean predicate over node variables. A *command* is a set of variable-assignments. A command of a node u can only update its own variables. On the other hand, u can read the variables of its neighbors. This classical communication model is called the *state model*.

An assignment of values to all variables in the system is called a *configuration*. Let Γ be the set of system configurations. A rule whose guard is **true** on one node u in some system configuration $\gamma \in \Gamma$ is said to be *enabled* on u in γ. Similarly, u is said to be enabled in γ. The atomic execution of a subset of enabled rules (at most one rule per node) results in a transition of the system from one configuration to another. This transition is called a *step*. A *run* of a distributed system is a maximal alternating sequence of configurations and steps. Maximality means that the execution is either infinite or its final configuration has no rule enabled.

2.2 Schedulers

The asynchronism of the system is modeled by an adversary (*a.k.a. scheduler*) which chooses, at each step, the subset of enabled nodes that are allowed to execute one of their rules during this step. Those schedulers can be classified according to their characteristics (like fairness, distribution, ...), and a taxonomy was presented in [15]. In this paper, we assume an *unfair asynchronous scheduler*. *Unfairness* means that even if a node u is continuously enabled, then u may never be chosen by the scheduler, unless u is the only enabled node. *Asynchronous* implies that during a computation step, if one or more nodes are enabled, then the scheduler chooses at least one (possibly more) of these enabled nodes to execute an action. This scheduler is the most challenging since no assumption is made on the subset of enabled nodes chosen by the scheduler at each step.

2.3 Self-stabilization

A predicate is a boolean function over the set of configuration Γ. A configuration $\gamma \in \Gamma$ *satisfies* some predicate R, if R evaluates to **true** for γ. Otherwise, γ *violates* R. We define a special predicate **true** as follows: for any $\gamma \in \Gamma$, γ satisfies **true**. We use the terms *closure* and *attractor* in the definition of stabilization. Given a predicate R, provided that the system starts from a configuration in Γ satisfying R, R is said to be *closed* for a certain run r, if every configuration of that run satisfies R. Given two predicates R_1 and R_2 over Γ, R_2 is an *attractor* for R_1 if every run that starts from a configuration that satisfies R_1 contains a configuration that satisfies R_2.

The specification SP_P of a problem P is a predicate over runs of the system, which describes a specific behavior of the system. A distributed algorithm $\mathcal{A}$ solves a problem P under a certain scheduler if every execution of $\mathcal{A}$ under that scheduler satisfies the specification of P. A distributed algorithm A is *self-stabilizing* for a specification SP_P if there exists a predicate R for P such that: (i) (*Convergence*) R is an attractor for **true**, and (ii) (*Closure*) Any run of A starting from a configuration satisfying R satisfies R.

2.4 DODAGs

We assume that each edge $\{u, v\} \in E$ is a bidirectional communication link provided with an orientation, from u to v, or from v to u, so that the resulting directed graph forms a *Destination-Oriented Directed Acyclic Graph* [20] (DODAG), *i.e.*, a Directed Acyclic Graph (DAG) with a unique root. Such orientation assumption is similar the one that assumes a consistent global orientation of a (virtual) ring, in which each node locally knows how to distinguish between its right and its left neighbor, or of a spanning tree, in which each node knows locally how to distinguish between neighbors that do and do not belong to the spanning tree, and among those that do belong to the tree, between its descendants and a single parent.

Consider an oriented edge from node u to node v of the DODAG, we said v is the parent of u and u is the child of v. Thus each node v has two sets of neighbors, the set of parents and the set of children. Note that unlike in tree topologies a node can have several parents.

2.5   Bit-by-Bit Communication of Identifier

In $\mathcal{S}$, nodes cannot communicate information to one single neighbor without using the fact that identifiers are, at least locally, unique. In other words, nodes must communicate their identifier stored in the immutable memory to their neighbors to address information to a unique neighbor. The size of an identifier is $\Theta(\log n)$ bits, while we aim at designing a sub-logarithmic algorithm. To reach a $\Theta(\log \log n)$ memory, [6] make the nodes communicate their identifier by smaller pieces, more precisely bit by bit. They define a function called $\mathtt{Bit}_v(i)$, which returns the position of the ith most significant bit equal to 1 in $\mathtt{id}_v$. In this paper, to achieve $\Theta(\log \log n)$ bits of memory per node, we use a similar $\mathtt{Bit}_v(i)$ function as [6].

3   Algorithm

Algorithm $\mathtt{Tok}^{\circlearrowleft}$ is based on a perpetual Depth-First traversal from the root r to the other nodes of the network. The classical DFS algorithm can be summarized as follows: first r has the token, then, the token is given to one unvisited neighbor v of r, which becomes r's child and r becomes v's parent. Node v then sends the token to one of its unvisited neighbors, and the token keeps being given from node to node until it reaches a node w such that w has no unvisited neighbors. At this point, node w sends the token back to its parent, which resumes the process of selecting unvisited neighbors. This traversal continues until all nodes in the network have been visited and the token is returned to the root node r. We focus on DODAGs, so a node v with the token selects the next node in the DODAG traversal among its children, rather than among all of its neighbors. From a local point of view, the tasks for a node v are as follows: receive the token from a node p, check whether it has any unvisited children, and if so, pass the token to one of them; if not, send the token back to its parent p.

So the self-stabilizing implementation of the DFS token circulation relies on several key concepts, including the presence or absence of the token, the identification of unvisited and visited neighbors, the selection of an unvisited neighbor, and the recording of the neighbor from which the token was received (i.e., the parent).

In order to prevent token loss or duplication, the circulation of unique token cannot occur in a single step and requires acknowledgment messages. A node v must inform node u that it is sending the token, and once u has confirmed receipt, v must delete the token. If v sends the token but does not verify that u has received it, the token may be lost. If u receives the token but does not verify that v has deleted the token, the token may be duplicated.

In this work, we consider the problem of token circulation in networks with sub-logarithmic memory available in each node. Storing identifiers in variables is not feasible, as a result, selecting the next unvisited child to receive the token becomes a non-trivial task that cannot be achieved in a single step. To address this, we propose a negotiation step where the node with the token interacts with its unvisited children. During this step, each child announces its identifier piece by piece, and the child with the largest identifier wins the negotiation. We also account for unvisited children who did not win the negotiation by proposing a mechanism to ensure that they will be visited later. Self-stabilizing algorithms must be able to detect and correct inconsistencies locally caused by transient faults. One fundamental idea in the design of $\text{Tok}^{\circlearrowright}$ is to minimize the frequency of node repairs. Firstly, the algorithm is already highly complex due to the numerous tasks that nodes must execute to achieve fair token circulation. Introducing additional rules to repair inconsistent situations would complicate both the algorithm and its proof. Secondly, in most cases, it is unnecessary to repair nodes as the token will eventually circulate in the network if the algorithm is resilient enough. The presence of the token and the rules for fair executions will allow the network to stabilize. However, there are some exceptions, i.e., repairing rules that are necessary for convergence. The global approach of $\text{Tok}^{\circlearrowright}$ is described by Algorithm 1, which describes the steps for a node v.

Algorithm 1: Local algorithm for a node v of DFS Token circulation

---

if *The token is present and some inconsistency is detected* **then**
| Release the token and consider that the token is in one of your descendants
else
| **if** *Received token from the parent* **then**
| | - Switch as a visited node;
| | - Wait for parent to release the token;
| **if** *Received token from a visited child* **then**
| | Wait for child to release the token;
| **if** *v has the token and nobody in its neighborhood has a token* **then**
| | **if** $\exists$ *unvisited children* **then**
| | | **while** $\exists$ *more than one unvisited child which are candidates to receive the token* **do**
| | | | - Ask all the candidates for a piece of their IDs;
| | | | - Publish the strongest piece;
| | | | - Wait for all the candidates which do not have the strongest piece to declare themselves losers;
| | | - Send the token to the winner;
| | | - Wait for the winner to receive the token;
| | | - Ask the losers to be ready for the next negotiation;
| | | - Wait for the update of the unvisited children;
| | | - Release the token;
| | **else**
| | | - Send the token to the parent;
| | | - Wait for the parent to receive the token;
| | | - Release the token;

3.1   Variables

Before presenting the rules of $\mathtt{Tok}^{\circlearrowright}$, let us first introduce the four variables it uses.

Visited and Unvisited Children. Nodes have to remember which of their children have already been visited in order to either send the token to unvisited children or to send it back to a parent. Classically, we implement this with a boolean variable *color*:

$$c_v \in \{\mathtt{red}, \mathtt{green}\}$$

Visited nodes are nodes that have the same color as the root. Nodes change from the state "unvisited" to "visited" (during a traversal initiated by the root) by changing their color upon receipt of the token, detailed below. The choice of the next child to be visited is based on the identifier of the unvisited children of the token holder: the chosen child is the one with the highest identifier. As a consequence, the circulation of the token always follows the same traversal path, depending on the relative order of the identifiers in the DODAG. Let us consider a node v with the token, and choose one of its unvisited neighbors u and sends the token to it, and so on to a node w surrounded by visited nodes. The node w sends the token back to its parent w_p which sends it to one of its unvisited children. The process is repeated until all nodes are visited, and the token returns to the node r. By construction the DFS token circulation is fair, since the unvisited nodes become visited exactly once. When the root r has no unvisited children, r switches its color to the other value (from red to green for example). Immediately, all the visited nodes become unvisited, for they now have the opposite color of r. This guarantees that this fair token circulation is perpetual. Only one operation, denoted by $c_v := \neg c_v$, is used in this variable, which is switching its value from **red** to **green** and vice versa.

State of the Token. To achieve atomicity of the operations required to maintain the network in a consistent state, we need to define a variable which takes eight different values.

$$\mathtt{tok}_v \in \{\bot, \bigstar, \bullet, \downdownarrows, \circlearrowleft, \downarrow, \circ, \uparrow\}$$

Let us give some explanations on those different values. The state $\underline{\mathtt{tok}_v = \bot}$ corresponds to nodes that are **away** from the current token circulation. For example, when the root has the token, all the other nodes are such that $\mathtt{tok}_v = \bot$. Reciprocally $\bot$ is a value that cannot be taken by the root, which would mean that no token is circulating in the network.

When a node that is not involved in the current token circulation receives the token from its parent, it simultaneously switches it color, and switches its token value to $\mathtt{tok}_v = \bigstar$. This value corresponds to the moment when a node **receives the token from its parent**, which is supposed to happen exactly once in each circulation round.

When the parent of v has finally left the token (recall that such operations are not atomic), v can **start negotiating** with its children to decide to which one it will send the token. To do that, it sets $\underline{\mathtt{tok}_v = \bullet}$, and keep that state until one winner is elected.

Once a winner u is elected $i.e.$ when all nodes of a different color than v have declared themselves losers, u excepted, v updates $\underline{\mathtt{tok}_v = \Downarrow}$ to **offer the token to one of its children**. After that, the child u that won the negotiation can update its own variable $\mathtt{tok}_u$ to ★.

After its child u took the token, v updates its variable to $\underline{\mathtt{tok}_v = \circlearrowleft}$, to let its children which lost the negotiation reset their variable in order to be **ready for the next negotiation**, when v will receive back the token from u.

When all the children of v have finished resetting their variable, v sets $\underline{\mathtt{tok}_v = \downarrow}$, which indicates that it is a node involved in a token circulation, but that **the token it had is below it**, to one of its descendants. After u has finished the token circulation below it, it sets $\mathtt{tok}_u = \uparrow$, to send the token back to v. If nothing wrong happened, v has only one child with a value $\mathtt{tok} \neq \perp$ at that moment, and therefore can take the token. To do so, it updates $\underline{\mathtt{tok}_v = \circ}$, and then **waits for u to effectively drop the token**, by setting $\mathtt{tok}_u = \perp$.

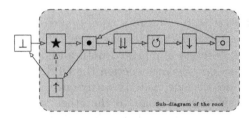

Fig. 1. Transition diagram for variable $\mathtt{tok}_v$.

After that, v restarts a negotiation, setting $\mathtt{tok}_v = \bullet$. After some time, v has no more children of a different color. When this happens, v **offers the token to its parent** w by setting $\mathtt{tok}_v = \uparrow$. Before effectively dropping the token, v waits that w acknowledges the token by updating its variable to $\mathtt{tok}_w = \circ$, but it also waits for all of its children to update their variable linked to the negotiation.

Indeed, in the current situation, all the children of v have successively won the token. As explained above, those nodes are prevented from taking one token of another color, to avoid livelocks. But now, the circulation of the token of v is terminated, so this becomes irrelevant. Worse, if nodes do not reset their variable to a value that allows them to negotiate, then they will all be ignored when a token from another color will come the next round. When all the children of v have reset their variable, v finally drops the token, and sets $\mathtt{tok}_v = \perp$.

Remark that for a node v with $\mathtt{tok}_v \notin \{\perp, \downarrow\}$, there is a strong pressure on the possible values of $\mathtt{tok}_u$, where u us a child of v, and most combinations are actually not supposed to occur in an execution. To summarize the Fig. 1 represents the order between the different states taken by the variable $\mathtt{tok}$.

Negotiation Variables: Bit-by-Bit Communication. To determine which of its children will receive the token, the parent progressively eliminates them, until only one remains. When it has exactly one child running for the token, it can finally declare that it gives the token. All the children that were eliminated ignore that message, and the one winner can take the token from its parent, and therefore no duplication of the token can occur.

The elimination process relies on the identifiers of the children, and on one additional variable which allows unvisited nodes to declare themselves as negotiating, or as losers of the negotiation. Step by step, the parent asks for the value of the i-th bit of the identifier of its children, and when all have answered, it reveals the biggest value it sees. All the nodes that have not announced this value consider themselves as eliminated until their parent allows them to negotiate again. This moment, when the parent allows its children who lost the election to negotiate again happens immediately after the winner of the election takes the token (Fig. 2).

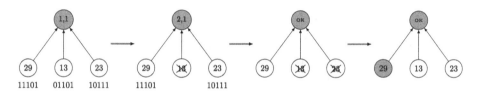

Fig. 2. Process of the designation of one child to give the token to. The parent reveals the biggest bit announced, then the children with small identifiers quit the negotiation, and in the end the winner takes the token and the losers are ready to negotiate again.

To negotiate for the token, nodes send bit-by-bit the value of their identifier to their parent. Two variables are required for that. The first one, ph, is used to communicate the position of the bit that is currently asked. It takes values between 1 and N, where N is the largest length of an identifier in the network, and thus $N \in O(\log n)$. The second variable is b, and is used to communicate the value of the bit that each child has, and for the parent to communicate the highest value it has seen. The variable b_v is used to communicate the values given by function Bit_v. It takes values in $\{0, 1, \perp\}$, where $\perp$ is the value that corresponds, for children, to the absence of a bit at that position (the identifier is too short), and, for the parent, to the request of the value of the bit. For simplicity, we denote the tuple (ph_v, b_v) by the notation id_v.

The general scheme is the following: first the parent sets $\text{id}_v = (1, \perp)$. Then all the children answer with the value of their first bit, by setting $\text{id}_u = (1, \text{Bit}_u(1))$. After all the children have answered, v sets $\text{id}_v = (1, b)$ where b is 1 if v saw a 1 in its children, and 0 otherwise. Then, all the children that do not have answered $(1, b)$ lose the negotiation. After all losers have left the negotiation, v sets $\text{id}_v = (2, \perp)$, and so on until only one node remains.

Negotiation Variables: State in the Negotiation Process. In addition to the identifier variables ph and b, we need a variable to remember where each node is in the negotiation process. This variable is denoted by play and takes four values:

$$\mathtt{play}_v \in \{\mathtt{P}, \mathtt{L}, \mathtt{W}, \mathtt{F}\}$$

Nodes v such that $\mathtt{play}_v = \mathtt{P}$ are the nodes that are available for a negotiation process with any of their parent that has a different color. They are the only nodes which can **participate** to a negotiation, the others have either already won, or lost. Note that P also stands for players.

When a node v **loses** the negotiation, it sets $\mathtt{play}_v = \mathtt{L}$, and thus becomes a loser, and is not involved in the current negotiation process anymore.

When a node v **wins** the negotiation and receives the token, it updates its color, its variable $\mathtt{tok}_v$, and it also updates $\mathtt{play}_v$ to W, to signify that it won a negotiation. After that, and until its parent sends the token back to its own parent, v won't negotiate with any of its parents. This guarantees that whatever the initial configuration of the system, v will not create oscillations between two tokens, and therefore no livelock.

However, the color is not sufficient to prevent livelocks or deadlocks caused by multiple circulating tokens. Indeed, even assuming only two tokens, one can design pathological configurations, and especially if the two tokens are of different colors. What we must absolutely prevent is situations in which one node alternatively takes a red token from one parent, sends it back later, then switches color and takes the green token from its other parent, sends it back, and then takes the red one again, and so on. With two such children, we can design scenarios where the tokens are in a livelock, and the circulation is blocked.

Note that nodes can neither simply ignore the token, for it would create a deadlock: both token would be blocked. To avoid such situations, we force a node that has received the token from one of its parents to not being available for a new token until this parent has sent the token back to its own parent. This requires being more subtle in how we define the election-related variables on the children.

As a consequence, we need the fourth value F to overcome a tricky situation. Recall that when a node u sends back the token to its parent w, it first waits for its children who won the negotiation (*i.e.* such that $\mathtt{play}_v = \mathtt{W}$) to reset their variable $\mathtt{play}_v$ to P, so that they will be able to negotiate during the next circulation round. But in triangle configurations, it might be that one such node v with $\mathtt{play}_v = \mathtt{W}$ has a common ancestor, with its parent u, as depicted in Fig. 3. In such a situation, v should not update $\mathtt{play}_v = \mathtt{P}$. Indeed if it does, then from w's perspective, it is exactly as if we never introduced the value W: it has one child who won the token and is at P. In particular, we can now create a livelock at the level of w.

But v can neither keep its value at W. That would create a deadlock with its parent u. Therefore, v **fakes** resetting its variable $\mathtt{play}_v$, by setting it to F, which is understood by its parent with $\mathtt{tok}_u = \uparrow$ as a P value, but both v and its grandparent w do recall that v has already had the token. This requires that as soon as possible, node v update its variable $\mathtt{play}_v = \mathtt{W}$, otherwise it could miss the opportunity to eventually reset its variable at P, notably if w sets $\mathtt{tok}_w = \uparrow$.

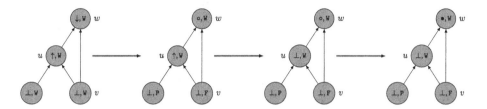

Fig. 3. Cleaning children before sending up the token in a triangle: $\mathtt{play}_v = \mathtt{F}$.

Although the alternating between two colors and the state $\mathtt{play}_v = \mathtt{W}$ share the common goal, which is to prevent a node to have several times the token, they are relevant in totally different situations. The alternation between two colors is built to prevent a node from receiving twice the same token, from parents that might be totally unrelated in the DODAG. On the other hand, the value W prevents a node from oscillating between two colors, from two different parents, which would block the circulation of the token.

3.2   Rules

The rules of $\mathtt{Tok}^{\circ}$ are grouped according to the task to which they relate: error handling, negotiation to choose the unvisited child, returning the token back to the parent, sending the token to a child. Figure 4 presents how the different rules follow each other during the execution of the algorithm, one rule, specific to the root is not presented in this section.

Error. As hinted above, some combinations of the variable $\mathtt{tok}$ should not occur on a node and its children. To repair such errors, the principle is that the parent trusts its child, in the sense that if both v and u believe they have a token, then v forgets its token, and repairs itself by setting $\mathtt{tok}_v = \downarrow$, acknowledging the fact that there is a token below it. The illegal pairs are detected by the predicate $\mathtt{Er}(v)$ and the correctness of the state of the token of node v is handled by Rule $\mathbb{E}_{\mathtt{TrustChild}}$. All the other rules of the algorithm suppose that node v is not in such an error. For homogeneity and readability, we constrained the other rules with $\neg\mathtt{Er}(v)$.

Negotiation. In a negotiation, two types of nodes are involved, the parent v which owns the token, and the unvisited children involved in the negotiation (we call these children "players" formally defined by the set $\mathtt{Players}(v)$).

Before describing the negotiation itself, let's explain how the nodes involved in the negotiation can leave it. The parent can leave the negotiation phase in two ways. The first is captured by Rule $\mathbb{R}_{\mathtt{OfferUp}}$, when the parent observes that the set of $players(v)$ is empty, which means that it has finished circulating below itself. As a consequence, v returns the token to its own parent ($\text{tok}_v = \uparrow$). The second possibility happens when node v has exactly one child u in $\text{Players}(v)$, after verifying some tricky problems, v can send the token to u, to achieve that, v puts $\text{tok}_v := \Downarrow$, thanks to Rule $\mathbb{R}_{\text{Give}}$. The first action that v takes, after u has actually received the token, is switching its variable tok to $\circlearrowright$ to inform its other unvisited children who lost the negotiation (a.k.a. the losers) that the negotiation phase is terminated, this is made by Rule $\mathbb{R}_{\text{NewPlay}}$. So each loser can change its variable $\text{play}_w = \text{P}$ and becomes players again when v will receive the token back (see Rule $\mathbb{R}_{\text{ReplayD}}$). Let us now consider the children $u \in \text{Players}(v)$, u quits the negotiation if it quits $\text{Players}(v)$. The node u can do this in three ways, u detects an inconsistent behavior in his neighborhood (see Rule $\mathbb{R}_{\text{FakeWin}}$), u wins negotiation (see Rule $\mathbb{R}_{\text{Win}}$), u loses in the negotiation (see Rule $\mathbb{R}_{\text{Lose}}$).

Now let us focus on the negotiation itself. When a node v has finished receiving the token (i.e., the parent of v has released the token $\text{tok}_{p(v)} = \downarrow$, Rule $\mathbb{R}_{\text{Drop}}$), it can begin the negotiation (execution of Rule $\mathbb{R}_{\text{Nego}}$). When all players have announced their binary value (see Rule $\mathbb{R}_{\text{NewBit}}$), v announces the highest value (see Rule $\mathbb{R}_{\text{maxPos}}$). After that, some players lose the negotiation and others remain players. If there is only one player left, the negotiation is over, otherwise v continues the negotiation (see Rule $\mathbb{R}_{\text{NewPh}}$).

Reception of the Token from a Child. Now consider the actions when a child u offers the token to its parent v by setting $\text{tok}_u = \uparrow$. The node v needs to be sure that it does not have any other children involved in a token circulation. If it has, then taking the token would lead to an inconsistent configuration, where one token is above an other one. Actually, u may have several parents related to each other, and only the deepest should take the token. If v can take the token, it executes the Rule $\mathbb{R}_{\text{Receive}}$ signifying to u that it has taken the token. Thus, when u is sure that v has taken the token, u can release the token and v can resume negotiating with its remaining players (Rule $\mathbb{R}_{\text{ReNego}}$).

End of the Circulation. After it has offered the token to its parent, a node v which is not the root should eventually drop the token, and set $\text{tok}_v = \bot$ (Rule $\mathbb{R}_{\text{Return}}$). Before that, it must check that all its children have correctly updated their variable from W to P to assure that when a token of the other color will come, all those nodes will be available for a negotiation phase.

In a similar situation, where all of its children have repaired themselves, the root does not drop the token, for it would simply destroy it. When the root is in such a situation, it means that the current circulation round is terminated, and one other can start. To achieve that the root changes its color, and thus if the circulation has already stabilized, the root has one color and all the node have the other color (*i.e.* all the other nodes are unvisited).

Let us consider a node v having parents that are sending the token higher and no other parent involved in a token circulation. If a node v is not ready to play again, then v cleans its variable play_v to be ready for the next circulation round (Rule $\mathbb{R}_{\text{ReplayUp}}$).

If v has some parents sending the token higher, and other parents which do not have terminated their token circulation, it must update its variable $\mathtt{play}_v = $ F (Rule $\mathbb{R}_{\mathtt{Fake}}$).

Finally, we must design a rule to cancel the effect of updating one's variable at F, to reset it at W, without Rule $\mathbb{R}_{\mathtt{ReWin}}$, the token could not be sent higher.

To Summarize. Figure 4 presents the scheme of an execution of $\mathtt{Tok}^\circlearrowleft$. It presents the update of the variables of one node v by the different rules of $\mathtt{Tok}^\circlearrowleft$.

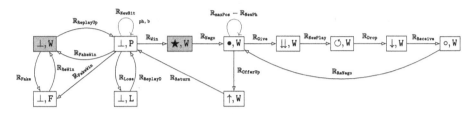

Fig. 4. Chain of the states taken by a node during the execution of the algorithm. The update of c_v is represented by color gray. Rules in blue corresponds to the negotiation. Rules in red correspond to the sending of a token from a child to a parent. Rules in black correspond to the sending of a token from a parent to a child.

4   Correctness

Let us draw here a roadmap of the proof of Theorems 1 and 2. This proof is subdivided into four parts. First, we establish that maximal executions of $\mathtt{Tok}^\circlearrowleft$ are infinite. Formally, we prove that in any configuration, there is at least one enabled node. This part of the proof is static, in the sense that we do not need to consider an execution of the algorithm but only isolated configurations. It essentially boils down to a syntactical analysis of the guards of the rules of $\mathtt{Tok}^\circlearrowleft$.

Although the previous does not need to consider executions, the two following parts of the proof are trickier. We aim at proving that eventually, one token fairly circulates in the network, in a self-stabilizing way. Therefore, we have to consider configurations in which several token exist in the network. To do so, we introduce a *ad hoc* structure which corresponds to the paths from the root to the tokens in the network, which is a DODAG. More precisely, it is a sub-DODAG of the DODAG representing the network. But there may exist some tokens that are not connected to the root by an admissible path. For these tokens, we also define this sub-DODAG structure, but we distinguish these tokens by saying that they are not pointed by the root. All the reasoning of the two following parts rely on these sub-DODAGs structures.

Secondly, we establish at once two crucial properties for the validity of $\mathtt{Tok}^\circlearrowleft$. The first property is that if a token is not pointed by the root of the network, then it can only circulate a finite number of times before being blocked, or deleted. This guarantees that no scheduler, even unfair, can create an execution in which the token pointed by the root is never activated. The second property is that even the token pointed by the root is restricted in its circulation, in the sense that it can only circulate a finite number of times before it reaches the root again, and the root "resets" it by the execution of a rule specific to the root, $\mathbb{R}^r_{\text{NewDFS}}$. The combination of these two properties, and of the liveness established previously leads to the guarantee that there is an infinity of finite circulations in any maximal execution of $\text{Tok}^\circlearrowleft$. To prove both properties, we define a potential function, which associates a positive value (a weight) to each token in any configuration. Then, we establish that any computing step decreases the weight of the token, unless the token is reset by an execution of $\mathbb{R}^r_{\text{NewDFS}}$. This establishes that a token cannot circulate indefinitely, since any step it takes makes a positive quantity decrease, unless this token is anchored at the root and there are an infinity of new circulations. Since we also prove that no token can be created, this guarantees that at some point, although several tokens may exist, only one is activated, the one which is pointed by the root.

Thirdly, we establish that $\text{Tok}^\circlearrowleft$ behaves correctly. By "behaves correctly" we mean that the token eventually fairly visit all the nodes of the network. To establish that, we largely use the previous results. In particular, we start the reasoning at a moment where the algorithm has already partly converged, in the sense that all the tokens except the one pointed by the root are not activated anymore. Then, we first establish some additional stability properties on the underlying structure of the DODAG which corresponds to the path from the root to the token. Then, after having properly defined what "to have the token" is, we establish that if one node receives the token from one of its parent, then all of its children that are able to receive the token will actually receive it at some point. On the other hand, we prove that if some children of one such node cannot receive the token, then after one or two additional circulations of the token, then these children will be able to receive it, and by the previous, will receive it. Therefore, circulation after circulation, the token visits nodes in deeper and deeper parts of the network. By induction, we finally prove that the token eventually fairly visits all the nodes of the network, indefinitely. In particular, this guarantees that all the other token were reached by the token pointed by the root, and merged with it.

Joined, these three first parts of the proof establish that $\text{Tok}^\circlearrowleft$ is a self-stabilizing algorithm for the fair token circulation in arbitrary DODAGs.

To prove the space optimality of $\text{Tok}^\circlearrowleft$, we use a similar approach to that [3] in which it is proven that if an algorithm $\mathcal{A}$ solves a problem P using $o(\log \log n)$ bits per node, then $\mathcal{A}$ solves P in anonymous networks, with the same settings (communication model, scheduler, topology...). We prove that even under a central strongly fair or synchronous scheduler, on a tree topology, and without the requirement of being self-stabilizing, no algorithm can solve the fair token circulation problem in anonymous network. The result established in [3] concludes to the space optimality of $\text{Tok}^\circlearrowleft$.

5   Conclusion

In this paper, we have presented $\text{Tok}^{\circlearrowleft}$, a memory-optimal self-stabilizing algorithm for the fair token circulation in arbitrary DODAGs under the unfair scheduler. Algorithm $\text{Tok}^{\circlearrowleft}$ works in the state model where no information is given on the neighbors, and requires only $\Theta(\log \log n)$ bits per node. To our knowledge, this is the first self-stabilizing algorithm achieving this space complexity in this model for graphs of arbitrary degree.

Of course, the construction of a DODAGs with the same space complexity remains an open question. Beyond the issue of DODAGs, our work is providing hope that it might be possible to design self-stabilizing token circulation algorithms for arbitrary networks with space-complexity $O(\log \log n)$ bits per node. The design of such algorithms requires to overcome at least two problems: the presence of more than one root, and the symmetry caused by the presence of cycles. The presence of multiple roots is an issue which may not be too dramatic, as it may be possible to let several tokens circulate, one per root, and to remove the tokens one by one until a single token remains. The symmetries caused by the presence of cycles appear to cause severe difficulties, and our current knowledge is insufficient to guarantee that a space-complexity of $O(\log \log n)$ bits per node can be achieved under such symmetries.

Such an algorithm with space-complexity $\Theta(\log \log n)$ bits per node, solving token circulation on arbitrary graphs would be a valuable toolbox which could be used to solve other problems, such as leader election, spanning tree construction, *etc.*

References

1. Blin, L., Boubekeur, F., Dubois, S.: A self-stabilizing memory efficient algorithm for the minimum diameter spanning tree under an omnipotent daemon. J. Parallel Distrib. Comput. **117**, 50–62 (2018)
2. Blin, L., Dubois, S., Feuilloley, L.: Silent MST approximation for tiny memory. In: Devismes, S., Mittal, N. (eds.) SSS 2020. LNCS, vol. 12514, pp. 118–132. Springer, Cham (2020). https://doi.org/10.1007/978-3-030-64348-5_10
3. Blin, L., Feuilloley, L., Le Bouder, G.: Optimal space lower bound for deterministic self-stabilizing leader election algorithms. Discret. Math. Theor. Comput. Sci. **25** (2023). https://doi.org/10.46298/dmtcs.9335
4. Blin, L., Le Bouder, G., Petit, F.: Optimal memory requirement for self-stabilizing token circulation. Technical report, HAL Open Science (2024). https://hal.science/hal-04448960
5. Blin, L., Potop-Butucaru, M., Rovedakis, S., Tixeuil, S.: A new self-stabilizing minimum spanning tree construction with loop-free property. In: Keidar, I. (ed.) DISC 2009. LNCS, vol. 5805, pp. 407–422. Springer, Heidelberg (2009). https://doi.org/10.1007/978-3-642-04355-0_43
6. Blin, L., Tixeuil, S.: Compact deterministic self-stabilizing leader election on a ring: the exponential advantage of being talkative. Distrib. Comput. **31**(2), 139–166 (2018). https://doi.org/10.1007/s00446-017-0294-2

7. Blin, L., Tixeuil, S.: Compact self-stabilizing leader election for general networks. J. Parallel Distrib. Comput. **144**, 278–294 (2020). https://doi.org/10.1016/j.jpdc.2020.05.019
8. Brown, G., Gouda, M., Wu, C.: Token systems that self-stabilize. IEEE Trans. Comput. **38**, 845–852 (1989)
9. Burns, J., Pachl, J.: Uniform self-stabilizing rings. ACM Trans. Program. Lang. Syst. **11**, 330–344 (1989)
10. Cournier, A., Devismes, S., Petit, F., Villain, V.: Snap-stabilizing depth-first search on arbitrary networks. Comput. J. **49**(3), 268–280 (2006)
11. Cournier, A., Devismes, S., Villain, V.: Light enabling snap-stabilization of fundamental protocols. ACM Trans. Auton. Adapt. Syst. **4**(1), 6:1–6:27 (2009)
12. Datta, A.K., Johnen, C., Petit, F., Villain, V.: Self-stabilizing depth-first token circulation in arbitrary rooted networks. Distrib. Comput. **13**(4), 207–218 (2000). https://doi.org/10.1007/PL00008919
13. Dijkstra, E.W.: Self-stabilizing systems in spite of distributed control. Commun. ACM **17**(11), 643–644 (1974). https://doi.org/10.1145/361179.361202
14. Dolev, S., Israeli, A., Moran, S.: Self-stabilization of dynamic systems assuming only read/write atomicity. Distrib. Comput. **7**, 3–16 (1993)
15. Dubois, S., Tixeuil, S.: A taxonomy of daemons in self-stabilization (2011). arXiv: 1110.0334
16. Ghosh, S.: An alternative solution to a problem on self-stabilization. ACM Trans. Program. Lang. Syst. **15**, 735–742 (1993)
17. Gouda, M.G., Haddix, F.F.: The stabilizing token ring in three bits. J. Parallel Distrib. Comput. **35**(1), 43–48 (1996). https://doi.org/10.1006/jpdc.1996.0066
18. Huang, S., Chen, N.: Self-stabilizing depth-first token circulation on networks. Distrib. Comput. **7**, 61–66 (1993)
19. Johnen, C., Beauquier, J.: Space-efficient distributed self-stabilizing depth-first token circulation. In: Proceedings of the Second Workshop on Self-stabilizing Systems, pp. 4.1–4.15 (1995)
20. Martin, J.J.: Distribution of the time through a directed, acyclic network. Oper. Res. **13**(1), 46–66 (1965). https://doi.org/10.1287/opre.13.1.46
21. Petit, F., Villain, V.: Optimal snap-stabilizing depth-first token circulation in tree networks. J. Parallel Distrib. Comput. **67**(1), 1–12 (2007)
22. Thubert, P., Richardson, M.: Routing for RPL (routing protocol for low-power and lossy networks) leaves. RFC 9010, RFC Editor (2021)
23. Winter, T., et al.: RPL: IPv6 routing protocol for low-power and lossy networks. RFC **6550**, 1–157 (2012)

Stand-Up Indulgent Gathering on Rings

Quentin Bramas[1], Sayaka Kamei[2], Anissa Lamani[1(✉)], and Sébastien Tixeuil[3]

[1] University of Strasbourg, ICube, CNRS, Strasbourg, France
`alamani@unistra.fr`
[2] Hiroshima University, Higashihiroshima, Japan
[3] Sorbonne University, CNRS, LIP6, IUF, Paris, France

Abstract. We consider a collection of $k \geq 2$ robots that evolve in a ring-shaped network without common orientation, and address a variant of the crash-tolerant gathering problem called the *Stand-Up Indulgent Gathering* (SUIG): given a collection of robots, if no robot crashes, robots have to meet at the same arbitrary location, not known beforehand, in finite time; if one robot or more robots crash on the same location, the remaining correct robots gather at the location of the crashed robots. We aim at characterizing the solvability of the SUIG problem without multiplicity detection capability.

Keywords: Stand-Up Indulgent Gathering · oblivious robots · fault-tolerance · discrete universe

1 Introduction

Mobile robotic swarms have recently gained significant interest within the Distributed Computing scientific community. For over two decades, researchers have aimed to characterize the precise assumptions that enable basic problems to be solved for robots represented as disoriented (each robot has its own coordinate system), oblivious (robots cannot remember past actions), and dimensionless points evolving in Euclidean space. One key assumption is the scheduling assumption [11], where robots can execute their protocol fully synchronized (FSYNC), completely asynchronously (ASYNC), or with a fairly chosen subset of robots scheduled for synchronous execution (SSYNC).

The *gathering* problem [13] serves as a benchmark due to its simplicity of expression (robots must gather at an unknown location in finite time) and its computational tractability (two SSYNC-scheduled robots cannot gather without additional assumptions).

As the number of robots increases, so does the probability of at least one failure. However, relatively few works consider the possibility of robot failures. One simple failure is the *crash* fault, where a robot unpredictably stops executing its protocol. In gathering, one must prescribe the expected behavior in the presence of crash failures. Two variants have been studied: *weak* gathering

This paper was supported by JSPS KAKENHI No. 23H03347, 23K11059, and ANR project SAPPORO (Ref. 2019-CE25-0005-1).

Y. Emek (Ed.): SIROCCO 2024, LNCS 14662, pp. 119–137, 2024.
https://doi.org/10.1007/978-3-031-60603-8_7

expects all correct (non-crashed) robots to gather regardless of crashed robot positions, while *strong* gathering (also known as stand-up indulgent gathering – SUIG) expects correct robots to gather at the unique crash location. In continuous Euclidean space, weak gathering is solvable in the SSYNC model [1,3,9,10], while SUIG (and its two-robot variant, stand-up indulgent rendezvous – SUIR) is only solvable in the FSYNC model [6–8].

A recent trend [11] has been to move from a continuous environment to a discrete one. In the discrete setting, robots can occupy a finite number of locations and move between neighboring locations. This neighborhood relation is conveniently represented by a graph whose nodes are locations, leading to the "robots on graphs" denomination. The discrete setting is better suited for describing constrained physical environments or environments where robot positioning is only available from discrete sensors [2]. From a computational perspective, the continuous and discrete settings are unrelated: on one hand, the number of possible configurations (i.e., robot positions) is much more constrained in the discrete setting than in the continuous setting (only a finite number of configurations exist in the discrete setting); on the other hand, the continuous setting offers algorithm designers more flexibility to solve problematic configurations (e.g., using arbitrarily small movements to break symmetry). To our knowledge, the only previous work considering SUIR and SUIG in a discrete setting is due to Bramas et al. [4]. They provide a complete characterization of the problems solvability for line-shaped networks. More specifically, they show that the problem is unsolvable in SSYNC, and provide a solution in FSYNC.

In this paper, we consider the discrete setting and aim to further characterize the solvability of the SUIR and SUIG problems when the visibility radius of robots is unlimited. In a set of locations whose neighborhood relation is represented by a ring-shaped graph, robots must gather at a single unknown location. Furthermore, if one or more robots crash at the same location, all robots must gather at that location. With respect to previous work [4], a ring topology induces a lot more possibility for symmetric situations, and a robot may no longer use the knowledge of being an edge robot (that is, being located at an extremal position with respect to all other robots on the line). We present the following results. With two robots, the SUIR problem is unsolvable in SSYNC, and solvable in FSYNC only when the initial configuration is not periodic and node-node symmetric. With three robots and more, impossibility results for SSYNC SUIR and FSYNC gathering (from a periodic or edge-edge symmetric configuration) extend to SUIG. Also, we prove that three robots cannot solve SUIG in FSYNC when the size of the ring is multiple of 3. On the positive side, we present a SUIG algorithm for more than three robots and an odd n in FSYNC, assuming an initial node-edge symmetric configuration that is not periodic. Among other difficulties, the key technique enabling the solution is to have at least two robots on distinct locations move in any situation (to withstand a crash), that can be of independent interest for further developing stand-up indulgent solutions.

The sequel of the paper is organized as follows: Sect. 2 formalizes our execution model, Sect. 3 presents results for the case of two robots, while Sect. 4 concentrates on the main case. Finally, concluding remarks are provided in Sect. 5.

2   Model

The rings we consider consist of n unlabeled nodes $u_0, u_1, u_2, \ldots, u_{n-1}$ such that u_i is connected to both $u_{((i-1) \mod n)}$ and $u_{((i+1) \mod n)}$.

Let $R = \{r_1, r_2, \ldots, r_k\}$ denote the set of $k \geq 2$ autonomous robots. Robots are assumed to be anonymous (i.e., they are indistinguishable), uniform (i.e., they all execute the same program, and do not make use of localized parameters), oblivious (i.e., they cannot remember their past actions), and disoriented (i.e., they have no common sense of direction). We assume that robots do not know k (the number of robots). In addition, they are unable to communicate directly, however, they have the ability to sense their environment, including the position of the other robots. Based on the configuration resulting of the sensing, they decide whether to move or to stay idle. Each robot r executes cycles infinitely many times: (i) first, r takes a snapshot of its environment to see the position of the other robots (LOOK phase), (ii) according to the snapshot, r decides to move or not to move (COMPUTE phase), and (iii) if r decides to move, it moves to one of its neighbor nodes depending on the choice made during the COMPUTE phase (MOVE phase). We call such cycles *LCM cycles*. We consider the *FSYNC* model in which at each time instant t, called *round*, each robot executes a LCM cycle synchronously with all other robots, and the *SSYNC* model, where a non-empty subset of robots chosen by an adversarial scheduler executes a LCM cycle synchronously, at each time instant t.

A node u is *occupied* if it hosts at least one robot, otherwise, u is *empty*. We say that there is a *tower* (or a *multiplicity*) on a node u if u hosts (strictly) more than one robot. The ability for the robots to identify nodes that host a tower is called *multiplicity detection*. When robots can detect a tower, we say that they have a *strong* multiplicity detection if they know the exact number of robots participating to the tower. Otherwise, they have a *weak* multiplicity detection capability. In this work, we assume that the robots have no multiplicity detection, *i.e.*, they cannot decide whether an occupied node is a tower.

During the execution, robots move and occupy some nodes, their positions form the *configuration* $C_t = (d(u_0), d(u_1), \ldots)$ of the system at time t, where $d(u_i) = 0$ if the node u_i is empty and $d(u_i) = 1$ if u_i is occupied. Initially, we assume that the configuration is distinct, *i.e.*, each node is occupied by at most one robot. For the analysis, a robot r may also denotes the node occupied by r.

A sequence of consecutive occupied nodes is a *block*. Similarly, a sequence of consecutive empty nodes is a *hole*.

The *distance* between two nodes u_i and u_j is the number of edges in a shortest path connecting them. The distance between two robots r_i and r_j is the distance between the two nodes occupied by r_i and r_j, respectively. We denote the distance between u_i and u_j (resp. r_i and r_j) $dist(u_i, u_j)$ (resp. $dist(r_i, r_j)$). Two robots or two nodes are *neighbors* if the distance between them is one.

An *algorithm* A is a function mapping the snapshot (obtained during the LOOK phase) to a neighbor node (or the same node should the robot decide not to move) destination to move to (during the MOVE phase). An *execution*

$\mathcal{E} = (C_0, C_1, \dots)$ of A is a sequence of configurations, where C_0 is an initial configuration, and every configuration C_{t+1} is obtained from C_t by applying A.

The *angle* between two nodes u_i and u_j is determined by the number of nodes in the path that connects them in either a clockwise or counterclockwise direction. This angle can be expressed as either the distance or its n-complement. For our analysis, we assume a fixed direction that is unknown to the robots and can be arbitrary unless otherwise specified. The rotation of an angle x is denoted by $\texttt{rot}(x)$, which maps each node to the x-th node in a specific direction. It's important to note that $\texttt{rot}(x)$ is equivalent to a rotation of angle $2x\pi/n$ radians in the same direction when mapping the n-node ring to a circle in Euclidean space with evenly spaced nodes. The image of node u_i after rotation $\texttt{rot}(x)$ is represented by $\texttt{rot}(x)(u_i)$.

An automorphism is a bijection f of nodes such that $f(u_i)$ and $f(u_j)$ are neighbors if and only if u_i and u_j are neighbors. In an n-ring, there are two types of automorphisms: rotational symmetries and reflections. These depend on whether the angles between nodes are preserved or not. The group of symmetries is the dihedral group D_n, which contains n rotational symmetries and n reflections. A reflection can be classified as an *edge-edge symmetry, node-edge symmetry*, or *node-node symmetry* depending on whether it has no fixed points, exactly one fixed point, or exactly two fixed points, respectively. The axis of a reflection is the imaginary line that passes through the fixed points and/or fixed edges. This axis corresponds to the axis of symmetry when the ring is mapped to the Euclidean plane. In the remainder of this text, a symmetry always refers to a reflection.

We say that a set S of nodes is *symmetric*, or contains an axis of symmetry, if there is a reflection f such that $f(S) = S$. We also say that a set S of nodes is *rotational symmetric* or *periodic* if there is a rotation f such that $f(S) = S$.

Configurations that have no tower are classified into three classes [12]. Let C be a configuration, and let S be the set of occupied nodes in C. Then, C is *symmetric* if S is symmetric. Also, C is *periodic* if there exists a non-trivial rotational symmetry f such that $f(S) = S$ (it is represented by a configuration of at least two copies of a sub-sequence). Otherwise, C is *rigid*.

We define the view of a robot r located on node u_i as the pair $\{S^+, S^-\}$ ordered in the lexicographic order where $S^+(t) = d(u_i), d(u_{i+1}), \dots, d(u_{i+n-1})$ and $S^-(t) = d(u_i), d(u_{i-1}), \dots, d(u_{i-(n-1)})$.

Problem Definition

A robot is said to be *crashed* at time t if it is not activated at any time $t' \geq t$. That is, a crashed robot stops execution and remains at the same position indefinitely. We call the node where robots crash *crashed node*. We assume that robots cannot distinguish the crashed node in their snapshots (i.e., they are able to see the crashed robots, but they remain unaware of their crashed status). A crash, if any, can occur at any round of the execution. Furthermore, if more than one crash occur, all crashes occur at the same location. In our model, since robots do not have multiplicity detection capability, a location with a single crashed robot and with multiple crashed robots are indistinguishable, and are thus equivalent. In the sequel, for simplicity, we consider at most one crashed robot.

We consider the *Stand Up Indulgent Gathering* (SUIG) problem. An algorithm solves the SUIG problem if, for any initial configuration C_0 with up to one crashed node, and for any execution $\mathcal{E} = (C_0, C_1, \dots)$, there exists a round t such that all robots (including the crashed one) gather at a single node, not known beforehand, for all $t' \geq t$. We call the special case with $k = 2$, the *Stand Up Indulgent Rendezvous* (SUIR) problem. Observe that in the case of multiple crashed nodes, the SUIG problem becomes unsolvable. Hence, in the sequel, we assume that if several robots crash, they crash on the same node. So, overall, there is at most one crashed node.

3   Stand Up Indulgent Rendezvous

In this section, we consider the case where two robots occupy initially distinct nodes. We first recall an impossibility result for gathering on rings as Theorem 1.

Theorem 1 ([12]). *The gathering problem is unsolvable in FSYNC on ring networks starting from a periodic or an edge-edge symmetric configuration, even with strong multiplicity detection, even with two robots.*

A direct consequence of this theorem is that SUIR is also impossible in the same setting (indeed, when there are no crashes, SUIR must guarantee rendezvous).

Next, we consider initial configurations that are node-edge symmetric (with n being odd).

Lemma 1. *On rings, starting from a node-edge symmetric configuration, there exists no deterministic algorithm that solves the SUIR problem by two oblivious robots without additional hypothesis, even in FSYNC.*

Proof. Assume for the purpose of contradiction that such an algorithm exists. As robots are oblivious and have no common sense of direction, they can either move towards each other or move away from each other.

Observe first that an even sized ring cannot be node-edge symmetric. So, we consider the case when the size of ring n is odd (then, every configuration is node-edge symmetric). When no fault occurs, the robots have to move towards the hole with an odd size to achieve rendezvous. However, if one of them crashes, whenever the other robot moves, it must apply the same behavior (move toward the hole with an odd size), and hence move back to its previous position. That is, the non-crashed robot moves back and forth between two nodes forever. As a result, the two robots never meet, a contradiction.   □

The remaining case to address is when the configuration is not periodic but node-node symmetric. Observe that such a symmetry only happens in even-sized rings. In this case, there is a simple strategy in FSYNC, which consists in making the robots move towards each other taking the shortest path (as the configuration is not periodic, there always exists a shortest path and a longest path between the two robots). So, if none of the two robots crashes, then the robots eventually meet on a single node (that is on the initial axis of symmetry). Otherwise, as a single robot moves, the smallest distance between the two robots becomes odd. However, the correct robot can deduce that the other has crashed, and hence continues to move towards the other robot through the shortest path. Hence, the distance between the two robots decreases at each round, and the two robots are eventually located on two adjacent nodes. Finally, to achieve the rendezvous, the non-crashed robot simply moves to the adjacent occupied node.

In the SSYNC model, SUIR remains impossible, even starting from a node-node symmetric configuration that is not periodic. Indeed, as both robots need to move at each time instant, if we assume a node-node symmetric configuration where a single robot r is activated, whatever the direction chosen by r, a node-edge symmetric configuration is reached. By Theorem 1 and Lemma 1, we can deduce:

Lemma 2. *On rings, starting from a node-node symmetric distinct configuration that is not periodic, there exists no SUIR algorithm in SSYNC, without additional hypotheses.*

Proof. Suppose for the sake of contradiction that there exists such an algorithm A. If the configuration is *only* node-node symmetric and not periodic, the distance between the two robots in both direction is always even, but they are different. Since A solves SUIR problem, robots cannot remain idle. Consider a schedule where from this configuration a single robot is activated, and thus moves in either direction. We reach a configuration in which the distance between the two robots in each direction is odd. Hence, the configuration is either periodic or edge-edge symmetric. By Theorem 1, algorithm A cannot solve SUIR from this configuration, a contradiction. □

Overall, SUIR is feasible in FSYNC in even-sized rings, starting from a non-periodic, node-node symmetric configuration, and impossible in all other settings.

4   Stand-Up Indulgent Gathering

In this section, we first extend existing impossibility results for SSYNC gathering and SSYNC SUIR to show that SUIG is not solvable in rings in SSYNC without additional hypothesis, which justifies our later assumption to only consider FSYNC.

In SSYNC, we can show that there always exists an execution where a configuration with only two occupied nodes is reached. As robots have no multiplicity detection, from such a configuration, it is known that the gathering is impossible. That is:

Lemma 3. *In rings, the SUIG problem is unsolvable in SSYNC without additional hypotheses.*

Proof. From Theorem 1, Lemmas 1 and 2, we know that the SUIR is not solvable in SSYNC. As we assume that robots do not have multiplicity detection, in the execution of every algorithm that solves the SUIG problem, no configuration shall contain exactly two occupied nodes. Indeed, the SSYNC adversary may activate all robots on the same node synchronously, making them behave as a single robot, leading to the previous impossibility case with two robots. Now, assume for the purpose of contradiction that there exists an algorithm A that solves the SUIG problem, and let C be the last configuration in the execution of A where the number of occupied nodes is strictly larger than 2 (i.e., the configuration reached just before gathering is achieved). As robots can only move by one edge at each round, the number of occupied nodes in C is 3. As we consider the SSYNC model, the adversary can activate only the robots located on a single node. That is, a configuration C' in which the number of occupied nodes is equal to 2 is reached. A contradiction.   □

Lemma 4. *There exists no algorithm in FSYNC that solves the SUIG problem by $k = 3$ oblivious robots when $n \mod 3 = 0$.*

Proof. Assume by contradiction that the Lemma does not hold and let consider the configuration shown in Fig. 1. As the robot that is on the axis of symmetry could be the faulty one, at least one of the two other robots should also be allowed to move. Since the configuration is symmetric, we cannot distinguish between the two other robots. That is both of them are allowed to move. If they move in the opposite side of the robot on the axis, eventually they become neighbors, and from there they can only exchange their positions. That is, in order to perform the gathering, the algorithm needs to have a rule making the two robots move towards the robot on the axis of symmetry. By moving a periodic configuration is reached. As robots cannot distinguish both of their sides of the ring, the two robots move back to their respective previous position. That is, the robots move back and forth forever. A contradiction.   □

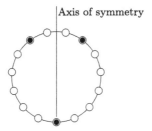

Fig. 1. An instance of an initial configuration with $k = 3$ when $n \mod 3 = 0$

As the gathering is known to be impossible from a periodic or an edge-edge symmetric configuration [12], we propose an algorithm which solves the SUIG problem on ring-shaped networks of size n by k robots such that $k > 3$ or $(k > 2 \wedge n \mod 3 \neq 0)$ assuming node-edge symmetric initial configurations. In the sequel, we use r to represent both a robot r and its occupied node.

Definition 1. *Let C be a configuration having a node-edge axis of symmetry L. The* target *node, denoted v_{target}, is the only node on L, i.e., the only node that is its own symmetric with respect to L. The* main *robots are the two robots that are closest to v_{target}, but not on v_{target}. The* secondary *robots are the two robots that are the farthest to v_{target} that have an empty adjacent node (See Fig. 2).*

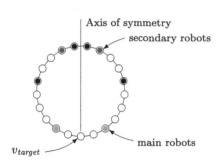

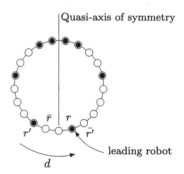

Fig. 2. An instance of a node-edge symmetric configuration

Fig. 3. An instance of a $\{r, r'\}$-quasi-node-edge symmetric configuration

In our algorithm, the goal is to move the main robots toward the target node. If the configuration remains node-edge symmetric, the same process continues so that all the robots reaches the target node. Nevertheless, in some cases, by moving, the main robots create a periodic configuration which makes the SUIG problem unsolvable. Hence, in this special case, together with the main robots, the secondary robots move as well.

During this process, if a non node-edge symmetric configuration is reached, then a robot has crashed and the configuration is quasi-symmetric, as defined below (See Fig. 3).

Definition 2. *Let C be a configuration and r, r' be occupied nodes by robots r, r' in C. We say C is $\{r, r'\}$-quasi-node-edge symmetric, with quasi-axis L, if $C \setminus \{r\} \cup \{\bar{r}'\}$ is node-edge symmetric of axis L, where $\bar{r} \notin C$ (resp. $\bar{r}' \notin C$) is the symmetric of r (resp. r') with respect to L and $r' \in C$ is at distance one from $\bar{r}$. The target node v_{target} of a quasi-axis is the middle node between r and $\bar{r}$ (there is only one middle node since n is odd). v_{target} is also the only node on L.*

From a symmetric configuration, a quasi-axis can be created by our algorithm in two cases: Either the two main robots are ordered to move and one of them is crashed, or the two secondary robots are ordered to move and one of them is crashed. We now define the gap distance associated with a quasi-axis, to distinguish between the two cases. Recall that, between two points in the ring, since n is odd, there are two paths, one is of odd length and one is of even length. In particular, the target node of a $\{r, r'\}$-quasi-axis of symmetry is located in the middle of the path of odd length between r and r'.

Definition 3. *Let C be $\{r, r'\}$-quasi-node-edge symmetric with target node v_{target}. If there is no robot in the path of odd length (between r and r'), or if only v_{target} is occupied, then the* gap *distance associated with $\{r, r'\}$ is the length of this path. Otherwise, the* gap *distance associated with $\{r, r'\}$ is the length of the path with even length.*

From this definition, we see that the parity of a quasi-axis determines whether the crashed robot is one of the main robots or not (if not, it is one of the secondary robots, see the algorithm for more details). That is, if the gap distance is odd, the crashed robot is one of the main robots, otherwise, one of the secondary robots.

Definition 4. *Let C be $\{r, r'\}$-quasi-node-edge symmetric with quasi-axis L, gap distance d, and the path P between r and r' of length d. The* leading robot *associated with $\{r, r'\}$ is the robot whose symmetric with respect to L is in P. If r is the leading robot, the* orientation *associated with $\{r, r'\}$ (or with the quasi-axis L) is the orientation of the ring such that $r = \mathbf{rot}(d)(r')$ (we also have $\mathbf{rot}(1)(r) = \bar{r'} \notin P$). We call the orientation such that $r' = \mathbf{rot}(d)(r)$ the* orientation opposite *to $\{r, r'\}$ (or with the quasi-axis L).*

Similarly we say that a configuration is r-*quasi-periodic* if, by moving r to an adjacent node, the configuration becomes periodic.

Let C be a possible initial configuration. The idea of the algorithm is to move the main robots towards v_{target}. If neither of the two main robots crashes, then the reached configuration remains symmetric and the main robots remain the ones to move. By contrast, if one of the two main robots crashes, then a quasi-node-edge-symmetric configuration C' is reached. Assume without loss of generality that C' is $\{r, r'\}$-quasi-node-edge symmetric with r as the leading robot. The strategy in this case is to move all the robots to their adjacent nodes with respect to the orientation opposite to $\{r, r'\}$. By doing so, the new configuration C'' is not only node-edge symmetric but also equivalent to the configuration reached if both main robots have moved. By repeating this process, robots move and eventually reach v_{target}. Observe that in some cases, by moving the main robots, a periodic configuration can be reached. To avoid this, robots compute the configuration to be reached when the main robots move. If it is periodic, then in addition to the main robots, the secondary robots move as well. Their destination is their adjacent free node towards v_{target} if it exists, otherwise, they move in the opposite direction of v_{target}.

Each move of our algorithm is executed on a configuration if the latter satisfies the move's condition. To define the conditions and define formally our solution, we use the following predicates:

- $NE(k)$: is the set of node-edge symmetric configurations having k occupied nodes.
- $QNE(k)$: is the set of quasi-node-edge symmetric configurations having k occupied nodes. We further partition this set depending on the number of quasi-axes and their orientation:
  - $QNE_{(a,b)}(k) \subset QNE(k)$ contains a quasi-axes having a given orientation and b quasi-axes with the opposite orientation.
  - $QNE_{(*,b)}(k) = \cup_{a \geq 0} QNE_{(a,b)}(k)$.
- $L5 \subset NE(5)$ (last five): is the set of node-edge symmetric configurations consisting of a block of size five.
- $L4 \subset NE(4)$ (last four): is the set of node-edge symmetric configurations consisting of four occupied nodes in blocks of size two separated by one empty node.
- $L4' \subset QNE(4)$ (last four prime): is the set of configurations consisting of four occupied nodes in a block of size three, followed by an empty node and a single occupied node.
- $L3 \subset NE(3)$ (last three): is the set of node-edge symmetric configurations consisting of three occupied nodes, two at distance 4 from each-other and one in the middle.
- $L3' \subset QNE(3)$ (last three prime): is the set of configurations consisting of two adjacent occupied nodes, followed by an empty node and a single occupied node.
- $L3'' \subset NE(3)$ (last three prime-second): is the set of configurations consisting of a block of size three.
- $L2 \subset NE(2)$ (last two): is the set of node-edge symmetric configurations consisting of two adjacent occupied nodes, and one of those two nodes only hosts crashed robots.
- P: is the set of periodic configurations.

We define three possible moves when the configuration is node-edge symmetric:

- M_{main}: the main robots are ordered to move towards the target node.
- $M_{secondary}$: the secondary robots move towards the target node if it is empty, or away from the target node otherwise.
- M_{L2}: all the robots move towards the other occupied node.

And two moves when the configuration is not node-edge symmetric, to recover after a crash:

- M_{same}: all the robots move in orientation opposite to the quasi-axis.
- M_{opp}: the two potentially crashed robots are ordered to move towards their respective target node.

The formal description of our solution is given in Algorithm 1. In the following sections, we show that, when executing Algorithm 1 starting from a configuration in $NE(k) \setminus P$ ($k > 3$), the gathering is achieved by the configuration transitions represented in Fig. 4, regardless of the presence or absence of crashed robots.

Let $\mathcal{C}$ be the set of configurations that are reachable by our algorithm. We write $C \rightsquigarrow C'$ to denote that C' is obtained from $C \in \mathcal{C}$ after one round of executing Algorithm 1. Similarly $C \overset{*}{\rightsquigarrow} C'$ denotes that C' is obtained from $C \in \mathcal{C}$ after a finite number of rounds executing Algorithm 1. To study different cases, we also write $C \overset{\text{no crash}}{\rightsquigarrow} C'$ (resp. $C \overset{\text{main crash}}{\rightsquigarrow} C'$, or $C \overset{\text{sec. crash}}{\rightsquigarrow} C'$) when each robot ordered to move reaches its destination (resp. one of the main robots is crashed, or one of the secondary robots is crashed).

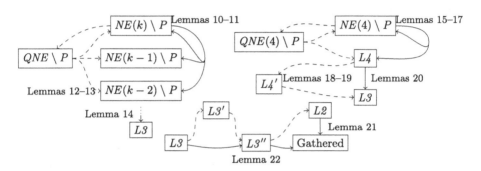

Fig. 4. Overview of the transitions between configurations. Dashed lines represent transitions that occur only when a crashed robot is ordered to move. Loops are shown to occur only a finite number of times, see the corresponding lemmas for more details.

Algorithm 1: SUIG algorithm on ring-shaped network

---
1 **if** $C \in L2$ **then**
2 | $\mathtt{M}_{L2}$
3 **else if** $C \in NE(k)$ **then**
4 | **if** $k \in \{4, 5\}$, $(n, k) \neq (7, 4)$, $C \notin L4 \cup L5$, and the main robots are adjacent to v_{target} **then**
5 | | $\mathtt{M}_{\text{secondary}}$ (See Fig. 5)
6 | **else if** Moving the main robots toward the target node does not create periodic configuration **then**
7 | | $\mathtt{M}_{\text{main}}$ (See Fig. 6)
8 | **else**
9 | | $\mathtt{M}_{\text{main}}$ and $\mathtt{M}_{\text{secondary}}$ (See Figs. 7 and 8)
10 **else if** $C \in QNE_{(*,0)}$ **then**
11 | $\mathtt{M}_{\text{same}}$ (See Fig. 9)
12 **else if** $C \in QNE_{(1,1)}$ **then**
13 | $\mathtt{M}_{\text{opp}}$ (See Fig. 10)

4.1 Overview of the Proof of Correctness

To prove the correctness of our algorithm, we first show that from a configuration with $k > 4$ occupied nodes, we reach in finite time a configuration with $k - 1$ or $k - 2$ occupied nodes (Subsect. 4.3). Repeating the argument, we eventually reach a configuration with either four or three occupied locations. Then, from three occupied locations, we show that after at most four rounds, the gathering

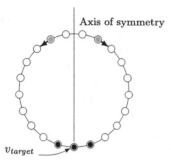

Fig. 5. Only the secondary robots are ordered to move towards the target node.

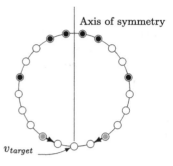

Fig. 6. The main robots are ordered to move towards the target node.

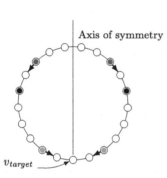

Fig. 7. The main robots and the secondary robots (to avoid reaching a periodic configuration) are ordered to move towards the target node.

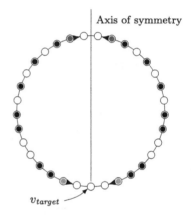

Fig. 8. The main robots are ordered to move towards the target node and, to avoid reaching a periodic configuration, the secondary robots move away from it (it is their only empty adjacent node).

is complete (Subsect. 4.5). When starting from four occupied nodes, the proof is different, but the result is the same (Subsect. 4.4).

During the execution, when the configuration C is node-edge symmetric, either 2 or 4 robots are ordered to move. If the moving robots are not crashed, the configuration remains node-edge-symmetric and the main robots eventually reach the target node. When one of these robots is crashed, the obtained configuration C' may not be node-edge symmetric anymore. In fact, this is the easiest case to consider because it implies that the configuration is quasi-node-edge symmetric, so all the robots can somehow realize that a crash occurred, and the algorithm can recover. Indeed, if all quasi-axes have the same orientation, then by moving all robots in the opposite orientation of that of the quasi-axis, the obtained configuration C'' is similar to the configuration obtained by moving the crashed robot towards its destination. So, the obtained configuration C'' is similar to the configuration obtained when no robot is crashed, i.e., $C \xrightarrow{\text{no crash}} C''$. If C' contains two quasi-axes with different orientation, then executing the move M_{opp} ensures that the obtained configuration is node-edge symmetric with a new axis of symmetry such that the main robots are correct. The last remaining case is when, after a crash, the configuration remains node-edge symmetric, so the robots do not realize that a crash occurred. In this case, we prove that the main robots in C' are correct, and we can apply a previous case. Due to space limitation, some proofs of the following sections are omitted but can be found in the companion technical report [5].

4.2 Geometric Results

Lemma 5. *Let C and C' be two periodic configurations with k occupied positions, and where more than $n/3$ consecutive nodes have the same state (occupied or empty) in both configurations, then $C = C'$.*

Lemma 6. *If a configuration C has two axes of symmetry, then the unique bisector line of these axes passing through a node is also an axis of symmetry.*

We know that if a secondary robot that is ordered to move is crashed, the resulting configuration is quasi-node-edge-symmetric and quasi-periodic. The next lemma states that in this case there exists no other quasi-axis of symmetry.

Lemma 7. *If a configuration with more than 3 occupied nodes has a $\{r_1, r_1'\}$-quasi-axis of symmetry with even gap distance, and is r_1-quasi-periodic, then it does not have another quasi-axis (regardless of its parity).*

Lemma 8. *Let $C \in NE(k) \setminus P$ with $k > 3$ occupied nodes, and $C \rightsquigarrow C'$, then $C' \notin P$, even if there exists a crashed robot.*

Lemma 9. *Let $C \in QNE(k) \setminus NE(k) \setminus P$ with $k > 3$. If all quasi-axes of C have odd gap distances, then, either*

– *all quasi-axes have the same orientation, or*
– *there are exactly two quasi-axes with opposite orientation.*

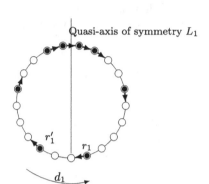

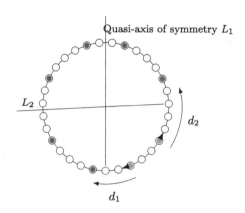

Fig. 9. When all the quasi-axis have the same orientation. All the robots are ordered to move (one of them is crashed).

Fig. 10. When there are two quasi-axes with opposite orientations. The two potentially crashed robots (orange robots) are ordered to move (one of them is crashed). (Color figure online)

4.3   From $NE(k)$, $k > 4$, to $L3$

In this section, first we show in Lemma 10, when no moving robot is crashed, the configuration remains node-edge-symmetric. In Lemma 11, we prove in this case that eventually the number of occupied nodes decreases. Then, we show that, when a crash occurs among the secondary robots, we recover in one round the same configuration that we would have reached if no crash occurred (Lemma 12). Finally, we show that, when a crash occurs among the main robots, we eventually reach a configuration where the main robots are correct (Lemma 13).

Lemma 10. *Let $C \in NE(k) \setminus P$, with $k > 4$ robots, and $C \rightsquigarrow C'$. If all moving robots reach their destination (no robot ordered to move is crashed), then $C' \in NE(k') \setminus P$, with $k' = k$, $k' = k - 1$, or $k' = k - 2$. Also, if $k' \leq 4$, then $C \in L5$ and $C' \in L3$.*

Proof. Clearly, the obtained configuration C' remains node-edge-symmetric, because the main (symmetric) robots are ordered to move towards the same target node located on the axis of symmetry, and the secondary (symmetric) robots might be ordered to move towards (or away) the target node. In the following, we show that C' is not periodic.

If only the main robots are ordered to move in C, then, from the condition of Line 6, we know that C' is not periodic. If they reach the target node, then $k' = k - 1$ or $k' = k - 2$ (depending whether the target node is occupied or not in C), otherwise $k' = k$. In the case $C \in L5$ ( ), only the main robots are ordered to move, and we obtain $C' \in L3$ ( ). If $k' \leq 4$, then it means the main robots have reached v_{target} (it is the only node where a multiplicity can be created). If v_{target} is empty in C, then $k' = k - 1$, but

k is even (C is node-edge-symmetric with no robot on the axis) and so $k \geq 6$, which implies $k' > 4$. So if $k' \leq 4$, it means v_{target} is occupied in C, $k = 5$ and $k' = k - 2 = 3$. When $k = 5$, the main robots are ordered to move towards the target only when the configuration is $L5$.

If only the secondary robots move in C, from the condition of Line 4, it means that either there are 4 occupied nodes (so C' is not periodic because n is odd) or there are 5 occupied nodes but 3 of them are adjacent, so C' is not periodic, and $k' = k$.

Otherwise, the main and the secondary robots move in C. In this case, assume by contradiction that C' is periodic. Since the secondary robots are ordered to move in C, then we know that C' is periodic if they stay idle, so C' is periodic and the configuration C'' obtained from C' by moving the two secondary robots to their original position is also periodic. Since C' is periodic with period at least 3 (because n is odd), then there are at least 6 occupied positions. By moving 2 robots to obtain C'', 4 occupied positions remain unchanged. Since $4 \geq 6/2$, by Lemma 5, we have the two configurations, C' and C'', are equal, which is a contradiction, so C' is not periodic. □

Lemma 11. *Let $C \in NE(k) \setminus P$ with $k > 4$ robots. If no moving robot crashes, then there exists $C' \in NE(k') \setminus P$, $C \overset{*}{\rightsquigarrow} C'$, with $k' = k - 1$ or $k' = k - 2$. Also, if $k' \leq 4$, then $C \in L5$ and $C' \in L3$.*

Lemma 12. *Let $C \in NE(k) \setminus P$ with $k > 4$ robots, and $C \rightsquigarrow C'$. If a secondary robot is crashed in C, then $C' \rightsquigarrow C''$ such that $C \overset{no \ crash}{\rightsquigarrow} C''$.*

Proof. When secondary robots are supposed to move and a secondary robot crashes in C, then the obtained configuration C' is quasi-periodic, and has a single quasi-axis of symmetry (Lemma 7). So it cannot be node-edge-symmetric (otherwise it would have another quasi-axis). So, after executing the algorithm for one more round, all the robots are ordered to move in the opposite orientation with respect to the quasi-axis. All the robots move except the crashed robot by applying M_{same}, so the obtained configuration C'' is similar to the configuration obtained by moving the crashed robot towards the target node, i.e., C'' is the configuration obtained from C if no robot is crashed. □

In the next Lemma, we show that, if a main robot is crashed, three cases can occur. Each case implies that the algorithm makes progress. The last two cases are easy to consider: (b) when the obtained configuration remains node-edge symmetric but the new main robots are correct, and (c) when it is not node-edge-symmetric, so it is quasi-node-edge symmetric and the robots detect that a crash occurred and recover (after one more round the configuration is the same as is no crash occurred). In the first case (a) however, we can only say that every time the crash of a main robot creates a new node-edge symmetric configuration where the crashed robot remains one of the main robots, the distance between two given robots decreases. This implies that this case can only occur a finite number of times. When the crashed robot is a main robot, then there are only two possible choice for the other main robot. If k is even, the other main robot is one of the two closest robots in both directions. If k is odd, the other main robot is one of the two second closest robots in both directions. In this Lemma, we denote by $M(C)$ the set of the two possible main robots defined above.

Lemma 13. *Let $C \in NE(k) \setminus P$ with $k > 4$ robots, and $C \rightsquigarrow C'$. If a main robot is crashed in C, then either (a) $C' \in NE(k) \setminus P$ and the distance between the two robots in $M(C)$ decreases, (b) $C' \in NE(k) \setminus P$ and the main robots are correct, or (c) $C' \overset{no\ crash}{\rightsquigarrow} C''$ such that either $C \overset{no\ crash}{\rightsquigarrow} C''$ or the main robots are correct in C''.*

Proof. As in Lemma 12, if the configuration C' is not node-edge symmetric, then either all the quasi-axes have the same orientation and move $\mathsf{M_{same}}$ is executed, or there are two quasi-axes with opposite orientation and move $\mathsf{M_{opp}}$ is executed. If move $\mathsf{M_{same}}$ is executed, the obtained configuration C'' is similar to the configuration obtained from C if no robot is crashed. If move $\mathsf{M_{opp}}$ is executed, the obtained configuration C'' has a new axis of symmetry where the main robots are correct. So, the case (c) holds.

If C' remains node-edge symmetric, then either the new main robots are correct (case (b)), or the correct main robot is one hop closer to the crashed robot (case (a)), hence to the other robot in $M(C)$ as well, so the distance between the two robots in $M(C)$ has decreased.   □

From Lemmas 10–13, we obtain the following results.

Lemma 14. *When executing Algorithm 1 from $C \in NE(k) \setminus P$ with $k > 4$ then, after a finite number of rounds, we obtain configuration $C' \in NE(k') \setminus P$, $k' = k - 1$ or $k' = k - 2$ even if there is a crashed robot. Also, if $k' \leq 4$, then $C \in L5$ and $C' \in L3$.*

4.4   From $NE(4)$ to $L3$

Starting from a configuration in $NE(4)$, we obtain a similar results: we prove that if no moving robot crashes, we reach configuration $L3$ (Lemmas 15, 16 and 20), and then we deal with cases where a crash occurs (Lemmas 17–19 and 20). However, here, we have one more case as the secondary robots might move not only to avoid a periodic configuration, but also to ensure that the only configuration reached in $NE(3)$ is $L3$. The case $n = 5, 7$ can be done by simple case analysis, so assume in this subsection that $n > 7$.

Lemma 15. *When executing Algorithm 1 from $C \in NE(4) \setminus P$ where the main robots are at distance more than 2, if the main robots do not crash, then, after a finite number of rounds, we obtain configuration $C' \in NE(4) \setminus P$, where the main robots are at distance 2.*

Lemma 16. *When executing Algorithm 1 from $C \in NE(4) \setminus P$, if the secondary robots are ordered to move and do not crash, then, after a finite number of rounds, we obtain configuration $C' \in L4$.*

Lemma 17. *When executing Algorithm 1 from $C \in NE(4) \setminus P$ where the main robots are at distance more than 2, if a main robot crashes, then, after a finite number of rounds, we obtain either a configuration in $L4$ or a configuration in $NE(4) \setminus P$ where the axis of symmetry changed from C.*

Proof. By the definition of C, the main robots are ordered to move. When a main robot crashes, either the obtained configuration C' is node-edge-symmetric or not. If C' is not node-edge symmetric, so is quasi-symmetric, and by applying M_{same} or M_{opp}, the obtained configuration C'' is in $NE(4) \setminus P$ where the distance between main robots decreased by 2. In the former case, if C' is node-edge-symmetric, then either the new pair of main robots is correct (not crashed) and the Lemma is proved, or the new pair still contain the crashed robot. From there, we apply the same reasoning and either we reach $L4$ ( ) or again the pair of main robots changes. If this occurs the crashed robot cannot be in the pair of main robots again, so the Lemma is proved. ☐

Lemma 18. *When executing Algorithm 1 from $C \in NE(4) \setminus P$ where the secondary robots are ordered to move, if the secondary robots are not adjacent, let C' be a configuration such that $C \xrightarrow{sec.\ crash} C'$. Then, $C' \rightsquigarrow C''$ such that $C \xrightarrow{no\ crash} C''$.*

Proof. Because the secondary robots are ordered to move in C, $k = 4$ and n is odd, $C \notin L4$ and the main robots are adjacent to v_{target}. Then the secondary robots are not adjacent to the main robots, and move towards v_{target}. Thus, in C'', the distance between secondary robots increased by 2 and between a secondary robot and a main robot decreased by 1 from C.

When a secondary robot crashes, C' is not node-edge symmetric (because the distance between the secondary robots is strictly greater than 2 as they were not adjacent in C) and is quasi-node-edge-symmetric and, after executing M_{same} or M_{opp}, the configuration is in $NE(4) \setminus P$ where the distance between secondary robots increased by 2 and the distance between a secondary robot and a main robot decreased by 1 from C, that is, it is C''.

Thus, the Lemma holds. ☐

Lemma 19. *When executing Algorithm 1 from $C \in NE(4) \setminus P$ where the secondary robots are ordered to move to obtain C', if the secondary robots are adjacent and one of them crashes, then, C' is in $NE(4) \setminus P$ where the main robots are correct and the secondary robots are not adjacent.*

Proof. If the secondary robots are ordered to move, this means that the main robots are adjacent to the target node. If the secondary robots are adjacent and one of them is crashed, then, after one round we obtain node-edge symmetric configuration C' where the two previous secondary robots are symmetric to the two previous main robots. In C', the new two secondary robots consist of a previous main robot and a previous secondary robot. Hence they are adjacent if the number of nodes is exactly 7. If there are 7 (or even 5) nodes, the secondary robots are not ordered to move as no periodic configuration can be created (condition at Line 4 does not apply). ☐

Lemma 20. *When executing Algorithm 1 from $C \in L4$, we reach a configuration in L3 in at most two rounds.*

4.5   From *L3* to the Gathered Configuration

Lemma 21. *When executing Algorithm 1 starting from a configuration in L2 and one occupied location only hosts crashed robots, a gathered configuration is reached in one round.*

Lemma 22. *When executing Algorithm 1 starting from a configuration in L3, a gathered configuration is reached in 4 rounds.*

From the previous Lemmas in Subsect. 4.3, 4.4 and 4.5, we obtain the following results.

Theorem 2. *Starting from a configuration $C \in NE(k) \setminus P$ (k > 3), Algorithm 1 solves the SUIG problem on ring-shaped networks without multiplicity detection in FSYNC.*

5   Concluding Remarks

We further characterized the solvability of the stand-up indulgent rendezvous and gathering problems on ring-shaped networks by Look-Compute-Move oblivious robots. A number of open questions are raised by our work:

- Is it possible to extend our possibility result for FSYNC SUIR in node-node symmetric configurations to general SUIG?
- Is it possible to solve FSYNC SUIG starting from a non-periodic, non-symmetric configuration?
- Aside from line-shaped networks (already studied by Bramas et al. [4]), is the problem solvable in other topologies?
- Can additional capabilities help the robots solve the problem in the cases we identified as impossible?

References

1. Agmon, N., Peleg, D.: Fault-tolerant gathering algorithms for autonomous mobile robots. SIAM J. Comput. **36**(1), 56–82 (2006)
2. Balabonski, T., Courtieu, P., Pelle, R., Rieg, L., Tixeuil, S., Urbain, X.: Continuous vs. discrete asynchronous moves: a certified approach for mobile robots. In: Atig, M.F., Schwarzmann, A.A. (eds.) NETYS 2019. LNCS, vol. 11704, pp. 93–109. Springer, Cham (2019). https://doi.org/10.1007/978-3-030-31277-0_7
3. Bouzid, Z., Das, S., Tixeuil, S.: Gathering of mobile robots tolerating multiple crash faults. In: IEEE 33rd International Conference on Distributed Computing Systems (ICDCS), pp. 337–346 (2013)
4. Bramas, Q., Kamei, S., Lamani, A., Tixeuil, S.: Stand-up indulgent gathering on lines. In: Dolev, S., Schieber, B. (eds.) SSS 2023. LNCS, vol. 14310, pp. 451–465. Springer, Cham (2023). https://doi.org/10.1007/978-3-031-44274-2_34

5. Bramas, Q., Kamei, S., Lamani, A., Tixeuil, S.: Stand-up indulgent gathering on rings (2024). arXiv:2402.14233

6. Bramas, Q., Lamani, A., Tixeuil, S.: Stand up indulgent rendezvous. In: Devismes, S., Mittal, N. (eds.) SSS 2020. LNCS, vol. 12514, pp. 45–59. Springer, Cham (2020). https://doi.org/10.1007/978-3-030-64348-5_4

7. Bramas, Q., Lamani, A., Tixeuil, S.: Stand up indulgent gathering. In: Gąsieniec, L., Klasing, R., Radzik, T. (eds.) ALGOSENSORS 2021. LNCS, vol. 12961, pp. 17–28. Springer, Cham (2021). https://doi.org/10.1007/978-3-030-89240-1_2

8. Bramas, Q., Lamani, A., Tixeuil, S.: Stand up indulgent gathering. Theor. Comput. Sci. **939**, 63–77 (2023). https://doi.org/10.1016/j.tcs.2022.10.015

9. Bramas, Q., Tixeuil, S.: Wait-free gathering without chirality. In: Scheideler, C. (ed.) SIROCCO 2014. LNCS, vol. 9439, pp. 313–327. Springer, Cham (2015). https://doi.org/10.1007/978-3-319-25258-2_22

10. Défago, X., Potop-Butucaru, M., Raipin-Parvédy, P.: Self-stabilizing gathering of mobile robots under crash or byzantine faults. Distrib. Comput. **33**, 393–421 (2020)

11. Flocchini, P., Prencipe, G., Santoro, N. (eds.): Distributed Computing by Mobile Entities, Current Research in Moving and Computing. LNCS, vol. 11340. Springer, Cham (2019). https://doi.org/10.1007/978-3-030-11072-7

12. Klasing, R., Markou, E., Pelc, A.: Gathering asynchronous oblivious mobile robots in a ring. Theoret. Comput. Sci. **390**(1), 27–39 (2008)

13. Suzuki, I., Yamashita, M.: Distributed anonymous mobile robots: formation of geometric patterns. SIAM J. Comput. **28**(4), 1347–1363 (1999)

In Search of the Lost Tree

Hardness and Relaxation of Spanning Trees in Temporal Graphs

Arnaud Casteigts[1] and Timothée Corsini[2]

[1] Department of CS, University of Geneva, Geneva, Switzerland
`arnaud.casteigts@unige.ch`
[2] LaBRI, CNRS, University of Bordeaux, Bordeaux INP, Bordeaux, France
`timothee.corsini@labri.fr`

Abstract. A graph whose edges only appear at certain points in time is called a temporal graph (among other names). These graphs are temporally connected if all pairs of vertices are connected by a path that traverses edges in chronological order (a temporal path). Reachability in these graphs departs from standard reachability; in particular, it is not transitive, with structural and algorithmic consequences. For instance, temporally connected graphs do not always admit spanning trees, i.e., subsets of edges that form a tree and preserve temporal connectivity among the nodes.

In this paper, we revisit fundamental questions about the loss of universality of spanning trees. To start, we show that deciding if a spanning tree exists in a given temporal graph is NP-complete. What could be appropriate replacement for the concept? Beyond having minimum size, spanning trees enjoy the feature of enabling reachability along the same underlying paths in both directions, a pretty uncommon feature in temporal graphs. We explore relaxations in this direction and show that testing the existence of bidirectional spanning structures (bi-spanners) is tractable in general. On the down side, finding *minimum* such structures is NP-hard even in simple temporal graphs. Still, the fact that bidirectionality can be tested efficiently may find applications, e.g. for routing and security, and the corresponding primitive that we introduce in the algorithm may be of independent interest.

Keywords: Temporal graphs · Temporal spanner · Spanning trees · Bidirectional paths

1 Introduction

Temporal graphs have gained attention as appropriate models to capture time-varying phenomena, ranging from dynamic networks (networks whose structure changes over the time) to dynamic interactions over static (or dynamic) networks. They have found applications in fields as various as transportation, social networks, biology, robotics, scheduling, and distributed computing. In its simplest version, a temporal graph can be represented as a labeled graph $\mathcal{G} = (V, E, \lambda)$ where V is a finite set of n vertices,

Supported by the French ANR, project ANR-22-CE48-0001 (TEMPOGRAL).

Y. Emek (Ed.): SIROCCO 2024, LNCS 14662, pp. 138–155, 2024.
https://doi.org/10.1007/978-3-031-60603-8_8

$E \subseteq V \times V$ is a finite set of directed or undirected edges (undirected, in this paper), and $\lambda : E \to 2^{\mathbb{N}}$ is a function that assigns one or several time labels to the edge of E, interpreted as presence times. More complex definitions exist, yet many basic features of temporal graphs are already captured in this simple model.

Reachability in temporal graphs is commonly defined in terms of *temporal paths*; i.e., paths which traverse the edges in chronological order. Reachability is not symmetrical: the fact that a node u can reach another node v does not imply that v can reach u (shorthand $u \rightsquigarrow v \not\Rightarrow v \rightsquigarrow u$). It is also not transitive ($u \rightsquigarrow v \wedge v \rightsquigarrow w \not\Rightarrow u \rightsquigarrow w$), with both structural and computational consequences.

Over the past two decades, a growing body of work has been devoted to better understanding temporal reachability, considered from various perspectives, e.g. k-connectivity and separators [19,25,28], connected components [2,4,6,29], feasibility of distributed tasks [3,9,15,26], schedule design [11,12], data structures [10,13,29,31], mitigation [21], shortest paths [13,16], enumeration [22], stochastic models [5,18], flows [1,30], and exploration [20,23,24,27].

One of the first questions, mentioned in the seminal work of Kempe, Kleinberg, and Kumar [28], concerns the existence of sparse spanning subgraphs, i.e., *temporal spanners*, defined as subgraphs of the input temporal graph that preserve temporal connectivity using as few edges as possible (one does not focus on distance preservation, as small spanners are not even guaranteed). Kempe *et al.* observe that spanning trees do not always exist in temporal graphs, and more generally, there exist temporally connected graphs with $\Theta(n \log n)$ edges that cannot be sparsified at all. Fifteen years later, Axiotis and Fotakis [4] establish a much stronger negative result, showing that there exist temporally connected graphs of size $\Theta(n^2)$ that cannot be sparsified. In some respects, this result killed the hope to define analogues of spanning trees in temporal graphs. Subsequent research focused on positive results for special cases. For example, if the input graph is a complete graph, then spanners with $O(n \log n)$ edges always exists [17]. If an Erdös-Rényi graph of parameters n and p is augmented with random time labels, then a nearly optimal spanner with $2n + o(n)$ edges exists asymptotically almost surely, as soon as the graph becomes temporally connected [18].

On the algorithmic side, Axiotis and Fotakis [4] and Akrida, Gąsieniec, Mertzios, and Spirakis [2] independently showed that minimizing the size temporal spanners is APX-hard. Special types of spanners were also investigated more recently, such as spanners with low stretch [8] and fault-tolerant spanners [7].

1.1 Contributions

In this article, we revisit one of the motivations of [28], questioning the (in)existence of spanning trees in a temporal graph, defined as a subset of edges that forms a tree and whose labels (inherited from the original graph) preserve temporal connectivity. Since these structures are not universal, a natural question is how hard it is to decide whether a spanning tree exists in a given temporal graph.

It is somewhat surprising that the answer to this question is still unknown more than two decades after the seminal work of Kempe *et al.* [28]. Our first result is to fill this gap by showing that deciding even such simple property is NP-complete. It is actually hard in both the strict and the non-strict setting, i.e., whether the labels along temporal paths must increase or only be non-decreasing. Interestingly, for many questions, these two settings turn out to be incomparable, which often creates some confusion in the temporal graph literature [14]. Here, we unify both settings by showing that the problem is difficult in the class of *proper* temporal graphs, where the distinction between strict and non-strict temporal paths vanishes. (We suggest to regard this type of reductions as a good practice, whenever it can be achieved, to make the results more general.)

Motivated by this negative result and the others mentioned above, we investigate new ways to relax spanning trees in temporal graphs. Spanning trees are not just optimal in size; they also guarantee that any two vertices can reach each other using the same underlying path, a pretty uncommon feature in temporal reachability. Relaxing trees in this direction, we investigate the concept of a bidirectional spanner (*bi-spanner*), i.e., one that preserves at least one bidirectional path among every pair of vertices when such paths exist.

We obtain both positive and negative results. On the positive side, we show that bi-spanners can be decided (and computed) in polynomial time. The algorithm relies on a more basic primitive, that finds a bidirectional path between a given pair of vertices (a *bi-path*). Then, the union of these paths over all pairs of vertices forms a bi-spanner. Taking the union here may seem pretty suboptimal in terms of size; however, we show that in some contrived cases, no better bi-spanners exist. Nonetheless, trying to mini-mize the size of such a spanner in general is a legitimate goal. We show that, unfortu-nately, this problem is NP-hard. It is even arguably harder than computing a spanning tree, since our reduction operates on *simple* temporal graphs (graphs whose edges have a unique label), a class where spanning trees can be decided efficiently.

Along these results, we make a number of structural observations related to tempo-ral spanning trees, temporal spanners, and bidirectional connectivity, which may be of independent interest. We also think that the bi-path primitive we introduce could find application in routing and security, as it allows two distant nodes to rely on a smaller number of intermediate nodes for communication.

1.2   Organization of the Paper

The paper is organized as follows. In Sect. 2, we define the main concepts and ques-tions investigated in the paper. In particular, we formulate the problems of TEMPORAL SPANNING TREE, BI-SPANNER, and k-BI-SPANNER. In Sect. 3, we characterize the landscape of tractability for TEMPORAL SPANNING TREE, showing that this problem is NP-complete (yet, tractable in very specific cases). In Sect. 4, we present an algo-rithm that solves BI-SPANNER for unrestricted size in polynomial time. Then, we show in Sect. 5 that k-BI-SPANNER is NP-complete even in simple temporal graphs (where TEMPORAL SPANNING TREE was tractable). We also briefly mention some research directions towards restrictions that could help the existence and the tractability of small bidirectional spanners. Section 6 concludes the paper with further remarks.

2   Definitions

Given a temporal graph $\mathcal{G} = (V, E, \lambda)$ defined as above, the static graph (V, E), undi-rected in this paper, is called the *footprint* of $\mathcal{G}$. We denote by $N(v) = \{u \in V \mid uv \in$

$E\}$ the neighbors of v in the footprint and $\Delta = \max_{v \in V}\{|N(v)|\}$ the maximum degree in the footprint. The terms of nodes and vertices are used interchangeably. The static graph $G_t = (V, E_t)$ where $E_t = \{e \in E \mid t \in \lambda(e)\}$ is the *snapshot* of $\mathcal{G}$ at time t. A pair (e, t) such that $e \in E$ and $t \in \lambda(e)$ is a *contact* in $\mathcal{G}$ (also called a temporal edge). The range of λ is the *lifetime* of $\mathcal{G}$, noted $\mathcal{T}$, of length τ. A *temporal path* (or journey) is a sequence of contacts $\langle (e_i, t_i) \rangle$ such that $\langle e_i \rangle$ is a path in G and $\langle t_i \rangle$ is non-decreasing (or increasing, for *strict* temporal paths). A graph $\mathcal{G}$ is *temporally connected* if there exists at least one temporal path between every ordered pair of vertices.

A temporal graph $\mathcal{G}' = (V', E', \lambda')$ is a *temporal subgraph* of $\mathcal{G}$, noted $\mathcal{G}' \subseteq \mathcal{G}$, if (V', E') is a subgraph of (V, E), $\lambda' \subseteq \lambda$, and λ' is restricted to E'. If $\mathcal{G}'$ is temporally connected, then it is a *temporal spanner* of $\mathcal{G}$. As explained above, temporal reachability is not transitive in general. However, it remains possible to compose two temporal paths when the first finishes before the second starts. A *pivot* in $\mathcal{G}$ is a vertex that every other node can reach *before* some time t, and that can reach all the other vertices after t. The existence of a pivot implies that of a temporal spanner of linear size. The converse does not hold, there exists linear-size spanners that do not rely on pivots.

Representation in Memory

There are several ways to represent a temporal graph in memory. One option is a sequence of snapshots, each specified as an adjacency matrix or adjacency list. A second option is as a list of triplets of the form (u, v, t). A third option is as a time-augmented adjacency list, i.e., an adjacency list whose nested entries contain a list of times. In the analysis of our bi-path algorithm, we assume that, when iterating over the neighbors of a given vertex, finding whether a label exists within a certain time range can be done in time $O(\tau)$, which is the case, e.g., with adjacency lists (this could even be done in time $O(\log \tau)$ in this case) or with sequences of adjacency matrices. For most of our other results, the representation does not matter, as one can convert them to one another in polynomial time.

2.1   Temporal Spanning Trees

A *temporal spanning tree* of $\mathcal{G}$ is a temporal spanner of $\mathcal{G}$ whose footprint is a tree. Clearly, such trees are not universal in temporally connected graphs, as illustrated by the graph in Fig. 1, whose underlying spanning trees all induce temporally disconnected graphs. Here, a temporal spanning tree would exist if, for example, the two edges labeled 1 had an additional label of value 3. We consider the following fundamental question:

Problem 1. TEMPORAL SPANNING TREE (TST)
Input: A temporal graph $\mathcal{G}$
Question: Does $\mathcal{G}$ admit a temporal spanning tree?

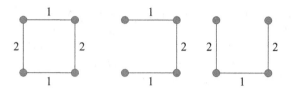

Fig. 1. A temporal graph that is temporally connected (left), in which all temporal subgraphs whose footprint is a tree are *not* temporally connected (middle and right, up to isomorphism).

2.2   Bidirectional Temporal Spanners

A bidirectional temporal path (bi-path, for short) between two vertices u and v is a couple (p_1, p_2) such that p_1 is a temporal path from u to v and p_2 is a temporal path from v to u that uses the same underlying path (reversed). A bidirectional temporal spanner (bi-spanner) is a spanner where all pairs of vertices can reach each other through a bi-path. We consider the following questions:

Problem 2. Bi-Spanner
Input: A temporal graph $\mathcal{G}$
Question: Does $\mathcal{G}$ admit a bidirectional temporal spanner?

Problem 3. k-Bi-Spanner
Input: A temporal graph $\mathcal{G}$, an integer k
Question: Does $\mathcal{G}$ admit a bidirectional temporal spanner with at most k edges?

Temporal spanning trees are particular cases of bi-spanners, as they require bi-paths between all pairs of vertices. However, this is not sufficient, as shown in Fig. 2, where all pairs can reach each other using bi-paths, but these paths cannot be mutualized in the form of a spanning tree. We will use this graph as a gadget in one of our reductions. Observe that this graph also has a pivot d.

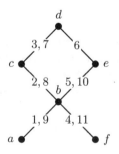

Fig. 2. A temporally connected graph with pivot vertex d and bidirectional paths between all vertices, but no temporal spanning tree.

2.3 Simple, Strict, Proper, Happy: Generality of Results

Temporal graphs can be restricted in many ways. Natural restrictions include the fact of being *simple*: every edge has a single presence time (λ is single-valued); and the fact of being *proper*: adjacent edges never have a label in common (i.e., every snapshot is a collection of matchings and isolated vertices). Temporal graphs that are both simple and proper are called *happy*. In addition, one may or may not impose that the temporal path are *strict* (already discussed).

A convenient fact about properness is that it makes the distinction between strict and non-strict vanish. Our main negative result establishes that TEMPORAL SPANNING TREE is NP-complete even in *proper* temporal graphs, thus this is true in both the strict and non-strict settings. Observe that TEMPORAL SPANNING TREE corresponds to k-BI-SPANNER for the particular case that $k = n - 1$, so we also get by the same result that k-BI-SPANNER is NP-complete. However, a stronger result is established in the paper, by showing that k-BI-SPANNER is also NP-complete in *simple* temporal graphs (where TEMPORAL SPANNING TREE is shown to be tractable). Since the number of edges and the number of time labels coincide in this class of temporal graphs, our result establishes that minimizing the number of labels instead of the number of edges would also be hard, a benefit of targeting *simple* temporal graphs. We insist that targeting well-chosen classes of temporal graphs in the reduction allows one to obtain several (otherwise different) results at once, this should thus be regarded as a good practice whenever possible. In fact, if the problem were hard in *happy* temporal graphs, then a single reduction would suffice to produce all the above results. However, this is not the case here, as both problems are easy for happy graphs. The reader is referred to [14] for a study of how these various settings impact the expressivity of temporal graphs in terms of reachability.

To summarize, the tractability landscape of the problems considered in this paper is depicted in Fig. 3.

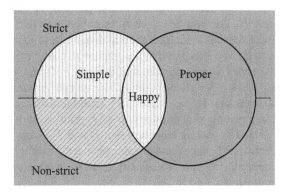

Fig. 3. Outline of the results. (*Plain red:* both TEMPORAL SPANNING TREE and k-BI-SPANNER are hard; *Vertical green:* both problems are tractable; and *Slanting blue:* only k-BI-SPANNER is hard. BI-SPANNER is always tractable.) (Color figure online)

3   TEMPORAL SPANNING TREES

In this section, we show that TEMPORAL SPANNING TREE is NP-complete in temporal graphs, whether the setting is strict or non-strict, through a reduction from SAT to this problem in *proper* temporal graphs. We also show, by simple arguments, that the problem is trivial to decide in the case of simple graphs.

3.1   NP-Completeness

The fact that TEMPORAL SPANNING TREE (TST) is in NP is straightforward: given an input graph $\mathcal{G}$ and a candidate subgraph $\mathcal{G}'$, it is easy to verify that (1) $\mathcal{G}' \subseteq \mathcal{G}$; (2) the footprint of $\mathcal{G}'$ is a tree; and (3) $\mathcal{G}'$ is temporally connected. The exact complexity of these steps may depend on the datastructure used for encoding the temporal graph. However, for all reasonable datastructures (including the ones discussed in Sect. 2), each step is clearly polynomial, using for example any of the algorithms from [13] for testing temporal connectivity.

We now prove that TST is NP-hard in *proper* temporal graphs (with the consequences discussed above), reducing from SAT.

Theorem 1. TEMPORAL SPANNING TREE is NP-hard in proper temporal graphs, even if every edge has only 2 time labels.

Proof. Let us start by discussing some properties of the temporal graph in Fig. 2, which will be used in the reduction. First, notice that the cycle in this graph is mandatory for temporal connectivity because if an edge on the right side is removed (de or be), then f can no longer reach d, while if an edge on the left side is removed, d can no longer reach a. Also recall that d is a pivot in this graph.

Let ϕ be a CNF formula with n variables and k clauses, we will transform ϕ into a proper temporal graph $\mathcal{G}$ such that $\mathcal{G}$ admits a temporal spanning tree if and only if ϕ is satisfiable. Without loss of generality, we assume that none of the clauses contain both a variable and its negation. For simplicity, we will use both negative and positive time labels, and possibly take some of them from the real domain. A suitable renormalization can be applied easily to convert this graph into a graph with positive time labels from $\mathbb{Z}$.

The graph $\mathcal{G}$ has a vertex x_i for each variable, a vertex c_j for each clause, and three special vertices (thus, $n + k + 3$ vertices in total). The special vertices are T (for True), F (for False), and B (for Base). These vertices play a role similar to c, e, and d in Fig. 2. In particular, B will be a pivot in $\mathcal{G}$. The edge of the footprint are built as follows. There is an edge between B and T, and an edge between B and F. Then, for each variable x_i, there is an edge between T and x_i and an edge between F and x_i. Finally, for each clause c_j where x_i appears (in plain or negated), there is an edge between x_i and c_j. The footprint of $\mathcal{G}$ is illustrated on Fig. 4a.

The time labels are as follows (see also Fig. 4b). Every edge is given two time labels, one positive (defined relative to a reference value $t^+ = n + k + 1$), and one negative (defined relative to $t^- = -t^+$). The time labels on BF are $-\epsilon$ and t^+ and those on BT are t^- and ϵ, with ϵ any positive value < 1. For each variable x_i, the labels on Tx_i are $t^- - i$ and i; the ones on Fx_i are $-i$ and $t^+ + i$. Finally, for each clause c_j where

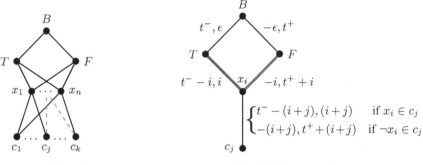

(a) Footprint of the gadgets.   (b) Labeling of a gadget.

Fig. 4. Reduction from SAT to TST.

x_i appears, we assign time labels to the edge $x_i c_j$ depending on whether x_i appears in plain or negated form in c_j, namely:

- If $x_i \in c_j$, the labels are $t^- - (i+j)$ and $(i+j)$. In this case, $x_i c_j$ is called a *positive edge*.
- If $\neg x_i \in c_j$, the labels are $-(i+j)$ and $t^+ + (i+j)$. In this case, $x_i c_j$ is called a *negative edge*.

The idea of the construction is that a spanning tree is equivalent to a valuation that satisfies ϕ. A spanning tree should break the cycles $TBFx_i$ for each variable x_i, we first show that the only edges that can be removed from such cycles are either Tx_i or Fx_i, for every i.

Claim 1. *Any temporal spanner of $\mathcal{G}$ contains both BT and BF in its footprint.*

Proof. Without BF, no journey exist from F to B. Indeed, for all x_i, the earliest time that F can reach T through x_i is i, which is larger than ϵ. Without BT, there is no journey from B to T, since for each x_i, the earliest time that B can reach x_i is $t^+ + i$, which is larger than i. □

Claim 1 implies that any temporal spanning tree of $\mathcal{G}$ requires the removal of either Tx_i or Fx_i for all i. Morally, this encodes the choice of an assignment for variable x_i in ϕ where Tx_i is kept if x_i is set to true and Fx_i is kept otherwise. Observe that there might exist some spanning trees without both Tx_i and Fx_i which means that the value of x_i does not impact the valuation of ϕ.

Next, we want to ensure that a clause is connected to B through a variable only if its valuation satisfies the clause.

Claim 2. *If $x_i c_j$ is a positive edge, then B does not share a bi-path with neither c_j nor x_i that crosses both F and the edge $x_i c_j$.*
If $x_i c_j$ is a negative edge, then B does not share a bi-path with neither c_j nor x_i that crosses both T and the edge $x_i c_j$.

Proof. – If x_ic_j is a positive edge, then no journey can be formed along vertices B, F, x_i, c_j, because the earliest time that B can reach x_i is $t^+ + i$, which is larger than $i + j$ since $t^+ > n + k$. Consider now that there is a journey J crossing the vertices $B, F, x_{i'}, c_j, x_i$, because of the previous case, $x_{i'}c_j$ is not a positive edge, so J will reach c_j at time $t^+ + (i' + j)$ which is greater than any label on x_ic_j, so J is not a journey.

– If x_ic_j is a negative edge, then no journey can be formed along vertices c_j, x_i, T, B, because the earliest time that c_j can reach T is i, which is greater than ϵ. Similarly, a journey J crossing vertices $x_i, c_j, x_{i'}, T, B$ cannot exists even if $c_jx_{i'}$ is positive as J would reach $x_{i'}$ at time $(i' + j)$ which is greater than the times on $Tx_{i'}$.   □

This claim means that there is no bi-path starting from B crossing both a negative edge and a positive edge. With these two claims, it is now possible to show that a spanning tree implies a valuation that satisfies ϕ.

Claim 3. *If $\mathcal{G}$ admits a temporal spanning tree, then ϕ is satisfiable.*

Proof. Suppose that $\mathcal{G}$ admits a temporal spanning tree $\mathcal{H}$, by Claim 1, $\mathcal{H}$ contains both BT and BF. Since $\mathcal{H}$ has no cycle, then for all x_i, $\mathcal{H}$ excludes at least one of Tx_i and Fx_i (possibly both, without consequence). By Claim 2, there is no bi-path from B crossing both positive and negative edges, which means that for each variable, only one valuation is kept, either through one of the edges Tx_i or Fx_i, or through another clause that was validated by a literal of the same sign. Thus, if $\mathcal{H}$ is a tree, then x_i can only have one value.

By construction, a positive edge is present in $\mathcal{G}$ if x_i is in c_j and a negative one if $\neg x_i$ is in c_j. Thus, c_j is in the tree only if one of its literals is true, and a temporal spanning tree $\mathcal{H}$ implies that the formula is satisfiable.   □

Knowing how to build a subgraph from a valuation of the formula, we show that such subgraph is indeed a temporal spanning tree through the next claim.

Claim 4. *If ϕ is satisfiable, then $\mathcal{G}$ admits a temporal spanning tree.*

Proof. Consider a satisfying assignment to ϕ. For each variable x, if x is set to true, remove Fx from $\mathcal{G}$ and all edges xc for each clause containing $\neg x$. If x is set to false, remove instead Tx and the links xc if c contains x in non-negated form. Finally for every clause c, pick one variable x such that xc is still in the graph and disconnect all other edges $x'c$ such that x is the only neighbor of c.

Remark that some variables may be disconnected from all clauses in such solution; however, this only means that removing the variable would still make the formula satisfiable.

Now, we need to prove that $\mathcal{H}$ is a temporal spanning tree.

1. The footprint of $\mathcal{H}$ is a tree: recall that the graph contains $|V| = n + k + 3$ vertices, where n is the number of variables and k the number of clauses.

   First, the edges BT and BF are kept. Moreover, for each variable, one chooses between True or False, keeping n edges, then, since the formula is True, we have that each clause is linked to exactly one of the variables, making k edges, for a total of $|V| - 1$ edges, the footprint of $\mathcal{H}$ is connected, so it must be a tree.

2. $\mathcal{H}$ is temporally connected: In order to prove that, we will prove that B is a pivot at time 0; in other words, we need to check that every vertex u may reach B at time ≤ 0 and be reached by B at time ≥ 0. If u is a clause vertex c_i, it can reach B by a journey along c_i, x_j, T or F (depending of the value of x_j), and finally B. If x_j is true, the journey will use times $t^- - (i + j)$, $t^- - j$, and finally time $t^- < 0$ from T to B. The backward journey from B to x_j will cross the same edges in reverse order at times ϵ, j, and $(i + j)$. If x_j is false, then starting from c_i towards B, the time labels will be $-(i + j)$, $-j$, and $-\epsilon$; the backward journey from B to x_j will use time labels $t^+, t^+ + j$, and $t^+ + i + j$. Now, any clause share a bidirectional journey with B, one towards it before time 0, and one from it after 0. Since all the of vertices are crossed by journeys from the clauses with respect to the time of the pivot, $\mathcal{H}$ is temporally connected. $\square$

Clearly, the size of $\mathcal{G}$ is polynomial in the size of ϕ. From Claims 3 and 4, ϕ is satisfiable if and only if $\mathcal{G}$ admits a temporal spanning tree. Furthermore, $\mathcal{G}$ is proper and each edge has 2 time labels, which concludes the proof. $\square$

3.2 Tractability in Simple Temporal Graphs

Here, we observe that temporal spanning trees do not exist, at all, in simple temporal graphs in the strict setting. In the non-strict setting, they may or may not exist, and this can be decided efficiently.

Lemma 1. *Simple temporal graphs do not admit spanning trees in the strict setting (if $n > 2$).*

Proof. Recall that a temporal spanning tree must offer bidirectional temporal paths (bi-paths) between all pairs of vertices. If the labeling is simple, then every edge has only one presence time, but since the strict setting imposes that the labels increase along a path, this implies that no path of length 2 or more can be used in both directions. $\square$

Lemma 2. TEMPORAL SPANNING TREE *can be solved in polynomial time in simple temporal graphs in the non-strict setting.*

Proof. By the same arguments as above, no path in the tree may rely on time labels that increase at some point along the path, as this would prevent bidirectionality (the temporal graph being simple). However, in the non-strict case, repetition of the time labels are allowed along a path. Thus, a temporal spanning tree exists if and only if a (standard) spanning tree exists in at least one of the snapshots, which can be tested (and found, if applicable) in polynomial time. $\square$

This completes the tractability landscape of TEMPORAL SPANNING TREE in the considered settings, as summarized by Fig. 3.

4 BI-SPANNERS of Unrestricted Size

In this section, we present a polynomial time algorithm that tests if a given temporal graph admits a bidirectional temporal spanner (a bi-spanner). The algorithm relies on a more basic primitive that finds, for a given node, all the bidirectional temporal paths (bi-paths) between this nodes and other nodes. If this algorithm, executed from each node, always reaches all the other nodes, then the union of the corresponding bi-paths forms a bi-spanner. Bi-spanners obtained in this way may be larger than needed. However, there exist pathological cases where no smaller bi-spanners exist, making this algorithm worst-case optimal (admittedly, these cases are extreme, which motivates the search for small solutions in general, as discussed in Sect. 5).

4.1 High-Level Description

Let s be the source node. The algorithm consists of updating recursively, for each node v, a set of triplets $B_v = \{(u_i, a_i, d_i)\}$, where $u_i \in N(v)$ and $a_i, d_i \in \mathcal{T}$, such that there exists a bi-path between s and v that arrives at v through its neighbor u_i before time a_i and departs from v through u_i after time d_i. Note that a_i needs not be smaller or larger than d_i, both direction of the bi-path being independent in time. The triplets are updated according to the following extension rule (see also Fig. 5).

Rule 1 (Extension rule). *Let (u_i, a_i, d_i) be a triplet at node v, let w be a neighbor of $v \neq u_i$. Let $a_i' = \min\{t \in \lambda(vw) \mid t \geq a_i\}$ and $d_i' = \max\{t \in \lambda(vw) \mid t \leq d_i\}$, or $\bot$ if the resulting set is empty. If $a_i' \neq \bot$ and $d_i' \neq \bot$, then (v, a_i', d_i') is a triplet at node w.*

Fig. 5. Example of the extension of a bi-path.

Lemma 3. *If the triplet (u_i, a_i, d_i) corresponds to a valid bi-path between s and v, and a triplet (v, a_i', d_i') is added by the extension rule to a neighbor w of v, then (v, a_i', d_i') corresponds to a valid bi-path between s and w.*

Proof. The existing triplet at v implies that a bi-path exists from s to v arriving at v before time a_i (included). If $a_i' \neq \bot$, then the extension rule guaranties that $a_i' \geq a_i$ and $a_i' \in \lambda(vw)$, thus the corresponding journey from s to v in this bi-path can be extended to w through edge vw. Likewise, if $b_i' \neq \bot$, then $b_i' \leq b_i$ and $b_i' \in \lambda(vw)$, so the journey from v to s can be preceded by a 1-hop journey from w to v. $\square$

Remark 1. It is sufficient to replace $\geq$ with $>$ and $\leq$ with $<$ in the extension rule for considering strict journeys instead of non-strict journeys.

Clearly, some triplets are strictly better than others. Namely, if a bi-path from the source arrives at an earliest time and departs back to it at a later time compared to second bi-path *through the same neighbor*, then all journeys using the corresponding times could be replaced by solution using the times of the first bi-path, plus extra waiting at the neighbor. Thus, it is useless to maintain the second triplet in the set. This is formalized through the following rule:

Rule 2 (Elimination rule). *Let (u_i, a_i, d_i) and (u'_i, a'_i, d'_i), with $u_i = u'_i$, be two triplets at node v, if $a_i \leq a'_i$ and $d_i \geq d'_i$, then (u_i, a'_i, d'_i) is removed from B_v.*

In the algorithm, we denote by $+$ the operation that consists of adding a new triplet to a set of triplets while taking into account this elimination rule. Then, we say that a node v has been *improved* by the addition of a triplet b if $B_v + b \neq B_v$. Observe that, despite the elimination rule, the set of triplets at a node may contain several triplets involving the same neighbor, if they are incomparable in time. Similarly, several triplets may co-exist with equal times if they involve different neighbors.

4.2   The Algorithm

The algorithm is quite compact, though its correctness and complexity are not immediate due to the fact that an edge (and even a contact on that edge) may be involved several times in the improvement of its endpoints. Initially, $B_v = \emptyset$ for all $v \neq s$ and $B_s = \{((\bot, -\infty, \infty))\}$. The algorithm starts with calling `improveNeighbors(s)`. Then, it improves the neighbors recursively until all the bi-paths from s have been computed.

Algorithm 1: Procedure `improveNeighbors()` that computes all the triplets from a source.

---

1 **Input**: current node u
2 **forall the** $(\text{-}, a_i, d_i) \in B_u$ **do**
3 $\quad$ **forall the** $v \in N(u)$ **do**
4 $\quad\quad$ $a'_i \leftarrow \min\{t \in \lambda(uv) \mid t \geq a_i\}$
5 $\quad\quad$ $d'_i \leftarrow \max\{t \in \lambda(uv) \mid t \leq d_i\}$
6 $\quad\quad$ **if** $B_v + (u, a'_i, d'_i) \neq B_v$ **then**
7 $\quad\quad\quad$ $B_v \leftarrow B_v + (u, a'_i, d'_i)$
8 $\quad\quad\quad$ `improveNeighbors(v)`

Theorem 2. *Algorithm 1 computes all bi-paths between s and the other vertices.*

Proof. First, let us prove by induction that at any step, for any vertex v, the triplets in B_v indeed correspond to bi-paths between s and v.

At the begining of the algorithm, all sets are empty except for B_s that contains $\{((\bot, -\infty, \infty))\}$, since s has a bi-path of length 0 with itself that can start and end at any time, this is a valid triplet.

Adding triplets to B_v is done through the $+$ operator that encapsulates the extension and elimination rules. Since the extension rule guarantees that the extended bi-path is valid if the previous one is (Lemma 3), this means that B_v contains only valid bi-paths.

Assume now that there exists a bi-path $b_{s,v}$ that was not computed by such algorithm, such bi-path can be seen as a sequence of triplets $((s, -\infty, \infty), (u_1, a_1, d_1), ..., (v, a_v, d_v))$ such that a_i are the arrival times to u_i of j_1 and d_i are the departure times from u_i of j_2.

By induction on the length of such bi-path: If $b_{s,v}$ has length 0, then it is computed by the algorithm since $\{((\bot, -\infty, \infty))\}$ is part of B_s. Suppose that $b_{s,v}$ has now length k and up to $(u_{k-1}, a_{k-1}, d_{k-1})$, it is part of a bi-path computed by the algorithm, i.e., there is a triplet (x, a'_{k-1}, d'_{k-1}) (where x is either u_{k-2} or $\bot$) in $B_{u_{k-1}}$ such that $a_{k-1} \geq a'_{k-1}$ and $d_{k-1} \leq d'_{k-1}$. Such triplet, by the time it was added to $B_{u_{k-1}}$, would imply a call to improveNeighbor(u_{k-1}), and by the extension rule, a triplet (u_{k-1}, a, d) in B_v such that $a = \min\{t \in \lambda(u_{k-1}v) \mid t \geq a'_{k-1}\}$ and $d = \max\{t \in \lambda(u_{k-1}v) \mid t \leq d'_{k-1}\}$.

Since a is the smallest $t \geq a'_{k-1}$ in $\lambda(u_{k-1}v)$ and $a_{k-1} \geq a'_{k-1}$, this means that the label a is either also the smallest label $\geq a_{k-1}$ or is a label even smaller than such label. Since $a_v \in \lambda(u_{k-1}v)$, that is $\geq a_{k-1}$ (otherwise j_1 would not be a journey), we have that $a \leq a_v$. By the similar argument, we also have $d \geq d_v$. Thus (u_{k-1}, a_v, d_v) is included into a triplet of B_v, and the bi-path $b_{s,v}$ is computed by the algorithm, a contradiction.   □

Theorem 3. *Algorithm 1 runs in polynomial time in the size of the input, namely in time $O(m\Delta^2\tau^4)$.*

Proof. By the elimination rule, for each vertex u, B_u contains less than $|N(u)| \cdot \tau$ triplets since given a neighbor and an arrival time, at most one departure time should be kept in B_u. Moreover, by the same rule, B_u can be improved at most $|N(u)| \cdot \tau^2$ times.

This means that the algorithm will be called at most $m \cdot \tau^2$ times. Each call of the function costs $|B(u)| \cdot |N(u)| \cdot O(\tau)$ where $O(\tau)$ upper bounds the cost of finding the minimum or maximum label in the given range, plus the cost of applying the elimination rule and comparing it, meaning that a call costs $O(|N(u)|^2 \cdot \tau^2)$ overall. Thus, the total running time is $O(m\Delta^2\tau^4)$.   □

Showing that BI-SPANNER was solvable in polynomial time was the main goal of this section, thus Algorithm 1 prioritizes simplicity over efficiency. It could be made slightly more efficient as follows.

Remark 2 (More efficient version). First, when a triplet $b = (u_i, a_i, b_i) \in B_u$ is considered for improving another neighbor v, the particular value of u_i does not matter. Thus, if there exists a better triplet $(-, a_j, b_j)$ whose times are more interesting, then b could be discarded in the loop. That way, the number of calls of the function would be reduced to τ^2 instead of $|N(u)|\tau^2$ while adding a verification similar to the elimination rule that can be done in $O(\tau)$ if B_v contains a specific set of pairs (a, d). Another improvement may consist of changing the input of the function in order to transmit only the new pair added to B_u, that way, the first loop of the algorithm is removed, dividing the running time by $|N(u)|\tau$. These improvements would result in a running time of $O(m\tau^3)$. Note that the size of all B_v, namely $O(m\tau)$, is a natural lower bound for the running time, in the current format of outputs.

5   BI-SPANNERS of Size k

In this section, we discuss several aspects of the size of bidirectional spanners. We start with a negative structural result regarding the minimum size of such spanners. This case being extreme, we investigate the problem of finding a bi-spanner of a certain size k. We already know, from the earlier sections, that this problem is hard, because spanning trees are a particular case. In fact, we strengthen this result by showing that k-BI-SPANNER is hard even in simple temporal graphs (where spanning trees were tractable). This motivates future work for identifying special cases by other means (e.g., specific parameters).

Recall, from the introduction, that sparse temporal spanners (let apart bidirectionality) are not guaranteed to exist in temporal graphs, as there exist temporal graphs with $\Theta(n^2)$ edges, all of which are critical for temporal connectivity [4]. However, if the footprint is a *complete* graph, then spanners of size $O(n \log n)$ exist whatever the labeling [17]. The following shows that no analogue result exists for bidirectional spanners.

Lemma 4. *Let $\mathcal{G} = (V, E, \lambda)$ be a happy clique, then $\mathcal{G}$ is bidirectionally temporally connected, but for every $e \in E, \mathcal{G} \setminus \{e\}$ is not.*

Proof. The fact that $\mathcal{G}$ is bidirectionally temporally connected is straightforward, as every edge in $\mathcal{G}$ forms a bi-path between its two endpoints and the footprint is a complete graph. Now, if an edge $\{u, v\} \in E$ is removed, then the only way u and v can reach each other is through a temporal path of length at least 2. Since the labeling is happy, the labels along such a journey must increase, thus the path is not reversible. □

Strictly speaking, Lemma 4 makes the algorithm of Sect. 4 worst-case optimal; however, for reasons that pertain to bidirectionality rather than to the algorithm itself, which is not satisfactory. A natural question is whether small bi-spanners can be found when they exist. In the following, we show that the answer is negative even in simple temporal graphs. We do so in the non-strict setting, since bi-spanner exist in the strict setting only if the footprint is complete (see Lemma 4), which is easy to test.

Theorem 4. k-BI-SPANNER *is NP-complete in the simple and non-strict setting.*

Proof. Given a candidate subgraph $\mathcal{G}' \subseteq \mathcal{G}$, it is easy to check whether $\mathcal{G}'$ uses at most k edges and whether it is bidirectionally connected (using the algorithm of Sect. 4), thus the problem is in NP. To show that this is NP-hard, we reduce from SET COVER, defined as follows: given a set of n elements $\mathcal{U}$ (the universe) and a collection $S = \{S_i\}$ of m subsets of $\mathcal{U}$ (whose union is $\mathcal{U}$), is there a union of at most k subsets that is equal to $\mathcal{U}$?

Let $\mathcal{I}$ be an instance of SET COVER, we construct a temporal graph $\mathcal{G} = (V, E, \lambda)$ with $n+m+3$ vertices such that $\mathcal{I}$ admits a solution of size k if and only if a bi-spanner of size $3n + 2m + 3 + k$ exists in $\mathcal{G}$. The vertices V are:

- one vertex u_i for each element of $\mathcal{U}$;
- one vertex s_i for each subset S_i;
- three auxiliary vertices v, v_1, v_2 used for connectivity.

Since $\mathcal{G}$ is simple and in the non-strict setting, each bidirectional journey must use edges from the same snapshot, thus the order of the time labels is irrelevant, what matters is only which sets of vertices are connected in a same snapshot. The edges are constructed as follows (see Fig. 6 for an illustration):

- s_i shares an edge at time i with v and with each $u_j \in S_i$;
- v_1 shares an edge at time $m+1$ with all $s_i \in S$ and with all $u_j \in \mathcal{U}$;
- v_2 shares an edge at time $m+2$ with v and with all $s_i \in S$;
- v_1 and v_2 share an edge with all the other nodes (and with each other) at unique arbitrary times.

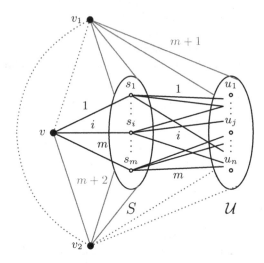

Fig. 6. The reduction, dotted edges have a distinct label from their neighborhood

Claim 5. $\mathcal{G}$ *is bidirectionally temporally connected.*

Proof. To start, observe that v_1 and v_2 both share an edge with all the other vertices and with each other, they are thus biconnected to all vertices. Moreover, each vertex of S is biconnected to v and to other vertices in S through v_2, and to all vertices of $\mathcal{U}$ through v_1. Vertices of $\mathcal{U}$ are also biconnected to each other through v_1. Finally, for each $u_j \in \mathcal{U}$, at least one s_i contains it, so u_j is biconnected to v at the corresponding time i. □

Now, observe that v_1 and v_2 are only biconnected to other vertices through bipaths of length 1, thus none of their incident edges can be removed. This implies that any bi-spanner must use at least these $2|V| - 3$ edges (twice $|V| - 1$, minus their common edge). These edges will always biconnect all the vertices in $\mathcal{U}$ with each other, all the vertices of S with each other, and all the vertices in $\mathcal{U}$ with all the vertices in S.

($\rightarrow$) If there is a solution of size k to SET COVER, then a bi-spanner of size $3n + 2m + k + 3$ can be built as follows: for each s_i selected in the solution, keep the edge vs_i (remove the others). For each u_j, keep an edge to one of the s_i that contains it and that is selected in the solution (remove the others). Then all u_j are still biconnected to v. This makes $k + n$ edges, plus the above required edges, which is $3n + 2m + k + 3$ in total.

($\leftarrow$) If a bi-spanner of size $3n + 2m + k + 3$ exists, then it contains at least the above required edges, so only $k + n$ edges overall are kept between v and vertices of S and between vertices of S and vertices of $\mathcal{U}$. The fact that each vertex in $\mathcal{U}$ remains biconnected to v implies that at least one edge is kept for each of them to a vertex of S, which makes at least n edges, and that the corresponding nodes in S still share an edge with v. There are at most k such edges, thus there exists a subset of S of size at most k that covers all the nodes in $\mathcal{U}$. $\square$

5.1 Further Restrictions (Open Questions)

Based on the above results, there is no straightforward settings where small bidirectional spanners exist and are easy to find at the same time. What type of restrictions may help? Intuitively, bi-paths enable some form of mutualization when they are *long*, as all pairs of intermediate vertices can also reach each other through the same bi-path. This motivates the following question:

Question 1. If the bi-path distance between pairs of vertices is large on average (over all the pairs), does this guarantee the existence of a small bidirectional spanner? Could such distances be large in a dense footprint?

Beyond structural questions, a natural approach for tractability is to explore parameters with respect to which the problem is fixed-parameter tractable.

Question 2. In what parameters are k-BI-SPANNER and TEMPORAL SPANNING TREE FPT?

As is now usual in temporal graphs, one could apply classical parameters to either the footprint or the snapshots. However, parameters which are themselves temporal may be more relevant. This sub-field of fixed-parameter tractability is mostly open.

6 Conclusion

In this paper, we answered a fundamental question in temporal graphs, namely, whether deciding the existence of a temporal spanning tree is tractable. It was already known that finding temporal spanners of minimum size is a hard problem. Unfortunately, we showed that even in the extreme case that the spanner is a tree, the problem is NP-complete. Our reduction rely on a proper temporal graph, and as such, applies to both strict and non-strict journeys. Motivated by this negative result, we explored different ways to relax spanning trees in temporal graphs. In particular, we showed that the property of bidirectionality without a prescribed size can be tested efficiently (it is even harder than the particular case of trees otherwise). Still, we believe that testing bidirectionality could prove useful in certain applications, and this adds to a small collection of non-trivial properties that remain tractable in temporal graphs. As already discussed, such applications might be related to routing or security. They also include transportation networks, where bus or train lines benefit from being operated along the same segments in both directions. Some questions remain open, in particular, what additional restrictions could force the existence of small bidirectional spanners, and what parameters could make their computation tractable.

References

1. Akrida, E.C., Czyzowicz, J., Gasieniec, L., Kuszner, Ł., Spirakis, P.G.: Temporal flows in temporal networks. J. Comput. Syst. Sci. **103**, 46–60 (2019)
2. Akrida, E.C., Gasieniec, L., Mertzios, G.B., Spirakis, P.G.: The complexity of optimal design of temporally connected graphs. Theory Comput. Syst. **61**(3), 907–944 (2017)
3. Altisen, K., Devismes, S., Durand, A., Johnen, C., Petit, F.: On implementing stabilizing leader election with weak assumptions on network dynamics. In: ACM Symposium on Principles of Distributed Computing, pp. 21–31 (2021)
4. Axiotis, K., Fotakis, D.: On the size and the approximability of minimum temporally connected subgraphs. In: 43rd International Colloquium on Automata, Languages, and Programming (ICALP), pp. 149:1–149:14 (2016)
5. Baumann, H., Crescenzi, P., Fraigniaud, P.: Parsimonious flooding in dynamic graphs. Distrib. Comput. **24**(1), 31–44 (2011)
6. Bhadra, S., Ferreira, A.: Complexity of connected components in evolving graphs and the computation of multicast trees in dynamic networks. In: Pierre, S., Barbeau, M., Kranakis, E. (eds.) ADHOC-NOW 2003. LNCS, vol. 2865, pp. 259–270. Springer, Heidelberg (2003). https://doi.org/10.1007/978-3-540-39611-6_23
7. Bilò, D., D'Angelo, G., Gualà, L., Leucci, S., Rossi, M.: Blackout-tolerant temporal spanners. In: Erlebach, T., Segal, M. (eds.) ALGOSENSORS 2022. LNCS, vol. 13707, pp. 31–44. Springer, Cham (2022). https://doi.org/10.1007/978-3-031-22050-0_3
8. Bilò, D., D'Angelo, G., Gualà, L., Leucci, S., Rossi, M.: Sparse temporal spanners with low stretch. In: 30th European Symposium on Algorithms (ESA). Schloss Dagstuhl-Leibniz-Zentrum für Informatik (2022)
9. Bramas, Q., Tixeuil, S.: The complexity of data aggregation in static and dynamic wireless sensor networks. In: Pelc, A., Schwarzmann, A.A. (eds.) SSS 2015. LNCS, vol. 9212, pp. 36–50. Springer, Cham (2015). https://doi.org/10.1007/978-3-319-21741-3_3
10. Brito, L.F.A., Albertini, M.K., Casteigts, A., Travençolo, B.A.N.: A dynamic data structure for temporal reachability with unsorted contact insertions. Soc. Netw. Anal. Min. **12**(1), 1–12 (2022)
11. Brunelli, F., Crescenzi, P., Viennot, L.: On computing pareto optimal paths in weighted time-dependent networks. Inf. Process. Lett. **168**, 106086 (2021). https://doi.org/10.1016/j.ipl.2020.106086
12. Brunelli, F., Crescenzi, P., Viennot, L.: Maximizing reachability in a temporal graph obtained by assigning starting times to a collection of walks. Networks **81**(2), 177–203 (2023)
13. Bui-Xuan, B., Ferreira, A., Jarry, A.: Computing shortest, fastest, and foremost journeys in dynamic networks. Int. J. Found. Comput. Sci. **14**(02), 267–285 (2003)
14. Casteigts, A., Corsini, T., Sarkar, W.: Invited paper: simple, strict, proper, happy: a study of reachability in temporal graphs. In: Devismes, S., Petit, F., Altisen, K., Di Luna, G.A., Fernandez Anta, A. (eds.) SSS 2022. LNCS, vol. 13751, pp. 3–18. Springer, Cham (2022). https://doi.org/10.1007/978-3-031-21017-4_1
15. Casteigts, A., Flocchini, P., Quattrociocchi, W., Santoro, N.: Time-varying graphs and dynamic networks. Int. J. Parallel Emergent Distrib. Syst. **27**(5), 387–408 (2012)
16. Casteigts, A., Himmel, A.-S., Molter, H., Zschoche, P.: Finding temporal paths under waiting time constraints. Algorithmica **83**(9), 2754–2802 (2021)
17. Casteigts, A., Peters, J.G., Schoeters, J.: Temporal cliques admit sparse spanners. In: 46th International Colloquium on Automata, Languages, and Programming (ICALP). LIPIcs, vol. 132, pp. 129:1–129:14. Schloss Dagstuhl - Leibniz-Zentrum für Informatik (2019)
18. Casteigts, A., Raskin, M., Renken, M., Zamaraev, V.: Sharp thresholds in random simple temporal graphs. In: 2021 IEEE 62nd Annual Symposium on Foundations of Computer Science (FOCS), pp. 319–326. IEEE (2022)

19. Conte, A., Crescenzi, P., Marino, A., Punzi, G.: Enumeration of s-d separators in DAGs with application to reliability analysis in temporal graphs. In: 45th International Symposium on Mathematical Foundations of Computer Science (MFCS). Schloss Dagstuhl-Leibniz-Zentrum für Informatik (2020)

20. Di Luna, G., Dobrev, S., Flocchini, P., Santoro, N.: Distributed exploration of dynamic rings. Distrib. Comput. **33**(1), 41–67 (2020)

21. Enright, J., Meeks, K., Mertzios, G.B., Zamaraev, V.: Deleting edges to restrict the size of an epidemic in temporal networks. In: 44th International Symposium on Mathematical Foundations of Computer Science (MFCS 2019). Schloss Dagstuhl-Leibniz-Zentrum fuer Informatik (2019)

22. Enright, J., Meeks, K., Molter, H.: Counting temporal paths. arXiv preprint arXiv:2202.12055 (2022)

23. Erlebach, T., Spooner, J.T.: Parameterized temporal exploration problems. In: 1st Symposium on Algorithmic Foundations of Dynamic Networks (SAND 2022). Schloss Dagstuhl-Leibniz-Zentrum für Informatik (2022)

24. Flocchini, P., Mans, B., Santoro, N.: On the exploration of time-varying networks. Theor. Comput. Sci. **469**, 53–68 (2013)

25. Fluschnik, T., Molter, H., Niedermeier, R., Renken, M., Zschoche, P.: Temporal graph classes: a view through temporal separators. Theor. Comput. Sci. **806**, 197–218 (2020)

26. Gómez-Calzado, C., Casteigts, A., Lafuente, A., Larrea, M.: A connectivity model for agreement in dynamic systems. In: Träff, J.L., Hunold, S., Versaci, F. (eds.) Euro-Par 2015. LNCS, vol. 9233, pp. 333–345. Springer, Heidelberg (2015). https://doi.org/10.1007/978-3-662-48096-0_26

27. Ilcinkas, D., Klasing, R., Wade, A.M.: Exploration of constantly connected dynamic graphs based on cactuses. In: Halldórsson, M.M. (ed.) SIROCCO 2014. LNCS, vol. 8576, pp. 250–262. Springer, Cham (2014). https://doi.org/10.1007/978-3-319-09620-9_20

28. Kempe, D., Kleinberg, J., Kumar, A.: Connectivity and inference problems for temporal networks. In: Proceedings of 32nd ACM Symposium on Theory of Computing (STOC), Portland, USA, pp. 504–513. ACM (2000)

29. Rannou, L., Magnien, C., Latapy, M.: Strongly connected components in stream graphs: computation and experimentations. In: Benito, R.M., Cherifi, C., Cherifi, H., Moro, E., Rocha, L.M., Sales-Pardo, M. (eds.) COMPLEX NETWORKS 2020. SCI, vol. 943, pp. 568–580. Springer, Cham (2021). https://doi.org/10.1007/978-3-030-65347-7_47

30. Vernet, M., Drozdowski, M., Pigné, Y., Sanlaville, E.: A theoretical and experimental study of a new algorithm for minimum cost flow in dynamic graphs. Discret. Appl. Math. **296**, 203–216 (2021)

31. Whitbeck, J., de Amorim, M.D., Conan, V., Guillaume, J.-L.: Temporal reachability graphs. In: Proceedings of the 18th Annual International Conference on Mobile Computing and Networking, pp. 377–388 (2012)

Near-Optimal Fault Tolerance
for Efficient Batch Matrix Multiplication
via an Additive Combinatorics Lens

Keren Censor-Hillel[1], Yuka Machino[2], and Pedro Soto[3(✉)]

[1] Technion, Haifa, Israel
ckeren@cs.technion.ac.il
[2] MIT, Cambridge, USA
yukam997@mit.edu
[3] Oxford University, Oxford, UK
Pedro.Soto@maths.ox.ac.uk

Abstract. Fault tolerance is a major concern in distributed computational settings. In the classic master-worker setting, a server (the master) needs to perform some heavy computation which it may distribute to m other machines (workers) in order to speed up the time complexity. In this setting, it is crucial that the computation is made robust to failed workers, in order for the master to be able to retrieve the result of the joint computation despite failures. A prime complexity measure is thus the *recovery threshold*, which is the number of workers that the master needs to wait for in order to derive the output. This is the counterpart to the number of failed workers that it can tolerate.

In this paper, we address the fundamental and well-studied task of matrix multiplication. Specifically, our focus is on when the master needs to multiply a batch of n pairs of matrices. Several coding techniques have been proven successful in reducing the recovery threshold for this task, and one approach that is also very efficient in terms of computation time is called *Rook Codes*. The previously best known recovery threshold for batch matrix multiplication using Rook Codes is $O(n^{\log_2 3}) = O(n^{1.585})$.

Our main contribution is a lower bound proof that says that any Rook Code for batch matrix multiplication must have a recovery threshold that is at least $\omega(n)$. Notably, we employ techniques from Additive Combinatorics in order to prove this, which may be of further interest. Moreover, we show a Rook Code that achieves a recovery threshold of $n^{1+o(1)}$, establishing a near-optimal answer to the fault tolerance of this coding scheme.

Keywords: Master-Worker Computation · Matrix Multiplication · Additive Combinatorics

1 Introduction

The master-worker computing paradigm has been extensively studied due to its ability to process large data-sets whose processing time is too expensive for

© The Author(s), under exclusive license to Springer Nature Switzerland AG 2024
Y. Emek (Ed.): SIROCCO 2024, LNCS 14662, pp. 156–173, 2024.
https://doi.org/10.1007/978-3-031-60603-8_9

a single machine. To address this, the main machine (master) may split the computation among m workers. Various tasks have been studied, such as matrix multiplication, gradient descent, tensors/multilinear computation, and more [14, 17,18,23,31,32,35,39,43–45,47,51–53].

A common issue in such settings is the need to cope with failures, so that faulty workers do not prohibit the completion of the computational task at hand. Indeed, faults are a major concern in many distributed and parallel settings, and coping with them has been studied for several decades [4,36].

In the master-worker setting, the main complexity measure that addresses the robustness of an algorithm to faults is its *recovery threshold*, defined as follows.

Definition of Recovery Threshold [52]. The *recovery threshold* of an algorithm in the master-worker setting is the number of workers the master must wait for in order to be able to compute its output.

A low recovery threshold means that the algorithm can tolerate a higher number of faults. For example, if the master splits the computation to m disjoint pieces and sends one to each worker then it cannot tolerate even a single failure and needs to wait for all m workers. If it splits the computation to $m/2$ pieces and sends each piece to 2 workers then it only needs to wait for $m-1$ machines, and thus can tolerate one failure. Notice that with 2 failures this approach does not work, as the 2 failed workers could be responsible for the same piece of computation. Indeed, such simple *replication* approaches are inefficient in terms of their robustness.

An efficient way for handling such faults is to use coding techniques. Coding has been an extremely successful technique in various distributed computations (see, e.g., [2,3,8,10,21,28]) which mostly address communication noise). In particular, coding schemes have been successfully proposed for the task of *matrix multiplication* in the master-worker setting, which is the focus of our work. Multiplying matrices is a vital computational task, which has been widely studied in distributed settings [9,11,12,16,26,27] and in parallel settings (see, e.g., [5,6,37] and many references therein).

Specifically, we address batch matrix multiplication in the master-worker setting, in which the master has n pairs of matrices, $\{(A_i, B_i)\}_{0 \le i \le n-1}$, whose n products $\{A_i \cdot B_i\}_{0 \le i \le n-1}$ it needs to compute. All matrices A_i have the same dimensions, as well as all matrices B_i.

To see why coding is useful, consider the above naïve approach for coping with faults by replication as applied for batch matrix multiplication. For simplicity, suppose that m is a multiple of n, i.e., $m = \lambda n$ for some integer λ. In the replication approach, the master sends each sub-task $A_i \cdot B_i$ to λ workers. It is straightforward to see that the recovery threshold here is $\text{Recovery}_{\text{replication}}(n) \ge (n-1)\lambda + 1 = m - \lambda + 1$, as λ failures could be of workers that are responsible for the same product, where for any given family $\mathcal{F}$ of algorithms for computing the n products, we denote by $\text{Recovery}_{\mathcal{F}}(n)$ the best recovery threshold of an algorithm in $\mathcal{F}$. In particular, this means that with simple replication, the recovery threshold is linear in the number m of machines.

However, it is well-known that with coding it is possible to do better. Consider the case $n = 2$. With replication and $m = 2\lambda$ workers, we have that $\texttt{Recovery}_{\text{replication}}(2) \geq \lambda + 1$. In contrast, consider the fault tolerance provided by the following simple coding scheme, in which we define $\tilde{A}(x) := A_0 + A_1 x$ and $\tilde{B}(x) := B_0 + B_1 x$. The master sends to each worker w the distinct values $\tilde{A}(x_w), \tilde{B}(x_w)$, and each worker w then returns the product $\tilde{A}(x_w) \cdot \tilde{B}(x_w)$ to the master. For any 3 workers u, v, w, it holds that

$$
\begin{bmatrix} x_u^0 & x_u^1 & x_u^2 \\ x_v^0 & x_v^1 & x_v^2 \\ x_w^0 & x_w^1 & x_w^2 \end{bmatrix} \begin{bmatrix} A_0 B_0 \\ A_0 B_1 + A_1 B_0 \\ A_1 B_1 \end{bmatrix} = \begin{bmatrix} \tilde{A}(x_u) \cdot \tilde{B}(x_u) \\ \tilde{A}(x_v) \cdot \tilde{B}(x_v) \\ \tilde{A}(x_w) \cdot \tilde{B}(x_w) \end{bmatrix},
$$

and since $x_u \neq x_v \neq x_w$ then the first matrix is invertible. Thus, the master can retrieve $A_0 \cdot B_0$ and $A_1 \cdot B_1$ using the results of any 3 workers, giving that $\texttt{Recovery}_{\text{coded}}(2) \leq 3$. In particular, the coding scheme achieves a fault tolerance of $\frac{m-3}{m} = 1 - \frac{3}{m}$, which tends to 100% as m grows. However, the replication scheme can only achieve a fault tolerance of $\frac{2\lambda - (\lambda+1)}{2\lambda} = \frac{1}{2} - \frac{1}{m}$, which is bounded by 50% even as m grows.

Rook Codes for Batch Matrix Multiplication. We consider the Rook Codes given by the following form, proposed in [44,45]. Denote by

$$
\tilde{A}(x) = \sum_{i \in [n]} A_i x^{p_i}, \quad \tilde{B}(x) = \sum_{j \in [n]} B_j x^{q_j} \tag{1}
$$

two matrix polynomials that are generated by the master, for some sequences of non-negative integers $P = (p_0, \ldots, p_{n-1})$ and $Q = (q_0, \ldots, q_{n-1})$. We denote:

$$
L_{P,Q} := |P + Q|.
$$

For $0 \leq w \leq m - 1$, the master sends to worker w the coded matrices $\tilde{A}(x_w)$ and $\tilde{B}(x_w)$, where $x_w \neq x_{w'}$ for every pair of different workers $w \neq w'$. Worker w then multiplies the two coded matrices and sends their product back to the master. Note that

$$
\tilde{A}(x) \cdot \tilde{B}(x) = \left(\sum_{i \in [n]} A_i x^{p_i} \right) \left(\sum_{j \in [n]} B_j x^{q_j} \right) = \sum_{i,j \in [n]} A_i B_j x^{p_i + q_j}.
$$

Note that while the degree of the above polynomial may be $\max\{P + Q\}$, the number of its non-zero coefficients is bounded by $L_{P,Q} = |P + Q|$, which may be significantly smaller. Thus, when the master receives back $L_{P,Q}$ point evaluations of $\tilde{A}(x) \cdot \tilde{B}(x)$, it can interpolate the polynomial. To be able to extract the coefficients $A_k \cdot B_k$ for all $k \in [n]$ out of the polynomial, we need $A_k \cdot B_k$ to be the only coefficient of $x^{p_k + q_k}$ in the polynomial. That is, the following property is necessary and sufficient.

Property 1 (The decodability property). For all $i, j, k \in [n]$, $p_k + q_k = p_i + q_j$ if and only if $i = k, j = k$.

Throughout the paper, we refer to codes given by Eq. 1 that satisfy Property 1 as Rook Codes for batch matrix multiplication in the master-worker setting. To summarize, given two sets of integers P, Q of size $|P| = |Q| = n$ for which Property 1 holds, we have that $\text{Recovery}_{\text{Rook-Codes}}(n) = \min_{P,Q}\{L_{P,Q}\}$ (the minimum is taken over all P, Q that satisfy Property 1). In order to derive the recovery threshold of Rook Codes for batch matrix multiplication, our goal is then to bound $L_{P,Q}$ for all such P, Q, or in other words, to bound $|P + Q|$.

How Robust Can Rook Codes Be? A simple Rook Code can be derived from the Polynomial Codes of [52], which are designed for multiplying a single pair of matrices. This would give $\tilde{A}(x) = \sum_{i \in [n]} A_i x^i$ and $\tilde{B}(x) = \sum_{j \in [n]} B_j x^{nj}$, i.e., we have $p_i = i$ and $q_j = nj$ for every $i, j \in [n]$. This would immediately bound the recovery threshold of Rook Codes by $\text{Recovery}_{\text{Rook-Codes}}(n) \leq ((n-1)+n(n-1))+1 = n^2$, which is already a significant improvement compared to the recovery threshold of replication since it does not depend on m but rather only on the size of the input. In [44], an intuition was given for why the task is non-trivial, in the form of a lower bound that says that with $P = (0, 1, 2, ..., n-1)$ (i.e., with $\tilde{A}(x) = \sum_{i \in [n]} A_i x^i$), any choice of Q has $L_{P,Q} \geq n^2/2$, which implies that in order to obtain a recovery threshold that is sub-quadratic in n, one cannot use such simple values for P.

But for other values of P, Q, Rook Codes can indeed do better, as shown by [44], who provided P and Q with $L_{P,Q} = O(n^{\log_2 3})$, implying that $\text{Recovery}_{\text{Rook-Codes}}(n) = O(n^{\log_2 3}) = O(n^{1.585})$. The parameters chosen for this are $P = Q = \{v_0 + v_1 \cdot 3 + v_2 \cdot 3^2 + ... + v_{\ell-1} \cdot 3^{\ell-1} \mid v_i \in \{0,1\}\}$, where $\ell = \log_2 n$, assuming that ℓ is an integer (an assumption that can easily be removed by standard padding). One can verify that this choice yields a Rook Code (i.e., that Property 1 holds), and that the recovery threshold is $\text{Recovery}_{\text{Rook-Codes}}(n) = O(3^{\ell}) = O(3^{\log_2 n}) = O(2^{\log_2 3 \cdot \log_2 n}) = O(n^{\log_2 3})$.

This significant progress still leaves the following question open:

Question: Are there Rook Codes for batch matrix multiplication in the master-worker setting with a recovery threshold that is linear in n?

1.1   Our Contribution

In this paper, we show that no Rook Code for batch matrix multiplication in the master-worker setting can get a strictly linear in n recovery threshold, but that we can get very close to such value.

Lower Bound. Our main contribution is showing a super-linear lower bound on the recovery threshold of Rook Codes for batch matrix multiplication in the master-worker setting.

Theorem 1 (Lower Bound). *Every Rook Code for batch matrix multiplication in the master-worker setting has a super-linear in n recovery threshold. That is,*

$$\text{Recovery}_{Rook\text{-}Codes}(n) = \omega(n).$$

Our main tool for proving Theorem 1 is the following theorem in additive combinatorics, whose proof is the key technical ingredient of our contribution (Sect. 3).

Theorem 2. *Let $P = \{p_0, p_1, \ldots, p_{n-1}\}$ and $Q = \{q_0, q_1, \ldots, q_{n-1}\}$ be finite subsets of the integers, such that $0 \in P$ and $0 \in Q$. Suppose that for all $0 \leq i, j, k \leq n - 1$, $p_k + q_k = p_i + q_j$ if and only if $i = k, j = k$. Then $|P + Q| = \omega(n)$.*

It is easy to see that Theorem 2 is sufficient for proving Theorem 1, as follows.

*Proof (**Proof of Theorem** 1 **given Theorem** 2).* Let $P = \{p_0, p_1, \ldots, p_{n-1}\}$ and $Q = \{q_0, q_1, \ldots, q_{n-1}\}$ be sets of integers that define a Rook Code for batch matrix multiplication in the master-worker setting. In particular, Property 1 holds, and hence by Theorem 2, we have that $|P + Q| = \omega(n)$. This implies that $L_{P,Q} = |P + Q| = \omega(n)$ and thus $\text{Recovery}_{Rook\text{-}Codes}(n) = \min_{P,Q}\{L_{P,Q}\} = \omega(n)$, as claimed.

Upper Bound. Our second contribution is that there exist Rook Codes for batch matrix multiplication in the master-worker setting that asymptotically get very close to the bound in Theorem 1. We prove the following (Sect. 4).

Theorem 3 (Upper Bound). *There exist Rook Codes for batch matrix multiplication in the master-worker setting with a recovery threshold of $ne^{O(\sqrt{\log n})}$. That is,*

$$\text{Recovery}_{Rook\text{-}Codes}(n) = n^{1+o(1)}.$$

1.2   Related Work

The use of coding for matrix-vector and vector-vector multiplication in the master-worker setting was first established concurrently in [18,35,52]. Shortly after, efficient constructions were given for general matrix-matrix multiplication [17,53]. In particular, [53] established the equivalence of the recovery threshold for general partitions to the tensor rank of matrix multiplication. Further, [46,47] considered coding for gradient descent in a master-worker setting for machine learning applications. Batch matrix multiplication was studied using Rook Codes [44,45], Lagrange Coded Computation (LCC) [51], and Cross Subspace Alignment (CSA) codes [31]. Follow-up works about LCC consider numerically stable extensions [23,43] and private polynomial computation [39]. There have also been follow-up works on CSA codes which include extending them to obtain security guarantees [14,32].

It is important to mention that LCC and CSA codes have recovery thresholds of $2n - 1$, and so, at a first glance, it may seem that LCC or CSA Codes already achieve a linear recovery threshold and thus outperform Rook Codes. However, an important aspect in which the Rook Codes that are considered in our work are powerful is in their computational complexity. The work in [44] supplies experimental evidence that the encoding step for Rook Codes is computationally more efficient. In Sect. 5, we give detailed descriptions of LCC and LCA codes, and provide some mathematical intuition as to why this may be the case.

In [19, 20], additive combinatorics is used to construct and analyze coded matrix multiplication algorithms. However, there are two key differences between these works and ours. The first is that they consider the special case of outer product of two block matrix vectors, which corresponds to batch matrix multiplication but with dependencies between the pairs of matrices which can be exploited; in particular, their constraint is simpler than the decodability property we need (Property 1). The second difference is that they consider security of the computation. Hence, our results are incomparable.

1.3 Roadmap

The following section shortly contains the notation for the additive combinatorics setup. Section 3 contains the proof of Theorem 2, which we proved above to imply the lower bound of Theorem 1. Section 4 proves the upper bound of Theorem 3. We conclude in Sect. 5 with a discussion of the computational efficiency that motivates our study of the recovery threshold of Rook Codes.

2   Notation

For two sets of integers A and B, the set $A + B$ is defined as $A + B := \{a + b \mid a \in A, b \in B\}$, and the set $A - B$ is defined as $A - B := \{a - b \mid a \in A, b \in B\}$. Note that when $A = B$ we still have the same definitions, so for example $A - A = \{a - b \mid a \in A, b \in A\}$, which is the set of all integers that are a subtraction of one element of A from another. For a set of integers A and an integer k, the set kA is defined as

$$kA := A + A + \cdots + A = \{a_0 + a_1 + \cdots + a_{k-1} \mid \forall 0 \leq i \leq k - 1, a_i \in A\}.$$

3   An $\omega(n)$ Lower Bound on the Recovery Threshold

In this section we prove that no Rook Code for batch matrix multiplication can achieve a strictly linear recovery threshold.

Theorem 1 (Lower Bound). *Every Rook Code for batch matrix multiplication in the master-worker setting has a super-linear in n recovery threshold. That is,*

$$\texttt{Recovery}_{Rook\text{-}Codes}(n) = \omega(n).$$

As proven in the introduction, the following theorem implies Theorem 1, and the remainder of this section is dedicated to its proof.

Theorem 2. *Let* $P = \{p_0, p_1, \ldots, p_{n-1}\}$ *and* $Q = \{q_0, q_1, \ldots, q_{n-1}\}$ *be finite subsets of the integers, such that* $0 \in P$ *and* $0 \in Q$. *Suppose that for all* $0 \leq i, j, k \leq n-1$, $p_k + q_k = p_i + q_j$ *if and only if* $i = k, j = k$. *Then* $|P + Q| = \omega(n)$.

Theorem 2 is closely related to Freiman's Theorem in [25] (see also [42, Theorem 1.3, Chapter 2]), a deep result in Additive Combinatorics which essentially states that a set P has a small *doubling constant* (i.e., the size of $|P+P|$ is only a constant multiple of $|P|$) if and only if P *is similar to* an arithmetic progression, for a well defined notion of similarity. Our theorem is comparable to Freiman's Theorem in the following sense. The contrapositive version of Theorem 2 states that if $|P+Q|$ is a constant multiple of $|P|$, then $P + Q$ *looks like* an arithmetic progression (in a different way compared to Freiman's similarity notion), in the sense that there exists i, j, k with $(i, j) \neq (k, k)$ such that $p_k + q_k = p_i + q_j$ (note that when $P = Q$, this implies that P contains a 3-term arithmetic progression, or 3-AP for short). In order to prove Theorem 2, we will invoke the following theorems from Additive Combinatorics.

Theorem 4 (Ruzsa's Triangle Inequality [48, **Lemma 2.6]).** *If* A, B, C *are finite subsets of integers, then*

$$|A||B - C| \leq |A - B||A - C|.$$

Theorem 5 (The Plünnecke-Ruzsa Inequality [38, **Theorem 1.2]).** *Let* A *and* B *be finite subsets of integers and let* K *be a constant satisfying the following inequality:*

$$|A + B| \leq K|A|.$$

Then for all integers $r, \ell \geq 0$,

$$|rB - \ell B| \leq K^{r+\ell}|A|.$$

We will also need the following lemma.

Lemma 1. *Let* $P, Q \subseteq \mathbb{Z}$ *such that* $|P| = |Q| = n$ *and* $|P + Q| \leq K|P| = Kn$ *for some constant* K. *Then* $|P + Q - P - Q| \leq K^7 n$.

Proof. First, using Ruzsa's Triangle Inequality from Theorem 4 with $A = P, B = P - P$, and $C = Q - Q$, we get that

$$|P + Q - P - Q| \leq |P - P + P||P + Q - Q|/|P|. \tag{2}$$

Next, using the Plünnecke-Ruzsa Inequality from Theorem 5, where $A = Q$, $B = P$, $r = 2, \ell = 1$ and the condition $|P + Q| \leq K|Q|$ (recall that $|P| = |Q|$), we get that

$$|P - P + P| = |2P - P| \leq K^{2+1}|Q| = K^3|P|. \tag{3}$$

Using symmetry of P and Q, we can similarly prove that $|Q - Q + Q| \leq K^3|Q|$. Hence, combining Eq. (2) and Eq. (3) together, we get that $|P + Q - P - Q|$ is bounded by

$$|P+Q-P-Q| \leq |P-P+P||P+Q-Q|/|P| \leq K^3|P||P+Q-Q|/|P| = K^3|P+Q-Q|.$$

Using Ruzsa's Triangle Inequality again, this time with $A = Q, B = -P$ and $C = Q - Q$, and then using the fact that $|Q + Q - Q| \leq K^3|Q|$, we get that:

$$|P + Q - Q| = |-P + Q - Q| \leq |P + Q||Q + Q - Q|/|Q| \leq |P + Q|K^3 \leq K^4 n.$$

Plugging this into the bound on $|P+Q-P-Q|$ gives that $|P+Q-P-Q| \leq K^3|P + Q - Q| \leq K^7 n$, as claimed.

Now we are ready to prove Theorem 2, for which we also use the following:

Theorem 6 (The Triangle Removal Lemma [48, Lemma 10.46]). *Let $G = G(V, E)$ be a graph which contains at most $o(|V|^3)$ triangles. Then it is possible to remove $o(|V|^2)$ edges from G to obtain a graph which is triangle-free.*

Overview: First we will assume for contradiction that there exists some constant K such that for all n, there exists P, Q with $|P| = |Q| = n$ such that $|P + Q| \leq Kn$ and P, Q satisfy Property 1. From this assumption, we will construct a tripartite graph $G = (V, E)$ with $V = (X, Y, Z)$ such that every edge is in exactly one triangle in G. Each of the three vertex sets will have size $N = |P + Q - P - Q|$ and the number of edges between each pair of sets is at least $Nn/4$. Because every edge is in exactly one triangle, we have that the number of triangles is exactly $|E|/3$, and thus it is $o(|V|^3)$. By using the Triangle Removal Lemma from Theorem 6, it follows that the number of edges that need to be removed to make the graph triangle-free is at most $o(|V|^2) = o(N^2)$. Since this naturally also bounds the total number of edges $|E|$, which is at least $3Nn/4$, we obtain that $3Nn/4 \leq o(|V|^2) = o(N^2)$ and hence $n = o(N)$. But from Lemma 1, N is bounded by a constant multiple of n, hence we get that $n = o(N) = o(n)$ which is a contradiction. The main technical challenge of this proof is in constructing a map between the set $|P+Q|$ and vertices of the graph in such a way so that the condition of Property 1, that $p_k + q_k = p_i + q_j$ if and only if $i = k, j = k$, translates into the condition that every edge is in exactly one triangle. Therefore, the first part of this proof will be about defining a map ψ with the desired properties, as presented in the proof of [42, Theorem 3.5, Chap. 2].

Proof (Proof of Theorem 2). Assume for contradiction that $|P+Q| = K|P|$. Let us call $N = |P + Q - P - Q|$.

First, we construct $\phi : \mathbb{Z} \to [q]$ for a prime $q > \max(P + Q - P - Q)$, such that if $x \in P + Q - P - Q$ and $\phi(x) = 0$, then $x = 0$ (we do not simply define $\phi(x) = x \mod q$, since we need ϕ to satisfy additional properties). We do this in the following way.

For each $\lambda \in [q]$, consider the map

$$\phi_\lambda : \mathbb{Z} \xrightarrow{\mathrm{mod}\ q} \mathbb{Z}/q\mathbb{Z} \xrightarrow{\times\lambda} \mathbb{Z}/q\mathbb{Z} \xrightarrow{(\mathrm{mod}\ q)^{-1}} [q].$$

It is important to note that $(\mathrm{mod}\ q)$ and $\times\lambda$ are group homomorphisms. Therefore, as the operation $\mathrm{mod}\ q$ restricted to $[q]$ is the identity map, ϕ_λ has the following property [42]:

Property 2. If $a_i, x_i \in \mathbb{Z}$ and $0 \leq \sum_{i=1}^n a_i \phi_\lambda(x_i) < q$ then $\sum_{i=1}^n a_i \phi_\lambda(x_i) = \phi_\lambda(\sum_{i=1}^n a_i x_i)$.

For each $x \neq 0$, $\phi_\lambda(x)$ takes on the values of $[q]$ with equal probability over all $\lambda \in [q]$. Therefore, for each $x \in (P + Q - P - Q) \setminus \{0\}$, the probability over all $\lambda \in [q]$ that N divides $\phi_\lambda(x)$ is at most $1/N$. As there are $N - 1$ elements in $(P + Q - P - Q) \setminus \{0\}$, by the union bound, we get:

$$\Pr_{\lambda \in [q]} (\forall x \in (P + Q - P - Q) \setminus \{0\}, N \text{ does not divide } \phi_\lambda(x))$$

$$\geq 1 - \left(\sum_{x \in (P+Q-P-Q)\setminus\{0\}} \Pr_{\lambda \in [q]} (N \text{ divides } \phi_\lambda(x)) \right)$$

$$\geq 1 - \frac{N-1}{N} = 1/N > 0.$$

This implies that there exists $\lambda \in [q]$ such that for all $x \in P + Q - P - Q \setminus \{0\}$ it holds that N does not divide $\phi_\lambda(x)$. We fix λ to be such a constant, and set $\phi = \phi_\lambda$. Now, we construct a set $T' \subseteq [n]$ so that we have the following property

Property 3. For all $i, j, k \in T'$, we have: $|\phi(p_k) - \phi(p_i) + \phi(q_k) - \phi(q_j)| \leq q$.

In order to construct T', we first pick one of the two intervals $[0, \frac{q-1}{2}]$ and $[\frac{q+1}{2}, q-1]$ to be I so that $T = \{k \in \{1 \ldots n\} : \phi(p_k) \in I\}$ and $|T| \geq n/2$. We then pick an interval J to be one of $[0, \frac{q-1}{2}]$ and $[\frac{q+1}{2}, q-1]$ so that $T' = \{k \in T : \phi(q_k) \in J\}$ and $|T'| \geq |T|/2 \geq n/4$. We now compose ϕ with $(\text{mod } N)$ to obtain a mapping ψ, as follows.

$$\psi : \mathbb{Z} \xrightarrow{\phi} \{0, 1 \ldots q-1\} \xrightarrow{\text{mod } N} \mathbb{Z}/N\mathbb{Z}.$$

Now, we construct a tripartite graph G whose vertex set is $V = X \dot\cup Y \dot\cup Z$, where X, Y, Z are all copies of $\mathbb{Z}/N\mathbb{Z}$. The edges of the graph are defined as follows.

- $(x, y) \in X \times Y$ is in $E(G)$ if and only if there exists $i \in T'$ such that $y - x = \psi(p_i)$.
- $(y, z) \in Y \times Z$ is in $E(G)$ if and only if there exists $i \in T'$ such that $z - y = \psi(q_i)$.
- $(x, z) \in X \times Z$ is in $E(G)$ if and only if there exists $i \in T'$ such that $z - x = \psi(p_i) + \psi(q_i) \bmod N$.

Note that $|V(G)| = |X| + |Y| + |Z| = 3N$, and $|E(G)| = 3N|T'| \geq \frac{3}{4}Nn$. If x, y, z form a triangle, then there exist i, j, k such that $y - x = \psi(p_i)$, $z - y = \psi(q_j)$, and $z - x = \psi(p_k) + \psi(q_k)$. We claim that in this case it must hold that $i = j = k$, which we prove as follows. Since $(z - y) + (y - x) = z - x$, we have that $\psi(p_k) + \psi(q_k) = \psi(p_i) + \psi(q_j)$, which means that $\psi(p_k) + \psi(q_k) - \psi(p_i) - \psi(q_j) = 0$. Since ψ applies a $(\text{mod } N)$ function to ϕ, this means that

N divides $\phi(p_k) + \phi(q_k) - \phi(p_i) - \phi(q_j)$. By Property 3 and our choice of T', it holds that $\phi(p_k) + \phi(q_k) - \phi(p_i) - \phi(q_j) \leq q$, and hence by Property 2, we have that $\phi(p_k) + \phi(q_k) - \phi(p_i) - \phi(q_j) = \phi(p_k + q_k - p_i - q_j)$. This implies that N divides $\phi(p_k + q_k - p_i - q_j)$, which only holds if $p_k + q_k - p_i - q_j = 0$ due to our choice of λ.

Since such a triangle implies $i = j = k$, this shows that each edge is in exactly one triangle hence, and so $Nn/4$ edges need to be removed to make G triangle free. The number of triangles is $Nn/4 = O(N^2) = o(N^3)$ and therefore, by the Triangle Removal Lemma from Theorem 6, one can remove at most $o(N^2)$ edges to make G triangle free. This implies that $Nn/4 = o(N^2)$, and hence $n = o(N)$. However, we proved in Lemma 1 that $N \leq K^7 n$, and hence $n \neq o(N)$ which is a contradiction. Therefore, our initial assumption does not hold, i.e., there is no constant K such that $|P + Q| \leq K|P|$ for all $n = |P|$. This implies that $|P + Q| = \omega(n)$ as claimed.

4   An $ne^{O(\sqrt{\log n})}$ Recovery Threshold Upper Bound

In this section, we show how to obtain Rook Codes for batch matrix multiplication in the master-worker setting with a near-optimal recovery threshold.

Theorem 3 (Upper Bound). *There exist Rook Codes for batch matrix multiplication in the master-worker setting with a recovery threshold of $ne^{O(\sqrt{\log n})}$. That is,*

$$\text{Recovery}_{Rook\text{-}Codes}(n) = n^{1+o(1)}.$$

To prove Theorem 3, we show that there exist sets P, Q with $|P + Q| = ne^{O(\sqrt{\log n})}$ such that the decodability property (Property 1) holds.

In order to show this, we use the following known construction of a set which does not contain any 3-term arithmetic progression, defined as follows.

Definition 1. *A set $A = \{a_1, \ldots, a_\ell\}$ is an ℓ-term arithmetic progression (ℓ-AP) if there is a value d such that $a_i = a_1 + (i - 1)d$, for all $2 \leq i \leq \ell$.*

Behrend [7] showed a construction of a set which does not contain any 3-AP, whose maximum element is bounded by a value which will be useful for us.

Theorem 7 (Behrend [7]). *For all n, there exist a set A of distinct positive integers of size $|A| = n$ which does not contain any 3-term arithmetic progressions (3-AP), and such that $\max_{a \in A}\{a\} = ne^{O(\sqrt{\log n})}$.*

Note that the proof of Theorem 7 shows that there exists a 3-AP-free subset A of $\{1, 2, \ldots N\}$ of size at least $N^{1 - \frac{2\sqrt{2 \log 2} + \epsilon}{\sqrt{\log N}}}$ elements. By choosing $n = N^{1 - \frac{2\sqrt{2 \log 2} + \epsilon}{\sqrt{\log N}}}$, we have that $\log n = \Theta(\log N)$. Therefore, expressing N in terms of n we get that $N = ne^{O(\sqrt{\log n})}$, hence for any n, we can construct a 3-AP-free set with n elements such that $\max_{a \in A}\{a\} = ne^{O(\sqrt{\log n})}$, which is the form of the theorem presented here. With this construction, we can easily prove our upper bound on the recovery threshold of Rook Codes, as follows.

Proof (Proof of Theorem 3). Let c be a suitable constant such that for each n, there is a set $A_n \subseteq [ne^{c\sqrt{\log n}}]$ of size n which contains no 3-AP, as obtained by Behrend's construction in Theorem 7. Denote $A_n = \{a_1, \ldots, a_n\}$. Let $P, Q = A$ where $p_i = q_i = a_i$. This choice of P and Q satisfies the decodability property (Property 1), because if $2p_i = p_k + q_j$ then p_k, p_i, q_j is a 3-AP, hence p_k, p_i, q_j must all be equal. Furthermore, $P + Q \subseteq \{1, 2 \ldots 2ne^{c\sqrt{\log n}}\}$, with a size of $ne^{O(\sqrt{\log n})}$. Therefore, $\text{Recovery}_{\text{Rook-Codes}}(n) = ne^{O(\sqrt{\log n})}$.

5   The Computational Efficiency of Rook Codes

We provide the mathematical background that could explain the experimental results of [44], suggesting that Rook Codes may have faster computational times.

Coding for Batch Matrix Multiplication. In [51], the method of Lagrange Coding Computation (LCC) has been suggested for batch matrix multiplication. In this approach, the master defines two polynomials $\tilde{A}(x)$ and $\tilde{B}(x)$, by invoking point evaluation over the input matrices. In more detail, $\tilde{A}(x)$ is defined to be the unique polynomial of degree at most $n - 1$, which is interpolated from n points $z_0, \ldots, z_{n-1}$ whose values are $A_0, \ldots, A_{n-1}$, respectively. That is, $\tilde{A}(z_i) = A_i$ for all $0 \le i \le n - 1$. Similarly, $\tilde{B}(x)$ is interpolated so that $\tilde{B}(z_i) = B_i$ for all $0 \le i \le n - 1$. Then, for $0 \le w \le m - 1$, the master sends to worker w the coded matrices $\tilde{A}(x_w)$ and $\tilde{B}(x_w)$, for some value x_w. Each worker w multiplies its two matrices and sends their product back to the master. Note that $(\tilde{A} \cdot \tilde{B})(x) = \tilde{A}(x) \cdot \tilde{B}(x)$ is a polynomial of degree at most $2n - 2$, for which $(\tilde{A} \cdot \tilde{B})(z_i) = A_i \cdot B_i$ for all $0 \le i \le n - 1$. Thus, once the master receives back $\tilde{A}(x_w) \cdot \tilde{B}(x_w)$ from $2n - 1$ workers, it can extract $A_i \cdot B_i$ for all $0 \le i \le n - 1$ by interpolating the polynomial $(\tilde{A} \cdot \tilde{B})(x)$ and evaluating it at $z_0, \ldots, z_{n-1}$. This yields $\text{Recovery}_{\text{Lagrange-Codes}}(n) \le 2n - 1$.

In [31], a method of Cross Subspace Alignment (CSA) Coding for batch matrix multiplication is suggested, in which rational functions are used. More precisely, for a vector $\vec{z} = (z_0, \ldots, z_{n-1})$ of n distinct values in a field, let $f_{\vec{z}}(x) = \prod_{i \in [n]}(z_i - x) = (z_0 - x) \cdot \ldots \cdot (z_{n-1} - x)$, and define the rational encoding functions as $\tilde{A}(x) = f_{\vec{z}}(x)\left(\sum_{i \in [n]} \frac{1}{z_i - x} A_i\right)$, and $\tilde{B}(x) = \sum_{i \in [n]} \frac{1}{z_i - x} B_i$. Then it holds that $(\tilde{A} \cdot \tilde{B})(x) = \tilde{A}(x)\tilde{B}(x) = \sum_{i \in [n]} \frac{c_i(\vec{z})}{z_i - x} A_i B_i + \sum_{i \in [n-1]} x^i N_i$, where each $c_i(\vec{z})$ does not depend on x and each N_i denotes a noise (matrix) coefficient that can be disregarded. The master computes $\tilde{A}(x_w)$ and $\tilde{B}(x_w)$ for some distinct x_w that satisfy $\{z_i\}_{i \in [n]} \cap \{x_w\}_{w \in [m]} = \emptyset$ and sends it to worker w, for each $0 \le w \le m - 1$. Each worker w then multiplies its two matrices and sends their product back to the master. Note that $(\tilde{A} \cdot \tilde{B})(x) = \tilde{A}(x)\tilde{B}(x)$ is a rational function with at most $2n - 1$ zeros and n poles[1], and so the master can perform

---

[1] For any $p = 2n-1$ points $\{(x_i, y_i)\}_{i \in [p]}$ and n numbers $\{(z_i)\}_{i \in [n]}$ such that $\{z_i\}_{i \in [v]} \cap \{x_k\}_{k \in [p]} = \emptyset$, there exists a unique rational function $h(x) = f(x)/g(x)$ such that $h(x_i) = y_i$, $g(z_i) = 0$, $\deg(f) = 2n - 1$, and $\deg(g) = n$. Notice that this is a natural generalization of polynomial interpolation (see, e.g., [31, Lemma 1]).

rational interpolation from the values sent to it from any $2n - 1$ workers. Then the master retrieves $A_i \cdot B_i$ for all $0 \leq i \leq n - 1$, by taking the coefficient of $c_i(\vec{z})/z_i - x$. This yields $\texttt{Recovery}_{\text{CSA-Codes}}(n) \leq 2n - 1$.

A Comparison with Rook Codes. As mentioned, at a first glance, it may seem that LCC or CSA Codes already achieve the desired linear recovery threshold. However, Rook Codes are powerful in their computational complexity, which motivates bounding their recovery threshold. To see this, let us be more specific about the parameters of the setting, as follows. For all $0 \leq i \leq n - 1$, denote by $\chi \times \zeta$ the dimensions of the matrix A_i and by $\zeta \times \upsilon$ the dimensions of B_i.

Encoding. For the encoding time, all three methods (Rook Codes, LCC, and CSA Codes) must use some fast multipoint evaluation algorithm. In particular, Rook Codes and LCC use multipoint evaluation of a polynomial, and CSA Codes use a multipoint evaluation of rational functions, i.e., Cauchy matrix multiplication [50]. All three methods can compute the m values $\{\tilde{A}(x_w), \tilde{B}(x_w)\}_{w \in [m]}$ in time $O((\chi\zeta + \zeta\upsilon) \max\{m, d\} \log^c \max\{m, d\}) = O((\chi + \upsilon)\zeta m \log^c m)$, where c is some constant and d is the degree of the polynomials $\tilde{A}(x), \tilde{B}(x)$ in Rook Codes and LCC, or the total number of their roots and poles in CSA [50].

Decoding in Rook Codes. For the decoding time, in Rook Codes, the master can decode in time $O(\chi\upsilon R \log^c(R))$, for interpolating a polynomial over $\chi \times \upsilon$ matrices out of $R = \texttt{Recovery}_{\text{Rook-Codes}}(n)$ points. This gives that the total computational time (for encoding and decoding) at the master is $O((\chi + \upsilon)\zeta m \log^c(m) + \chi\upsilon R \log^c(R))$. This decoding time is much smaller than the encoding time in some cases, for example when $\chi = \upsilon$ and $\zeta = \omega(\chi R \log^c R/(m \log^c m))$. Since in such cases the encoding is the more expensive procedure, this motivates delegating the encoding to the workers in scenarios in which the master should do as little computation as possible (for example, when it is responsible for more than a single task). This is done by sending all matrices A_i and B_i for $0 \leq i \leq n - 1$ to all workers, and having each worker w compute $A(x_w), B(x_w)$ (the values in P, Q can be either also sent by the master or be agreed upon as part of the algorithm, depending on the setup). While this clearly has an overhead in terms of communication, in some cases this overhead can be very small compared to the original schemes in which encoding takes place at the master – as an example, if all workers are fully connected then the master can send A_i, B_i to a different worker w_i for each i, and each w_i can forward these two matrices to all other workers concurrently.

The benefit of delegating the encoding to the workers is that it allows the master to avoid any computation for encoding, and each worker w only has to evaluate the polynomial at its own x_w value, which needs at most $(\chi + \upsilon)\zeta n + O(n \log^{1/2}(n))$ multiplications (see Sect. 5.1). The $O(n \log^{1/2}(n))$ term does not depend on χ, υ, ζ and thus, is negligible when the matrices have large dimensions.

Decoding in LCC and LCA. The crucial difference between Rook Codes and LCC or CSA Codes, is that in Rook Codes, worker w can locally encode $\tilde{A}(x_w), \tilde{B}(x_w)$ using division-free algorithm (i.e., an algorithm using only $+$ or

$\cdot$, but not $/$); this stands to LCC and CSA Codes, which must perform divisions (by definition). Furthermore, it is not known how to efficiently delegate the encoding procedure to the workers in LCC since the encoding functions are $\tilde{A}(x) = \sum_{i \in n} A_i \prod_{j \in [n] \setminus \{i\}} \frac{x - z_j}{z_i - z_j}$ and $\tilde{B}(x) = \sum_{i \in n} B_i \prod_{j \in [n] \setminus \{i\}} \frac{x - z_j}{z_i - z_j}$ and, in particular, since there are $O(n^2)$ many products of the form $1/(z_i - z_j)$, any naïve algorithm must take quadratic time, and converting it to coefficient form for using Horner's scheme (see Sect. 5) incurs a $O(n \log^c n)$ overhead. Thus, the encoding step for LCC has complexity $O((\chi + \upsilon)\zeta n \log^c n)$, by [50, Theorem 10.10]. Moreover, divisions are known to be more expensive (for both LCC and CSA) [29,50], thus they increase the overhead at each worker. In particular, the current theoretical bound (using Newton iterations) on division in terms of multiplication for number fields is 3 [29]. Notice that evaluating the encoding function $\tilde{B}(x) = \sum_{i \in [n]} B_i \frac{1}{z_i - z}$ at a point x_w is very unlikely to use less than $n(\chi + \upsilon)\zeta$ divisions since even if we give it a common denominator, i.e., $\tilde{B}(x) = (\prod_{i \in [n]} \frac{1}{z_i - z}) \sum_{i \in [n]} B_i (\prod_{i \in [n]} z_i - z) \frac{1}{z_i - z}$, we would now have a super-linear increase in multiplications in the numerator. Furthermore, the evaluation of $\tilde{B}(x_w)$ is equivalent to computations involving Cauchy matrices, an old computational problem that has been studied extensively and conjectured to have a super-linear lower bound in the number of points plus poles (see [50]). Furthermore, division-free algorithms are often preferred because they tend to avoid issues of division-by-0 and false-equality-with-0 [41].

To summarize, there are settings in which Rook Codes are preferable over LCC and LCA codes, which motivates bounding their recovery threshold to obtain high robustness to faults.

5.1   Bounding Encoding Time of Our Rook Codes Construction

We provide here the promised analysis of the encoding time of our construction.

Lemma 2. *Let $p_0 < p_1 < ... < p_{n-1}$ and $q_0 < q_1 < ... < q_{n-1}$ be increasing orderings of the elements in P and Q, respectively, and let $\delta(P, Q)$ equal the number of multiplications needed to compute the values $\{x^{p_i - p_{i-1}}\}_{i \in [n]} \cup \{x^{q_i - q_{i-1}}\}_{i \in [n]}$, where $p_{-1} = q_{-1} = 0$. Then, given A_i, B_i for all i, each worker w can locally encode its values $\tilde{A}(x_w)$ and $\tilde{B}(x_w)$ using at most $\delta(P, Q) + (\chi + \upsilon)\zeta n$ multiplications.*

Proof. Each worker w uses Horner's rule [15] to compute $\tilde{A}(x_w), \tilde{B}(x_w)$. There are at most n non-zero coefficients in $\tilde{A}(x), \tilde{B}(x)$, thus Horner's rule takes the form $\tilde{A}(x) = x^{p_0}(A_0 + x^{p_1 - p_0}(A_1 + x^{p_2 - p_1}(A_2 + ...)))$ for $\tilde{A}(x)$ (similarly for $\tilde{B}(x)$), where the values $c_{w,p,k} = x_w^{p_k - p_{k-1}}$ and $c_{w,q,k} = x_w^{q_k - q_{k-1}}$ can all be computed using successive squaring for at most $\delta(P, Q)$ multiplications. More precisely, we have that $f_{n-1}(x_w) := \tilde{A}(x_w)$ can be recursively computed as $f_{i+1}(x_w) := c_{w,p,n-1-i}(A_{n-1-i} + f_i(x_w))$, where $f_0 = x_w^{p_{n-1} - p_{n-2}} A_{n-1}$. Therefore, computing $\tilde{A}(x_w)$ can be done using at most $\delta(P, Q) + (\chi + \upsilon)\zeta n$ multiplications and $(\chi + \upsilon)\zeta n$ additions. A similar argument applies for computing $\tilde{B}(x_w)$.

The following lemma bounds $\delta(P, Q)$, so that plugging this bound in Lemma 2 gives our claimed time for encoding of our Rook Codes.

Lemma 3. $\delta(P, Q) = O(n \log^{1/2}(n))$.

Due to space limitation, the proof of the above is deferred to the full version of this paper [13]

6   Discussion

This paper analyzes batch matrix multiplication in the master-worker setting that is both efficient and robust, by showing nearly matching upper and lower bounds for the recovery threshold of Rook Codes.

Our lower bound in Theorem 1 is surprising since, in the classical case of polynomial codes [30, 49] (or Reed-Solomon codes [40]), the communication rate is the same whether or not one defines the codewords as evaluations or as the coefficients of a polynomial. However, in the case of Rook codes and LCC we have that the recovery threshold (the measure of complexity that we care about here for batch matrix multiplication) is different between the evaluation codewords (Rook coding) and coefficient codewords (LCC).

We state here some further research directions.

Constructing 3-AP Free Sets. In the above setting, it is assumed that the number of pairs of matrices n is fixed. One may consider the non-uniform case, in which n is given as a parameter. In such a setting, the master would have to construct the appropriate 3-AP free sets according to n as part of its computation. For this, one would need a construction that is also computationally efficient. It may be an interesting direction for further research and, in particular, the construction of [22] could turn out useful. The latter constructions is also better in lower order terms of its size.

Possible Improvements of the Lower Bound. In additive combinatorics, problems such as finding 3-AP free sets are known as tri-colored sum-free problems [1, 24, 34]. Although these are similar to 3-AP free sets, the bound achieved is stronger for 3-AP free sets than for tri-colored sum free sets. If $A \subset [N]$ is a 3-AP free set, it is proven that $|A| = O(\max(A) \exp(-\Omega((\log N)^\beta)))$ for some constant $\beta > 0$ [33]. However, such strong qualitative bounds are not yet found for tri-colored sum free sets. In the language of [54], one reason for this is that the equality $a + b = 2c$, which is characteristic of the 3-AP problem, is preserved even after a, b, c is translated, whereas the equality $p + q = r$, which is characteristic of the tri-colored sum-free set is not preserved after translation (i.e., $(p + t) + (q + t) \neq (r + t)$). The proof for 3-AP free sets exploits this property by iteratively selecting sub-sequences of A, then scaling and translating this set to increase the density $|A| / \max\{A\}$ of the set. It is an open question whether the lower bound can indeed be increased to exactly match our upper bound.

Acknowledgements. This project has received funding from the European Union's Horizon 2020 research and innovation programme under grant agreement no. 755839, and from ISF grant 529/23. We thank Yufei Zhao for his advice for an alternate and more concise proof of Theorem 2.

References

1. Aaronson, J.: A connection between matchings and removal in abelian groups (2016)
2. Aggarwal, A., Dani, V., Hayes, T.P., Saia, J.: Multiparty interactive communication with private channels. In: Robinson, P., Ellen, F. (eds.) Proceedings of the 2019 ACM Symposium on Principles of Distributed Computing, PODC 2019, Toronto, ON, Canada, 29 July–2 August 2019, pp. 147–149. ACM (2019). https://doi.org/10.1145/3293611.3331571
3. Alon, N., Braverman, M., Efremenko, K., Gelles, R., Haeupler, B.: Reliable communication over highly connected noisy networks. Distrib. Comput. **32**(6), 505–515 (2019). https://doi.org/10.1007/s00446-017-0303-5
4. Attiya, H., Welch, J.L.: Distributed Computing - Fundamentals, Simulations, and Advanced Topics. Wiley Series on Parallel and Distributed Computing, 2nd edn. Wiley (2004)
5. Azad, A., et al.: Exploiting multiple levels of parallelism in sparse matrix-matrix multiplication. SIAM J. Sci. Comput. **38**(6) (2016). https://doi.org/10.1137/15M104253X
6. Ballard, G., Druinsky, A., Knight, N., Schwartz, O.: Hypergraph partitioning for sparse matrix-matrix multiplication. ACM Trans. Parallel Comput. **3**(3), 18:1–18:34 (2016). https://doi.org/10.1145/3015144
7. Behrend, F.A.: On sets of integers which contain no three terms in arithmetical progression. Proc. Natl. Acad. Sci. **32**(12), 331–332 (1946). https://doi.org/10.1073/pnas.32.12.331. https://www.pnas.org/doi/abs/10.1073/pnas.32.12.331
8. Censor-Hillel, K., Cohen, S., Gelles, R., Sela, G.: Distributed computations in fully-defective networks. In: Milani, A., Woelfel, P. (eds.) PODC 2022: ACM Symposium on Principles of Distributed Computing, Salerno, Italy, 25–29 July 2022, pp. 141–150. ACM (2022). https://doi.org/10.1145/3519270.3538432
9. Censor-Hillel, K., Dory, M., Korhonen, J.H., Leitersdorf, D.: Fast approximate shortest paths in the congested clique. Distrib. Comput. (2020). https://doi.org/10.1007/s00446-020-00380-5
10. Censor-Hillel, K., Gelles, R., Haeupler, B.: Making asynchronous distributed computations robust to noise. Distrib. Comput. **32**(5), 405–421 (2019). https://doi.org/10.1007/s00446-018-0343-5
11. Censor-Hillel, K., Kaski, P., Korhonen, J.H., Lenzen, C., Paz, A., Suomela, J.: Algebraic methods in the congested clique. Distrib. Comput. **32**(6), 461–478 (2019). https://doi.org/10.1007/s00446-016-0270-2
12. Censor-Hillel, K., Leitersdorf, D., Turner, E.: Sparse matrix multiplication and triangle listing in the congested clique model. Theor. Comput. Sci. **809**, 45–60 (2020). https://doi.org/10.1016/j.tcs.2019.11.006
13. Censor-Hillel, K., Machino, Y., Soto, P.: Near-optimal fault tolerance for efficient batch matrix multiplication via an additive combinatorics lens. CoRR abs/2312.16460 (2023). https://doi.org/10.48550/ARXIV.2312.16460

14. Chen, Z., Jia, Z., Wang, Z., Jafar, S.A.: GCSA codes with noise alignment for secure coded multi-party batch matrix multiplication. IEEE J. Sel. Areas Inf. Theory **2**(1), 306–316 (2021). https://doi.org/10.1109/JSAIT.2021.3052934

15. Cormen, T.H., Leiserson, C.E., Rivest, R.L., Stein, C.: Introduction to Algorithms, 3rd edn. MIT Press (2009). http://mitpress.mit.edu/books/introduction-algorithms

16. Drucker, A., Kuhn, F., Oshman, R.: On the power of the congested clique model. In: Proceedings of the ACM Symposium on Principles of Distributed Computing (PODC), pp. 367–376 (2014). https://doi.org/10.1145/2611462.2611493

17. Dutra, S., Bai, Z., Jeong, H., Low, T.M., Grover, P.: A unified coded deep neural network training strategy based on generalized polydot codes, pp. 1585–1589 (2018). https://doi.org/10.1109/ISIT.2018.8437852

18. Dutta, S., Fahim, M., Haddadpour, F., Jeong, H., Cadambe, V., Grover, P.: On the optimal recovery threshold of coded matrix multiplication. IEEE Trans. Inf. Theory 278–301 (2020)

19. D'Oliveira, R.G.L., El Rouayheb, S., Heinlein, D., Karpuk, D.: Degree tables for secure distributed matrix multiplication. IEEE J. Sel. Areas Inf. Theory **2**(3), 907–918 (2021). https://doi.org/10.1109/JSAIT.2021.3102882

20. D'Oliveira, R.G.L., El Rouayheb, S., Karpuk, D.: Gasp codes for secure distributed matrix multiplication. IEEE Trans. Inf. Theory **66**(7), 4038–4050 (2020). https://doi.org/10.1109/TIT.2020.2975021

21. Efremenko, K., Kol, G., Saxena, R.R.: Noisy beeps. In: Emek, Y., Cachin, C. (eds.) PODC 2020: ACM Symposium on Principles of Distributed Computing, Virtual Event, Italy, 3–7 August 2020, pp. 418–427. ACM (2020). https://doi.org/10.1145/3382734.3404501

22. Elkin, M.: An improved construction of progression-free sets. In: Charikar, M. (ed.) Proceedings of the Twenty-First Annual ACM-SIAM Symposium on Discrete Algorithms, SODA 2010, Austin, Texas, USA, 17–19 January 2010, pp. 886–905. SIAM (2010). https://doi.org/10.1137/1.9781611973075.72

23. Fahim, M., Cadambe, V.R.: Numerically stable polynomially coded computing. IEEE Trans. Inf. Theory **67**(5), 2758–2785 (2021). https://doi.org/10.1109/TIT.2021.3050526

24. Fox, J., Lovász, L.M.: A tight bound for green's arithmetic triangle removal lemma in vector spaces. In: Klein, P.N. (ed.) Proceedings of the Twenty-Eighth Annual ACM-SIAM Symposium on Discrete Algorithms, SODA 2017, Barcelona, Spain, Hotel Porta Fira, 16–19 January, pp. 1612–1617. SIAM (2017). https://doi.org/10.1137/1.9781611974782.106

25. Freiman, G.: Foundations of a structural theory of set addition. American Mathematical Society, Translations of Mathematical Mono graphs, vol. 37 (1973)

26. Le Gall, F.: Further algebraic algorithms in the congested clique model and applications to graph-theoretic problems. In: Gavoille, C., Ilcinkas, D. (eds.) DISC 2016. LNCS, vol. 9888, pp. 57–70. Springer, Heidelberg (2016). https://doi.org/10.1007/978-3-662-53426-7_5

27. Gupta, C., Hirvonen, J., Korhonen, J.H., Studený, J., Suomela, J.: Sparse matrix multiplication in the low-bandwidth model. In: Agrawal, K., Lee, I.A. (eds.) SPAA 2022: 34th ACM Symposium on Parallelism in Algorithms and Architectures, Philadelphia, PA, USA, 11–14 July 2022, pp. 435–444. ACM (2022). https://doi.org/10.1145/3490148.3538575

28. Haeupler, B., Wajc, D., Zuzic, G.: Network coding gaps for completion times of multiple unicasts. In: Irani, S. (ed.) 61st IEEE Annual Symposium on Foundations

of Computer Science, FOCS 2020, Durham, NC, USA, 16–19 November 2020, pp. 494–505. IEEE (2020). https://doi.org/10.1109/FOCS46700.2020.00053

29. Hasselström, K.: Fast division of large integers: a comparison of algorithms. Master's thesis (2003)
30. Huffman, W.C., Pless, V.: Fundamentals of error-correcting codes (2003). https://api.semanticscholar.org/CorpusID:60284659
31. Jia, Z., Jafar, S.A.: Cross subspace alignment codes for coded distributed batch computation. IEEE Trans. Inf. Theory **67**(5), 2821–2846 (2021). https://doi.org/10.1109/TIT.2021.3064827
32. Jia, Z., Jafar, S.A.: On the capacity of secure distributed batch matrix multiplication. IEEE Trans. Inf. Theory **67**(11), 7420–7437 (2021). https://doi.org/10.1109/TIT.2021.3112952
33. Kelley, Z., Meka, R.: Strong bounds for 3-progressions (2023)
34. Kleinberg, R., Sawin, W., Speyer, D.: The growth rate of tri-colored sum-free sets. Discret. Anal. (2018). https://doi.org/10.19086/da.3734. https://doi.org/10.19086%2Fda.3734
35. Lee, K., Lam, M., Pedarsani, R., Papailiopoulos, D., Ramchandran, K.: Speeding up distributed machine learning using codes. IEEE Trans. Inf. Theory 1514–1529 (2018)
36. Lynch, N.A.: Distributed Algorithms. Morgan Kaufmann (1996)
37. Milakovic, S., Selvitopi, O., Nisa, I., Budimlic, Z., Buluç, A.: Parallel algorithms for masked sparse matrix-matrix products. In: Lee, J., Agrawal, K., Spear, M.F. (eds.) PPoPP 2022: 27th ACM SIGPLAN Symposium on Principles and Practice of Parallel Programming, Seoul, Republic of Korea, 2–6 April 2022, pp. 453–454. ACM (2022). https://doi.org/10.1145/3503221.3508430
38. Petridis, G.: New proofs of Plünnecke-type estimates for product sets in groups. Combinatorica **32**(6), 721–733 (2012). https://doi.org/10.1007/s00493-012-2818-5
39. Raviv, N., Karpuk, D.A.: Private polynomial computation from Lagrange encoding. IEEE Trans. Inf. Forensics Secur. **15**, 553–563 (2020). https://doi.org/10.1109/TIFS.2019.2925723
40. Reed, I.S., Solomon, G.: Polynomial codes over certain finite fields. J. Soc. Ind. Appl. Math. **8**(2), 300–304 (1960). https://doi.org/10.1137/0108018
41. Rote, G.: Division-free algorithms for the determinant and the Pfaffian: algebraic and combinatorial approaches. In: Alt, H. (ed.) Computational Discrete Mathematics. LNCS, vol. 2122, pp. 119–135. Springer, Heidelberg (2001). https://doi.org/10.1007/3-540-45506-X_9
42. Ruzsa, I.: Sumsets and structure. Combinatorial Number Theory and Additive Group Theory (2009)
43. Soleymani, M., Mahdavifar, H., Avestimehr, A.S.: Analog Lagrange coded computing. IEEE J. Sel. Areas Inf. Theory **2**(1), 283–295 (2021). https://doi.org/10.1109/JSAIT.2021.3056377
44. Soto, P., Fan, X., Saldivia, A., Li, J.: Rook coding for batch matrix multiplication. IEEE Trans. Commun. **70**, 3641–3654 (2022). https://doi.org/10.1109/TCOMM.2022.3165201
45. Soto, P., Li, J.: Straggler-free coding for concurrent matrix multiplications. In: 2020 IEEE International Symposium on Information Theory (ISIT), pp. 233–238 (2020). https://doi.org/10.1109/ISIT44484.2020.sl20
46. Soto, P.J., Ilmer, I., Guan, H., Li, J.: Lightweight projective derivative codes for compressed asynchronous gradient descent. In: Chaudhuri, K., Jegelka, S., Song, L.,

Szepesvari, C., Niu, G., Sabato, S. (eds.) Proceedings of the 39th International Conference on Machine Learning. Proceedings of Machine Learning Research, vol. 162, pp. 20444–20458. PMLR (2022). https://proceedings.mlr.press/v162/soto22a.html

47. Tandon, R., Lei, Q., Dimakis, A.G., Karampatziakis, N.: Gradient coding: avoiding stragglers in distributed learning. In: ICML, pp. 3368–3376 (2017)

48. Tao, T., Vu, V.H.: Additive Combinatorics. Cambridge Studies in Advanced Mathematics. Cambridge University Press (2006). https://doi.org/10.1017/CBO9780511755149

49. Van Lint, J.H.: Introduction to Coding Theory, vol. 86. Springer, Cham (1998)

50. Von Zur Gathen, J., Gerhard, J.: Modern Computer Algebra. Cambridge University Press, Cambridge (2013)

51. Yu, Q., Li, S., Raviv, N., Kalan, S.M.M., Soltanolkotabi, M., Avestimehr, A.S.: Lagrange coded computing: optimal design for resiliency, security, and privacy. In: The 22nd International Conference on Artificial Intelligence and Statistics (AISTATS). Proceedings of Machine Learning Research, vol. 89, pp. 1215–1225. PMLR (2019). http://proceedings.mlr.press/v89/yu19b.html

52. Yu, Q., Maddah-Ali, M.A., Avestimehr, A.S.: Polynomial codes: an optimal design for high-dimensional coded matrix multiplication. In: Proceedings of the 31st International Conference on Neural Information Processing Systems (NIPS), pp. 4406–4416 (2017)

53. Yu, Q., Maddah-Ali, M.A., Avestimehr, A.S.: Straggler mitigation in distributed matrix multiplication: fundamental limits and optimal coding. IEEE Trans. Inf. Theory **66**(3), 1920–1933 (2020). https://doi.org/10.1109/TIT.2019.2963864

54. Zhao, Y.: Graph Theory and Additive Combinatorics: Exploring Structure and Randomness. Cambridge University Press, Cambridge (2023)

Deterministic Leader Election
for Stationary Programmable Matter
with Common Direction

Jérémie Chalopin[ID], Shantanu Das[ID], and Maria Kokkou[(✉)][ID]

Aix Marseille Univ, CNRS, LIS, Marseille, France
{jeremie.chalopin,shantanu.das,maria.kokkou}@lis-lab.fr

Abstract. Leader Election is an important primitive for programmable
matter, since it is often an intermediate step for the solution of more complex problems. Although the leader election problem itself is well studied
even in the specific context of programmable matter systems, research
on fault tolerant approaches is more limited. We consider the problem
in the previously studied Amoebot model on a triangular grid, when the
configuration is connected but contains nodes the particles cannot move
to (e.g., obstacles). We assume that particles agree on a common direction (i.e., the horizontal axis) but do not have chirality (i.e., they do not
agree on the other two directions of the triangular grid). We begin by
showing that an election algorithm with explicit termination is not possible in this case, but we provide an implicitly terminating algorithm that
elects a unique leader without requiring any movement. These results
are in contrast to those in the more common model with chirality but
no agreement on directions, where explicit termination is always possible but the number of elected leaders depends on the symmetry of the
initial configuration. Solving the problem under the assumption of one
common direction allows for a unique leader to be elected in a stationary and deterministic way, which until now was only possible for simply
connected configurations under a sequential scheduler.

Keywords: programmable matter · leader election · distributed
algorithms · Amoebot · stationary · common direction

1 Introduction

Programmable Matter (PM) refers to a distributed system consisting of a large
number of constant–memory computational entities (called *particles*) evolving
in a geometric environment and acting collaboratively in order to accomplish
a given task. We consider the task to be the well–known Leader Election (LE)
problem, that is, electing a unique leader among the particles. Particles in PM
systems do not have unique identifiers or a global sense of direction, making
it difficult to break the symmetry and elect a unique leader. In some contexts,
particles may be unable to move. For example, this can be so as to maintain

Y. Emek (Ed.): SIROCCO 2024, LNCS 14662, pp. 174–191, 2024.
https://doi.org/10.1007/978-3-031-60603-8_10

a configuration or due to the presence of obstacles (such as foreign objects) in the system. Hence, we study the case of particles electing a leader without any movement, proposing a so-called *stationary* algorithm. Leader Election without movement has been studied in the literature in the context of PM (e.g., [2,10]), however, it is not always possible to deterministically elect a unique leader. As a simple example consider a synchronous system of three particles forming a triangle in a triangular grid. Without additional assumptions, there is no deterministic algorithm that elects a unique leader. Moreover, even if the particles have chirality (i.e., a common notion of clockwise and counterclockwise directions) a unique leader cannot be obtained for every initial configuration by existing algorithms, such as [2,11]. Providing a method to elect a unique leader can be useful for problems where a task cannot be performed by multiple leaders but a single leader would make the problem solvable. For example, in [10] it is proven that k leaders in a k–symmetric initial configuration of particles cannot perform shape formation without additional capabilities, unless the target shape is also k–symmetric. We show that adding agreement on one direction instead of chirality suffices to elect a unique leader. We give an implicitly terminating algorithm where a unique particle eventually switches its state to leader and no particle ever changes state again, but particles never know whether the algorithm has terminated. We prove that explicit termination is not possible for particles agreeing on one direction without any additional capabilities. However, complimentary to the case of particles with common chirality, a unique leader is eventually elected in our setting even under an asynchronous but fair scheduler.

1.1   Related Work and Motivation

The concept of PM was introduced and formalized in [19]. Since then, various very distinct models such as [9,13,14,20] have been proposed. We use the Amoebot model (Sect. 2), introduced in [9] and updated in [8]. Leader Election has been studied in both a two dimensional setting, as we describe below, and in three dimensions in [4,16]. All LE algorithms for deterministic 2D settings, including the one in this paper, assume that particles operate in a geometric environment (i.e., triangular grid), have constant memory and are activated by a fair scheduler. However, each of the existing algorithms uses at least two additional assumptions or capabilities. We list the different options for those model choices here and underline the ones used in our work: *(i)* using a randomized algorithm to break symmetries (e.g., [7]) or a deterministic algorithm *(ii)* electing $k \in \{1, 2, 3, 6\}$ leaders in k-symmetric configurations (e.g., [10]) or a unique leader regardless of the configuration *(iii)* a simply connected configuration (e.g., [15]) or a configuration containing holes *(iv)* particles with (e.g., [2]) or without common chirality which is a common sense of rotational orientation *(v)* particles that can move from a node to a neighbouring one (e.g., [12]) or stationary particles and *(vi)* a sequential scheduler (e.g., [11]) which activates one particle at a time and the particle finishes its action or an asynchronous scheduler which simultaneously activates any number of particles. It is worth noting that using the concurrency control framework presented in

Table 1. Deterministic LE in two-dimensions.

Paper	Leaders	Simply Connected Particles	Chirality	Movement	Seq. Scheduler	One Direction	Dimension
[15]	1	✓	✓	✗	✓	✗	2D
[12]	1	✗	✗	✓	✓	✗	2D
[11]	1	✗	✓	✓	✓	✗	2D
[10]	3	✓	✗	✗	✗	✗	2D
[11]	6	✗	✓	✗	✓	✗	2D
[2]	6	✗	✓	✗	✗	✗	2D
[4]	1	✓	✗	✗	✓	✗	2D & 3D
This Work	1	✗	✗	✗	✗	✓	2D

[8], algorithms which terminate under a sequential scheduler and satisfy certain conditions can be transformed into equivalent algorithms that work in the concurrent setting using *locks*. However, even if we remove the sequential scheduler assumption in previous stationary LE algorithms given the results of [8], current stationary algorithms cannot elect a unique leader without additional capabilities. We summarise the results and assumptions of previous work in Table 1.

We are interested in determining the minimum capabilities that particles need to deterministically elect a unique leader in a 2D system. To the best of our knowledge the case of particles agreeing on one direction instead of having chirality or movement capabilities has not been studied before. Therefore, an additional motivation was to determine the differences between assuming chirality and assuming common direction. We show that the difficulties that arise from each of the assumptions are different and the results that can be obtained are complementary to each other. In previous work like [2] it was shown that assuming chirality we can get an explicitly terminating algorithm, but up to six leaders are elected depending on the symmetry of the configuration. We show here that agreement on one direction allows for a single leader to be elected, however, an algorithm with explicit termination is no longer possible (Sect. 3). Hence, perhaps surprisingly, common chirality and common direction are not directly comparable with respect to which one is more general.

From a practical perspective, we find scenarios where a large number of particles can autonomously detect the intended common direction more natural, whereas, agreement on chirality is potentially a less intuitive capability to implement. For example, particles can be placed at a slightly tilted plane in order to collectively agree on a common direction, similarly to the setting in [3]. Alternatively, a large enough source of light placed at a sufficiently long distance can provide a common direction for light detecting particles. A setting with some similarities to this case was described in [18].

1.2 Our Contributions

We provide the first result on deterministically electing a unique leader without using movement or chirality, assuming that the particles agree on one direction

and that the initial configuration is connected. We show that explicitly terminating LE is not possible for this case (Sect. 3) but we give an algorithm that
elects a unique leader in $O(n^3)$ rounds, where n is the number of particles in
the system (Sect. 5). Our results are complementary to the case of particles only
agreeing on chirality, such as in [2].

Our algorithm (Sect. 5) uses a message passing procedure to elect intermediate leaders on boundaries. In Sect. 4 we show that if a pair of particles on a
boundary can *detect* each other's chirality, they can differentiate between the
boundaries in which they participate. We then give a message passing procedure
for such particles, so that a message originating at a particle on some boundary
B is only received by particles also on B. Once intermediate leaders are elected,
the system is encoded as a set of trees, each rooted at one of the leaders, and the
trees are compared and merged until only one root remains. Encoding parts of
a network as trees in order to compare them, is a technique that is used in both
mobile agent computing (e.g., [6]) and in PM systems (e.g., [10]). However, here
we mark the endpoints of the edges where a comparison occurs in the encoding
of the tree, adding a dynamic aspect to the system encoding.

Due to space constraints, some proofs, the pseudocode and a complexity
discussion have been omitted but can be found in the full version [5].

2   Model and Preliminaries

Let G_Δ be an infinite regular triangular grid. We use two coordinates, x and y, to
describe the relative position of nodes in the grid. Suppose $v_1 = (x_1, y_1)$ and v_2
are adjacent nodes of G_Δ. If v_2 is on the East (resp. West) of v_1, $v_2 = (x_1+2, y_1)$
(resp. $v_2 = (x_1 - 2, y_1)$). If v_2 is on the North East (resp. South West) of v_1,
$v_2 = (x_1 + 1, y_1 + 1)$ (resp. $v_2 = (x_1 - 1, y_1 - 1)$). Finally, if v_2 is on the North
West (resp. South East) of v_1, $v_2 = (x_1 - 1, y_1 + 1)$ (resp. $v_2 = (x_1 + 1, y_1 - 1)$).
The particle system consists of a finite, connected subset of G_Δ, such that each
of the nodes in this subset contains exactly one particle. We refer to nodes
occupied by particles as *particle nodes* and to nodes not occupied by particles as
non–particle nodes. Each particle has six ports associated with a *port number* in
$\{0, \ldots, 5\}$, corresponding to each edge leading to a neighbouring node. We define
ports 0 (East) and 3 (West) to be common for all particles. The remaining ports
are labelled in a circular way such that port i and port $i + 1$ (mod 6) lead to
neighbouring nodes and the local labelling of the ports is known to the particle.
We refer to ports with a port number in $\{0, 1, 5\}$ (resp. $\{2, 3, 4\}$) as *right* (resp.
left) ports. For directions that are not 0 or 3, the port numbers are not consistent,
even among neighbours. The particles have constant memory and they cannot
move, but they can communicate with particles at distance one by exchanging
messages. A message, m, sent by an active particle p to a neighbouring particle
p' is received when p' is activated and both p and p' know the ports m is sent and
received through. We assume that particles are activated by a *fair asynchronous*
scheduler so that every particle is activated infinitely often and it is possible but
not necessary that all particles are activated simultaneously.

We call an edge uv a *local boundary* if there exists a non–particle node o that is a common neighbour of u and v and we write (uv, o) to denote the local boundary. A sequence of local boundaries $(u_1 u_2, o_1), (u_2 u_3, o_2), \ldots, (u_k u_1, o_k)$ is a *boundary* when moving around u_i from u_{i-1} to u_{i+1} only non–particle nodes starting with o_{i-1} and ending with o_i are encountered. Notice that it is possible that $o_{i-1} = o_i$. For any (uv, o) we say that u and v are neighbours on the local boundary. We call particles that are on at least one boundary and are connected to each other through their 0 and 3 ports, a *horizontal path*. An edge between two particles, p_a and p_b, on a horizontal path in which p_a and p_b do not have a common neighbouring particle, is called a *dark blue edge* (DBE). An edge between p_a and p_b in a horizontal path where p_a and p_b have one common neighbouring particle is called a *light blue edge* (LBE). All other edges are *grey edges*. A subsystem of particles connected by grey edges and LBEs but not by DBEs, is called a *grey component*. An example is shown in Fig. 1.

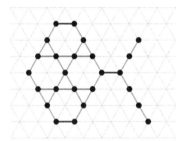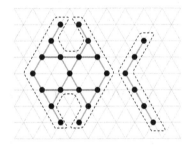

Fig. 1. Example of grey, dark blue and light blue edges in two grey components connected by a DBE. The two grey components without any of the DBEs are traced in the second subfigure. (Color figure online)

Remark 1. Any pair of particles, $\{p, p'\}$, connected by a grey edge can detect whether they have common chirality.

Each particle looks at the port leading to the other particle of the pair. If the grey edge that is incident to port $i \in \{1, 2, 4, 5\}$ of p and incident to port $i' \in \{1, 2, 4, 5\}$ of p' satisfies $i + 3 \pmod{6} = i'$ and $i' + 3 \pmod{6} = i$ the particles know they agree on chirality, otherwise they know they disagree.

Remark 2. Any pair of particles, $\{p, p'\}$, connected by an LBE can detect whether they have the same chirality.

Each of p, p' is connected to a common neighbour, q, by a grey edge. From Remark 1, $\{p, q\}$ (resp. $\{p', q\}$) know whether they have common chirality. So q can inform p (resp. p') whether it has common chirality with p' (resp. p).

Remark 3. Any pair of particles, $\{p, p'\}$, connected by a DBE cannot locally detect whether they have the same chirality.

Contrary to Remark 1, it is always the case that $i+3 \pmod 6 = i'$ and $i'+3$ (mod 6) $= i$, so this cannot be used to detect chirality. Contrary to Remark 2, p and p' do not have common neighbours and by the problem definition they cannot move to communicate information like in [10,12]. So p and p' cannot locally detect whether they have common chirality. The difference in whether particles on horizontal edges can detect having common chirality (Remarks 2 and 3) is why we split horizontal edges into DBEs and LBEs.

3   Explicit Termination Impossibility

We prove that an explicitly terminating LE algorithm is not possible without imposing any restrictions on the system, using the same basic indistinguishability argument as in [1]. Although we assume a synchronous scheduler, the same result holds for sequential activations by activating all equivalent particles (as defined in Fig. 2) sequentially instead of simultaneously. Our proof only addresses the deterministic case, however, it can be shown that no probabilistic algorithm solves the problem by applying the method of [17].

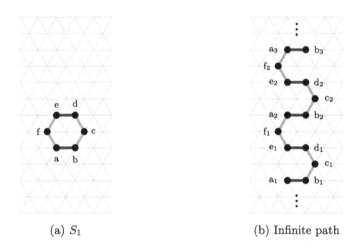

(a) S_1 (b) Infinite path

Fig. 2. Systems where all equivalent particles have the same local information cannot locally differentiate between the configurations.

Theorem 1. *There does not exist a terminating algorithm that solves LE under a fair synchronous scheduler when the initial configuration can contain holes and the particles cannot move, even if all particles agree on a direction.*

Proof. Let S_1 be the particle system in Fig. 2a and S_{inf} be the infinite particle system presented in Fig. 2b. Each occupied node v_i in S_{inf} is mapped to an occupied node v in S_1, such that $v_i \mapsto v$ for $v \in \{a, b, c, d, e, f\}$ and $i \in \mathbb{Z}$. We assume that all particles b_i, c_i or d_i (resp. a_i, e_i or f_i) in S_{inf} have common (resp. opposite) chirality as the particle they are mapped to in S_1. Locally, the neighbourhood at distance one of $\{a_i, b_i, \ldots, f_i\} \in S_{inf}$ is the same as the neighbourhood at distance one of $\{a, b, \ldots, f\} \in S_1$, so particles in S_1 and S_{inf} have the same input and local information. Let us assume that there exists an algorithm $\mathcal{A}$ solving Terminating LE in all connected configurations. Let us further assume that $\mathcal{A}$ solves Terminating LE in S_1 after s synchronous steps. Let u be a node in S_{inf} and let S_2 be the subgraph induced in S_{inf} by all particles at distance $k \geq (2*6+2) + (2*s+2)$ from u. Now let us consider a synchronous execution of $\mathcal{A}$ in S_2. By construction, there must exist at least two particles in S_2, q and q', that are in the same state as the elected particle in S_1 after s steps of $\mathcal{A}$ such that both q and q' are at distance at least $s + 1$ from each endpoint of S_2. Therefore, $\mathcal{A}$ elects at least two leaders in S_2 and there cannot exist any algorithm solving Terminating LE.

4 Message Forwarding on Boundaries

Existing algorithms such as [2,11] use chirality to move messages along boundaries. Although in this work we assume that particles do not have common chirality, we define a method to substitute the chirality assumption in existing algorithms in order to use techniques from previous work. The following procedure forwards messages on boundaries of grey components and we prove that a message originating at a particle in some boundary B is received by particles on B and it is not received by particles not on B.

Message Forwarding on Boundaries: Let p_s, p_1 and p_2 be three consecutive particles on boundary B and say that p_s holds a message it will send to p_1. Call the common non–particle neighbour of p_s and p_1 on B, o_1 and the common non–particle neighbour of p_1, p_2 on B, o_2. When p_s sends the message to p_1 it attaches the label of the port of p_1 leading to o_1 using the orientation of p_1. Call that label $boundary\text{–}label$. When p_1 receives the message from p_s through port z it reads the $boundary\text{–}label$ attached to the message and sets $next\text{–}particle$ to be the first particle to be reached by following the cyclic port ordering $< z, boundary\text{–}label, \ldots >$, which in this case we have called p_2. p_1 detects whether it has common chirality with p_2, using the method described in Remark 1 if (p_1, p_2) is a grey edge or the method described in Remark 2 if (p_1, p_2) is an LBE. Call the port of p_1 leading to p_2 port x, the port of p_2 leading to p_1 port y, the port of p_1 leading to o_2 port i and the port of p_2 leading to o_2 port j. If p_1 and p_2 have common chirality and $i = x - 1 \pmod 6$ (resp. $i = x + 1 \pmod 6$), p_1 calculates $j = y + 1 \pmod 6$ (resp. $j = y - 1 \pmod 6$). Equivalently, if p_1 and p_2 do not have common chirality and $i = x - 1 \pmod 6$ (resp. $i = x + 1 \pmod 6$), p_1 calculates $j = y - 1 \pmod 6$ (resp. $j = y + 1 \pmod 6$). Finally, p_1 sets $boundary\text{–}label$ to j and sends the message to p_2.

Theorem 2. *Let $\{p, p'\}$ be a pair of particles neighbouring on at least one boundary of a grey component. If both p and p' use the above procedure, any message passing from p to p' always remains on the same boundary.*

Proof. Let p be the particle from which p' receives the message. If p is the particle the message originates from, p decides the boundary, B, on which the message is sent. From the definition of a *local boundary*, we know that p and p' share a common non–particle neighbour, say o. Due to the common direction, if p is connected to p' through a right port (resp. left port), p' is connected to p through a left port (resp. right port). Furthermore, if (p, p') is grey and p is connected to o through a left port (resp. right port), p' is also connected to o through a left port (resp. right port). Particles connected by an LBE know they only have one common non–particle neighbour. Since o is a common neighbour it must be reached by a port at distance one from the port connecting p to p'. The type of edge (i.e., grey or LBE), whether o is reached by a left or right port and the port of p' leading to p are known to p. Moreover, the handedness of p' can be computed by Remarks 1 and 2. Therefore, p can compute the port of p' leading to o. If p is not the particle the message originates from, p learns the boundary the message is forwarded on by receiving the port label leading to the common non–particle neighbour defining the boundary when receiving m by its predecessor on the boundary, in the same way as p' learns the boundary in the case where the message originates from p and the proof remains the same.

5   Algorithm Description

The system consists of a number of grey components connected by DBEs. Each grey component elects a local leader. In each round, every component attempts to *"compete"* with two neighbouring components, where *"neighbouring"* refers to components connected by a DBE. The result of a competition for each component is *win*, *lose* or *draw*. The first two results occur when the competing components detect a difference between them. In this case, the components merge creating a new component consisting of all the particles of the two previous components and the local leader of the winning component. The result of a competition is a draw when two components cannot detect a difference between them, for example when a component is competing with itself. In this case, the components remain separate and do not compete on that DBE again, unless at least one of them merges with another component. We prove that when no competition on any DBE is possible anymore, all particles belong to one common component with one leader. We address each part of the algorithm separately.

5.1   Grey Leader Election

We begin by electing a unique leader in every grey component. Each particle sees its neighbourhood at distance one and differentiates between grey, light blue and dark blue edges. For a DBE uv connecting two particles, p at u and p' at v, p (resp. p') removes p' (resp. p) from its list of neighbours during Grey LE and treats v (resp. u) as a node not occupied by a particle. In existing stationary LE algorithms (e.g., [2]) and algorithms that present intermediate tools for stationary particles (e.g., [11,12]), it is assumed that particles have common chirality which is used to exchange information along boundaries. For k–symmetric configurations, those algorithms elect $k \in \{1, 2, 3, 6\}$ leaders in symmetric positions on the outer boundary. To facilitate the presentation of our algorithm, we say that we use the algorithm presented in [2] in this step, however, any algorithm for stationary particles that terminates electing 1,2,3 or 6 heads in symmetric positions on the outer boundary could be used. In [2] common chirality is used to move information along boundaries. We substitute the assumption of common chirality in [2] in the parts of the system where this is possible (i.e., grey components) by the method in Sect. 4. However, since the particles do not agree on chirality we execute the algorithm of [2] in both directions on each boundary of each grey component simultaneously, electing $k \in \{1, 2, 3, 6\}$ heads that know the number of elected particles for each direction of the outer boundary. Particles in inner boundary competitions learn that they are not elected when the competition on the unique outer boundary terminates and the procedure described in Sect. 5.2 begins.

Each head calculates an ID for itself (the port number $a \in \{0, \dots, 5\}$ leading to the particle's neighbour on the boundary in dir) and an ID_i for the i–th head (where $i = \{1, \dots, k-1\}$) on the boundary, $ID_i = a + i * \frac{6}{k} \pmod 6$, using the local port numbering. We impose the ordering $ID = 0 > ID = 3 > ID = \{1, 5\} > ID = \{2, 4\}$ and only the head with the maximum ID remains a head. Notice that particles with different chirality might not agree on the IDs but since the left and right directions are common, if a head h computes $ID_h \in \{0, \{1, 5\}, 3, \{2, 4\}\}$ for its own ID, all other heads on the boundary also compute ID_h as the ID of h. Eventually, two heads remain per boundary after winning the dir and the dir' competitions.

Each of the two remaining heads sends a competition message in the direction it was elected. Either both messages are received by a common particle or by two neighbouring particles. In the first case, the single particle picks one of the heads based on its local orientation. In the latter case, p_{c_1}, p_{c_2} know which is the rightmost particle by looking at the ports connecting them. The rightmost particle then sends an *elected* message to the head that reached it and the leftmost particle sends a *not–elected* message to the remaining head. After this step, one leader remains and it is on the outer boundary. The leader then initiates the *Tree Construction* phase of the algorithm and particles located on inner boundaries learn that LE in the component has terminated.

Correctness. We prove the correctness of Grey Leader Election.

Lemma 1. *After executing the LE algorithm of [2] in both directions for every boundary, 1,2,3 or 6 leaders are elected on symmetric positions in the outer boundary for each direction and each head knows the direction of the competition in which it was elected and the number k of elected heads in that direction.*

Each head on the outer boundary can compute a set of distinct IDs, each corresponding to another head on the boundary, with respect to its own coordinate system, even in the setting of [2]. However, in that case it is possible that different particles locally compute the same ID for distinct heads in the boundary. We show that by adding agreement on one direction, the heads do not compute the same set of IDs corresponding to different particles.

Lemma 2. *Using the locally computed identifiers of all heads on the boundary for a given direction, all heads choose the same particle to remain a head, without communicating with each other. Eventually, only one head remains in each direction on the outer boundary.*

Proof. Due to the particles not having chirality, the sets of IDs computed by different heads may be different. The calculation of 0 or 3 as the ID for any head is consistent among all heads since 0 and 3 are common directions. Suppose that for $k \in \{3, 6\}$ there does not exist a particle with $ID = 0$ or $ID = 3$. Using the formula for calculating IDs, $ID_i = a + i * \frac{6}{k} \pmod 6$, the ID calculated for at least one of the heads is either 0 or 3 and we immediately get a contradiction. For $k = 2$ it is possible that no head with $ID \in \{0, 3\}$ exists. This time, we use the fact that port 1 is mapped either to port 1 or port 5 (similarly for port 5), depending on the chirality of the particles and the same holds for ports 2 and 4. A head, h, calculating $a_h \in \{1, 5\}$ (resp. $a_h \in \{2, 4\}$) for its ID calculates $a_{h'} \in \{2, 4\}$ (resp. $a_{h'} \in \{1, 5\}$) for the other head, h'. In the same setting h' computes $a_{h'} \in \{2, 4\}$ (resp. $a_{h'} \in \{1, 5\}$) for its own ID and $a_h \in \{1, 5\}$ (resp. $a_h \in \{2, 4\}$) for the ID of h. Thus $a \in \{1, 5\}$ corresponds to only one of the two heads, and it is elected. Therefore, in all cases one head remains on the outer boundary for each direction in each grey component.

Lemma 3. *For a boundary B where exactly one leader remains for each direction, the leaders can always reach either one common node or a pair of neighbouring nodes occupied by particles. The reached particle(s) can differentiate between the leaders and elect a unique leader for the boundary.*

From the above lemmas, we obtain the following.

Theorem 3. *After the execution of the Grey Leader Election algorithm, a unique leader is elected on the outer boundary of every grey component.*

5.2   Tree Construction

We encode each grey component so that leaders in different grey components, connected by DBEs, can compare their components. We create a spanning tree of the component, using a standard technique from the distributed computing literature, in which the root is the unique leader of the component. Each particle in the tree computes a label encoding its neighbourhood. By following the reasoning in [10], we encode the neighbourhood of a particle p using six characters, each corresponding to the node reachable through port $i \in \{0, \ldots, 5\}$. Each character in the label has one of the following values.

- *P*: Node occupied by the parent of *p*.
- *C*: Node occupied by a child of *p*.
- *D*: Node connected to *p* through a DBE. Note that if port[0] is encoded as D the DBE is outgoing while if port[3] is encoded as D the DBE is incoming.
- *E*: Node not occupied by a particle.
- *N*: Neighbour particle of *p* that is not a neighbour on the tree or a DBE.

Each particle knows its parent and its children, as well as the information that is locally observed from its immediate neighbourhood (i.e., non–particle or particle nodes and incoming or outgoing DBEs). Notice that the characters in the label are mutually exclusive during Tree Construction, so it cannot be the case that more than one characters in {P,C,D,E,N} are needed to represent a neighbour of *p*. The correctness of the tree construction phase comes directly from [10].

5.3   Component Competition

In this section, we describe how particles in different components are compared. We begin by describing how the trees defined in Sect. 5.2 are traversed in order to be compared to adjacent trees through the directed DBEs connecting them. Then we discuss how components choose which DBEs to compete on and how components merge after a comparison, if merging is possible. Finally, we show that comparisons eventually stop and that only one leader remains in the system when no more comparisons are made. We refer to each iteration of the set of procedures in this section (i.e., Tree Traversal, Tree Comparisons and Merging Components) by a component as a *round*.

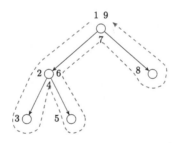

Fig. 3. An example of cyclic–DFS demonstrating the traversal of a tree represented as a ring. The numbers denote the order in which each node is visited.

Tree Traversal: Each node orders its children in increasing order with respect to its port numbers so that the ordering shown in Fig. 3 can be obtained. We call the ring calculated in such a way "cyclic–DFS", also known as the Euler tour of the tree. Notice that the five node tree of Fig. 3 is represented by a nine–node ring and that in the traversal order of Fig. 3, the root is the first as well as the last node. The root marks its first label as *root–label* and its last label as the *last–node*. We later use the ring encoding to move information in the form of labels and tokens between particles of the same tree.

Encoding a Tree as a Ring: Each node of the tree is split into $c + 1$ virtual nodes called *agents*, where c is the number of children of that node in the tree. Each agent represents one node of the equivalent ring. Let $root = v_1, v_2, \ldots, v_n$ be the nodes of the tree encoding of the component. We write $v_i.agent_j$ for $i \in \{1, \ldots, n\}$ and $j \in \{1, \ldots, 7\}$ to describe the j-th time node v_i is visited in cyclic DFS. Equivalently, $v_i.agent_j$ also denotes that $j - 1$ children of v_i have been visited in the traversal. In "Label Forwarding Cycle" we show how to move information using the ring. The "Label Forwarding Cycle" procedure can be started by any arbitrary agent, say *p.agent*. The procedure finishes when the label of a target node, call it target–label, reaches *p.agent*.

Label Forwarding Cycle (LFC): *p.agent* simultaneously sends its label to the *previous–node* on the ring and pulls the label, l, of the *next–node*, on the ring. When some $p'.agent \neq p.agent$ holds two labels, it forwards the oldest label to the previous–node. When $p'.agent$ holds one label but the previous–node does not hold any label, $p'.agent$ forwards the label it holds to the previous–node. In all other cases, particles do nothing. While l is not the *target–label*, *p.agent* waits until the next–node gets a new label and then pulls the label from the next–node and forwards its current label to the previous–node. When l is the target–label, *p.agent* sends a termination–message around the tree through its previous–node. Each particle that receives the termination–message, forwards it if it holds one label or holds it while it holds more than one labels. When the termination–message returns to *p.agent*, the label forwarding procedure finishes.

The label of an agent r at distance i from *p.agent* in the ring reaches *p.agent* once each particle between r and *p.agent* has performed $2i$ steps. This is because $i - 1$ particles between r and *p.agent* need to forward their label before r sends its own and i more steps are needed to reach *p.agent*.

Tree Comparisons: Let C be a grey component. Initially, all incoming and outgoing DBEs are marked as *not–compared*. C marks as *chosen* the first outgoing DBE uv encountered in the cyclic–DFS traversal of the tree encoding C that is not already marked as *compared* if one exists, such that $u \in C$ and $v \in C'$, where C' is also a component. Of the incoming DBEs that are marked as chosen by components at distance one, C marks as *chosen* the first one that is encountered in the cyclic–DFS traversal of the tree encoding it, if one exists. If there is at least one incoming edge, C makes a comparison on the incoming edge it selected. If the outgoing edge selected by C is also selected by a neighboring component as an incoming edge, C makes a comparison on this edge. Thus, C performs at most two comparisons at the same time. It is possible that no outgoing or incoming DBEs exist or that all outgoing DBEs are already marked as *compared*, so C does not make any comparisons. When C selects an incoming DBE for a comparison it messages all other incoming DBEs selected by neighbouring components that they do not participate in the current competition and the corresponding edges remain *not–compared*. The two endpoints of each edge on which a comparison is performed also encode being part of an edge participating in a comparison as part of their label. For example a label ECEDEP is changed to ECE<u>D</u>EP as long as the DBE is used for a comparison. When the edge is no longer used, the label is changed back to ECEDEP.

Let p_{db} be the endpoint of a DBE selected for a competition. $p_{db}.agent_1$ initiates an LFC, using *root–label* as the target label. If two comparisons are made simultaneously, an LFC is done separately but in parallel for the incoming and the outgoing comparisons. Each node in the tree holds two copies of the labels, one for each of the forwarding procedures. When initiating the LFC, $p_{db}.agent_1$ marks whether the forwarding cycle it initiates is for the comparison on an incoming DBE or an outgoing DBE. When the LFC finishes, p_{db} marks itself as *ready* and waits for the remaining endpoint of the DBE it is competing on, p'_{db}, to also mark itself as *ready*. When p_{db} and p'_{db} are both marked as *ready*, they each restart the LFC mechanism and compare each of the labels they receive during LFC to the label received by the other endpoint of the DBE, as follows. If the trees differ in the label of some particle, for the first pair of particles with different labels, take the first position of the label in which a difference is detected, j. We arbitrarily define the ordering $P > C > D > \mathbf{D} > E > N$. We say that the component that has the smaller symbol in position j, *loses* the comparison, or equivalently, that it is a *losing* component. If the trees of the compared components have the same labels during the comparison but one tree has more nodes, the component with less particles *loses*. Each competing component waits until all comparisons in which it participates finish before participating in a merge. Every node in the tree knows the number of comparisons the component participates in due to the LFCs. When a particle incident to a DBE finishes a comparison, it sends a message to the root through its parent with the result of the comparison. A component, C, that loses one comparison to another component C', merges with C' using the procedure in "Merging Components". If C loses two comparisons, it picks the component reached by the outgoing DBE it lost to, call it C', and only merges with C' using the procedure in "Merging Components". Each edge on which either a merge is performed or the components are detected to be equivalent is marked as *compared*. When two components merge, the DBE on which the merge was performed becomes *compared* and all DBEs that were marked as compared in a previous round become *not–compared*. When an endpoint of an edge is marked as *not–compared*, the remaining endpoint of the edge also marks the edge as *not–compared*.

Merging Components: After a component C' loses, it merges with the component that won, C. Let uv, such that $u \in C$ and $v \in C'$, be the DBE on which the merge is executed. The particle at v initiates the merge by sending a message to its parent. The merging message is moved through the tree by each particle marking itself as in–merge and forwarding the message to its parent until the root is reached. When the root receives the merging message, it sets its state to *follower*, marks its in–merge neighbour as its parent by switching its corresponding port in the label to P and sends the merging message back to the in–merge neighbour. Each in–merge particle that receives the merge–message for a second time, marks the node from which it received the message as its child (i.e., switches P to C in its label), marks its remaining in–merge neighbour as its parent by switching its corresponding port in the label to P, forwards the merge–message to its in–merge neighbour and unmarks itself as in–merge. When v receives back the merge–message it marks the node from which it received the message as its child and u as its parent in its label. Finally, when v changes its label to mark u as a parent (i.e., v changes $\mathbf{D}$ in its label to P), u also marks v as its child (i.e., u changes $\mathbf{D}$ in its label to C), the merge is complete and the new component CC' is ready to begin a new comparison. When the merge is complete, u informs the root by sending a merge–complete message to its parent. The merge–complete message is then forwarded by each particle that receives it to its parent until the root is reached. After a merge, CC' marks all endpoints of DBEs (i.e., particles with D in their label) as "finished" and waits until the remaining endpoint for each of the DBEs is also marked as finished. When all neighbouring components have finished the current comparison, CC' begins executing "Tree Traversal" again, until no more edges marked as not–compared remain. While CC' only has *compared* edges it waits until an edge becomes not–compared as described in "Tree Comparisons". A component participating in a merge as a losing (resp. winning) component might be simultaneously participating in a merge on a different DBE as a winning (resp. losing or winning) component. A component participates in at most one merge as *losing*, so the exchanged messages in the above procedure are unique. Additionally, a component simultaneously participating in two merges as *winning* does so only through the two particles incident to the chosen DBEs and the resulting changes in the tree are local changes in the respective positions of the labels of those particles. It is possible that a component consisting of only one particle that is only connected to components consisting of one particle each, is a *winning* component in both merges. In this case, the single particle of the component implements both winning merges as described above, by recording the merges in different positions of its own label. If the DBE neighbour of a winning component marks itself as *finished* instead of merging with the winning component, it is inferred that the losing component merged on a different DBE. The particle of the winning component incident to the DBE on which a comparison is made also marks itself as *finished*, marks the DBE as *not–compared* and informs the root that the merge is complete through its parent. Finally, a component participating in two simultaneous merges both as losing and as winning, participates in both merges independently since even if the particle changing its label for the winning merge participates in the losing merge, that must be through different ports and as a result, different parts of its label are changed for each of the merges.

Correctness. The omitted part of the proof consists of the following steps. The encoding of a competing component is not altered during the execution of LFC. This ensures that each of the particles incident to an active DBE of a component receives the same encoding during a comparison. A comparison

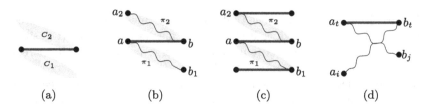

Fig. 4. Visualisation of the setting described in the proof of Theorem 4

between two components always terminates in *win*, *lose* or *draw*. A component wins (resp. draws) on a DBE if and only if the component competing on the same edge loses (resp. draws). Progress is always ensured during the execution of the algorithm since either the number of competing components decreases or the number of DBEs on which comparisons can be performed decreases. Eventually no DBEs on which a competition can be performed exist and when this happens only one component with one leader remains (Theorem 4).

Notice that since a component participates in at most two simultaneous comparisons the component might win both comparisons, lose one comparison and win one comparison, win or lose one comparison and draw on the other, draw in both comparisons or lose both comparisons. In the first three cases, the changes corresponding to different merges of the component are independent of each other. A component drawing in both competitions only marks edges as compared without changing the encoding of the component. When losing both comparisons, the root of the component only participates in one merge and the second merge is aborted. After a comparison between two components, either the components merge or they do not merge but the DBE on which the competition is made is marked as *compared*.

We define a *rightmost DBE ab* to be a DBE such that there does not exist a particle b' that is an endpoint of another DBE with $x_{b'} > x_b$. We further define a *topmost* DBE ab to be a rightmost DBE such that there does not exist a b' that is an endpoint of another rightmost DBE with $y_{b'} > y_b$. We call an edge connecting particles in the same component (resp. different components) *internal* (resp. *external*). A DBE where a comparison is being made is *active*. We can prove that for every comparison, it is not possible that an internal and an external DBE of a component are both active (Lemma 4). In each round performed by a component, either an external DBE is chosen and a merge occurs in the system or an internal edge is chosen and it becomes *compared*. Consequently the system stabilizes in a final configuration where all DBEs are compared. We will show that this is possible only if there is only one component.

Lemma 4. *Let e be an internal DBE of some component C and let e' be a different DBE of C. For every comparison it is not possible that e and e' are both active.*

Theorem 4. *Starting from any number of grey components encoded as trees, after executing the above algorithm only one tree with one leader, remains.*

Proof. Let us suppose that there exists at least one DBE in the final configuration which remains external after the execution of the algorithm. Out of the edges that remain external in the final configuration, take e to be the rightmost topmost one. This means that e connects two components of the final configuration which did not merge. Call the configuration on which e was marked as *active* for the last time $\mathcal{F}$ and call the components connected by e in $\mathcal{F}$, C_1 and C_2. Notice that all edges that are external and incident to C_1 or C_2 in $\mathcal{F}$ are also external in the final configuration, otherwise, $\mathcal{F}$ is not the last configuration in which e is *active*. Out of the active external DBEs connecting C_1 and C_2 in $\mathcal{F}$ take $e = ab$ to be a rightmost DBE such that $a \in C_1$ and $b \in C_2$. The edge ab and the two components are those in Fig. 4a. In our algorithm components with different encodings merge, so the tree encodings of C_1 and C_2 (call them T_1 and T_2) must be the same. Since a and b are both marked as endpoints of an active DBE in $\mathcal{F}$, when a (resp. b) is compared to its equivalent particle in T_2 (resp. T_1), the equivalent particle a_2 (resp. b_1), must also be the endpoint of an active DBE. So there must be a path π_1 in T_1 (resp. π_2 in T_2) connecting a (resp. b) to b_1 (resp. a_2) (Fig. 4b). Due to Lemma 4 we know that the DBEs incident to a_2 and b_1 (Fig. 4c) cannot be internal since they would not be active simultaneously with ab. We further know that π_1 must be equivalent to π_2 since we have assumed that the encodings of C_1 and C_2 are the same. Since the paths are equivalent, the horizontal shift between a, b_1 and a_2, b must also be equal. We then have $x_a - x_{b_1} = x_{a_2} - x_b$ or equivalently $x_a + x_b = x_{a_2} + x_{b_1}$. Since ab is a rightmost DBE, it must be that $x_{a_2} \leq x_a$ and $x_{b_1} \leq x_b$. Suppose $x_{a_2} < x_a$. Then to maintain the above equality, it must be that $x_{b_1} > x_b$ which is a contradiction, so we must have $x_{a_2} = x_a$. Similarly, we also get $x_{b_1} = x_b$. We have hence proved that T_1 (resp. T_2) is incident to two active, external, rightmost DBEs that are connected to each other by paths in T_1 and T_2 in $\mathcal{F}$.

Out of all rightmost DBEs in $\mathcal{F}$, we now take $a_t b_t$ to be the topmost DBE and following the argumentation of the previous paragraph, there must exist a path from b_t to a lower particle a_i (i.e., $y_{a_i} < y_{b_t}$) and a path from a_t to a lower particle b_j (i.e., $y_{b_j} < y_{a_t}$). Take the paths starting at the a particles: a_t and a_i. Since all particles by definition agree on the left and right directions, every time the path starting at a_t moves right (resp. left) the path starting at a_i also moves right (resp. left) or equivalently the paths starting at rightmost a particles start from the same x coordinate and always maintain the same x coordinates. However, the path starting at a_t (resp. a_i) must eventually move down (resp. up) to reach b_j (resp. b_t) that is not (resp. that is) at the topmost DBE. Since the two equal paths start from the same x coordinate and $y_{b_t} > y_{b_j}$, the paths must cross (Fig. 4d). However, if the paths cross there is a path connecting a_t to a_i so a_t and a_i must be in the same tree encoding of a component. We have defined a particles to be endpoints to active outgoing DBEs and we know that there is only one active outgoing DBE in each component, so this setting is not possible and C_1, C_2 must have different tree encodings. Hence a merge occurs either between C_1 and C_2 (which we have assumed not to be the case) or between one of C_1, C_2 and a neighbouring component. In the latter case, e

must become *active* in some future round since the components have changed, which contradicts the assumption that $\mathcal{F}$ is the last time e is active. In either case, there cannot exist a DBE which is external in the final configuration so the system eventually consists of only one component with one leader.

6   Conclusion and Open Problems

We presented an algorithm that deterministically solves the LE problem for stationary particles that agree on one direction. We showed that the difficulties in this case are complementary to those of particles with common chirality. In the former case a unique leader can be elected by an implicit termination algorithm, whereas, in the latter case an explicit termination algorithm is possible but up to six leaders are elected. Notice that stationary particles agreeing on a common direction cannot agree on chirality even after LE, as chirality cannot be communicated along DBEs. A natural next step is determining minimal sets of properties of an initial configuration that would allow a stationary, deterministic and explicitly terminating LE algorithm that elects a unique leader. For example, in the case of particles agreeing on a common direction, if the system does not contain DBEs, Grey LE solves the problem. If only one DBE exists in the system, due to the common direction, one of the endpoints of the DBE can be elected. Can a terminating LE algorithm be found for any constant number of DBEs? An alternative, is to study the problem in systems where the number of DBEs is not constant but each grey component has at most one incoming and at most one outgoing DBE. Another assumption is to restrict to configurations where each particle is in at most one or at most two boundaries, instead of the three boundaries that we consider here.

Acknowledgments. This work has been partially supported by ANR project DUCAT (ANR-20-CE48-0006)

Disclosure of Interests. The authors have no competing interests to declare that are relevant to the content of this article.

References

1. Angluin, D.: Local and global properties in networks of processors (extended abstract). In: STOC, pp. 82–93. ACM (1980)
2. Bazzi, R.A., Briones, J.L.: Stationary and deterministic leader election in self-organizing particle systems. In: Ghaffari, M., Nesterenko, M., Tixeuil, S., Tucci, S., Yamauchi, Y. (eds.) SSS 2019. LNCS, vol. 11914, pp. 22–37. Springer, Cham (2019). https://doi.org/10.1007/978-3-030-34992-9_3
3. Becker, A., Demaine, E.D., Fekete, S.P., Habibi, G., McLurkin, J.: Reconfiguring massive particle swarms with limited, global control. In: Flocchini, P., Gao, J., Kranakis, E., Meyer auf der Heide, F. (eds.) ALGOSENSORS 2013. LNCS, vol. 8243, pp. 51–66. Springer, Heidelberg (2014). https://doi.org/10.1007/978-3-642-45346-5_5

4. Briones, J.L., Chhabra, T., Daymude, J.J., Richa, A.W.: Invited paper: asynchronous deterministic leader election in three-dimensional programmable matter. In: ICDCN, pp. 38–47. ACM (2023)
5. Chalopin, J., Das, S., Kokkou, M.: Deterministic leader election for stationary programmable matter with common direction (2024). arXiv:cs.DC/2402.10582
6. Das, S., Flocchini, P., Nayak, A., Santoro, N.: Effective elections for anonymous mobile agents. In: Asano, T. (ed.) ISAAC 2006. LNCS, vol. 4288, pp. 732–743. Springer, Heidelberg (2006). https://doi.org/10.1007/11940128_73
7. Daymude, J.J., Gmyr, R., Richa, A.W., Scheideler, C., Strothmann, T.: Improved leader election for self-organizing programmable matter. In: Fernández Anta, A., Jurdzinski, T., Mosteiro, M.A., Zhang, Y. (eds.) ALGOSENSORS 2017. LNCS, vol. 10718, pp. 127–140. Springer, Cham (2017). https://doi.org/10.1007/978-3-319-72751-6_10
8. Daymude, J.J., Richa, A.W., Scheideler, C.: The canonical Amoebot model: algorithms and concurrency control. Distrib. Comput. 36(2), 159–192 (2023)
9. Derakhshandeh, Z., Dolev, S., Gmyr, R., Richa, A.W., Scheideler, C., Strothmann, T.: Brief announcement: Amoebot - a new model for programmable matter. In: SPAA, pp. 220–222. ACM (2014)
10. Di Luna, G.A., Flocchini, P., Santoro, N., Viglietta, G., Yamauchi, Y.: Shape formation by programmable particles. Distrib. Comput. 33(1), 69–101 (2020)
11. Dufoulon, F., Kutten, S., Moses, W.K., Jr.: Efficient deterministic leader election for programmable matter. In: PODC, pp. 103–113. ACM (2021)
12. Emek, Y., Kutten, S., Lavi, R., Moses, W.K., Jr.: Deterministic leader election in programmable matter. In: ICALP. LIPIcs Leibniz International Proceedings in Informatics, vol. 132, pp. 140:1–140:14. Schloss Dagstuhl - Leibniz-Zentrum für Informatik (2019)
13. Fekete, S.P., Gmyr, R., Hugo, S., Keldenich, P., Scheffer, C., Schmidt, A.: CADbots: algorithmic aspects of manipulating programmable matter with finite automata. Algorithmica 83(1), 387–412 (2021)
14. Feldmann, M., Padalkin, A., Scheideler, C., Dolev, S.: Coordinating amoebots via reconfigurable circuits. J. Comput. Biol. 29(4), 317–343 (2022)
15. Gastineau, N., Abdou, W., Mbarek, N., Togni, O.: Distributed leader election and computation of local identifiers for programmable matter. In: Gilbert, S., Hughes, D., Krishnamachari, B. (eds.) ALGOSENSORS 2018. LNCS, vol. 11410, pp. 159–179. Springer, Cham (2019). https://doi.org/10.1007/978-3-030-14094-6_11
16. Gastineau, N., Abdou, W., Mbarek, N., Togni, O.: Leader election and local identifiers for three-dimensional programmable matter. Concurr. Comput. Pract. Exp. 34(7) (2022)
17. Itai, A., Rodeh, M.: Symmetry breaking in distributed networks. 88, 60–87 (1990)
18. Savoie, W., et al.: Phototactic supersmarticles. Artif. Life Robot. 23(4), 459–468 (2018)
19. Toffoli, T., Margolus, N.: Programmable matter: concepts and realization. Int. J. High Speed Comput. 5(2), 155–170 (1993)
20. Woods, D., Chen, H., Goodfriend, S., Dabby, N., Winfree, E., Yin, P.: Active self-assembly of algorithmic shapes and patterns in polylogarithmic time. In: ITCS, pp. 353–354. ACM (2013)

Mutual Visibility in Hypercube-Like Graphs

Serafino Cicerone[1], Alessia Di Fonso[1(✉)], Gabriele Di Stefano[1],
Alfredo Navarra[2], and Francesco Piselli[2]

[1] Università degli Studi dell'Aquila, 67100 L'Aquila, Italy
{serafino.cicerone,alessia.difonso,gabriele.distefano}@univaq.it
[2] Università degli Studi di Perugia, 06123 Perugia, Italy
alfredo.navarra@unipg.it, francesco.piselli@unifi.it

Abstract. Let G be a graph and $X \subseteq V(G)$. Then, vertices x and y of G are X-visible if there exists a shortest x, y-path where no internal vertices belong to X. The set X is a mutual-visibility set of G if every two vertices of X are X-visible, while X is a total mutual-visibility set if any two vertices from $V(G)$ are X-visible. The cardinality of a largest mutual-visibility set (resp. total mutual-visibility set) is the mutual-visibility number (resp. total mutual-visibility number) $\mu(G)$ (resp. $\mu_t(G)$) of G. It is known that computing $\mu(G)$ is an NP-complete problem, as well as $\mu_t(G)$. In this paper, we study the (total) mutual-visibility in hypercube-like networks (namely, hypercubes, cube-connected cycles, and butterflies). Concerning computing $\mu(G)$, we provide approximation algorithms for both hypercubes and cube-connected cycles, while we give an exact formula for butterflies. Concerning computing $\mu_t(G)$ (in the literature, already studied in hypercubes), we provide exact formulae for both cube-connected cycles and butterflies.

1 Introduction

Problems about sets of points in the Euclidean plane and their mutual visibility have been investigated for a long time. For example, in [20] Dudeney posed the famous *no-three-in-line* problem: finding the maximum number of points that can be placed in an $n \times n$ grid such that there are no three points on a line. Beyond the theoretical interest, solutions to these types of geometric/combinatorial problems have proved useful in recent years in the context of swarm robotics. The requirement is to define algorithms that allow autonomous mobile robots to change in a finite time their configuration in the plane so they can see each other (see, e.g., [17, 28, 30]). When robots are constrained to move on graphs, then Cycles, Grids, Hypercubes, Butterfly and other well-structured

The work has been supported in part by the European Union - NextGenerationEU under the Italian Ministry of University and Research (MUR) National Innovation Ecosystem grant ECS00000041 - VITALITY - CUP J97G22000170005, and by the Italian National Group for Scientific Computation (GNCS-INdAM).

Y. Emek (Ed.): SIROCCO 2024, LNCS 14662, pp. 192–207, 2024.
https://doi.org/10.1007/978-3-031-60603-8_11

topologies represent intriguing challenges due to their high level of symmetry (see, e.g., [1, 12, 14–16, 19, 21] where robots are required to gather all at one vertex). Furthermore, such topologies play a central role in computer network architectures, especially for parallel and distributed computations.

Mutual visibility in graphs with respect to a set of vertices has been recently introduced and studied in [18] in terms of the existence of a shortest path between two vertices not containing a third vertex from the set. The visibility property is then understood as a kind of non-existence of "obstacles" between the two vertices in the mentioned shortest path, which makes them "visible" to each other. Formally, let G be a connected and undirected graph, and $X \subseteq V(G)$ a subset of the vertices of G. Two vertices $x, y \in V(G)$ are X-visible if there exists a shortest x, y-path where no internal vertex belongs to X. X is a mutual-visibility set if its vertices are pairwise X-visible. The cardinality of a largest mutual-visibility set is the mutual-visibility number of G, and it is denoted by $\mu(G)$. Computing one of such largest sets, referred to as a μ-set of G, solves the so-called MUTUAL-VISIBILITY problem. In [18], it is also shown that computing $\mu(G)$ is an NP-complete problem, but there exist exact formulae for the mutual-visibility number of special graph classes like paths, cycles, blocks, cographs, grids, and in [8] also for distance-hereditary graphs. In the subsequent works [10] and [11], exact formulae have been derived also for both the Cartesian and the Strong product of graphs. The contributions presented in those initial works showed several interesting connections with other mathematical contexts. For instance, the mutual-visibility is related to the general position problem in graphs (see, e.g., [22, 26, 29] and references therein) and to the context of swarm robotics (cf. [4–6]).

In [11], a natural extension of the mutual-visibility has been proposed. Formally, $X \subseteq V(G)$ is a total mutual-visibility set of G if every two vertices x and y of G are X-visible. A largest total mutual-visibility set of G is a μ_t-set of G, and its cardinality is the total mutual-visibility number of G denoted as $\mu_t(G)$. Of course, $\mu_t(G) \leq \mu(G)$. In [9] it is shown that also computing $\mu_t(G)$ is an NP-complete problem.

The most recent works in the literature on mutual-visibility concern total mutual-visibility. This setting is clearly more restrictive, but it surprisingly turns out to become very useful when considering networks having some Cartesian properties in the vertex set, namely, those ones of product-like structures (cf. [11]). In [23], in fact, the total mutual-visibility number of Cartesian products is bounded and several exact results proved. Furthermore, a sufficient and necessary condition is provided for asserting when $\mu_t(G) = 0$. In [24], the authors give several bounds for $\mu_t(G)$ in terms of the diameter, order and/or connected domination number of G. They also determine the exact value of the total mutual-visibility number of lexicographic products. The total mutual-visibility of Hamming graphs (and, as a byproduct, for hypercubes) is studied in [3]. Finally, the mutual and the total visibility problems are studied for graphs of diameter two in [13] showing bounds and/or closed formulae.

Results. In this work, we study the mutual-visibility and the total mutual-visibility of hypercube-like graphs. In particular, we provide approximated solutions for hypercubes and cube-connected cycles, whereas we obtain optimal results for butterflies. An extended version of this paper can be found in [7].

Outline. Concerning the organization of the paper, in the next section we provide all the necessary notation and preliminary concepts. The subsequent Sects. 3, 4, and 5 are specialized for presenting - in order - results about (total) mutual visibility for hypercubes, cube-connected cycles, and butterflies. Concluding remarks are provided in Sect. 6.

2   Notation and Preliminaries

In this work, we consider undirected and connected graphs. We use standard terminologies from [2], some of which are briefly reviewed here.

Given a graph G, $V(G)$ and $E(G)$ are used to denote its vertex set and its edge set, resp., and $n(G)$ is used to represent the size of $V(G)$. Whenever G is clearly known by the context, we denote $n(G)$ simply by n. If $u, v \in V(G)$ are adjacent, $(u, v) \in E(G)$ represents the corresponding edge. If $X \subseteq V(G)$, then $G[X]$ denotes the subgraph of G *induced* by X, that is the maximal subgraph of G with vertex set X.

The usual notation for representing special graphs is adopted. K_n is the *complete graph* with n vertices, $n \geq 1$, that is the graph where each pair of distinct vertices are adjacent. P_n represents any *path* $(v_1, v_2, \ldots, v_n)$ with $n \geq 2$ distinct vertices where v_i is adjacent to v_j if $|i - j| = 1$. Vertices $v_2, \ldots, v_{n-1}$ are all *internal* of P_n. A *cycle* C_n, $n \geq 3$, in G is a path $(v_0, v_1, \ldots, v_{n-1})$ where also $(v_0, v_{n-1}) \in E(G)$. Two vertices v_i and v_j are *consecutive* in C_n if $j = (i + 1) \bmod n$ or $i = (j + 1) \bmod n$. The *distance* function on a graph G is the usual shortest path distance.

Two graphs G and H are *isomorphic* if there exists a bijection $\varphi : V(G) \to V(H)$ such that $(u, v) \in E(G) \Leftrightarrow (\varphi(u), \varphi(v)) \in E(H)$ for all $u, v \in V(G)$. Such a bijection φ is called *isomorphism*. Given G and H, the *Cartesian product* $G \square H$ has the vertex set $V(G) \times V(H)$ and the edge set $E(G \square H) = \{((g, h), (g', h')) : (g, g') \in E(G) \text{ and } h = h', \text{ or, } g = g' \text{ and } (h, h') \in E(H)\}$.

If H is a subgraph of G, H is said to be *convex* if all shortest paths in G between vertices of H actually belong to H. Concerning convex subgraphs, we recall the following useful statement:

Lemma 1. [18] *Let H be a convex subgraph of any graph G, and X be a mutual-visibility set of G. Then, $X \cap V(H)$ is a mutual-visibility set of H.*

Finally, note that binary strings are used as vertex labels or components of vertex labels for the classes of graphs studied in this paper (hypercubes, cube-connected cycles, and butterflies). For a binary string $x = x_0 x_1 \cdots x_i \cdots x_{d-1}$ with d bits, position 0 corresponds to the leftmost bit, and position $n - 1$ to the rightmost bit. Sometimes, we interpret these strings as (binary) numbers. We use $x(i)$ to denote the binary string obtained from x by complementing the bit in position i. We will abbreviate "the vertex with label x" to "vertex x".

3   Hypercube

The d-dimensional hypercube Q_d is an undirected graph with vertex set $V(Q_d) = \{0,1\}^d$, and two vertices are adjacent if and only if the two binary strings differ by exactly one bit, that is, the Hamming distance[1] of the two binary strings is 1. Hypercube Q_d can also be recursively defined in terms of the Cartesian product of two graphs as follows:

- $Q_1 = K_2$,
- $Q_d = Q_{d-1} \,\square\, K_2$, for $d \geq 2$.

This makes clear that Q_d can be also seen as formed by two subgraphs both isomorphic to Q_{d-1} (cf. Fig. 1).

3.1   An Upper Bound for $\mu(Q_d)$

Here, we first analyze specific optimal solutions for the Mutual-Visibility problem in hypercubes of small dimension $d \leq 5$ and then we provide an upper bound holding for any $d \geq 6$.

First of all, we can claim the following lemma.

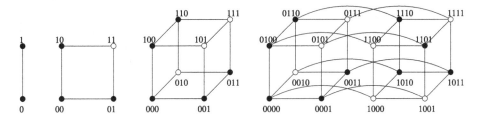

Fig. 1. A representation of the hypercube Q_d for $d = 1, 2, 3$ and 4. In each graph, the black vertices form a μ-set.

Lemma 2. *Each subgraph Q' of Q_d that is isomorphic to $Q_{d'}$, $d' < d$, is convex.*

Proof. Consider a hypercube Q' subgraph of Q_d that is isomorphic to $Q_{d'}$ for some $d' < d$. By considering two vertices x and y of Q', we have to show that there does not exist a shortest path between x and y passing through a vertex $z \in V(Q_d) \setminus V(Q')$. Since the distance between two vertices in a hypercube is governed by the Hamming distance, the bits that differ between the labels associated with x and y concern only Q'. In fact, there cannot be a shortest path that makes use of a vertex $z \in V(Q_d) \setminus V(Q')$ as, by construction, the corresponding bit leading to (the dimension of) z is different from the one in the labels of both x and y. □

---

[1] The Hamming distance between two strings of equal length is the number of positions at which the corresponding symbols are different.

Corollary 1. $\mu(Q_d) \leq 2\mu(Q_{d-1})$, $d \geq 2$.

Proof. The proof simply follows by recalling that Q_d can be obtained by the Cartesian product of $Q_{d-1} \square K_2$, for $d \geq 2$, i.e., by suitably connecting two hypercubes of dimension $d - 1$. Therefore, by Lemma 2 the claim holds.   □

We now consider all the dimensions $d \leq 5$, one by one, for which we can provide optimal solutions.

From results provided in [18], we know that $\mu(Q_1) = \mu(K_2) = 2$ and that $\mu(Q_2) = \mu(C_4) = 3$. For Q_3 (that contains 8 vertices) there exists exactly one μ-set of size 5, up to isomorphisms. By referring to Fig. 1, the optimal solution is provided by the set $\{000, 001, 100, 110, 011\}$. It can be observed that there does not exist any solution with more than 5 vertices. By contradiction, assume that X, with $|X| \geq 6$, is a μ-set for Q_3. Then, for each subcube Q' of Q_3 isomorphic to Q_2 we have $|V(Q') \cap X| < 4$ (in fact, Q' is a convex subgraph isomorphic to a cycle). Hence, the two copies of Q_2 must have 3 elements each in X. To avoid having a cycle C_4 whose elements are all in X, the two vertices not in X must be antipodal; but, this implies that there exists another pair of antipodal vertices that are in X which are not mutually-visible.

For Q_4 (that contains 16 vertices), Corollary 1 implies that each μ-set can have at most 10 elements, i.e., 5 vertices per each subcube of dimension 3. Since the μ-set for Q_3 is unique up to isomorphisms, we have few possibilities to combine the two subcubes. By a computer-assisted exhaustive search, we have that there does not exist a solution with 10 vertices selected, whereas we can provide a mutual-visibility set with 9 vertices selected. By referring to Fig. 1, the optimal solution is provided by the set $\{0000, 0001, 0100, 0110, 0011, 1101, 1010, 1011, 1110\}$. Furthermore, the obtained solution is unique up to isomorphisms.

Concerning Q_5, since $\mu(Q_4) = 9$, by Lemma 2 we can obtain a mutual-visibility set with at most 18 vertices. Again, by means of a computer-assisted case-by-case analysis, we have that the optimal and unique (up to isomorphisms) solution is of size 16. The optimal solution is provided by the set $\{00000, 00001, 00100, 00110, 00011, 01101, 01010, 01011, 01110, 10101, 10111, 11000, 11001, 11100, 11110, 11011\}$.

For $d \geq 6$, we can state the next corollary that provides an upper bound to $\mu(Q_d)$. In particular, since we have shown that for Q_5 any μ-set contains 2^4 elements, i.e., exactly half of the vertices, then the next claim simply follows by Corollary 1:

Corollary 2. $\mu(Q_d) \leq 2^{d-1}$, $d \geq 5$.

3.2   Lower Bound and an Approximation Algorithm

Here, we first provide a lower bound for $\mu(Q_d)$. Then, we derive an approximation algorithm for the mutual-visibility problem in hypercubes.

Theorem 1. $\mu(Q_d) \geq \binom{d}{\lfloor \frac{d}{2} \rfloor} + \binom{d}{\lfloor \frac{d}{2} \rfloor + 3}$, $d \geq 1$.

Proof. Let X_p be the subset of $V(Q_d)$ containing all the vertices that are at distance p from the vertex labeled with all zeroes. By construction, and by the Hamming distance, the elements of X_p are all the vertices whose labels contain exactly p 1's. Hence, it is easy to find a shortest path between two of the selected vertices, say x and y that are at a distance j from each other. It suffices to detect the j differences among the labels associated with the two vertices. Then, by first replacing (one per step) the 1's present in x but not in y with 0's and then similarly replacing the 0's with 1's, equals to determine a shortest path between x and y. Such a path is shortest because at each step makes a change in the direction of the destination, i.e., its length is exactly j. Furthermore, along the chosen path, there are no vertices in X_p as the number of 1's is always smaller than p until the destination.

The number of vertices with a fixed number p of 1's is $\binom{d}{p}$, which is maximized for $p = \lfloor \frac{d}{2} \rfloor$.

For $d < 5$, the size of $X_{\lfloor \frac{d}{2} \rfloor}$ is $\binom{d}{\lfloor \frac{d}{2} \rfloor}$ and assumes the values $1, 2, 3, 6$ for $d = 1, 2, 3, 4$. These values are less than the corresponding optimal values of $\mu(Q_d)$ shown above. As in these cases $\binom{d}{\lfloor \frac{d}{2} \rfloor + 3} = 0$, the statement holds.

For $d \geq 5$, we can choose a larger set of vertices, still guaranteeing mutual visibility. Given a hypercube Q_d, we define a set X as follows:

- Let $p = \lfloor \frac{d}{2} \rfloor$;
- $X = X_p \cup X_{p+3}$.

Let $x \in X_p$ and $y \in X_{p+3}$. There are shortest path between x and y whose internal vertices are labeled with $p+1$ and $p+2$ 1's only. Indeed, a shortest path from a vertex labeled with p 1's to one labeled with $p + 3$ 1's can involve only internal vertices obtained by first changing two 0's in 1's and then alternating one 1 in 0 and one 0 in 1 until the last step. By the number of 1's in each label of the internal vertices we are guaranteed that x and y are mutually visible. If $x, y \in X_p$, to show the mutual visibility we can find a shortest path with vertices not in X as described above, whereas if $x, y \in X_{p+3}$ the vertices not in X in a shortest path between x and y, can be obtained by first replacing (one per step) the 0's present in x but not in y with 1's and then similarly replacing the 1's with 0's. □

An immediate consequence of the above theorem is the existence of an approximation algorithm for the mutual-visibility problem in the context of hypercubes, as stated by the next two corollaries.

Corollary 3. *There exists an algorithm for the* MUTUAL-VISIBILITY *problem on a hypercube* Q_d *providing a solution of size greater than* $\frac{2^d}{\sqrt{\frac{\pi}{2} d}}$.

Sketchproof. For $d \leq 5$, we have already shown we can provide the optimal solution. For $d \geq 6$, we show that the procedure defined in the proof of Theorem 1 (for defining the μ-set X for Q_d) provides the requested algorithm. In fact, by exploiting the Stirling's approximation [31], we obtain $|X| > \frac{2^d}{\sqrt{\frac{\pi}{2} d}}$. □

From Corollaries 2 and 3, we obtain

Corollary 4. *There exists an $O(\sqrt{d})$-approximation algorithm for the* MUTUAL-VISIBILITY *problem on a hypercube Q_d.*

We remind that in a hypercube Q_d of n vertices, $n = 2^d$, i.e., $d = \log n$. Consequently, the approximation provided by the above theorem for an n-vertex hypercube can be expressed as $O(\sqrt{\log n})$.

About the total mutual visibility on hypercubes, from [3, Theorem 3], it turns out that $\mu_t(Q_d) \geq \frac{2^{d-2}}{d(d+1)}$ which is asymptotically much smaller than the bound $\mu(Q_d) > \frac{2^d}{\sqrt{\frac{\pi}{2}d}}$ provided in the proof of Corollary 3.

4   Cube-Connected Cycles

A cube-connected cycle of order $d \geq 3$ (denoted CCC_d) can be defined as an undirected graph formed by a set of $d \cdot 2^d$ vertices labelled $[\ell, x]$, where ℓ is an integer between 0 and $d-1$ and x is a binary string of length d. Two vertices $[\ell, x]$ and $[\ell', x']$ are adjacent if and only if: (i) either $x = x'$ and $(\ell - \ell') \mod d = 1$, (ii) or $\ell = \ell'$ and $x' = x(\ell)$. In this last case, x and x' differ exactly for the bit in position ℓ.

The edges satisfying condition (i) are referred to as *cycle edges*, those satisfying condition (ii) are instead referred to as the *hypercube edges*.

Note that, the removal of all hypercube edges produces a graph with 2^d components, each of which is isomorphic to a cycle C_d. For this reason, each cycle C_d in CCC_d which does not include any hypercube edge is referred to as a *supervertex* of the (embedded) hypercube.

Furthermore, contracting all the cycle edges in CCC_d will produce a graph with 2^d vertices isomorphic to a hypercube Q_d. Consequently, the smallest CCC_d is in fact defined for $d = 3$.

Natural Routing in CCC_d. Let $u = [\ell, x]$ and $v = [\ell', x']$ be two vertices of a CCC_d, and $H = \{p_0, \dots, p_{h-1}\}$ be the positions in which the binary strings representing x and x' differ. By [32], the distance between u and v equals $h + k$, with k being the number of edges in the shortest walk on a cycle C_d starting at index ℓ and ending at index ℓ' which includes a visit to every vertex with an index in the set H. Then, a *natural routing* from u to v is to traverse the shortest walk while traversing the hypercube edges each time a vertex with an index in the set H is met for the first time.

Lemma 3. *Each subgraph Q' composed by the supervertices of a CCC_d, isomorphic to $Q_{d'}$, $d' < d$, induces a subgraph of CCC_d that is convex.*

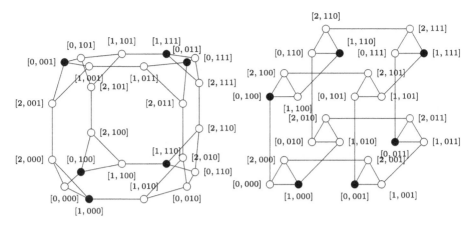

Fig. 2. *(left)* A representation of a CCC_3 graph, black vertices represent a μ-set for CCC_3. *(right)* An alternative representation of a CCC_3.

Proof. By Lemma 1, we have that the shortest paths between vertices in $Q_{d'}$ are confined within $Q_{d'}$ itself. Consider now the subgraph obtained by restoring each supervertex with the corresponding cycle C_d along with its edges within $Q_{d'}$. By the natural routing, any shortest path among two vertices u and v of the defined subgraph makes use of the hypercube edges only from vertices with index in the set H. Since all such edges are included in the obtained subgraph, the claim holds. ☐

Lemma 4. $\mu(CCC_d) \leq 3 \cdot 2^{d-2}$, $d \geq 3$.

Proof. The proof follows by first observing that for $d = 3$, the optimal solution provides $\mu(CCC_3) = 3 \cdot 2^{3-2} = 6$. In fact, we have obtained this result by a computer-assisted exhaustive search, and the optimal solution found is shown in Fig. 2. Then, by Lemma 3, we have that, as d increases, $\mu(CCC_d)$ can double at most. ☐

Theorem 2. $\mu(CCC_d) \geq 2^{\lceil \frac{d}{2} \rceil}$, $d \geq 3$.

Proof. Given a cube connected cycle CCC_d, $d \geq 3$, we define a set X according to the following procedure:

- Insert into X any vertex $v = [0, x]$ with a number of 1's bounded by $\lceil \frac{d}{2} \rceil$ in the $\lceil \frac{d}{2} \rceil$ most significant bits composing x.

In doing so, and reminding the natural routing in a CCC_d, we have that the shortest path between two vertices v_1 and v_2 belonging to X is well defined by considering the shortest walk in C_d that from 0 reaches p_{h-1} and then comes back to 0 either proceeding backward or forward, it depends on the distances. The choice of X guarantees to traverse at most d cycle edges, i.e., an entire C_d at most, or just a portion back and forth. Since all vertices in X have their first coordinate 0, all the traversed edges within any encountered C_d as well as all hypercube edges included in the shortest path never meet other vertices in X.

Summarizing, we have that any two vertices $[0, x]$ and $[0, y]$ in X are mutually visible since the shortest path forced by the chosen set X ensures to never encounter a vertex $[0, z]$ for any $z \notin \{x, y\}$. By construction, $|X| = 2^{\lceil \frac{d}{2} \rceil}$. □

From Lemma 4 and Theorem 2 we obtain

Corollary 5. *There exists a* $3 \cdot 2^{\lfloor \frac{d}{2} \rfloor - 2}$*-approximation algorithm for the* MUTUAL-VISIBILITY *problem on a cube-connected cycle* CCC_d.

We remind that in a cube-connected cycle CCC_d of n vertices, $n = d \cdot 2^d$, i.e., $d = \log n - \log d = \log n - o(\log n)$. Consequently, the approximation provided by the above theorem for an n-vertex cube-connected cycle can be expressed as $O(\sqrt{n})$.

4.1   Total Mutual Visibility

In [23], a characterization of all graphs having total mutual-visibility zero is provided. The characterization is based on the notion of *bypass vertex*: a vertex u of G is a bypass vertex if u is not the middle vertex of a convex P_3 of G. Let $\mathrm{bp}(G)$ be the number of bypass vertices of G.

Theorem 3. *[23, Theorem 3.3] Let G be a graph with $n(G) \geq 2$. Then $\mu_t(G) = 0$ if and only if $\mathrm{bp}(G) = 0$.*

This result directly provides the value of $\mu_t(CCC_d)$.

Corollary 6. $\mu_t(CCC_d) = 0$, $d \geq 3$.

Proof. Let u be a vertex of CCC_d and consider a path P_3 given by (v', u, v''), where v' and v'' belong to different cycles of the graph. It can be observed that such a path is convex. Since it is well-known that CCC_d is a vertex-transitive graph, then each vertex u is not a bypass vertex and hence $\mathrm{bp}(CCC_d) = 0$. □

5   Butterfly

A d-dimensional butterfly $BF(d)$ is an undirected graph with vertices $[\ell, c]$, where $\ell \in \{0, 1, \ldots, d\}$ is the *level* and $c \in \{0, 1\}^d$ is the *column*. The vertices $[\ell, c]$ and $[\ell', c']$ are adjacent if $|\ell - \ell'| = 1$, and either $c = c'$ or c and c' differ precisely in the ℓ-th bit. $BF(d)$ has $d + 1$ levels with 2^d vertices at each level, which gives $(d+1) \cdot 2^d$ vertices in total. The vertices at levels 0 and d have degree 2 whereas the rest have degree 4. $BF(d)$ has two standard graphical representations, namely normal and diamond representations (see Fig. 3). For further details one may refer to [25].

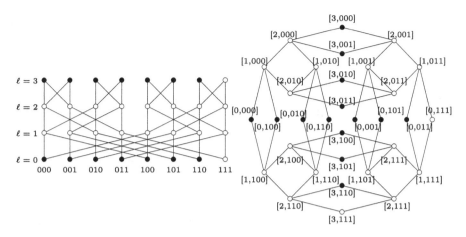

Fig. 3. Normal representation and diamond representation of $BF(3)$, black vertices represent a μ-set for $BF(3)$.

Some additional notation is required: let $A_i = \{[\ell, i] \mid \ell \in \{0, 1, \ldots, d\}\}$ be the vertex set forming the i-th column of the butterfly, and let $L_j = \{[j, c] \mid c \in \{0, 1\}^d\}$, be the vertex set forming its j-th level. Note that, $BF(d)$ can be partitioned into two copies of $BF(d-1)$ (that we denote as $BF'(d-1)$ and $BF''(d-1)$) and L_0 (cf. Fig. 4). Notice that both $BF'(d-1)$ and $BF''(d-1)$ are convex subgraphs of $BF(d)$.

Natural Routing in $BF(d)$. Consider the case in which starting from a vertex $[0, i]$, it is necessary to reach a vertex $[d, j]$ along a shortest path. A *natural routing* for such a task simply requires comparing the corresponding bits of i and j starting from the leftmost: if they coincide, the edge along the current column is traversed, otherwise, the edge for changing column is used. Symmetrically, a routing from $[d, i]$ toward $[0, j]$ can be obtained in a similar way but comparing the corresponding bits of i and j starting from the rightmost. This leads to the following useful property:

$\mathcal{P}$: Let $u = [\ell', i]$ and $v = [\ell'', j]$ be two vertices, $\ell_{\min} = \min\{\ell', \ell''\}$ and $\ell_{\max} = \max\{\ell', \ell''\}$. Each shortest path between u and v is comprised between levels $\min\{k' - 1, \ell_{\min}\}$ and $\max\{k'', \ell_{\max}\}$, where k' and k'' are the positions of the first and last bit, resp., in which i and j differ.

As special cases for this property we get: $(\mathcal{P}_1)$ if from $[0, i]$ it is necessary to reach $[0, j]$, then a shortest path reaches level k iff the k-th bits of i and j differ; $(\mathcal{P}_2)$ if from $[d, i]$ it is necessary to reach $[d, j]$, then a shortest path reaches level $d - k$ iff the $(d - k)$-th bit of i and j differ.

Lemma 5. *Let X be a mutual-visibility set of $BF(d)$. Then, $|X \cap A_i| \leq 2$ for each $i \in \{0, 1, \ldots, d\}$.*

Proof. According to Lemma 1, since A_i is a convex subgraph, A_i is isomorphic to a path graph, which mutual-visibility number is two.   □

Lemma 6. $X = (L_0 \cup L_d) \setminus \{[0, 111 \ldots 1], [d, 111 \ldots 1]\}$ *is a mutual-visibility set of $BF(d)$.*

Proof. Consider $u = [0, i]$ and $v = [d, j]$. A shortest path that makes u and v X-visible is given by the natural routing from u to v.

Consider now $u = [0, i]$ and $v = [0, j]$. According to Property $\mathcal{P}_1$, if the last bit of i and j does not differ, then there exists a shortest path whose interior vertices belong all at levels $1, 2, \ldots, d-1$, and this property trivially makes the two vertices X-visible. If the last bit of i and j differs, then their distance is $2d$ and a shortest path that makes the two vertices X-visible can be defined by composing the natural routing from u to $[d, 111 \ldots 1]$ with the natural routing from v to $[d, 111 \ldots 1]$.

Consider the last case in which $u = [d, i]$ and $v = [d, j]$. According to Property $\mathcal{P}_2$, if the first bit of i and j does not differ, then there exists a shortest path whose interior vertices belong all at levels $1, 2, \ldots, d-1$, and this property trivially makes the two vertices X-visible. If the first bit of i and j differs, then their distance is $2d$ and a shortest path that makes the two vertices X-visible can be defined by composing the natural routing from $[0, 111 \ldots 1]$ to u with that from $[0, 111 \ldots 1]$ to v. □

Lemma 7. $\mu(BF(d)) \leq 2^{d+1} - 2$, $d \geq 2$.

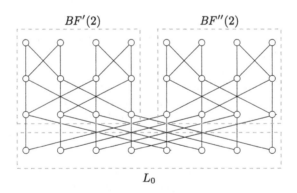

Fig. 4. Partition of $BF(3)$ into $BF'(2)$, $BF''(2)$, and L_0.

Sketchproof. By contradiction, assume the existence of a mutual-visibility set of $BF(d)$ such that $|X| = 2^{d+1} - 1$. According to Lemma 5, one column (say A_τ) contains exactly one element of X while every other column has two. The proof proceeds by considering the following two cases, according to $|X \cap L_0|$.

$|X \cap L_0| \geq 2$. In such a case, by the cardinality of X it follows that there is one element of X at level 0 for each column different from A_τ, and, if also A_τ has an element of X at level 0, then X cannot be a mutual-visibility set. In case A_τ does not have an element of X at level 0, we get that each other column must have a vertex of X at level d. This last property leads to show that X cannot be a mutual-visibility set.

$|X \cap L_0| \leq 1$. In this case, we consider the set X' (X'', resp.) containing all vertices of X that are contained in $BF'(d-1)$ ($BF''(d-1)$, resp.). Since the cardinality of both X' and X'' is at least $2^d - 1$, then we can recursively apply this proof to both the subgraphs. Then, either the first case will eventually apply, or we end the recursion with the terminal situation given by a subgraph isomorphic to $BF(2)$ that shares with the original set X seven or eight elements, at most two of such elements per column, and at most one element at the lower level. Up to isomorphisms, there are a few such configurations, and each of them has elements of X that are not in mutual visibility.   □

Since the mutual-visibility set provided by Lemma 6 contains $2^{d+1} - 2$ elements, then from Lemma 7 we have:

Theorem 4. $\mu(BF(d)) = 2^{d+1} - 2$, $d \geq 2$.

5.1   Total Mutual-Visibility

Given $BF(d)$, let L_0' (L_d', resp.) be the subset of L_0 (L_d, resp.) containing all vertices $[0, i]$ ($[d, i]$, resp.) fulfilling one of the following conditions, where i is interpreted as a binary number:

- i is odd and $i \leq 2^{d-1}$,
- i is even and $i > 2^{d-1}$.

Figure 5 shows $L_0' \cup L_d'$ in $BF(3)$.

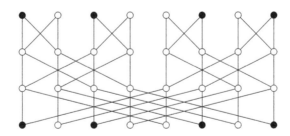

Fig. 5. Visualization of $L_0' \cup L_d'$ as computed in $BF(3)$.

Lemma 8. $X = L_0' \cup L_d'$ is a total mutual-visibility set of $BF(d)$, $d \geq 1$.

Proof. Let u and v be two distinct vertices of $BF(d)$. From the proof of Lemma 6 we easily derive that u and v are X-visible in each of the following three cases: (1) $u = [0, i]$ and $v = [0, j]$, (2) $u = [0, i]$ and $v = [d, j]$, and (3) $u = [d, i]$ and $v = [d, j]$. Then, it remains to consider three additional situations.

(4) $u = [0, i]$ and $v = [\ell, j]$, with $\ell \notin \{0, d\}$. If $d(u, v) \leq d$, then u and v are X-visible. Otherwise there is a u, v-shortest path passing through $u' = [d, j]$. If u' is not occupied by a robot, then u and v are X-visible, otherwise there exists $u'' = [d, j']$, with $d(u', u'') = 2$, that is not occupied such that u'' belongs to a u, v-shortest path, therefore u and v are X-visible.

(5) $u = [d, i]$ and $v = [\ell, j]$, with $\ell \notin \{0, d\}$. It is symmetric to the previous case.

(6) $u = [\ell', i]$ and $v = [\ell'', j]$, with $\ell', \ell'' \notin \{0, d\}$. There are some subcases.

(a) $i = j$. The statement trivially holds since a u, v-shortest path not passing through vertices of X is contained in column A_i;

(b) $i \neq j$, and u and v are located on the same sub-butterfly. Without loss of generality, assume that both u and v are in $BF'(d - 1)$ and $i < j$. Since u and v are in the same sub-butterfly, each u, v-shortest path does not touch level 0 of $BF(d)$.

If i and j are both even, then, according to Property $\mathcal{P}$, we get that each u, v-shortest path does not touch level d of $BF(d)$ unless u or v are at that level. Since u and v are not at level d by hypothesis, it follows that u and v are X-visible. When both i and j are odd, their binary strings do not differ in the last position, and hence from Property $\mathcal{P}$ we still get that each u, v-shortest path does not touch level d; again, this guarantees the mutual-visibility between the two vertices.

If i is odd and j is even, then, according to Property $\mathcal{P}$, we get that each u, v-shortest path must touch level d of $BF(d)$, regardless of the level of u an v; in this case, it can be observed that such a shortest path can pass through the vertex $[d, i]$. Conversely, if i is even and j is odd, the requested shortest path can pass through either $[d, i + 1]$ or $[d, j]$.

(c) $i \neq j$, and u and v are located on different sub-butterflies. Without loss of generality, assume u in $BF'(d-1)$ and v in $BF''(d-1)$. If i is odd, then it can be easily observed that a shortest path from u to v can be traversed by first reaching the vertex $[1, j]$ and then $[0, j']$ in the other sub-butterfly (note that this vertex is not in X by definition). Now, the remaining part of the u, v-shortest path can be identified as in the case (4) above. If i is even, the requested shortest path can be defined by the same strategy, but now it is necessary to reach the vertex $[0, j]$ first and then $[1, j']$ in the other sub-butterfly. The remaining part of the u, v-shortest path can be identified as in the case (6.a) or (6.b). □

Theorem 5. $\mu_t(BF(d)) = 2^d$, $d \geq 1$.

Proof. Let X be any total mutual-visibility set of $BF(d)$. If $u = [0, i]$ and $v = [d, j]$, according to the natural routing of the butterfly, there exists a unique u, v-shortest path in $BF(d)$. Consequently, each vertex in this u, v-shortest path cannot belong to X. In general, this leads to the property that each vertex $[\ell, i]$ such that $\ell \notin \{0, d\}$ does not belong to X.

Consider now two vertices $u = [d, i]$ and $v = [d, i+1]$, with i even. They belong to a convex subgraph of the butterfly isomorphic to a cycle C_4, and hence they cannot both belong to X otherwise the other pair of vertices in that cycle are not in mutual visibility. Symmetrically, vertices $u = [0, i]$ and $v = [0, i + 2^{d-1}]$, i even, cannot both belong to X as they belong to a convex subgraph C_4.

From the arguments above it follows that in X cannot exist any vertex with degree four. Moreover, only half of the vertices at level d and at level 0 can belong to X. Hence, $\mu_t(BF(d)) \leq 2^d$. The statement follows by observing that the total mutual-visibility set provided by Lemma 8 contains exactly 2^d elements. □

6   Conclusions

In this paper, we have studied the mutual-visibility and the total mutual-visibility in hypercubes, cube-connected cycles, and butterflies. While for any butterfly $BF(d)$ we were able to provide exact formulae to calculate both $\mu(BF(d))$ and $\mu_t(BF(d))$, for the other topologies we were able to identify approximation algorithms. These results, together with those obtained in [3] on Hamming graphs, suggest that the study of mutual visibility properties on such topologies seems to be particularly complex. This of course suggests further investigations within such topologies, like for instance on wrapped butterflies or Fibonacci cubes [27].

Furthermore, we aim at studying distributed algorithms for swarm of robots to bring them in configurations where mutual visibility is satisfied. First results have been shown in [5] for robots moving on Trees and Square Grids. Hypercube-like graphs, represent a challenging environment as the high level of symmetry of the topologies does not provide useful means for the resolution of the problem.

References

1. Bose, K., Kundu, M.K., Adhikary, R., Sau, B.: Optimal gathering by asynchronous oblivious robots in hypercubes. In: Gilbert, S., Hughes, D., Krishnamachari, B. (eds.) ALGOSENSORS 2018. LNCS, vol. 11410, pp. 102–117. Springer, Cham (2019). https://doi.org/10.1007/978-3-030-14094-6_7
2. Brandstädt, A., Le, V.B., Spinrad, J.P.: Graph Classes: A Survey. SIAM, Philadelphia (1999)
3. Bujtá, C., Klavžar, S., Tian, J.: Total mutual-visibility in hamming graphs (2023). https://arxiv.org/abs/2307.05168
4. Cicerone, S., Di Fonso, A., Di Stefano, G., Navarra, A.: The geodesic mutual visibility problem for oblivious robots: the case of trees. In: 24th International Conference on Distributed Computing and Networking, ICDCN 2023, pp. 150–159. ACM (2023). https://doi.org/10.1145/3571306.3571401
5. Cicerone, S., Di Fonso, A., Di Stefano, G., Navarra, A.: The geodesic mutual visibility problem: oblivious robots on grids and trees. Pervasive Mob. Comput. **95**, 101842 (2023). https://doi.org/10.1016/j.pmcj.2023.101842. https://www.sciencedirect.com/science/article/pii/S1574119223001001
6. Cicerone, S., Di Fonso, A., Di Stefano, G., Navarra, A.: Time-optimal geodesic mutual visibility of robots on grids within minimum area. In: Dolev, S., Schieber, B. (eds.) SSS 2023. LNCS, vol. 14310, pp. 385–399. Springer, Cham (2023). https://doi.org/10.1007/978-3-031-44274-2_29

7. Cicerone, S., Di Fonso, A., Di Stefano, G., Navarra, A., Piselli, F.: Mutual visibility in hypercube-like graphs (2023). https://arxiv.org/abs/2308.14443
8. Cicerone, S., Di Stefano, G.: Mutual-visibility in distance-hereditary graphs: a linear-time algorithm. In: Fernandes, C.G., Rajsbaum, S. (eds.) Proceedings of the XII Latin-American Algorithms, Graphs and Optimization Symposium, LAGOS 2023, Huatulco, Mexico, 18–22 September 2023. Procedia Computer Science, vol. 223, pp. 104–111. Elsevier (2023). https://doi.org/10.1016/J.PROCS.2023.08.219
9. Cicerone, S., Di Stefano, G., Drožđek, L., Hedžet, J., Klavžar, S., Yero, I.G.: Variety of mutual-visibility problems in graphs. Theor. Comput. Sci. **974**, 114096 (2023). https://doi.org/10.1016/j.tcs.2023.114096
10. Cicerone, S., Di Stefano, G., Klavžar, S.: On the mutual visibility in cartesian products and triangle-free graphs. Appl. Math. Comput. **438**, 127619 (2023). https://doi.org/10.1016/j.amc.2022.127619
11. Cicerone, S., Di Stefano, G., Klavžar, S., Yero, I.G.: Mutual-visibility in strong products of graphs via total mutual-visibility (2022). https://arxiv.org/abs/2210.07835
12. Cicerone, S., Di Stefano, G., Navarra, A.: Gathering robots in graphs: the central role of synchronicity. Theor. Comput. Sci. **849**, 99–120 (2021). https://doi.org/10.1016/j.tcs.2020.10.011
13. Cicerone, S., Di Stefano, G., Klavžar, S., Yero, I.G.: Mutual-visibility problems on graphs of diameter two (2024). https://arxiv.org/abs/2401.02373
14. D'Angelo, G., Di Stefano, G., Klasing, R., Navarra, A.: Gathering of robots on anonymous grids and trees without multiplicity detection. Theor. Comput. Sci. **610**, 158–168 (2016)
15. D'Angelo, G., Di Stefano, G., Navarra, A.: Gathering on rings under the look-compute-move model. Distrib. Comput. **27**(4), 255–285 (2014)
16. D'Angelo, G., Navarra, A., Nisse, N.: A unified approach for gathering and exclusive searching on rings under weak assumptions. Distrib. Comput. **30**(1), 17–48 (2017)
17. Di Luna, G.A., Flocchini, P., Chaudhuri, S.G., Poloni, F., Santoro, N., Viglietta, G.: Mutual visibility by luminous robots without collisions. Inf. Comput. **254**, 392–418 (2017). https://doi.org/10.1016/j.ic.2016.09.005
18. Di Stefano, G.: Mutual visibility in graphs. Appl. Math. Comput. **419**, 126850 (2022). https://doi.org/10.1016/j.amc.2021.126850
19. Di Stefano, G., Navarra, A.: Optimal gathering of oblivious robots in anonymous graphs and its application on trees and rings. Distrib. Comput. **30**(2), 75–86 (2017)
20. Dudeney, H.E.: Amusements in Mathematics. Nelson, Edinburgh (1917)
21. Klasing, R., Kosowski, A., Navarra, A.: Taking advantage of symmetries: gathering of many asynchronous oblivious robots on a ring. Theor. Comput. Sci. **411**, 3235–3246 (2010)
22. Klavzar, S., Neethu, P.K., Chandran, S.V.U.: The general position achievement game played on graphs. Discret. Appl. Math. **317**, 109–116 (2022). https://doi.org/10.1016/j.dam.2022.04.019
23. Klavžar, S., Tian, J.: Graphs with total mutual-visibility number zero and total mutual-visibility in cartesian products. Discussiones Mathematicae Graph Theory (2023). https://doi.org/10.7151/dmgt.2496
24. Kuziak, D., Rodríguez-Velázquez, J.A.: Total mutual-visibility in graphs with emphasis on lexicographic and cartesian products (2023). https://arxiv.org/abs/2306.15818

25. Manuel, P.D., Abd-El-Barr, M.I., Rajasingh, I., Rajan, B.: An efficient representation of Benes networks and its applications. J. Discret. Algorithms **6**(1), 11–19 (2008). https://doi.org/10.1016/j.jda.2006.08.003

26. Manuel, P.D., Klavzar, S.: The graph theory general position problem on some interconnection networks. Fundam. Informaticae **163**(4), 339–350 (2018). https://doi.org/10.3233/FI-2018-1748

27. Navarra, A., Piselli, F.: Mutual-visibility in Fibonacci cubes. In: Barolli, L. (ed.) AINA 2024. LNDECT, vol. 199, pp. 22–33. Springer, Cham (2024). https://doi.org/10.1007/978-3-031-57840-3_3

28. Poudel, P., Aljohani, A., Sharma, G.: Fault-tolerant complete visibility for asynchronous robots with lights under one-axis agreement. Theor. Comput. Sci. **850**, 116–134 (2021). https://doi.org/10.1016/j.tcs.2020.10.033

29. Prabha, R., Devi, S.R., Manuel, P.: General position problem of butterfly networks (2023). https://arxiv.org/abs/2302.06154

30. Sharma, G., Vaidyanathan, R., Trahan, J.L.: Optimal randomized complete visibility on a grid for asynchronous robots with lights. Int. J. Netw. Comput. **11**(1), 50–77 (2021)

31. Stirling, J., Holliday, F.: The Differential Method: Or, A Treatise Concerning Summation and Interpolation of Infinite Series. E. Cave (1749)

32. Van, D., Marilynn, W., Quentin, L., Stout, Q.: Perfect dominating sets on cube-connected cycles. Congr. Numer. **97**, 51–70 (1993)

Non-negotiating Distributed Computing

Carole Delporte-Gallet[1]([⊠]), Hugues Fauconnier[1], Pierre Fraigniaud[1],
Sergio Rajsbaum[1,2], and Corentin Travers[3]

[1] IRIF, Université Paris Cité and CNRS, Paris, France
`cd@irif.fr`
[2] Instituto de Matematicas, Universidad Nacional Autónoma de México,
Mexico City, Mexico
[3] LIS, Aix-Marseille Université, Marseille, France

Abstract. A recent trend in distributed computing aims at designing
models as simple as possible for capturing the inherently limited computing and communication capabilities of insects, cells, or tiny technological
artefacts, yet powerful enough for solving non trivial tasks. This paper
is contributing further in this field, by introducing a new model of distributed computing, that we call *non-negotiating*. In the non-negotiating
model, a process decides *a priori* what it is going to communicate to the
other processes, before the computation starts. Thus, the information a
process sends does not depend on what the process hears from others
during an execution. We consider non-negotiating distributed computing in the read/write shared memory model in which processes are asynchronous and subject to crash failures. We show that non-negotiating
distributed computing is *universal*, in the sense that it is capable of
solving any colorless task solvable by an unrestricted full-information
algorithm in which processes can remember all their history, and send
it to the other processes at any point in time. To prove this universality
result, we present a non-negotiating algorithm for solving multidimensional approximate agreement, with arbitrary precision $\epsilon > 0$.

1 Introduction

1.1 Context and Objective

Standard models of distributed computing assume a set of $n \geq 2$ processes
exchanging information via some communication medium. The communication
media may be of very different kinds, from static to dynamic networks, and
from message-passing to shared memory, to mention just a few, under various failure assumptions. Several textbooks study such models, e.g. [6,31,32].
There has been however, much interest recently in models to study systems
of limited capabilities. This is for instance the case of, e.g., sensor networks,
for which space or energy considerations limit the computing power of each
device. Biological systems like a flock of birds, a school of fish, or a colony
of cells or insects, attracted a lot of interest from the distributed computing

This work is funded in part by the ANR project DUCAT ANR-20-CE48-0006.

Y. Emek (Ed.): SIROCCO 2024, LNCS 14662, pp. 208–225, 2024.
https://doi.org/10.1007/978-3-031-60603-8_12

community recently (see, e.g., [1,2,15,19]). Extremely weak models have thus been introduced, where, for instance, the processes are assumed to be finite-state automata [3], or the communication bandwidth may be drastically limited, e.g., from typically $O(\log n)$ bits in the CONGEST model [31] to $O(1)$ bits in the STONE-AGE model [13] — the BEEPING model [7] (see also [8,10,22]) even assumes that processes are capable to communicate only by either emitting a "beep" or remaining silent. Many of these works considered synchronous failure-free distributed computing. Others considered asynchronous models, including work assuming noisy communication channels [14], work on self-stabilizing systems in which processes are subject to *transient* failures [21], and a few other contributions to asynchronous shared-memory computing, e.g., to understand epigenetic cell modification [33].

This paper introduces the study of a new type of limitation that we call *non-negotiating distributed computing*. In all previous models we are aware of, the contents of a message sent by a process may depend on the messages it has previously received. Even in beeping models, a process decides in each round to emit a beep or remains silent based on the beeps it has heard in the previous rounds. We consider non-negotiating computing, in the sense that the contents of all communication sent by a process is fixed before the execution starts. Messages sent can contain input values, but cannot depend on what a process has heard from others during an execution. Is it possible to do useful non-negotiating distributed computing in an asynchronous system?

We show that the answer is *yes*, in the standard wait-free model in which n processes communicate by writing and reading in a shared memory, and processes are subject to permanent *crash* failures. The shared memory consists of an array of n single-writer/multi-reader registers (SWMR), one per process. The processes are asynchronous, and up to $n-1$ of them may crash. Algorithms in this model are *wait-free* because a process can never wait for another process to perform an action, as the latter can crash.

1.2   Non-negotiating Distributed Computation

We propose a very weak asynchronous crash failure model. After writing its input to the shared memory, each process just repeats a loop consisting of (1) writing a private counter in its shared register, (2) reading all the counters currently present in the other registers, and (3) incrementing its private counter, until some stopping condition is fulfilled. Once the process stops, it decides an output value, and terminates. The pseudo-code of a process is presented in Algorithm 1.

We call this model *non-negotiating*, because each process decides what is going to say (through a sequence of write operations), before the computation starts. The i-th write operation of a process is simply i, it cannot write a value that depends on what it has read so far.

1.3   Results

In a nutshell, we show that in fact, a very large class of problems can be solved in the non-negotiating model. Moreover, we establish a universality result: any

Algorithm 1. Non-Negotiating Distributed Algorithm for Input-Output Tasks

1: **write** *myinput*   ▷ Private input written (once) in shared memory
2: *mycounter* ← 1   ▷ Initialization of private counter
3: **repeat**
4:     **write** *mycounter*   ▷ Private counter written in shared memory
5:     **read** all counters   ▷ Counters are read sequentially, in arbitrary order
6:     *mycounter* ← *mycounter* + 1   ▷ Private counter incremented
7: **until** test({*counters*})   ▷ Testing the stopping condition
8: **read** all inputs   ▷ Inputs are read sequentially, in arbitrary order
9: **decide** output({*counters*}, {*inputs*})   ▷ Computing the output

---

colorless task that is solvable wait-free by an unrestricted algorithm, is solvable wait-free by a non-negotiating algorithm. This result is established by showing how to solve multidimensional approximate agreement for an arbitrary small precision $\epsilon > 0$, using a non-negotiating algorithm, and then proving a theorem that shows how to solve any colorless task using the multidimensional approximate agreement algorithm. Recall that in *multidimensional approximate agreement*, processes start with private input values in $\mathbb{R}^d$, and are required to output values in the convex hull of the inputs that are at most ϵ apart.

Detailed Results. We start by considering *uniform* non-negotiating algorithms, that is, an even more restricted non-negotiating model, where processes execute the same number of rounds, k, fixed a priori. For $n = 2$ processes, we show that there is a uniform non-negotiating algorithm for one-dimensional approximate agreement, for any arbitrary small ϵ, using a sufficiently large value of k (Theorem 1). Always, both processes execute exactly k rounds of the algorithm, and stop.

We however show that there is no uniform non-negotiating algorithms for approximate agreement for more than two processes (Theorem 3). Remarkably, to expose its full computational power, a non-negotiating algorithm, for $n > 2$, must use, in addition to counters, the ability of a process to stop early, before executing k writes, as implied by our first main technical result: we present a non-negotiating algorithm solving multidimensional approximate agreement, for $n \geq 2$ processes (Theorem 4). It is remarkable that a process may convey information to other processes by stopping, since in a wait-free setting a process crash or stop is indistinguishable from the process just being slow.

Our second main contribution is to show that our non-negotiating algorithm for multidimensional approximate agreement can be used to solve any colorless task that is solvable by an unrestricted, full-information algorithm (Theorem 5). Our reduction to multidimensional agreement is in fact general, using our non-negotiating algorithm as a black box, and is also of independent interest.

Discussion. The class of tasks that we consider are called *colorless*, because they can be specified by sets of possible inputs to the processes, and, for each one, a set of legal sets of outputs, without referring to process IDs. Colorless tasks have been thoroughly studied, e.g., [24], as they include many of the standard tasks considered in the context of distributed computing. Note that, even for just three processes, it is undecidable whether a colorless task with finite inputs is wait-free solvable [18, 25].

The main result of the area is the Asynchronous Computability Theorem [28] characterizing asynchronous read/write shared memory task solvability in terms of topology. There is also a corresponding characterization theorem for colorless tasks [26, 27]. These and all subsequent results [24] assumed an unrestricted model: unbounded size registers and *full-information* protocols where a process remembers all its past, and includes all of it in each write operation to the shared memory [24]. We show in this paper, that for colorless tasks, a full-information protocol can be replaced by writing a counter as in a non-negotiating algorithm. We stress however, in terms of time complexity, our approximate agreement algorithm is exponentially slow with respect to full-information algorithms [29]. We do not know if this is unavoidable, except for *uniform* algorithms, for which we prove a corresponding lower bound for the case of two processes (Theorem 2).

The approximate agreement algorithms we design assumes *fixed inputs*: a process always starts with the same fixed input value. But our reduction in Sect. 4 shows how to solve multidimensional approximate agreement for any (finite) set of input values, because such a task is colorless [24]. Our non-negotiating multidimensional agreement algorithm is of independent interest. Indeed, since consensus is not wait-free solvable [16, 23], approximate agreement (that is, 1-dimensional approximate agreement) has been thoroughly studied since in the early days of the field [11]. The multidimensional approximate agreement version, where processes start with values in $\mathbb{R}^d$, has also received recently much attention, e.g., [4, 17, 20]. It was considered in the context of message-passing Byzantine failures systems [30] as well as in shared memory systems with crash failures [27]. Connections between approximate agreement, distributed optimization, and machine learning have also been recently identified [12].

Organization of the Paper. We present the results for two processes in Sect. 2. Our non-negotiating algorithm solving multidimensional agreement is described in Sect. 3. The fact that this algorithm can be used to solve any task that is solvable by an unrestricted algorithm is established in Sect. 4. Section 5 concludes the paper. Due to lack of space, some proof details are omitted. They can be found in the long version [9].

2   Uniform Non-negotiated Approximate Agreement

In a *uniform* algorithm, processes always execute the same number of rounds. We present a uniform approximate agreement algorithm for two processes in Sect. 2.1. The proof provides intuition about the algorithm of Sect. 3 for any number of processes, but the intuition is only partial, since we will show that actually, for $n > 2$ processes, there is no approximate agreement uniform algorithm

We stress that both in this section and in Sect. 3 we design fixed inputs approximate agreement algorithms, in other words, assuming each process always starts with the same input. Approximate agreement for arbitrary (finite) inputs can be formally specified as a colorless task [24], and hence, the results in Sect. 4 show that they can be solved by a non-negotiating algorithm.

2.1   A Uniform Algorithm for 2-Processes Approximate Agreement

We assume two processes, p_1 starts with input value 0 and process p_2 starts with input value 1. In every execution of an algorithm, each non-crashed process therefore has to decide a value in the interval $[0, 1]$, such that if only $p_i, i \in \{1, 2\}$ participates (takes steps), it decides its own input, and in any case, the decided values are at most ϵ apart.[1]

Theorem 1. *For every $\epsilon > 0$, there exists a uniform non-negotiating algorithm that solves ϵ-approximate agreement for two processes.*

Algorithm 2 is used to prove this theorem. Processes execute k rounds, where k depends on ϵ, that is, the algorithm solves $\epsilon = \frac{1}{2k+1}$-approximate agreement.

Algorithm 2 . Uniform non-negotiating algorithm for 2-processes $\frac{1}{2k+1}$-approximate agreement in $[0, 1]$. Code for p_i, $i \in \{1, 2\}$ with input $i - 1$.

1: $c_i \leftarrow 0$; $view_i[r] \leftarrow 0, 1 \leq r \leq k$
2: **for** $r = 1, \ldots, k$ **do**
3:      $c_i \leftarrow c_i + 1$
4:      write(c_i) to $R[i]$ ▷ $R[i]$ is p_i's register
5:      $view_i[r] \leftarrow$ read($R[3 - i]$)   ▷ read other process register
                                                                  ▷ store value read in the array $view_i$

6: **let** d be $\begin{cases} k \text{ if } view_i[k] = 0 \\ \text{such that } (view_i[d] \leq k - d) \wedge (view_i[d + 1] \geq k - d) \text{ otherwise} \end{cases}$

7: **decide** $y_i = \begin{cases} \frac{2(k-d)}{2k+1} & \text{if } i = 1 \\ \frac{2d+1}{2k+1} & \text{if } i = 2 \end{cases}$

Each process $p_i, i \in \{1, 2\}$ has two local variables: a local counter c_i and an array $view_i$. The counter c_i records the local progress of p_i: it is incremented each time p_i repeats its **for** loop. The local array $view_i$ stores the successive values of the other process counter, as read by p_i. That is, for $r \in \{1, \ldots, k\}$, $view_i[r]$ is reserved to store the value returned by its r-th read operation, of the other process' register. After k iterations, process p_i decides a value y_i based on k and its $view_i$ (at line 7).

The decision of process p_i depends on the index d_i such that $view_i[d_i] + d_i \leq k$ and $view_i[d_i + 1] + d_i \geq k$. It is easy to see that such an index d_i can always be found since the function $r \rightarrow view_i[r] + r$ is increasing, and p_i performs k

---

[1] Following previous papers, we consider inputs $0, 1$ for concreteness, but a similar algorithm can be designed with inputs $(1, 0)$, $(0, 1)$ as in Sect. 3.

rounds. Moreover, in the case when both processes terminate the algorithm, the indexes d_1 and d_2 of p_1 and p_2 are closely related, $d_1 = k - d_2$ or $d_1 = k - d_2 - 1$, as we prove (Lemma 1). Therefore, the decision values y_1 and y_2 are at most $\frac{1}{2k+1}$ apart, because $v_1 = \frac{2(k-d_1)}{2k+1}$ and $v_2 = \frac{2d_2+1}{2k+1}$, see line 7.

The validity requirement is trivially satisfied: in case only one process participates, say p_1, its array $view_1$ contains only 0's at the end of the algorithm, and the index d_1 is consequently equal to k. p_1 thus decides $\frac{2(k-d_1)}{2k+1} = 0$, as required. Similarly, in a solo execution p_2 decides 1. When both processes participate, decisions are in the range $[0, 1]$.

The main technical ingredient is first to show that the arrays $view$ obtained by the processes when the algorithm terminates can be partitioned into disjoint classes $C_1, C_2, \ldots$ defined as follows.

Definition 1. *Let $view$ be an array of k integers and let $d \geq 0$ be an integer. We say that $view$ is in C_d if and only if:*

- *For $r \in \{1, \ldots, d\}$, $view[r] \in \{0, .., k - d\}$ and*
- *For $r \in \{d + 1, \ldots, k\}$, $view[r] \in \{k - d, \ldots, k\}$.*

In particular, $[k, k, \ldots, k]$ belongs to the class C_0, and $[0, 0, \ldots, 0]$ belongs to the class C_k. Second, we establish that when both processes terminate the algorithm, their array $view_1$ and $view_2$ belong to two related classes:

Lemma 1. *In every execution where p_i and p_j terminate their algorithm, we have $view_i \in C_d \implies (view_j \in C_{k-d-1}) \vee (view_j \in C_{k-d})$*

This is illustrated Fig. 1. In the case where both processes terminate, the views of both processes are always in "neighbour" classes. As processes chooses their output based on the index of the class their view belongs to, this ensures that outputs are at most $\epsilon = \frac{1}{2k+1}$ apart.

$$C_k \text{---} C_0 \text{---} C_{k-1} \text{---} C_1 \text{---} \ldots \ldots \ C_2 \text{---} C_{k-2} \text{---} \ C_1 \text{---} C_{k-1} \text{---} \ C_0 \text{---} C_k$$

Fig. 1. An edge between C_i and C_j represents an execution of the two processes ending in these classes, where red is for p_1 and blue for p_2. Thus, in one extreme p_1 decides 0 and in the other p_2 decides 1, and in adjacent vertices decisions are at most $\frac{1}{2k+1}$ apart. (Color figure online)

An intuition behind the algorithm can be obtained by representing executions in a 2-dimensional grid, as illustrated in Fig. 2. Starting from the lower left corner, an execution is depicted by moving 1 unit to the right each time p_1 executes a read or write operation, and moving up 1 unit each time p_2 executes an operation. In this example $k = 5$, and each process executes an alternating sequence of 5 write and 5 read operations. Three executions are depicted, all three indistinguishable to p_1, and hence with the same $view_1$. However, p_2 distinguishes the three executions, and thus it has a different view in each one.

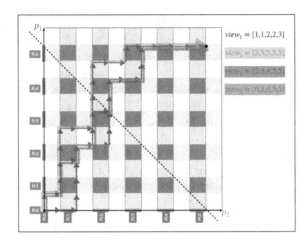

Fig. 2. Three executions for $k = 5$, where p_1 has the same view $[1, 1, 2, 2, 3]$, while p_2 has different views. A horizontal arrow represents a read or write operation by process p_2 while a vertical one, an operation by process p_1. Arrows on the edge of a bi-colored red/blue cell correspond to read operations. For p_1 (respectively, for p_2), output of a read depends solely on the row index (respectively column index) of the bi-colored cell is on the edge of. (Color figure online)

The diagonal $x \rightarrow k - x$ is also depicted (black dotted line.). Every path representing the execution crosses the diagonal. Notice that the last two executions (highlighted in green and purple) cross the diagonal at the same point, but not the third one. The intuition is that the decision of the process depends on when (it becomes aware from its view that) the execution has reached the diagonal. In the yellow execution, p_2 knows that the crossing point has been reached after its second read, while this happens only after its third read in the last two executions.

2.2   Inherent Slowness of Uniform Non-negotiating Computing

We show that our uniform two-processes algorithm is essentially optimal with regards to the number of rounds k. Thus, although it is exponentially slower than unrestricted algorithms such as [29], this is unavoidable. Indeed, we establish that $\Omega(\frac{1}{k})$ is a lower bound on the agreement parameter ϵ of any uniform non-negotiating algorithm that stops after k rounds, for any k:

Theorem 2. *For any $\epsilon < \frac{1}{4k-1}$, there is no uniform k-non negotiating algorithm for two processes that implements ϵ-agreement.*

The main idea is to consider the subset of executions of a uniform k-non negotiating algorithm that have an immediate snapshot schedule, following [5]. Such an execution is defined by a sequence of concurrency classes, $c_1, c_2, \ldots$, where each c_i consists of a non-empty subset of $\{p_1, p_2\}$. In the corresponding execution, processes in c_i write their counters (in an arbitrary order), and then they read each other counters (in an arbitrary order). A sequence of concurrency classes $C = c_1, c_2, \ldots, c_\ell$ is *complete* if each process appears in exactly k concurrency classes, i.e., it is an execution of a *uniform* k-non negotiating algorithm. Notice that the number ℓ of concurrency classes satisfies $k \leq \ell \leq 2k$.

Some immediate snapshot executions are represented in Fig. 3, where the horizontal axis corresponds to the operations executed by p_1 and the vertical axis by p_2. Thus, a path on the grid defines an interleaving of the operations of the two processes. Arrows represent concurrency classes. Table (a) represents the fully synchronous execution of 4 concurrency classes, where every concurrency class c_i is equal to $\{p_1, p_2\}$, while Table (h) represents a fully sequential execution of 8 concurrency classes, where first p_1 executes and then p_2, so the first 4 concurrency classes c_i are equal to $\{p_1\}$, and the other 4 concurrency classes are equal to $\{p_2\}$. Notice that a full concurrency class $\{p_1, p_2\}$ is represented as a diagonal arrow, for the two write operations, and for the two read operations, since both commute: no process distinguishes in which order the operations were performed. Full details of the argument are in [9]. But in brief, the idea is to

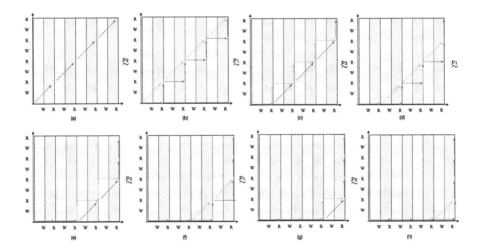

Fig. 3. Some immediate snapshot executions of a uniform k-non negotiating algorithm, $k = 4$, and their indistinguishability relations.

count the number of executions from the fully sequential to the fully concurrent. In more detail, there is a path of $4k - 1$ executions (vertices), with C_1 as the central vertex, and whose endpoints are the fully sequential executions, C', C'', and each two consecutive executions are indistinguishable to one process. This claim implies that the best ϵ-agreement that can be achieved is $\frac{1}{4k-1}$. This is because in each two consecutive executions one process decides the same value.

3 Non-negotiated Multidimensional Approximate Agreement

In this section, we show how to solve multidimensional approximate agreement in the non-negotiating model. As in the previous section, we concentrate on a *fixed inputs* version of *multidimensional approximate agreement* in which each process starts with a fixed input value. It will be used in Sect. 4 to solve *any* colorless task. Let $\epsilon > 0$. Each process $p_i, i \in \{1, \dots, n\}$, $n \geq 2$, starts with input $\mathbf{x}_i = (0, \dots, 0, 1, 0, \dots, 0) \in \{0, 1\}^n$ where the unique 1 stands at the i-th coordinate. Each participating process p_i must output a vector $\mathbf{y}_i \in \mathbb{Q}^n$ such that, if $I \subseteq \{1, \dots, n\}$ denotes the set of participating processes, then (1) for every non-faulty process $p_i \in I$, $\mathbf{y}_i \in \mathsf{Hull}(\{\mathbf{x}_i \mid i \in I\})$, and (2) for every two non-faulty processes $p_i, p_j \in I$, $\|\mathbf{y}_i - \mathbf{y}_j\|_2 \leq \epsilon$. Here, for a set S of points in $\{0, 1\}^n$, $\mathsf{Hull}(S)$ denotes the convex hull of S. Note that since, for every i, process p_i always starts with input $\mathbf{x}_i$, it does not need to write $\mathbf{x}_i$ in memory.

3.1 No Uniform Algorithm Can Solve Multidimensional Approximate Agreement

We first show that actually no uniform algorithm can solve multidimensional approximate agreement (for arbitrarily small ϵ).

Theorem 3. *Let $n \geq 3$. For $\epsilon < \sqrt{\frac{n-2}{n-1}}$, there is no uniform non-negotiating algorithm that implements multidimensional approximate agreement for n processes.*

Proof. Let $\mathcal{A}$ be a n-processes uniform non-negotiating algorithm that implements multidimensional ϵ-agreement for some $\epsilon, 0 \leq \epsilon < 1$. Recall that in our fixed inputs version, the input of each process p_i is the point $\mathbf{u}_i = (0, \dots, 0, 1, 0, \dots, 0)$ whose all coordinates are 0, except the i-th which is 1.

In a uniform non-negotiating algorithm, in every execution, every process performs k iterations of the repeat loop before deciding (unless it fails). The constant k may depend on ϵ, but it is independent of the execution.

The proof is based on an indistinguishability argument. For $i, 1 \leq i \leq n-1$, let e_i be an execution such that:

- Process p_i performs first its k iterations and decides,
- Then processes $p_j, j \in \{1, \dots, n-1\} \backslash \{i\}$ perform their k iterations of the repeat loop. The order in which these processes take steps does not matter. For example, they may perform their k iterations sequentially in increasing index order.
- Finally, process p_n performs its k iterations of the for loop of the algorithm.

Let us first observe that for each $i, 1 \leq i \leq n-1$, process p_i cannot distinguish e_i from an execution in which it is the only participating process. It thus decides $\mathbf{u}_i$ in execution e_i.

Second, note that executions $e_1, \dots, e_{n-1}$ are indistinguishable for process p_n. Indeed, in each of them, the state of the shared memory is $[k, \dots, k, 0]$

when p_n starts executing its algorithm as every process has written the final value, k, of its counter before p_n has taken even one step. Hence, in each execution $e_i, 1 \le i \le n-1$, the successive views of the counters for process p_n is $[k, \ldots, k, 1], \ldots, [k, \ldots, k, k]$. p_n thus decides the same point $\mathbf{y}$ in $e_1, \ldots, e_{n-1}$.

By the first observation, it must be the case that $||\mathbf{y} - \mathbf{u}_i||_2 \le \epsilon$ for all $i, 1 \le i \le n-1$. The coordinates of the point $\mathbf{c}$ closest to $\mathbf{u}_1, \ldots, \mathbf{u}_{n-1}$ (that is, the centroid of $\mathbf{u}_1, \ldots, \mathbf{u}_{n-1}$) are $(\frac{1}{n-1}, \ldots, \frac{1}{n-1}, 0)$. As $||\mathbf{c} - \mathbf{u}_i||_2 = \sqrt{\frac{n-2}{n-1}}$ for every $i, 1 \le i \le n-1$, it follows that $\epsilon \ge \sqrt{\frac{n-2}{n-1}}$. $\square$

3.2 Not Uniform General Multidimensional Approximate Agreement Algorithm

We now present our general multidimensional approximate agreement algorithm, which in light of the previous impossibility, is not uniform.

Theorem 4. *For every $n \ge 2$, and for every $\epsilon > 0$, Algorithm 3 with $k = \frac{3n}{\epsilon}$ solves multidimensional ϵ-approximate agreement for n processes.*

Algorithm 3. Non-negotiating Algorithm for multidimensional approximate agreement. Code for process $p_i, i \in \{1, \ldots, n\}$, starting with input $\mathbf{x}_i$. R is a shared array of n integers initialized to 0.

1: $c_i \leftarrow 0$ ▷ Counter of p_i
2: $view_i \leftarrow [0, \ldots, 0]$ ▷ $view_i[j]$ contain last read value of p_j's counter
3: **repeat**
4:     $c_i \leftarrow c_i + 1$
5:     write(c_i) to $R[i]$ ▷ $R[i]$ is the shared register of p_i
6:     $view_i \leftarrow$ collect(R)   ▷ instruction collect reads all registers in arbitrary order
7: **until** $\sum_{j=1}^{n} view_i[j] \ge k$ ▷ stop when the sum of counters reaches threshold k
8: **decide** $\mathbf{y}_i = \frac{\sum_{j=1}^{n} (view_i[j] \cdot \mathbf{x}_j)}{\sum_{j=1}^{n} view_i[j]}$

The rest of the section is dedicated to the proof of Theorem 4. In Algorithm 3, the accuracy ϵ of the agreement is controlled through parameter $k = O(\frac{n}{\epsilon})$. A process stops incrementing its counter when it observes that the sum of the counters is larger than the threshold k (line 7). In other words, each process looks for the first iteration d at which the sum of the counters of the processes is at least k. In each iteration of the repeat-loop, the value of the counters, as observed by process p_i, are stored in the local variable $view_i$, and are interpreted as the coordinates of a point $\mathbf{d}_i \in \mathbb{R}^n$ where

$$\mathbf{d}_i = \sum_{j=1}^{n} (view_i[j] \cdot \mathbf{x}_j).$$

The algorithm stops at process p_i when $\mathbf{d}_i$ is far enough from the origin $(0, \ldots, 0)$, that is, when it is on the other side of the hyperplane H_k with respect to the origin — see Fig. 4. H_k is the hyperplane that contains the n points $(k, 0, \ldots, 0), \ldots, (0, \ldots, 0, k)$. The coordinates of $\mathbf{d}_i$ are then scaled down to produce a decision $\mathbf{y}_i$ that belongs to the hyperplane H_1 (line 8). As we shall prove in Lemma 3, scaling down the points $\mathbf{d}_i$ ensures validity as the resulting points $\mathbf{y}_i$ are in the convex hull of the unit vectors corresponding to the participating processes. For agreement, we shall prove in Lemma 6 that the decisions $\mathbf{y}_i, i \in I$, all lie in a same L_2-ball of diameter $O(\frac{n}{k})$.

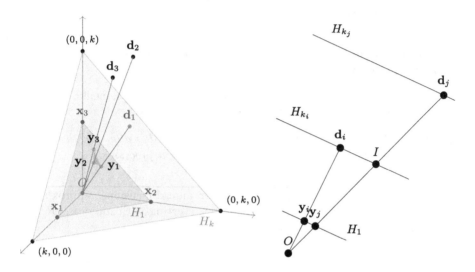

Fig. 4. Points $\mathbf{d}_i, i \in \{1, 2, 3\}$, and outputs $\mathbf{y}_i$ for $n = 3$.

Fig. 5. The points used in the proof of Lemma 6

We first show that Algorithm 3 terminates.

Lemma 2. *In every execution, every process decides after a finite number of iterations.*

Proof. Let p_i be a process. Suppose for contradiction that there is an infinite execution in which p_i does not fail, and does not decide. This means that p_i performs infinitely many iterations of the repeat-loop. In particular, in iteration k, the value of the counter of process p_i is k, which is the value that p_i writes to shared memory in this iteration. Hence, the view it obtains after reading the memory in this iteration is such that $view_i[i] = k$. Since the initial value of each register is 0 and each process writes only positive values to its register, we get that $\sum_{1 \leq j \leq n} view_i[j] \geq k$. Therefore, process p_i exits the repeat-loop, and, as it does not fail, decides at line 8. □

Next, we show that the validity condition of multidimensional ϵ-agreement is satisfied.

Lemma 3. *In every execution, any decision is in the convex hull of the inputs of the participating processes.*

Proof. Let e be an execution of Algorithm 3, let P be the set of participating processes in e, and let $p_i \in P$ be a process that decides in e. We denote by $\mathbf{y}_i$ its decision. It is required that any decision $\mathbf{y}_i$ belongs to the convex hull of the inputs of the participating processes, that is, $\mathbf{y} = \sum_{j \in P} \lambda_j \mathbf{x}_j$ where, for each $j \in P$, $\lambda_j \geq 0$ and $\sum_{j \in P} \lambda_j = 1$. Let $view_i$ be the result of the last collect by process p_i before it decides. Note that for each $p_j \notin P$, $view_i[j] = 0$ as a non-participating process never writes to shared memory, and all registers are initialized to 0. By Line 8, the coordinates of the output decision $\mathbf{y}_i$ are $\frac{1}{\sum_j view_i[j]} (view_i[1], \ldots, view_i[n])$. Therefore, the coordinates are all positive, and their sum is 1. Since, in addition, $view_i[j] = 0$ for every $j \notin P$, it follows that $\mathbf{y}_i \in \mathsf{Hull}(\{\mathbf{x}_j : j \in P\})$. □

We now establish a couple of technical lemmas that will be used for proving that the outputs are close to each other. Their proof can be found in [9].

Lemma 4. *Let e be an execution, and let $i \in \{1, \ldots, n\}$ such that process p_i is correct in e. Let $view_i$ be the last view of p_i before p_i decides. Then*

$$k \leq \sum_{1 \leq j \leq n} view_i[j] \leq k + n - 1.$$

The previous lemma establishes a bound on the sum of the components of the final views of the processes. This sum cannot be too far away from the threshold value k. In the next lemma, we examine each component individually, and we show that each component j cannot be to far from some value c_j.

Lemma 5. *Let e be a finite execution in which at least one process decides. For each process p_j, $j \in \{1, \ldots, n\}$, let c_j be the last value of its counter as written to its register, with $c_j = 0$ if process p_j does not participate in e. For every process p_i that decides there exists n non-negative integers $\delta_i^1, \ldots, \delta_i^n$ such that $\sum_{\ell=1}^n \delta_i^\ell \leq n - 1$, and $\mathbf{d}_i = \sum_{\ell=1}^n (c_\ell - \delta_i^\ell) \cdot \mathbf{x}_\ell$.*

We now have all the ingredients to show that multidimensional ϵ-agreement is solved by Algorithm 3.

Lemma 6. *Let us assume that, in some execution of Algorithm 3, p_i and p_j decide $\mathbf{y}_i$ and $\mathbf{y}_j$, respectively. Then $\|\mathbf{y}_i - \mathbf{y}_j\|_2 \leq \frac{3(n-1)}{k}$.*

Proof (Sketch of the proof). We denote by $view_i$ and $view_j$ the last counters collected on the memory obtained by p_i and p_j, respectively. If $view_i = view_j$, then $\mathbf{y}_i = \mathbf{y}_j$, and the lemma follows. Let us suppose that $view_i \neq view_j$. Let k_i (respectively, k_j) be the sum of the components of $view_i$ (respectively, $view_j$). Without loss of generality, we assume that $k_i \leq k_j$. For every integer $\ell > 0$, let H_ℓ be the hyperplane that contains the n points $(\ell, 0, \ldots, 0), \ldots, (0, \ldots, 0, \ell)$.

That is, $H_\ell = \{\mathbf{v} = (v_1, \ldots, v_n) \mid \sum_{1 \leq \lambda \leq n} v_\lambda = \ell\}$. Note that $\mathbf{d}_i \in H_{k_i}$ as it is the point whose coordinates are $(view_i[1], \ldots, view_i[n])$.

The line (Od_i), where $O = (0, \ldots, 0)$ is the origin, intersects H_k in a single point that we denote by I — See Fig. 5. Indeed, $\mathbf{d}_j$ is either in H_{k_i}, in which case $I = \mathbf{d}_j$, or on the opposite side of H_k. By line 8 in Algorithm 3, $\mathbf{y}_i$ stands on the line (Od_i). Similarly, $\mathbf{y}_i$ stands on the line (Od_j), and, by definition, I also stands on this line. Therefore, points O, $\mathbf{d}_i$, $\mathbf{d}_j$, $\mathbf{y}_i$, $\mathbf{y}_j$, and I are co-planar. To bound the distance between $\mathbf{y}_i$ and $\mathbf{y}_j$, we are going to use the Intercept Theorem on the triangles (Od_iI) and (Oy_iy_j).

The main steps of the proofs are as follows (See [9] for the complete proof):

- We show, using Lemma 5, that we can bound the L_2-norm $||I - \mathbf{d}_i||_2$ from above by a quantity that does not depend on k,

$$||\mathbf{d}_i - I||_2 \leq ||\mathbf{d}_i - I||_1 \leq 3(n-1) \tag{1}$$

- Finally, we use the Intercept Theorem to bound the distance between the decision $\mathbf{y}_i$ and $\mathbf{y}_j$. $O, \mathbf{y}_i, \mathbf{d}_i, \mathbf{y}_j$ and I are co-planar. Hyperplanes H_1 and H_{k_i} have the same direction, and contain $\mathbf{y}_i, \mathbf{y}_j$ and $\mathbf{d}_i, I$, respectively. Therefore, the lines $(\mathbf{y}_i\mathbf{y}_j)$ and $(\mathbf{d}_iI)$ are parallel. From the Intercept Theorem, we have

$$\frac{||\mathbf{y}_j - \mathbf{y}_i||_2}{||I - \mathbf{d}_i||_2} = \frac{||\mathbf{y}_i - O||_2}{||\mathbf{d}_i - O||_2}$$

Now, the decision of process p_i is $\mathbf{y}_i = \frac{\mathbf{d}_i}{\sum_{1 \leq \ell \leq n} view_i[\ell]} = \frac{\mathbf{d}_i}{k_i}$. Therefore,

$$||\mathbf{y}_j - \mathbf{y}_i||_2 = \frac{||\mathbf{y}_i||_2}{||\mathbf{d}_i||_2}||I - \mathbf{d}_i||_2 = \frac{1}{k_i}\frac{||\mathbf{d}_i||_2}{||\mathbf{d}_i||_2}||I - \mathbf{d}_i||_2 \leq 3\frac{(n-1)}{k_i} \leq 3\frac{(n-1)}{k}.$$

The penultimate inequality comes from Eq. (1), and the last inequality follows from the fact that the sum k_i of the components in the last collect $view_i$ of process p_i is at least k.   □

4   Universality of Multidimensional Approximate Agreement

In this section, we show that multidimensional approximate agreement with fixed inputs is complete, in the sense that any problem that is solvable wait-free can be solved by merely solving multidimensional ϵ-agreement for an appropriate setting of ϵ, and inferring the outputs of the problem directly from the solution to multidimensional ϵ-agreement.

To formally define the notion of "problem", it is convenient to adopt the terminology of algebraic topology (see, e.g., [24]). Recall that a *simplicial complex* $\mathcal{K}$ with vertex set V is a collection of non-empty subsets of V containing each singleton $\{v\}$, $v \in V$, and closed by inclusion, i.e., if $\sigma \in \mathcal{K}$ then $\sigma' \in \mathcal{K}$ for every non-empty $\sigma' \subseteq \sigma$. Each set in $\mathcal{K}$ is called a *simplex*. A *task* is then defined as a triple $\varPi = (\mathcal{I}, \mathcal{O}, \varDelta)$ where $\mathcal{I}$ and $\mathcal{O}$ are simplicial complexes, and $\varDelta : \mathcal{I} \to 2^{\mathcal{O}}$ is the input-output specification. That is, every vertex of $\mathcal{I}$ (resp., of $\mathcal{O}$) is a possible input value (resp., output value), and every $\sigma \in \mathcal{I}$ (resp., $\tau \in \mathcal{O}$) represents a collection of legal input configurations (resp., output configurations) of the system. In other words, if $\sigma = \{x_1, \ldots, x_k\}$ belongs to $\mathcal{I}$, then it is legal that $n \geq k$ processes conjointly start with this set of inputs, i.e., some processes start with input x_1, some others with input x_2, etc. Similarly, if $\tau = \{y_1, \ldots, y_k\}$ belongs to $\mathcal{O}$ then it is legal for a set of $n \geq k$ processes to conjointly output τ, i.e., some processes output y_1, while some others output y_2, etc. Finally, for every $\sigma \in \mathcal{I}$, $\varDelta(\sigma)$ is a sub-complex of $\mathcal{O}$ specifying the set of legal outputs for σ, i.e., any set of processes with input configuration σ can collectively output any simplex $\tau \in \varDelta(\sigma)$.

For instance, n-dimensional ϵ-agreement with fixed inputs is the task $\varPi = (\mathcal{I}, \mathcal{O}, \varDelta)$ with

$$\mathcal{I} = \left\{ \{\mathbf{x}_i \mid i \in I\} \mid (I \neq \varnothing) \wedge (I \subseteq \{1, \ldots, n\}) \right\}$$

where, for each $i \in \{1, \ldots, n\}$, $\mathbf{x}_i$ is the n-dimensional vector $(0, \ldots, 0, 1, 0, \ldots, 0)$ with the 1 at the i-th coordinate,

$$\mathcal{O} = \left\{ \{\mathbf{y}_j \mid j \in J\} \in \mathcal{P}(\mathbb{Q}^n) \mid (\varnothing \neq J \subseteq \{1, \ldots, n\}) \wedge (\forall i, j \in J, \|\mathbf{y}_i - \mathbf{y}_j\|_2 \leq \epsilon) \right\}$$

and, for every $\sigma = \{\mathbf{x}_i \mid i \in I\} \in \mathcal{I}$, and every $\tau = \{\mathbf{y}_j \mid j \in J\} \in \mathcal{O}$, we have

$$\tau \in \varDelta(\sigma) \iff \mathbf{y}_j \in \mathsf{Hull}(\{\mathbf{x}_i \mid i \in I\}) \text{ for every } j \in J.$$

The following theorem essentially states that an algorithm for multidimensional ϵ-agreement can be used to solve any (solvable) task.

Theorem 5. *Let $n \geq 2$, and let $\varPi = (\mathcal{I}, \mathcal{O}, \varDelta)$ be a task solvable by n processes. There exists $\epsilon > 0$ such that, for every input $\sigma = \{x_i \mid i \in I\} \in \mathcal{I}$ for $\varPi$, with $\varnothing \neq I \subseteq \{1, \ldots, n\}$, if $\{\mathbf{y}_i \mid i \in I\}$ denotes any solution of multidimensional ϵ-agreement whenever process p_i starts with input $\mathbf{x}_i$ for every $i \in I$, then every process $p_i, i \in I$ can compute locally from $\mathbf{y}_i$ an output y_i for $\varPi$ such that $\{y_i \mid i \in I\} \in \varDelta(\sigma)$.*

Proof. By the colorless wait-free computability theorem [24, 27], since $\varPi = (\mathcal{I}, \mathcal{O}, \varDelta)$ is a task solvable by n processes, there exists an integer $T \geq 0$ and a simplicial map[2]

$$f : \mathcal{B}^T(\mathcal{I}) \to \mathcal{O}$$

where $\mathcal{B}^T(\mathcal{I})$ is the simplicial complex obtained by applying T times the barycentric subdivision operator to $\mathcal{I}$ (see Fig. 6). Moreover, this map f agrees with the input-output specification $\varDelta$ of the task, that is, for every $\sigma \in \mathcal{I}$,

$$f(\mathcal{B}^T(\sigma)) \subseteq \varDelta(\sigma),$$

i.e., every simplex $\sigma' \in \mathcal{B}^T(\sigma)$ is mapped by f to a simplex $f(\sigma')$ of $\varDelta(\sigma)$.

---

[2] Recall that a map f from the vertex set of a complex $\mathcal{K}_1$ to the vertex set of a complex $\mathcal{K}_2$ is simplicial if, for every $\sigma \in \mathcal{K}_1$, $f(\sigma) \in \mathcal{K}_2$. Such a map is therefore a map $f : \mathcal{K}_1 \to \mathcal{K}_2$, mapping every simplex of $\mathcal{K}_1$ to a simplex of $\mathcal{K}_2$.

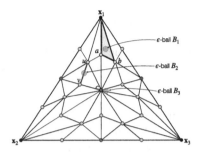

Fig. 6. Applying two times the barycentric subdivision operator to $\sigma_\epsilon = \{\mathbf{x}_1, \mathbf{x}_2, \mathbf{x}_3\}$. Formally, the figure displays the canonical geometric realization of $\mathcal{B}^2(\sigma_\epsilon)$, and the outer triangle represents the frontier of the convex hull of the three points $\mathbf{x}_1, \mathbf{x}_2, \mathbf{x}_3$.

Let us denote by $\Pi_\epsilon = (\mathcal{I}_\epsilon, \mathcal{O}_\epsilon, \Delta_\epsilon)$ the multidimensional ϵ-agreement task. Recall that we mean here the fixed inputs version, for which every $i \in \{1, \dots, n\}$, process p_i can only start with the point $\mathbf{x}_i$. That is, the input complex of multidimensional ϵ-agreement is simply $\mathcal{I}_\epsilon = \{\mathbf{x}_1, \dots, \mathbf{x}_n\}$. Now, recall that, by definition of multidimensional ϵ-agreement, the outputs $\mathbf{y}_j, j \in J$, will be at mutual distance at most ϵ, and will stand in the convex hull of the input simplex $\sigma = \{\mathbf{x}_i \mid i \in J\}$.

A first observation is that, by picking ϵ sufficiently small, any ball of radius ϵ in the convex hull of σ can be mapped to a simplex of $\mathcal{B}^T(\sigma)$. For instance, in Fig. 6, the ball B_1 is included in the simplex $\{a, b, \mathbf{x}_1\}$ of $\mathcal{B}^2(\sigma)$. Therefore, each $\mathbf{y}_j$ in B_1 can be mapped to any of the three points a, b, or $\mathbf{x}_1$, e.g., to the closed point. Instead, the ball B_2 intersects a face of $\mathcal{B}^2(\sigma)$, namely the edge $\{u, v\}$. In this case, each $\mathbf{y}_j$ in B_2 can be mapped to any of the two points u or v, e.g., the closest. A ball of radius ϵ may however intersect many different faces, just like the ball B_3 does in Fig. 6. However, for ϵ sufficiently small, this may occur only for balls that are close to a single vertex (c in the case of the ball B_3). Therefore, each $\mathbf{y}_j$ in B_3 can simply be mapped to this vertex. This mapping is denoted by

$$g : \mathcal{O}_\epsilon \to \mathcal{B}^T(\mathcal{I}_\epsilon)$$

A second observation is that there is a canonical one-to-one correspondence between any input simplex $\sigma = \{x_i \mid i \in I\} \in \mathcal{I}$ of the task Π and the input simplex $\sigma_\epsilon = \{\mathbf{x}_i \mid i \in I\} \in \mathcal{I}_\epsilon$ of multidimensional ϵ-agreement. Therefore, the same holds for their barycentric subdivisions. Let us denote by

$$h_\sigma : \mathcal{B}^T(\sigma) \to \mathcal{B}^T(\sigma_\epsilon)$$

this one-to-one map. Note that, for every face σ' of σ, $h_{\sigma'}$ coincides with h_σ restricted to $\mathcal{B}^T(\sigma')$.

We have now all the ingredients for establishing the theorem. Let us first explain how the generic non-negotiating algorithm (Algorithm 1 of Sect. 1.2) is instantiated by each process. Every process p_j starts by writing its input for task Π (line 1 of the generic algorithm). It then solves multidimensional ϵ-agreement (with fixed input $\mathbf{x}_j$) using the multidimensional approximate agreement algorithm (Algorithm 3). Following Algorithm 3, this consists in performing $O(n/\epsilon)$ iterations of the **repeat** loop of the generic algorithm. p_j finally reads all inputs for task Π previously written (line 8 of the generic algorithm.).

By reading the inputs for Π, process p_j thus collects an input simplex $\sigma = \{x_i \mid i \in I\} \in \mathcal{I}$. From σ, p_i infers $\sigma_\epsilon = \{\mathbf{x}_i \mid i \in I\}$ which is its view of the input simplex of the multidimensional approximate agreement. Let $\mathbf{y}$ be the output of p_j in multidimensional ϵ-agreement. Using the map $g : \mathcal{O}_\epsilon \to \mathcal{B}^T(\mathcal{I}_\epsilon)$, process p_j computes

$$\mathbf{z} = g(\mathbf{y}) \in \mathcal{B}^T(\mathcal{I}_\epsilon).$$

Given $\mathbf{z}$, σ, and σ_ϵ, process p_j can then use the one-to-one map $h_\sigma : \mathcal{B}^T(\sigma) \to \mathcal{B}^T(\sigma_\epsilon)$ to compute

$$z = h_\sigma^{-1}(\mathbf{z}) \in \mathcal{B}^T(\sigma).$$

Finally, given $z \in \mathcal{B}^T(\sigma)$, process p_j just outputs $y = f(z)$ where $f : \mathcal{B}^T(\mathcal{I}) \to \mathcal{O}$ is the aforementioned simplicial map whose existence is guaranteed by the wait-free computability theorem.

To show correctness, let $I \subseteq \{1, \ldots, n\}$ be the set of (correct) participating processes, with input simplex $\{x_i \mid i \in I\}$. These processes solve multidimensional ϵ-agreement with input $\sigma_\epsilon = \{\mathbf{x}_i \mid i \in I\}$, and output $\{\mathbf{y}_j \mid j \in J\} \in \Delta_\epsilon(\sigma_\epsilon)$. Thanks to the mapping g, these output values are mapped to a simplex $\tau \in \mathcal{B}^T(\sigma_\epsilon)$, which is, thanks to h_σ, in one-to-one correspondence with a simplex $\tau' \in \mathcal{B}^T(\sigma)$. Since f is simplicial, the latter simplex is mapped to a simplex $\tau'' \in \mathcal{O}$. In fact, since f solves the task Π, and since $\tau' \in \mathcal{B}^T(\sigma)$, we necessarily have $\tau'' \in \Delta(\sigma)$ as f agrees with Δ. This completes the proof. □

The following is a direct consequence of Theorem 5, merely because multidimensional ϵ-agreement with process p_i starting with input $\mathbf{x}_i$ for every $i \in \{1, \ldots, n\}$ can be solved using a non-negotiating algorithm.

Corollary 1. *Let $n \geq 2$, and let $\Pi = (\mathcal{I}, \mathcal{O}, \Delta)$ be a task solvable by n processes with an unrestricted algorithm. There exists a non-negotiating algorithm solving Π for n processes.*

5   Conclusion

We have introduced a very restricted form of distributed wait-free shared memory computing, and shown that nevertheless, it is universal, in the sense that it is capable of solving any colorless task that is wait-free solvable under no restrictions on the algorithm. The result is achieved by proving a general result about multidimensional approximate agreement: a black box that solves this task for arbitrary $\epsilon > 0$, can be used to solve any (wait-free solvable) colorless task.

Our results open several interesting avenues for future research. They uncover the remarkable power of asynchronous read/write shared memory, that allows processes to communicate to each other information, through the timing of their read/write operations, which is not under their control. Our non-negotiating algorithm for ϵ-multidimensional agreement has step complexity $O(n/\epsilon)$ which is exponentially slower than standard shared memory algorithms. Although we proved a lower bound for two processes uniform algorithms, in general we don't know if this slowdown is unavoidable.

The class of tasks we have considered are *colorless* [24], which includes consensus, approximate agreement, set agreement, and many others. It remains an open question if our non-negotiating model can be used to solve general tasks, most notably, renaming [6].

Acknowledgments. We thank the anonymous reviewers for their nice comments that help improve the presentation of that paper.

References

1. Afek, Y., Alon, N., Barad, O., Hornstein, E., Barkai, N., Bar-Joseph, Z.: A biological solution to a fundamental distributed computing problem. Science **331**(6014), 183–185 (2011)
2. Ancona, B., Bajwa, A., Lynch, N.A., Mallmann-Trenn, F.: How to color a French flag - biologically inspired algorithms for scale-invariant patterning. In: Kohayakawa, Y., Miyazawa, F.K. (eds.) LATIN 2021. LNCS, vol. 12118, pp. 413–424. Springer, Cham (2020). https://doi.org/10.1007/978-3-030-61792-9_33
3. Angluin, D., Aspnes, J., Diamadi, Z., Fischer, M.J., Peralta, R.: Computation in networks of passively mobile finite-state sensors. Distrib. Comput. **18**(4), 235–253 (2006)
4. Attiya, H., Ellen, F.: The step complexity of multidimensional approximate agreement. In: 26th International Conference on Principles of Distributed Systems, OPODIS. LIPIcs, vol. 253, pp. 6:1–6:12 (2022)
5. Attiya, H., Rajsbaum, S.: The combinatorial structure of wait-free solvable tasks. SIAM J. Comput. **31**(4), 1286–1313 (2002)
6. Attiya, H., Welch, J.: Distributed Computing: Fundamentals, Simulations, and Advanced Topics, vol. 19. Wiley, Hoboken (2004)
7. Cornejo, A., Kuhn, F.: Deploying wireless networks with beeps. In: Lynch, N.A., Shvartsman, A.A. (eds.) DISC 2010. LNCS, vol. 6343, pp. 148–162. Springer, Heidelberg (2010). https://doi.org/10.1007/978-3-642-15763-9_15
8. Davies, P.: Optimal message-passing with noisy beeps. In: 42th ACM Symposium on Principles of Distributed Computing (PODC), pp. 300–309 (2023)
9. Delporte-Gallet, C., Fauconnier, H., Fraigniaud, P., Rajsbaum, S., Travers, C.: Non-negotiating distributed computing. https://hal.science/hal-04470425
10. Delporte-Gallet, C., Fauconnier, H., Rajsbaum, S.: Communication complexity of wait-free computability in dynamic networks. In: Richa, A., Scheideler, C. (eds.) SIROCCO 2020. LNCS, vol. 12156, pp. 291–309. Springer, Cham (2020). https://doi.org/10.1007/978-3-030-54921-3_17
11. Dolev, D., Lynch, N.A., Pinter, S.S., Stark, E.W., Weihl, W.E.: Reaching approximate agreement in the presence of faults. J. ACM **33**(3), 499–516 (1986)
12. El-Mhamdi, E.M., Guerraoui, R., Guirguis, A., Hoang, L.N., Rouault, S.: Genuinely distributed byzantine machine learning. Distrib. Comput. **35**(4), 305–331 (2022)
13. Emek, Y., Wattenhofer, R.: Stone age distributed computing. In: 32nd ACM Symposium on Principles of Distributed Computing (PODC), pp. 137–146 (2013)

14. Feinerman, O., Haeupler, B., Korman, A.: Breathe before speaking: efficient information dissemination despite noisy, limited and anonymous communication. Distributed Comput. **30**(5), 339–355 (2017)
15. Feinerman, O., Korman, A.: The ANTS problem. Distrib. Comput. **30**(3), 149–168 (2017)
16. Fischer, M.J., Lynch, N.A., Paterson, M.: Impossibility of distributed consensus with one faulty process. J. ACM **32**(2), 374–382 (1985)
17. Függer, M., Nowak, T., Schwarz, M.: Tight bounds for asymptotic and approximate consensus. J. ACM **68**(6), 46:1–46:35 (2021)
18. Gafni, E., Koutsoupias, E.: Three-processor tasks are undecidable. SIAM J. Comput. **28**(3), 970–983 (1999)
19. Gelblum, A., Fonio, E., Rodeh, Y., Korman, A., Feinerman, O.: Ant collective cognition allows for efficient navigation through disordered environments. eLife **9**, e55195 (2020)
20. Ghinea, D., Liu-Zhang, C., Wattenhofer, R.: Multidimensional approximate agreement with asynchronous fallback. In: 35th ACM Symposium on Parallelism in Algorithms and Architectures (SPAA), pp. 141–151 (2023)
21. Giakkoupis, G., Ziccardi, I.: Distributed self-stabilizing MIS with few states and weak communication. In: 42nd ACM Symposium on Principles of Distributed Computing (PODC), pp. 310–320 (2023)
22. Hella, L.: Weak models of distributed computing, with connections to modal logic. Distrib. Comput. **28**(1), 31–53 (2015)
23. Herlihy, M.: Wait-free synchronization. ACM Trans. Program. Lang. Syst. **13**(1), 124–149 (1991)
24. Herlihy, M., Kozlov, D.N., Rajsbaum, S.: Distributed Computing Through Combinatorial Topology. Morgan Kaufmann, Burlington (2013)
25. Herlihy, M., Rajsbaum, S.: The decidability of distributed decision tasks. In: Proceedings of the Twenty-Ninth Annual ACM Symposium on the Theory of Computing (STOC), pp. 589–598. ACM (1997)
26. Herlihy, M., Rajsbaum, S.: The topology of distributed adversaries. Distrib. Comput. **26**(3), 173–192 (2013)
27. Herlihy, M., Rajsbaum, S., Raynal, M., Stainer, J.: From wait-free to arbitrary concurrent solo executions in colorless distributed computing. Theor. Comput. Sci. **683**, 1–21 (2017)
28. Herlihy, M., Shavit, N.: The topological structure of asynchronous computability. J. ACM **46**(6), 858–923 (1999)
29. Hoest, G., Shavit, N.: Toward a topological characterization of asynchronous complexity. SIAM J. Comput. **36**(2), 457–497 (2006)
30. Mendes, H., Herlihy, M., Vaidya, N.H., Garg, V.K.: Multidimensional agreement in byzantine systems. Distrib. Comput. **28**(6), 423–441 (2015)
31. Peleg, D.: Distributed Computing: A Locality-Sensitive Approach. Society for Industrial and Applied Mathematics (2000)
32. Raynal, M.: Fault-Tolerant Message-Passing Distributed Systems - An Algorithmic Approach. Springer, Cham (2018)
33. Taubenfeld, G.: Anonymous shared memory. J. ACM **69**(4) (2022)

Better Sooner Rather Than Later

Anaïs Durand[1], Michel Raynal[2(✉)], and Gadi Taubenfeld[3]

[1] LIMOS, Université Clermont Auvergne CNRS UMR 6158, Aubière, France
[2] IRISA, CNRS, Inria, Univ Rennes, 35042 Rennes, France
tgadi@runi.ac.il
[3] Reichman University, 4610101 Herzliya, Israel

Abstract. This article unifies and generalizes fundamental results related to n-process asynchronous crash-prone distributed computing. More precisely, it proves that for every $0 \leq k \leq n$, assuming that process failures occur only before the number of participating processes bypasses a predefined threshold that equals $n - k$ (a participating process is a process that has executed at least one statement of its code), an asynchronous algorithm exists that solves consensus for n processes in the presence of f crash failures *if and only if $f \leq k$*. In a very simple and interesting way, the "extreme" case $k = 0$ boils down to the celebrated FLP impossibility result (1985, 1987). Moreover, the second extreme case, namely $k = n$, captures the celebrated mutual exclusion result by E.W. Dijkstra (1965) that states that mutual exclusion can be solved for n processes in an asynchronous read/write shared memory system where any number of processes may crash (but only) before starting to participate in the algorithm (that is, participation is not required, but once a process starts participating it may not fail). More generally, the possibility/impossibility stated above demonstrates that more failures can be tolerated when they occur earlier in the computation (hence the title).

Keywords: Adopt/commit · Asynchronous read/write system · Concurrency · Consensus, Contention · Mutual exclusion · Process participation · Process crash · Time-constrained crash failure · Simplicity

1 Introduction

1.1 Two Fundamental Problems in Distributed Computing

On the Nature of Distributed Computing. Parallel computing aims to track and exploit data independence in order to obtain efficient algorithms: the decomposition of a problem into data-independent sub-problems is under the control of the programmer. The nature of *distributed computing* is different; namely, distributed computing is the science of cooperation in the presence of adversaries (the most common being asynchrony and process failures): a set of predefined processes, each with its own input (this is not on the control of the programmer) must exchange information in order to attain a common goal. The two most famous distributed computing problems are *consensus* and *mutual exclusion*.

ⓒ The Author(s), under exclusive license to Springer Nature Switzerland AG 2024
Y. Emek (Ed.): SIROCCO 2024, LNCS 14662, pp. 226–237, 2024.
https://doi.org/10.1007/978-3-031-60603-8_13

The Consensus Problem. The consensus problem was initially introduced in the context of synchronous message-passing systems in which some processes are prone to Byzantine failures [9, 11]. Consensus is a one-shot object providing the processes with a single operation denoted propose(). This operation takes an input parameter and returns a value. When a process invokes propose(v), we say it proposes the value v. If propose() returns the value v', we say that it decides v'. The following set of properties defines consensus. A *faulty* process is a process that commits a failure (crash in our case, i.e., an unexpected premature and definitive stop). An *initial failure* is a process crash that occurs before the process starts participating [18]. A process that is not faulty is said to be *correct*.

- Validity. If a process decides value v, then v was proposed by some process.
- Agreement. No two processes decide different values.
- Termination. If a correct process invokes propose(v) then it decides on a value.

A fundamental result related to consensus in asynchronous crash-prone systems where processes communicate by reading and writing atomic registers only is its impossibility if even only one process may crash [10] (read/write counterpart of the famous FLP result stated for asynchronous message-passing systems [4]).

The Mutual Exclusion Problem. Mutual exclusion is the oldest and one of the most important synchronization problems. Formalized by E.W. Dijkstra in the mid-sixties [1], it consists of building what is called a lock (or mutex) object, defined by two operations, denoted acquire() and release(). The invocation of these operations by a process p_i follows the following pattern: "acquire(); *critical section*; release()", where "critical section" is any sequence of code. It is assumed that, once in the critical section, a process eventually invokes release(). A mutex object must satisfy the following two properties.

- Mutual exclusion: No two processes are simultaneously in their critical section.
- Deadlock-freedom progress condition: If there is a process p_i that has a pending operation acquire() (i.e., it invoked acquire() and its invocation is not terminated) and there is no process in the critical section, there is a process p_j (maybe $p_j \neq p_i$) that eventually enters the critical section.

A fundamental result related to mutual exclusion in asynchronous fault-free systems where processes communicate by reading and writing atomic registers only is that mutual exclusion can be solved for any finite number of processes even when (process) participation is not required [1].

Observation 1. *In a shared memory system with no failures and where participation is not required, mutual exclusion is solvable if and only if consensus is solvable.*

The proof is straightforward. To solve consensus using mutual exclusion, we can simply let everybody decide on the proposed value of the first process to enter the critical section. To solve mutual exclusion using consensus, the processes can participate in a sequence of consensus objects to decide on the next process to enter the critical section.

1.2   Contention-Related Crash Failures

The Notion of λ-Constrained Crash Failures. Consensus can be solved in crash-prone (read/write or message-passing) synchronous systems. So, an approach to solve consensus in crash-prone asynchronous systems consists in capturing a "logical time notion" that can be exploited to circumvent the consensus impossibility. In this article, the notion of time is captured by the increasing number of processes that started participating in the consensus algorithm (a process becomes *participating* when it accesses the shared memory for the first time).[1] Crash failures in such a context have given rise to the notion of λ-*constrained crash failures* (introduced in [17], where they are named *weak* failures). Then, they have been investigated in [2,3]. The idea consists in allowing some number k of processes to crash only while the current number of participating processes has not bypassed some predefined threshold denoted λ, see Fig. 1.

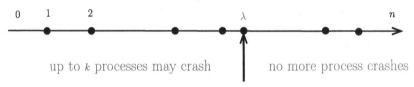

Fig. 1. Asynchronous λ-constrained crash failures

Observation 2 : A computability equivalence. The following observation follows immediately from the definitions concerning process participation, initial crashes, and λ-constrained crash failures. From a computability point of view, the three following statements are equivalent (each implies the two others).

It is possible to solve consensus and mutual exclusion in an asynchronous system

- in a fault-free system where participation is not required, or
- in the presence of any number of initial failures, or
- in the presence of any number of 0-constrained crash failures.

Motivation: Why study λ-constrained failures? As discussed and demonstrated in [2, 17], the new type of λ-constrained failures enables the design of algorithms that can tolerate several traditional "any-time" failures plus several additional λ-constrained failures. More precisely, assume that a problem can be solved in the presence of t traditional (i.e., any-time) failures but cannot be solved in the presence of $t + 1$ such failures. Yet, the problem might be solvable in the presence of $t_1 \leq t$ "any-time" failures plus t_2 λ-constrained failures, where $t_1 + t_2 > t$.

Adding the ability to tolerate λ-constrained failures to algorithms that are already designed to circumvent various impossibility results, such as the Paxos algorithm [8] and indulgent algorithms in general [6,7], would make such algorithms even more

---

[1] Let us remind that such a process participation assumption is implicit in all asynchronous message-passing systems.

robust against possible failures. An indulgent algorithm never violates its safety property and eventually satisfies its liveness property when the synchrony assumptions it relies on are satisfied. An indulgent algorithm which in addition (to being indulgent) tolerates λ-constrained failures may, in many cases, satisfy its liveness property even before the synchrony assumptions it relies on are satisfied.

When facing a failure-related impossibility result, such as the impossibility of consensus in the presence of a single faulty process (discussed earlier [4, 10]) one is often tempted to use a solution that guarantees no resiliency at all. We point out that there is a middle ground: tolerating λ-constrained failures enables to tolerate failures some of the time. Notice that traditional t-resilient algorithms also tolerate failures only some of the time (i.e., as long as the number of failures is at most t). After all, *something is better than nothing*. As a simple example, a message-passing algorithm is described in [4], which solves consensus despite asynchrony and up to $t < n/2$ processes crashes if these crashes occur initially (hence no participating process crashes).

1.3 Computational Model

Our model of computation consists of a collection of n asynchronous deterministic processes that communicate by atomically reading and writing shared registers. A process can read or write at each atomic step, but not both. A register that can be written and read by any process is a multi-writer multi-reader (MWMR) register. If a register can be written by a single (predefined) process and read by all, it is a single-writer multi-reader (SWMR) register. Asynchrony means that there is no assumption on the relative speeds of the processes. Each process has a unique identifier. The only type of failure considered in this paper is a process *crash* failure. As already said, a crash is a premature halt. Thus, until a process possibly crashes, it behaves correctly by executing its code. The following known observation implies that an impossibility results proved for the shared memory model also holds for such a message-passing system.

Observation 3. *A shared memory system that supports atomic registers can simulate a message-passing system that supports send, receive, and even broadcast operations.*

The proof is straightforward. The simulation is as follows. With each process p, we associate an unbounded array of shared registers which all processes can read from, but only p can write into. To simulate a broadcast (or sending) of a message, p writes to the next unused register in its associated array. When p has to receive a message, it reads the new messages from each process.

1.4 Contributions and Related Work

The article unifies and generalizes fundamental results about the mutual exclusion and consensus problems. To this end, it states and proves the following theorem.

Theorem 1 (Main result). *For every $0 \leq k \leq n$, an algorithm exists that solves consensus for n processes in the presence of f $(n-k)$-constrained crash failures if and only if $f \leq k$.*

There are two special cases that are of special interest.

- The first special case, when $k = 0$, indicates that in the presence of any number of n-constrained crash failures, not even a single failure can be tolerated. This implies the celebrated impossibility results (from 1985 and 1987) which states that consensus cannot be solved by n processes in an asynchronous message-passing or read/write shared memory system in which even a single process may crash at any time [4, 10]. Here, we use Observation 3 that a shared memory system can simulate a message passing system.
- The second special case, when $k = n$ implies that consensus can be solved for n processes in an asynchronous read/write shared memory system in the presence of any number of 0-constrained crash failures. This result, together with Observation 1 and Observation 2, implies the celebrated result by E.W. Dijkstra (from 1965), which originated the field of distributed computing, that mutual exclusion can be solved for n processes in an asynchronous read/write shared memory fault-free system where (process) participation is not required [1].

It is shown in [2, 17], among other results, that consensus can be solved (1) despite a single process crash if this crash occurs before the number of participating processes bypasses $\lambda = n - 1$; and (2) despite $k - 1$ process crashes, where $k > 1$, if these crashes occur before the number of participating processes bypasses $\lambda = n - k$. The main question left open in [2, 17] is whether this possibility result is tight.

Our main result, as stated in Theorem 1, shows that the answer to this open question is negative and proves a new stronger result which is shown to be tight. Furthermore, two cumbersome and complicated consensus algorithms were presented to prove the above results [2, 17]. These algorithms are based on totally different design principles, and the following question was posed as an open problem in [2]: "Does it exist a non-trivial generic consensus algorithm that can be instantiated for any value of $k \geq 1$?"[2]. Our result answers this second question positively.

To prove the if direction part in the proof of Theorem 1, a rather simple and elegant consensus algorithm is presented. This new algorithm is based on two underlying (read/write implementable) objects, namely a crash-tolerant adopt-commit object [5] and a not-crash-tolerant deadlock-free acquire-restricted mutex object (a mutex object without a release operation [15, 16]). We show that the proposed algorithm is optimal in the λ-constrained crash failures model.

Finally, contention-related crash failures were also investigated in [3] in a model where processes communicate by accessing shared objects which are computationally stronger than atomic read/write registers.

2   The Consensus Algorithm

This section proves the "if direction" of Theorem 1.

---

[2] "Non-trivial generic" means here that the algorithm must not be a case statement with different sub-algorithms for different values of k.

Theorem 2 (If direction). *For every* $0 \leq k \leq n$, *an algorithm exists that solves consensus for* n *processes in the presence of* f $(n - k)$-*constrained crash failures* **if** $f \leq k$.

To prove this theorem, we present below a consensus algorithm tolerating k λ-constrained failures, where $\lambda = n - k$. The processes, denoted $p_1, p_2, ..., p_n$, execute the same code. It is assumed that proposed values are integers and that the default value $\perp$ is greater than any integer.

2.1   Shared and Local Objects Used by the Algorithm

Shared Objects. The processes cooperate through the following shared objects (which can be built on top of asynchronous read/write systems, the first one in the presence of any number of crashes, the second one in failure-free systems, but as we will see, the access to this object will be restricted to correct processes only).

- $INPUT[1..n]$ is an array of atomic single-writer multi-reader registers. It is initialized to $[\perp, ..., \perp]$. $INPUT[i]$ will contain the value proposed by p_i.
- DEC is a multi-writer multi-reader atomic register, the aim of which is to contain the decided value. It is initialized to $\perp$ (a value that cannot be proposed).
- AC is an adopt/commit object. This object, which can be built in asynchronous read/write systems prone to any number of process crashes, was introduced in [5]. It provides the processes with a single operation (that a process can invoke only once) denoted ac_propose(). This operation takes a value as an input parameter and returns a pair $\langle tag, v \rangle$, where $tag \in \{\texttt{commit}, \texttt{adopt}\}$ and v is a proposed value (we say that the process decides a pair). The following properties define the object.
  - *Termination.* A correct process that invokes ac_propose() returns from its invocation.
  - *Validity.* If a process returns the pair $\langle -, v \rangle$, then v was proposed by a process.
  - *Obligation.* If the processes that invoke ac_propose() propose the same input value v, only the pair $\langle \texttt{commit}, v \rangle$ can be returned.
  - *Weak agreement.* If a process decides $\langle \texttt{commit}, v \rangle$ then any process that decides returns the pair $\langle \texttt{commit}, v \rangle$ or $\langle \texttt{adopt}, v \rangle$.

  Let us remark that if, initially, a process executes solo ac_propose(v), it returns the value v, and, if any, all later all invocations of ac_propose() will return v. The same occurs if (initially) a set of processes invoke ac_propose() with the same value v: the adopt/commit object will always return v. Wait-free implementation of the adopt-commit object are described in [5, 13].
- ARM is a one-shot *acquire-restricted* deadlock-free mutex object, i.e., a mutex object that provides the processes with a single operation denoted acquire() (i.e., a mutex object without release() operation). One-shot means that a process can invoke acquire() at most once.

  Let us observe that as there is no release() operation, only one process can return from its invocation of acquire(). The other processes that invoked acquire() never terminate their acquire() operation. The ARM object will be used to elect a process when needed in specific circumstances.

As we will see, the proposed consensus algorithm allows only correct processes to invoke the acquire() operation. So any algorithm implementing a failure-free deadlock-free mutex algorithm (or a read/write-based leader election algorithm) can be used [15]. Such space efficient algorithms exist, that use only log n atomic read/write registers [16].

Local Objects. Each process p_i manages four local variables denoted $input_i[1..n]$, val_i, res_i and tag_i. Their initial values are irrelevant.

2.2 An Informal Description of the Algorithm

We present below the algorithm for process p_i. Recall that there are at most k λ-constrained crash failures, where $\lambda = n - k$.

1. p_i first deposits its proposed value in_i in $INPUT[i]$.
2. p_i repeatedly reads the $INPUT[1..n]$ array until $INPUT[1..n]$ contains at least $n-k$ entries different from their initial value $\perp$. Because at most k processes may crash, and the process participation assumption, this loop statement eventually terminates.
3. p_i computes the smallest value deposited in the array $INPUT[1..n]$ and sets val_i to that value.
4. p_i champions the value in val_i for it to be decided. To this end, it uses the underlying wait-free adopt/commit object; namely, it invokes AC.ac_propose(val$_i$) from which it obtains a pair $\langle tag_i, res_i \rangle$.
5. Once p_i's invocation of the adopt-commit object terminates, there are two possible cases,
    - if tag_i = commit, due to the weak agreement property of the object AC, no value different from res_i can be decided. Consequently, p_i writes res_i in the shared register DEC and returns res_i as the agreed upon consensus value, and terminates.
    - if tag_i = adopt, p_i continues to the next step below.
6. Notice that if p_i arrives here, it must be the case that process participation is above $n - k$, and hence no process will fail from that point in time. So, p_i continually checks whether $DEC \neq \perp$ and, in parallel, starts participating in the single-shot *mutex* object.
7. the value of DEC as the agreed-upon consensus value and terminates.
8. If p_i enters the critical section, it writes res_i in the shared register DEC, returns res_i as the agreed-upon consensus value, and terminates.

Notice that if process p_i terminates in step 5, and process p_j terminates in step 8, then, due to the weak agreement property of the object AC it must be the case that $res_i = res_j$.

2.3 A Formal Description and Correctness Proof

Algorithm 1 describes the behavior of a process p_i. The statement return(v) returns the value v to the invoking process and terminates its execution of the algorithm. The idea that underlies the design of this algorithm is pretty simple, namely:

- Failure-prone part: Exploitation of the *participating processes* assumption to benefit from the adopt-commit object AC (Lines 1–5) and try to decide from it.
- Failure-free part: Exploitation of the λ-*constrained failures* assumption (Lines 6–8) to ensure that, if the adopt-commit object does not allow processes to decide, the decision will be obtained from the acquire-restricted mutex object, whose invocations occur in a failure-free context (crashes can no longer occur when processes access ARM).

---

operation propose(in$_i$) **is**
(1) $INPUT[i] \leftarrow in_i$;
(2)   **repeat** $input_i[1..n] \leftarrow$ asynchronous non-atomic reading of $INPUT[1..n]$
              until $\big(input_i[1..n]$ contains at most $k \perp \big)$ **end repeat**;
(3) $val_i \leftarrow$ min$\big($values deposited in $input_i[1..n]\big)$;
(4) $\langle tag_i, res_i \rangle \leftarrow AC$.ac_propose(val$_i$);
(5)   **if** $(tag_i =$ commit$)$ **then** $DEC \leftarrow res_i$; return(DEC) **end if**;
(6)   Launch in parallel the local thread T;
(7)   wait$\big(DEC \neq \perp\big)$; kill(T); return(DEC).

thread T is
(8) ARM.acquire(); **if** $DEC = \perp$ **then** $DEC \leftarrow res_i$ **end if**.

Algorithm 1: Consensus tolerating k λ-constrained failures, where $\lambda = n - k$

Lemma 1 (Validity). *A decided value is a proposed value.*

Proof. A process decides either on Line 5 or 7. Whatever the line, it decides the value of the shared register DEC, which was previously assigned a value that has been deposited in a local variable res_i (Line 5 or 8). The only place where a local variable res_i is updated is Line 4, and it follows from the validity property of the adopt-commit object that this value is the proposed value val_j of some process p_j. Since val_j is the minimum value seen by p_j in $INPUT$ (Line 3) that contains only the input values of the processes (and maybe some $\perp$ values that are, by definition, greater than any input variables), val_j contains the proposed value of some process. $\square_{Lemma\ 1}$

Lemma 2. *If, when a process p_i exits at Line 2 at time t, at least $n - k + 1$ entries of INPUT are different from $\perp$, then p_i is a correct process and no more crash occurs after time t.*

Proof. If there is a time t at which at least $n - k + 1$ entries of $INPUT$ are different from $\perp$, it follows that the number of participating processes is greater than $n - k$. It then follows from the λ-constrained crash failures no process crashes after time t and p_i is a correct process. $\square_{Lemma\ 2}$

Lemma 3. *If a process p_i executes Line 6, it is a correct process.*

Proof. Let p_i be a process that executes Line 6. If at least $n - k + 1$ entries of $INPUT[1..n]$ were different from $\perp$ when p_i exited Line 2, it follows from Lemma 2 that p_i is a correct process. So, let us consider the case where, when p_i exited Line 2, exactly $n - k$ entries of $INPUT[1..n]$ were different from $\perp$.

Recall that by obligation property, if the processes that invoke ac_propose() propose the same input value v, only the pair $\langle \texttt{commit}, v \rangle$ can be returned. Thus, since process p_i did not obtain $tag_i = \texttt{commit}$ at Line 4, it must be that some other process proposed, at Line 4, a value different than the value proposed by p_i. This implies that the minimum value computed by p_i at Line 3, is (1) different than the minimum value computed by some other process, say process p_j, at Line 3, and (2) that process p_j computed this minimum value at Line 3 before p_i reached Line 6.

Consequently, the set (of size $n - k$) of non-$\perp$ entries in $input_i$ at the time when p_i has exited Line 2 must be different than the set (of size at least $n - k$) of non-$\perp$ entries in $input_j$ at the time when p_j has exited Line 2, from which it follows that when the last of p_i and p_j exited Line 4, there were at least $n - k + 1$ participating processes. Thus, by Lemma 2, p_i is a correct process. $\qquad \Box_{Lemma\ 3}$

Lemma 4 (Termination). *Every correct process decides.*

Proof. Correct processes are required to participate, and there are no more than k crashes (model assumption). Thus, at least $n - k$ processes eventually write their input value into $INPUT$ and, thus, no process remains stuck in the loop at Line 2.

Since the adopt-commit object is wait-free, the invocation of $AC.\texttt{ac\_propose}(\texttt{val}_i)$ at Line 4 always terminates. If a correct process p_i obtains the pair $\langle \texttt{commit}, v \rangle$ when it invokes $AC.\texttt{ac\_propose}(\texttt{val}_i)$ at Line 4, it assigns $v \neq \perp$ to the shared register DEC and then decides. Any process that obtains the tag $\texttt{adopt}$ will later decide at Line 7.

When no process p_i obtains the pair $\langle \texttt{commit}, v \rangle$ at Line 5, or when every process that obtains a pair $\langle \texttt{commit}, v \rangle$ crashes before updating the shared register DEC, DEC will not be updated at Line 5.

In such a case by Lemma 3, every correct process launches in parallel its local thread T (Line 6). By the deadlock-freedom property of ARM, some process, say process p_k, will eventually enter the critical section (Line 8). Process p_k then assigns $v \neq \perp$ to DEC. Again, all other processes will be able to decide with their threads T (Line 7). $\qquad \Box_{Lemma\ 4}$

Lemma 5 (Agreement). *No two processes decide different values.*

Proof. We consider two cases. The first is when some process p_i obtains the pair $\langle \texttt{commit}, v \rangle$ from the invocation of $AC.\texttt{ac\_propose}(\texttt{in}_i)$ at Line 4. In this case, due to the weak agreement property of the adopt-commit object, all the processes that return from this invocation obtain a pair $\langle -, v \rangle$. It follows that the local variables res_j of every correct process p_j contains v. As only the content of the shared variable DEC or a local variable res_j can be decided by p_j, only the value v can be decided.

The second case is when no process p_i obtains the pair $\langle \texttt{commit}, - \rangle$. In this case, when a process p_i decides, this occurs at Line 7. By Lemma 3, p_i is correct (and also all the processes that cross Line 6 are correct) and launched its local thread T. So, only correct processes launch their threads T. Due to the deadlock-freedom property of mutex, one and only one of them, say process p_j, terminates its invocation of ARM.acquire() and imposes rec_j as the decided value. $\square_{Lemma\ 5}$

Theorem 2 follows from Algorithm 1, Lemma 1, Lemma 4, and Lemma 5.

3   Optimality of the Algorithm

This section proves the "only if direction" of Theorem 1. We point out that the impossibility result we give below was essentially already presented in [2, 17]. Our proof is an adaptation of the proof from [2, 17].

Theorem 3 (Only if direction). *For every $0 \leq k \leq n$, an algorithm exists that solves consensus for n processes in the presence of f $(n - k)$-constrained crash failures **only if** $f \leq k$.*

Proof. To prove the only if direction, we have to show that, in the context of process participation and λ-constrained crash failures, with $\lambda = n - k$, there is no read/write registers-based algorithm that solves consensus while tolerating $(k + 1)$ λ-constrained crash failures. To this end, assume to the contrary that for some k such that $n > k + 1$, and $\lambda = n - k$ that there is a read/write-based algorithm A that tolerates $k + 1$ λ-constrained crash failures.

Given an execution of A, let us remove any set of k processes by assuming they crashed initially. It then follows from the contradiction assumption that algorithm A solves consensus in a system of $n' = n - k$ processes. However, in a system of $n' = n - k$ processes, the number of participating processes is always smaller or equal to n', from which follows that, in such an execution, n'-constrained crash failures are crashes that occur at any time, i.e., these crashes are not constrained by some timing assumption. It follows that A may be used to generate a read/write-based consensus algorithm for $n' - k$ processes that tolerates one crash failure that can occur at any time. This contradicts the known impossibility of consensus in the presence of asynchrony and even a single crash failure, presented in [4, 10]. $\square_{Theorem\ 3}$

4   Discussion

Better Sooner than Later in General. There are many reasons why it is better for failures to occur sooner rather than later. For example, identifying failures early in the software development life cycle helps save valuable time and resources. When failures are detected early in, the necessary actions can be taken promptly to mitigate or address the issue. Early failures offer a chance to iterate and optimize, increasing the chances of success in subsequent attempts. It also provides ample time to recover and redirect efforts toward alternative solutions.

Better Sooner than Later in this Article. In this article, we have identified yet another reason why it is better for failures to occur sooner rather than later: in the context of asynchronous distributed algorithms, more failures can be tolerated when it is a priori known that they may occur earlier in the computation. That is, we have demonstrated a tradeoff between the number of failures that can be tolerated and the information about how early they may occur. In the two extreme cases, if failures may occur only initially, then both mutual exclusion and consensus can be solved in the presence of any number of (initial) failures; while when failures may occur at any time, then it is impossible to solve these problems even in the presence of a single (any time) failure. More generally, for every $0 \leq k \leq n$, if it is known that failures may occur only before the number of participating processes bypasses a predefined threshold that equals $n - k$, then it is possible to solve consensus for n processes in the presence of up to k failures, but not in the presence of $k + 1$ failures.

On Simplicity. The proposed algorithm is simple. This does not mean that the problem was simple! As correctness, simplicity is a first class citizen property. Simplicity, as it captures the essence of a problem, makes its understanding easier. As said by A. Perlis (the very first Turing Award), *"Simplicity does not precede complexity, but follows it."* [12].

Finally, let us notice that the following question has recently been addressed in [14]: *Are consensus and mutex the same problem?* It is worth noticing that the present paper adds a new relation linking mutex and consensus when considering the notions of participating processes and failure timing.

References

1. Dijkstra, E.W.: Solution of a problem in concurrent programming control. Commun. ACM **8**(9), 569 (1965)
2. Durand, A., Raynal, M., Taubenfeld, G.: Contention-related crash failures: definitions, agreement algorithms, and impossibility results. Theoret. Comput. Sci. **909**, 76–86 (2022)
3. Durand, A., Raynal, M., Taubenfeld G.: Reaching agreement in the presence of contention-related crash failures. Theor. Comput. Sci. 966-967, 12 (2023)
4. Fischer, M.J., Lynch, N.A., Paterson, M.S.: Impossibility of distributed consensus with one faulty process. J. ACM **32**(2), 374–382 (1985)
5. Gafni, E.: Round-by-round fault detectors: unifying synchrony and asynchrony. In: Proceedings of 17th ACM Symposium on Principles of Distributed Computing (PODC), pp. 143–152. ACM Press (1998)
6. Guerraoui R.: Indulgent algorithms. In: Proceedings 19th Annual ACM Symposium on Principles of Distributed Computing (PODC 2000), pp. 289–297. ACM Press (2000)
7. Guerraoui, R., Raynal, M.: The information structure of indulgent consensus. IEEE Trans. Comput. **53**(4), 453–466 (2004)
8. Lamport, L.: The part-time parliament. ACM Trans. Comput. Syst. **16**(2), 133–169 (1998)
9. Lamport, L., Shostak, R., Pease, M.: The Byzantine generals problem. ACM Trans. Program. Lang. Syst. **4**(3), 382–401 (1982)
10. Loui, M., Abu-Amara, H.: Memory requirements for agreement among unreliable asynchronous processes. Adv. Comput. Res. **4**, 163–183 (1987)

11. Pease, M., Shostak, R., Lamport, L.: Reaching agreement in the presence of faults. J. ACM **27**, 228–234 (1980)
12. Perlis, A.: Epigrams on programming. ACM Sigplan **17**(9), 7–13 (1982)
13. Raynal, M.: Concurrent Programming: Algorithms, Principles and Foundations, p. 515. Springer, Heidelberg (2013)
14. Raynal, M.: Mutual exclusion vs consensus: both sides of the same coin? Bull. Eur. Assoc. Theoret. Comput. Sci. (EATCS), **140**, 14 (2023)
15. Raynal, M., Taubenfeld, G.: A visit to mutual exclusion in seven dates. Theoret. Comput. Sci. **919**, 47–65 (2022)
16. Styer, E., Peterson, G.L.: Tight bounds for shared memory symmetric mutual exclusion problems. In: Proceedings 6th ACM Symposium on Principles of Distributed Computing (PODC), pp. 117–191. ACM Press (1989)
17. Taubenfeld, G.: Weak failures: definition, algorithms, and impossibility results. In: Podelski, A., Taïani, F. (eds.) Proceedings 6th International Conference on Networked Systems. LNCS, vol. 11028, pp. 269–283. Springer, Cham (2018). https://doi.org/10.1007/978-3-030-05529-5_4
18. Taubenfeld, G., Katz, S., Moran, S.: Initial failures in distributed computations. Int. J. Parallel Prog. **18**(4), 255–276 (1989)

Highly-Efficient Persistent FIFO Queues

Panagiota Fatourou[1,2(✉)], Nikos Giachoudis[1], and George Mallis[1,2]

[1] FORTH ICS, Heraklion, Greece
{faturu,ngiachou}@ics.forth.gr
[2] University of Crete, Rethymno, Greece
csd4165@csd.uoc.gr

Abstract. In this paper, we study the question whether techniques employed, in a conventional system, by state-of-the-art concurrent algorithms to avoid contended hot spots are still efficient for recoverable computing in settings with Non-Volatile Memory (NVM). We focus on concurrent FIFO queues that have two end-points, head and tail, which are highly contended.

We present a persistent FIFO queue implementation that performs a pair of persistence instructions per operation (enqueue or dequeue). The algorithm achieves to perform these instructions on variables of low contention by employing Fetch&Increment and using the state-of-the-art queue implementation by Afek and Morrison (PPoPP'13). These result in performance that is up to 2× faster than state-of-the-art persistent FIFO queue implementations.

Keywords: Non-volatile memory (NVM) · NVM-based computing · Persistence · Recoverable algorithms · Recoverable data structures · Concurrent data structures · FIFO queue · Lock-freedom · Persistence cost analysis

1 Introduction

Non-Volatile Memory (NVM) has been proposed as an emerging memory technology to provide the persistence capabilities of secondary storage at access speeds that are close to those of DRAM. In systems with NVM, concurrent algorithms can be designed to be *recoverable* (or *persistent*), by persisting parts of the data they use. This enables to restore their states after system-crash failures. In this direction, a vast amount of papers that appear in the literature propose persistent implementations for a wide collection of concurrent data structures, including stacks [9,22], queues [9,11,24], heaps [9], and many others. Moreover, a lot of papers have focused on designing general transformations [1,2,9] and universal constructions [3,4,9,21] that can be applied on many conventional

Supported by the Hellenic Foundation for Research and Innovation (HFRI) under the "Second Call for HFRI Research Projects to support Faculty Members and Researchers" (project number: 3684).

Y. Emek (Ed.): SIROCCO 2024, LNCS 14662, pp. 238–261, 2024.
https://doi.org/10.1007/978-3-031-60603-8_14

concurrent data structures to get their persistent analogs. (This list of references is by no means exhaustive.)

In this paper, we study the question whether techniques employed, in a conventional system, by state-of-the-art concurrent algorithms, to avoid contended hot spots, are still efficient for persistent computing in settings with NVMs. We focus on concurrent queues that have two end-points, head and tail, which are highly contended. Accesses to the queue's endpoints result in bottlenecks, making most existing concurrent queue implementations non-scalable. Several papers have proposed techniques for reducing this cost. One approach uses software combining [5–8,14,17], a technique that attempts to reduce synchronization by having a thread, the *combiner*, to collect and apply operations (of the same type) on each of the endpoints. Other papers [5,7,10,19] use *Fetch&Increment* (*FAI*) to avoid hotspots; *FAI* is an *atomic* primitive which increases by one the value of a shared variable it takes as a parameter, and returns its previous value. Avoiding hotspots using *FAI*, when designing a concurrent FIFO queue, resulted in LCRQ, the state-of-the-art concurrent queue implementation, for which it has been experimentally shown [19] to significantly outperform the previously-proposed state-of-the-art algorithm [6], which was based on software-combining.

The performance power of software combining in persistence has been studied in [6]. The combining-based persistent FIFO queue implementation [6] has been shown to outperform all other persistent queue implementations, specialized or not. As specialized algorithms take into consideration the semantics of the data structure to make design decisions in favor of performance, this was kind of surprising. However, the specialized persistent queues that are proposed in the literature [11,24] are not based on state-of-the-art queue implementations [9,19].

In this paper, we examine whether the use of *FAI* is a more promising approach for designing persistent queues than employing combining. Specifically, we focus on the state-of-the-art queue implementation (LCRQ) [19], and study the correctness and performance implications of different persistence choices to make it persistent. We end up with a persistent FIFO queue implementation, called PERLCRQ, that performs much better than any previous such implementation.

LCRQ is inspired by the simple idea of using an infinite array, Q (which initially contains $\bot$), and two *FAI* objects, Tail and Head (initially 0). An enqueuer executes a *FAI* on Tail to get an index i of the array where the new item should be enqueued. Then, it performs a GET&SET to swap the element to be inserted with the current value of $Q[i]$ (which in a successful enqueue it should be $\bot$). Similarly, a dequeuer executes *FAI* on Head, to get the array index from where it will dequeue an item by swapping the value of that element with $\top$. This way, each element of the array has at most one enqueuer that tries to insert an element in it and at most one dequeuer that tries to dequeue from it. This simple algorithm, which we call IQ, is not practical, as it requires an array of infinite size. Moreover, it may result in a livelock with an enqueuer and a dequeuer always interfering with one another without ever making progress. CRQ [19] follows similar ideas as IQ but it uses a circular array (of bounded size) to store the new items. This introduces many more cases where synchronization is needed between enqueuers and dequeuers, but results in a more practical algorithm. To solve the live-lock problem (and the space limitation of the static array), LCRQ employs several CRQ instances, connected in a linked list to form a queue as in the well-known lock-free queue implementation in [18]. (Sect. 3 presents the details of IQ, CRQ and LCRQ.)

Our effort focuses on making LCRQ persistent with the least possible *persistence cost* (Sect. 4), i.e. without paying a high performance overhead to support persistence, in periods of time where no failures occur. A shared variable can be persisted by executing a pair of persistence instructions: `pwb`, which requests the system to flush back the current value of a shared variable to NVM and works in a non-blocking way, and `psync`, that blocks until flushing has been completed for all preceeding `pwb` s. Persistence can easily be supported on top of an existing concurrent algorithm by persisting all shared variables [16] every time they are updated. However, this technique would have excessive performance overhead. On the other hand, the simple approach of persisting only `Head` (`Tail`) every time a *FAI* is executed on it, results also in excessive performance cost as it violates two major persistence principles crucial for performance [1]: a) it executes many persistence instructions per operation, and b) it applies persistence instructions on highly contended shared variables accessed by all threads.

We focused first on achieving persistence with a single pair of persistence instructions per operation, which is optimal. Starting from IQ, we managed to persist only once per operation. This required careful design of the algorithms' *recovery functions* (the recovery function of an implementation is executed by the system, after a crash, to bring the data structure in a coherent state). It also required long argumentation to prove that the proposed algorithms are indeed correct (i.e., they satisfy *durable linearizability* [16]). We consider these to be main contributions of the paper, as the persistence part of the algorithms themselves (without considering the recovery functions) had to be maintained as simple and minimal as possible for achieving our performance goals. We next observed that even when we persist only once, we cannot beat the state-of-the-art combining based approaches [9], if persistence instructions are executed on `Head` and `Tail` that are highly contended. So, we came up with persistent variants of the algorithms, where an enqueue operation *op* persists only the last value it writes into Q. Moreover, for IQ, we came up with PERIQ, where dequeue operations persist also only the last value they write into Q. As each position of the array is accessed by only two threads, both persistence principles mentioned above are respected by PERIQ. This results in low performance overhead.

Respecting the persistence principles [1], when designing PERCRQ (the persistent version of CRQ) was more challenging. A position i of Q can now be accessed by any operation that reads $i + kR$ in `Tail` (or `Head`), where $k \geq 0$ is an integer and R is the size of the circular array. This leads to the necessity to persist `Head` every time an item is dequeued to ensure durable linearizability. However, as explained previously, doing so has considerable performance overhead. To avoid this cost, we introduce the following technique. Each thread maintains a local copy of Head. Every time it is about to complete a dequeue operation, it persists the local copy of Head (instead of its shared version, on which *FAI* is executed to get an index in Q). Since the local copies of Head are single-writer, single-reader variables, this significantly reduces the cost of persistence (Sect. 5, Figs. 2, 3). We believe that this technique is of independent interest and can be used to get persistent versions of many other state-of-the-art concurrent algorithms.

Our experimental analysis (Sect. 5) shows that PERLCRQ is at least 2x faster than its best competitor, PBQUEUE [9]. We also provide experiments to support all our claims above.

Performing a single pair of persisting instructions per operation in PERIQ introduces some recovery cost. (The *recovery cost* is the time needed to execute the recovery function.) The *recovery function* of PERIQ has to scan Q until it finds a streak of n subsequent $\perp$ values. This might end up to be expensive, as bigger parts of Q has to be scanned as the time is progressing. On the other hand, we may choose to trade performance at normal execution time (where no failures occur) by periodically persisting Head and Tail. This illustrates an interesting tradeoff between the performance at normal execution time and the recovery cost: better performance at normal execution time results in higher cost at recovery (and vice versa). It is an interesting problem to study whether this trade-off appears when designing the persistence versions of other state-of-the-art techniques for ensuring synchronization and for designing concurrent data structures in a conventional (non-NVM) setting. We believe that this trade-off is inherent in many cases.

We provide experiments (Sect. 5) to simulate failures which allows us to measure the recovery cost. To the best of our knowledge, none of the papers on persistent data structures in shared memory systems provide experiments to measure the recovery cost.

Summarizing, the main contributions of this paper are the following:

(**1**) We present PERLCRQ, a persistent implementation of a FIFO queue that is much faster than existing state-of-the-art persistent queue implementations. PERLCRQ promotes the idea of starting from the state-of-the-art implementation of a concurrent data structure (DS) to come up with an efficient persistent implementation of it. This departs from the approach followed in previous specialized persistent FIFO queue implementations [11,24].

(**2**) We illustrate an interesting tradeoff between the persistence cost of an algorithm at normal execution time and its recovery cost.

(**3**) We provide a framework to simulate failures and measure the recovery cost.

(**4**) Our experimental analysis shows that a) PERLCRQ is at least two times faster than its competitors, and that b) respecting the persistence principles presented in [1] is indeed crucial for performance.

(**5**) Based on our experiments, we propose a new technique for reducing the persistence cost, namely using local copies of highly-contended shared variables by each thread and persisting them instead. This technique transfers part of the persistence cost from normal execution time to recovery, and is of independent interest.

2   Preliminaries

We consider a system where n asynchronous threads run concurrently and communicate using shared objects. In addition to read/write objects, that support
atomic *reads* and *writes* on/to the state of the object, the system also provides
the following *atomic* primitives: a) FETCH&INCREMENT(V) (*FAI*), which adds
1 to V's current value and returns the value that V had before the increment, b)
GET&SET(V, v'), which stores value v' into V and returns V's previous value, c)
COMPARE&SWAP(V, v_{old}, v_{new}) (*CAS*), which checks the value of V and if it is
v_{old}, it changes its value to v_{new} and returns `true`, otherwise it makes no change
and returns `false`. A variant of *CAS* is CAS2($V, (v_0^{old}, v_1^{old}), (v_0^{new}, v_1^{new})$),
which operates atomically on an array V of two elements. Specifically, it checks
if $V[0] == v_0^{old} \land V[1] == v_1^{old}$ and if this is true it changes the value of V to
$\{v_0^{new}, v_1^{new}\}$ and returns `true`; otherwise it returns `false` without changing V.
Finally, TEST&SET(B) changes the value of a shared bit B to 1 and returns
its value before the change; it comes together with RESET(B), which resets the
value of B to 0. In the pseudocodes in later sections, the names of shared objects
start with a capital letter. We assume the *Total Store Order* (*TSO*) model, where
writes become visible in program order.

The system's memory is comprised of a non-volatile part and its volatile
components (registers, caches, DRAM). We study persistence under the *full-
system crash failure* (or *system-wide crash failure*) model [16]. When a failure
occurs, the values of all variables stored in volatile memory are lost and are
reset to their initial values at recovery time. On the contrary, any data that
was written back (*persisted*) to NVM persist (i.e., they are available at recovery
time).

We assume *explicit epoch persistency* [16], where a write-back to the persistent
memory is initiated through a specific instruction called *persistent write-back*
(pwb); pwb is asynchronous, i.e., the order in which pwbs take effect may not be
preserved inherently. To enforce ordering when necessary, a `pfence` instruction
ensures that all pwb instructions performed before the `pfence`, are realized before
those that follow it. However, `pfence` is also asynchronous, so it does not provide
any guarantee that the pwbs that preceded it have occured by the time of a crash.
A `psync` instruction causes a thread to block until all prior pwb instructions have
been realized (i.e., the data they flush have indeed been written back to the
NVM). We collectively refer to pwb, `pfence`, and `psync` as the set of *persistence
instructions*. The *persistence cost* of an operation's instance (an execution) is
the time overhead incurred from executing its persistence instructions.

We assume that a recoverable implementation has an associated *recovery
function* (for the system as a whole). After a system failure, the recovery function
is invoked to bring the system in a consistent state. Then, the system may
(asynchronously) recover failed threads, which start by invoking new operations.
The *recovery cost* of an implementation is the time needed to execute its recovery
function. We usually measure it by taking the average among several runs, in
each of which the recovery function is executed at the end after a system-wide
failure (see details in Sect. 5).

Consider an execution α, and let $c_1, c_2, \ldots$ be crash events in α, in order. We divide α into *epochs* $E_1, E_2, \ldots$. The first epoch starts at C_0 and ends with c_1, if C_1 exists, otherwise E_1 is the entire execution. For each $k > 0$, epoch E_k starts with the event following c_{k-1} and ends with c_k, if c_k exists in E_k. If c_k does not exist, E_k is the suffix of α after c_{k-1}.

An operation *op* has an *invocation* and may have a *response*; *op*'s *execution interval* starts with its invocation and ends with its response (if it exists) or it is the suffix of an execution starting with its invocation (otherwise). An implementation is *lock-free*, if in every infinite execution it produces, it holds that if the number of failures is finite, an infinite number of operations respond.

In a conventional setting where threads never recover after a crash, an execution α is *linearizable* [15] if there exists a sequential execution σ which contains all completed operations in α (and possibly some of the uncompleted ones), such that a) the response of every operation in α is the same as that of the corresponding operation in σ, and b) the order of operations in σ respects the partial order induced by the execution intervals of operations in α. An algorithm is *linearizable*, if all the executions it produces are linearizable.

We now consider a setting where failed threads may recover. Briefly, an execution is *durably linearizable* [16], if the state of the object after every crash reflects all the changes performed by the operations that have completed by the time of the crash (and possibly by some of the uncompleted operations). Our goal is to design durably linearizable implementations of FIFO queues with low persistence cost.

3   LCRQ: A Brief Description

LCRQ [19] uses multiple instances of a static circular FIFO queue implementation, called CRQ. CRQ is based on a simple but impractical algorithm, which we call IQ and we discuss it first.

The IQ Algorithm. In IQ (black lines in Algorithm 1), the queue is represented as an array (Q) of infinite size, initialized with $\perp$ in all of its cells, and two variables Head and Tail, initially 0. Head stores the index of Q's item that is going to be dequeued next, whereas Tail stores the index of the next available Q's cell to store a new item.

Enqueue uses *FAI* to read Tail and get the current available index of Q. It then attempts to store the new item there using GET&SET. If the GET&SET returns $\perp$, the enqueue is successful. Otherwise, it re-tries by applying the steps described above. A dequeue uses *FAI* to identify the next item to be removed by reading Head. It then attempts to read that item and replace it with $\top$ by using GET&SET. If GET&SET returns a not $\perp$ item, the dequeue is successful and returns it. If GET&SET returns $\perp$ and Head > Tail, the dequeue returns EMPTY (i.e., the queue is empty). Otherwise, these steps are repeated until either some non-$\perp$ value or EMPTY is returned.

The IQ algorithm is very simple but unrealistic, because of the infinite table it uses. Another issue of the IQ algorithm is that if a dequeue operation continuously swaps ⊤ in cells that an enqueue operation is trying to access, we would have a *livelock*.

The CRQ Algorithm. To address the infinite array issue, CRQ (see black lines of Algorithm 3) implements $\mathcal{Q}$ as a circular array, of size R. As in IQ, Head and Tail are implemented as *FAI* objects, whose values are incremented indefinitely. Each enqueue (dequeue) reads the value i in Tail (Head). Consider an enqueue (dequeue) operation that reads index i by executing *FAI* on Tail (Head) and stores (removes) an element in (from) $\mathcal{Q}[i \bmod R]$. We say that the *index* of the operation is i. A pair of enqueue and dequeue operations that both have index i are called *matching operations*, and we denote them by enq_i and deq_i.

Every cell in the array now stores a 3-tuple object (s, i, v), containing the *safe bit*, an *index*, and the *value* of the item it stores; its initial value is $(1, u, \perp)$, where u is the actual index of each cell of the array. If the node's value is $\perp$ then that node is *unoccupied*, otherwise it is *occupied*. An enqueue with index i and value v tries to store the triplet $(1, v, i)$ in $\mathcal{Q}[i \bmod R]$, if this array cell is unoccupied. CRQ ensures that only the dequeue with index i can dequeue this item.

The use of a circular array results in several synchronization challenges between enqueuers and dequeuers. Specifically, CRQ copes with the following cases: a) a dequeue *deq* finds an element of $\mathcal{Q}$ it accesses to store an index less than or equal to the index it uses and be unoccupied (*empty transition*), and b) *deq* finds $\mathcal{Q}$'s element occupied by an item that an enqueue other than its matching one has inserted (*unsafe transition*). A *transition* is a change of the value of an array cell. A dequeue operation *deq* performs a *dequeue transition* if it successfully executes the CAS2 in line 34, changing the value of item $\mathcal{Q}[t \bmod R]$ from (s, t, v) to $(s, t + R, \perp)$. In order to have that transition, t should match the index i of the cell, and the cell must be also occupied (line 32). *deq* executes an *empty transition* if it arrives before the enqueue that gets index t stores its item in the cell, and finds the cell unoccupied. Then, the CAS2 is called (line 41) to change the index of that cell to $t + R$. This prevents an *enq* that reads i in Tail to add its item in this array element. This is needed to ensure linearizability, as there will be no dequeuer to dequeue this item (a dequeuer has already *exhausted* index i). Finally, *deq* performs an *unsafe transition*, if it arrives while the cell is occupied with index $t - kR$, where $k > 0$. Then, *deq* executes the CAS2 of line 38 to change the safe bit to 0. This way it informs later enqueuers to be careful, as the items they want to store in this position may never be removed. An enqueuer has to check whether Head $> i$ and if this is so, the dequeuer may have already passed from this position, so they try the insertion with the next available index.

Algorithm 1. Persistent IQ (PERIQ). Ignore code in blue to get the pseudocode for IQ.

```
1: function ENQUEUE(x: Item)              15:         pwb(Q[h]); psync()
2:     while true do                      16:         return EMPTY
3:         t ← FETCH&INCREMENT(Tail)
4:         if GET&SET(Q[t], x) == ⊥ then  17: function RECOVERY
5:             pwb(Q[t]); psync()         18:     count⊥ = 0
6:             return OK                   19:     while count⊥ < n do
                                          20:         if Q[Tail] == ⊥ then count⊥ =
7: function DEQUEUE                              count⊥ + 1
8:     while true do                      21:         else count⊥ = 0
9:         h ← FETCH&INCREMENT(Head)      22:         Tail ← Tail + 1
10:        x ← GET&SET(Q[h], ⊤)          23:     Tail ← Tail − n + 1
11:        if x ≠ ⊥ then                 24:     Head ← Tail
12:            pwb(Q[h]); psync()         25:     while Q[Head] ≠ ⊤ and Head ≥ 0 do
13:            return x                    26:         Head ← Head − 1
14:        if Tail ≤ h + 1 then           27:     Head ← Head + 1
```

Before attempting its insertion, a thread executing an ENQUEUE(crq, x), that reads t in Tail (line 4), checks that: a) the index i of cell $Q[t \bmod R]$ is at most as big as the value t (line 14) (otherwise, the enqueue operation tries the next potentially available cell, as another operation with index higher than t has already processed this item), b) the safe bit s is equal to 1 or Head is at most t (line 14) (the second condition ensures that the dequeue with index t has not started its execution yet, so the signal through the safe bit does not concern it). An *enqueue transition* succeeds when the CAS2 operation in line 14 succeeds.

Linearizability of IQ and CRQ are discussed in [19]. CRQ implements a *tantrum* queue, i.e. a FIFO queue with relaxed semantics: any enqueue operation may return a special value CLOSED (which identifies that the array is full or that it faces a livelock when trying to insert a new element in it); as soon as an enqueue returns CLOSED on an instance of CRQ, all other enqueue operations on the same CRQ instance should also return CLOSED.

The LCRQ Implementation. The main idea of LCRQ (see black lines of Algorithm 5) is to use a linked list of nodes, each representing an instance of CRQ. The linked list together with two pointers (**First** and **Last**) to its endpoints, implement a FIFO queue, which closely follows the lock-free FIFO queue implementation by Michael and Scott [18]. When an enqueue on the current instance of CRQ returns CLOSED, it creates a new CRQ instance containing the item to be inserted and appends it to the list. When a dequeue on the CRQ instance pointed by First, returns EMPTY, it dequeues this node by moving First to point to the next CRQ node in the list. LCRQ is a linearizable implementation of a FIFO queue [19].

4 Persistent FIFO Queue Algorithm

4.1 Persistent IQ

PERIQ (Algorithm 1) works like IQ and persists just the last write into Q of each operation. This is optimal, but also it ensures low persistence cost due to low contention [1], as IQ ensures that each element of $\mathcal{Q}$ is accessed by at most two threads. Note that persisting Head and Tail, instead, could also work, but it would be more expensive in terms of performance, since these variables are accessed by all threads and thus they are highly contended (see Sect. 5, Fig. 6).

On the other hand, avoiding writing back Head and Tail, adds complexity, and thus also performance overhead, to the recovery function. To recover Tail, PERIQ searches the array for the first continuous streak of n unoccupied cells, where n is the total number of threads in the system. Note that when a crash happens, some of the active enqueuers may already have inserted their items into $\mathcal{Q}$ (and their write-backs to NVM has taken effect) but others might not. However, all the unoccupied positions between two occupied are taken from enqueue operations that are active at the time of the crash. Since there are n processes running concurrently, it is easy to see that $l < n$. Thus, there are at most $n - 1$ unoccupied cells between two occupied. Because of this, when the recovery function in PERIQ finds the first streak of n unoccupied cells in $\mathcal{Q}$, it sets Tail to the first cell of that streak.

To find Head, PERIQ (Algorithm 1) starts from Tail and traverses towards the beginning of $\mathcal{Q}$ until it first sees the value $\top$. Head is set to point to the element on the right of this $\top$ in $\mathcal{Q}$ (or to 0 if such an element does not exist). This way, there does not exist an element with the value $\top$ between Head and Tail.

We now discuss why PERIQ ensures durable linearizability. We denote by enq_j the enqueue operation that reads j in Tail (line 3) in its last iteration of the while loop (line 2) and successfully inserts an item x_j in position $\mathcal{Q}[j]$. A dequeue operation *matches* enq_j if it reads j in Head (line 9) and replaces x_j with $\top$ in $\mathcal{Q}[j]$ (line 10) . We denote by deq_j the matching dequeue operation of enq_j. Any dequeue that reads $j' > j$ in Head (line 9) and replaces a value $\notin \{\top, \bot\}$ with $\top$ in $\mathcal{Q}[j']$ (line 10) is called a *following* dequeue to deq_j. An operation *belongs to* an epoch E_k, if it performs a *FAI* in E_k. A dequeue operation is *successful* if it replaces a value $x \notin \{\top, \bot\}$ with $\top$ in $\mathcal{Q}$ (line 10). We say that an enqueue operation is *persisted* if it has executed the GET&SET (line 4), the return value is $\bot$, and the write-back of the new value has taken effect. A dequeue operation is *persisted*, if it has executed the GET&SET (line 10), read a value $x \neq \bot$, and the write-back of $\top$ has taken effect. Persisted operations terminate if the system does not crash.

We linearize an enqueue enq that belongs to an epoch E_k, if either enq is persisted in E_k, or there exists a matching dequeue deq to enq which is persisted in E_k; enq is linearized at the last *FAI* it executes. We linearize a dequeue deq of E_k, if either it is persisted in E_k, or there exist a matching enqueue to deq that has been included in the linearization and a following dequeue to deq of E_k is persisted. If a dequeue returns EMPTY, it is linearized when it reads Tail on line 14. The linearization points of successful dequeues are assigned in the same order as the linearization points of their matching enqueues, so that the FIFO property holds. Algorithm 2 assigns linearization points to PERIQ in a formal (algorithmic) way.

Algorithm 2. PERIQ linearization procedure

Execution α and set of epochs $\mathcal{E} := \{E_1, E_2, \ldots\}$ in α
1: LinQ auxiliary infinite array
2: $head(\text{LinQ})$ accessor to head index of LinQ
3: $tail(\text{LinQ})$ accessor to tail index of LinQ

4: **for** each epoch $k = 1, 2, \ldots$ **do**
5:    **for** each event $j = 1, 2, \ldots$ in epoch E_k **do**
6:       **if** $e_j = \langle t \leftarrow \text{FETCH\&INCREMENT(Tail)}\rangle$ **then**
7:          Let $enq(x)$ be the enqueue that executes e_j
8:          **if** e_j is the last *FAI* of $enq(x)$ and $enq(x)$ is persisted
             or $\exists$ matching deque $\in E_k$ for $enq(x)$, which is persisted **then**
9:             LinQ$[t] \leftarrow x$
10:            $tail(\text{LinQ}) \leftarrow t + 1$
11:            Linearize that enqueue operation at e_j
12:         **else if** e_j is a Tail read by $\langle deq : per \rangle$ such that
             Tail $\leq$ Head at the time of the read **then**
13:            Linearize that dequeue operation at e_j
14:         **while** $head(\text{LinQ}) < tail(\text{LinQ})$ **do**
15:            $h \leftarrow \min\{i : \text{LinQ}[i] \neq \bot\}$
16:            Let $deq()$ be the dequeue whose *FAI* returns h
17:            **if** $deq()$ not active in $e_1, \ldots, e_j$ of E_k **then**
18:               **break**
19:            **if** $deq()$ is persisted **or** $\exists$ a following persisted dequeue to $deq()$ in E_k **then**
20:               $x \leftarrow \text{LinQ}[h]$
21:               LinQ$[h] \leftarrow \bot$
22:               $head(\text{LinQ}) \leftarrow h + 1$
23:               Linearize that dequeue operation at e_j

Roughly speaking, all enqueue and dequeue operations that have been persisted by the time of the crash are linearized. This is needed to ensure durable linearizability, as such operations may have also returned. We linearize all active non-persisted enqueue operations that have matching dequeue operations which are persisted. Since all persisted dequeue operations are linearized, linearizing also their matching enqueues (even if they are not persisted) is necessary for correctness. We linearize an active non-persisted dequeue *deq* only if there is a matching enqueue operation *enq* that has been included in the linearization and a following dequeue *deq'* to *deq* that is persisted in E_k (and thus it is linearized). This is necessary because in the absence of the linearization of *deq*, the linearization of *deq'* would violate the FIFO property of the queue (given that *enq* has been linearized). Note that if we do not assign a linearization point to an enqueue and its matching dequeue, the FIFO property of the rest operations is not violated. So, we do not linearize *deq*, if there is no following dequeue that is linearized in E_k.

4.2 Persistent CRQ

PERCRQ (Algorithm 3) builds upon CRQ, so we focus on how to support persistence on top of CRQ. As in PERIQ, our goal is to execute a single pwb instruction in each operation. An enqueue operation, *enq*, either returns CLOSED (line 9 or 22), or it returns OK (line 16). As shown in Algorithm 3, we managed to insert a single pair of pwb-psync instructions in each of these cases. Apparently, the new value written by *enq* has to be written back to NVM before it completes, to

Algorithm 3. PERCRQ, code for thread p_i. Ignore code in blue to get the pseudocode for CRQ. Instances of $\mathcal{Q}$, Head, and Tail refer to $crq.\mathcal{Q}$, $crq.$Head and $crq.$Tail.

```
 1: function ENQUEUE(crq, x)
 2:     closedFlag : integer, initially 0
 3:     while true do
 4:         (cb, t) ← FETCH&INCREMENT(Tail)
 5:         if cb == 1 then // closed bit is set
 6:             if closedFlag == 0 then
 7:                 pwb(Tail); psync()
 8:                 closedFlag ← 1
 9:             return CLOSED
10:         e ← Q[t mod R] // get item
11:         v ← e.val // get item's value
12:         (s, i) ← (e.s, e.idx) // get safe bit
            and index
13:         if v == ⊥ then
14:             if i ≤ t and (s == 1 or Head ≤ t)
            and
                  CAS2(Q[t mod R], (s, i, ⊥), (1, t, x))
            then
15:                 pwb(Q[t mod R]); psync()
16:                 return OK
17:         h ← Head // read Head
18:         if t − h ≥ R or STARVING then
19:             TEST&SET(Tail.cb)
20:             pwb(Tail); psync()
21:             closedFlag ← 1
22:             return CLOSED

23: function DEQUEUE(crq)
24:     while true do
25:         Head_i ← FETCH&INCREMENT(Head)
26:         h ← Head_i
27:         e ← Q[h mod R] // read item
28:         while true do
29:             v ← e.v // read item's value
30:             (s, i) ← (e.s, e.idx)
31:             if i > h then goto line 43
32:             if v ≠ ⊥ then
33:                 if i == h then
34:                     if          CAS2(Q[h mod
R], (s, h, v), (s, h + R, ⊥)) then // dequeue
transition
35:                         pwb(Head_i); psync()
36:                         return v
37:                 else
38:                     if          CAS2(Q[h mod
R], (s, i, v), (0, i, v)) then // unsafe tran-
sition
39:                         goto line 43
40:                 else
41:                     if CAS2(Q[h mod R], (s, i, ⊥), (s, i +
R, ⊥)) then // empty transition
42:                         goto line 43
43:         (cb, t) ← Tail
44:         if t ≤ h + 1 then
45:             pwb(Head_i); psync()
46:             FIXSTATE(crq)
47:             return EMPTY

48: function FIXSTATE(crq)
49:     while True do
50:         h ← FETCH&ADD(crq.head, 0)
51:         t ← FETCH&ADD(crq.tail, 0)
52:         if crq.tail ≠ t then
53:             continue
54:         if h ≤ t then
55:             return
56:         if CAS(crq.tail, t, h) then
57:             return

58: function RECOVERY
59:     R: size of circular array
60:     Head ← max_{i=1...n} Head_i
61:     (cb, t) ← Tail
62:     Tail ← (cb, 0)
63:     for i ← 0;  i < R;  i++ do
64:         if  Q[i].val  ≠  ⊥  and  Tail  <
Q[i].idx + 1 then
65:             Tail ← Q[i].idx + 1
66:         else if  Q[i].val  ==  ⊥  and
Q[i].idx ≥ R then
67:             if Tail < Q[i].idx − R + 1 then
68:                 Tail ← Q[i].idx − R + 1
69:     if Head > Tail then Tail ← Head //
Empty queue
70:     else// Head ≤ Tail
71:         max ← Head
72:         for i ← Head; i mod R ≠ Tail mod
R; i++ do
73:             if Q[i mod R].val == ⊥ and
Q[i mod R].idx − R > max then
74:                 max ← Q[i mod R].idx − R + 1
75:         Head ← max
76:         min ← Tail
77:         for i ← Head; i mod R ≠ Tail mod
R; i++ do
78:             if Q[i mod R].val  ≠  ⊥  and
Q[i mod R].idx  <  min  and  Q[i mod
R].idx ≥ Head then
79:                 min ← Q[i mod R].idx
80:         if min < Tail then Head ← min
81:     for i ← Head − 1; i mod R ≠ Tail mod
R; i-- do
82:         Q[i mod R] ← (1, i + R, ⊥)
83:     for i = 0; i < R; i ← i+1 do Q[i].s ← 1
```

ensure durable linearizability. So, the pair of pwb-psync in line 15 is necessary. In order to respect the semantics of a tantrum queue, if *enq* returns CLOSED, all other enqueues on this instance of CRQ that will be linearized after *enq* should also return CLOSED. Thus, PERCRQ has to ensure that the state of Tail is CLOSED after recovery. This means that we need to persist the CLOSED value in Tail, as soon as it is written there. This justifies the pwb-psync pair of line 20. However, a pair of pwb-psync is also needed in line 7, to avoid the following bad scenario: Assume that a thread performs the *TAS* of line 19 and becomes slow before executing the pwb instruction of line 20. Afterwards, another thread reads in Tail the value 1 for the closed bit and returns CLOSED in line 9 without persisting the closed bit. Then, if a crash occurs, at recovery time, Tail will not be closed anymore. So, it may happen that a subsequent enqueue operation returns OK, violating the semantics of a tantrum queue.

To reduce the persistence cost, we use a known technique [1,9], which uses a flag (*closedFlag*) to indicate whether executing the persistence instructions can be avoided.

Consider now a dequeue operation, *deq*. Assume first that *deq* has executed the *CAS* of line 34 (dequeue transition), successfully, and thus, at the absence of a crash, it will return in line 36. Our first effort was to follow the same approach as in PERIQ, and persist the triplet written by *deq* into $\mathcal{Q}$. Scenario 1 explains why persisting just this triplet is not enough.

Scenario 1. *Consider a circular array $\mathcal{Q}$ of size $R = 5$ and let its state consist of the following triplets: $\{(1, x_0, 0), (1, x_1, 1), (1, x_2, 2), (1, x_8, 8), (1, \perp, 4)\}$ (see Fig. 1a for a graphical illustration). Assume that the values of these triplets are stored in NVM. If a crash occurs, at recovery time, Algorithm 3 cannot distinguish between the following two cases: (a) the enqueue with index 8 (enq$_8$) occurs after the enqueue with index 3 (enq$_3$) and its matching dequeue (deq$_3$) have been completed. (b) enq$_8$ stores its item in $\mathcal{Q}[3]$ before enq$_3$ stores its own item there and no dequeue operations (at all) have been executed. In both cases we linearize enqueues enq$_0$, enq$_1$, enq$_2$ and enq$_8$ (as they are persisted operations). However, in the first case, we have to also linearize enq$_3$ and dequeues deq$_0$, deq$_1$, deq$_2$, and deq$_3$, to ensure the FIFO property. Thus, at recovery time Head should have a value at least 4. Note that in the second case, only enqueue operations are executed, hence at recovery time Head should be equal to 0. Since these two cases are indistinguishable to PERCRQ, RECOVERY cannot choose a correct value for Head.*

From the above, we figured out that the value of Head should be written back to NVM. However, we made the following observation: as long as the value of Head is written back to NVM, writing the triplet of line 34 back is unnecessary. If *deq* reads h in Head, then by writing Head back, no dequeue after the crash gets index h or less, and *deq* is considered to be persisted.

Writing Head back to NVM only in a successful dequeue is not enough. To argue that the pair of pwb-psync of line 45 is needed, assume it is omitted from the code. Then, there is an execution which results in a state where before the crash the last linearized operation is a dequeue which returns EMPTY, whereas after the crash we can have a dequeue operation that returns some item without any enqueue operation to be linearized between these two dequeues. This violates the FIFO property. Note that after any pwb-psync instruction pair, an operation will return in the absence of a crash. This gives us the optimal one pwb-psync instruction pair for each operation.

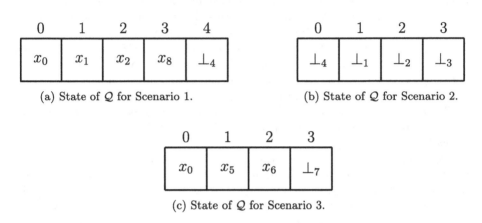

(a) State of $\mathcal{Q}$ for Scenario 1.

(b) State of $\mathcal{Q}$ for Scenario 2.

(c) State of $\mathcal{Q}$ for Scenario 3.

Fig. 1. Circular array states. On top of the array are the array indexes. Each cell has a value; the subscript corresponds to the index field value.

We now describe the recovery function for PERCRQ. To find Tail, we traverse the entire array (line 63) and search for an index that is at least as large as the maximum index recorded in each occupied cell of $\mathcal{Q}$ (lines 64-65). This is needed to ensure that all items that have been written back before the crash will be dequeued. We have to also consider unoccupied elements of $\mathcal{Q}$, because they can result from pairs of enqueues-dequeues which are linearized[1]. Hence Tail should get a bigger index. To achieve this, Tail gets the maximum index of unoccupied elements (minus R), which is greater than its current value (lines 66-68).

We next focus on Head, which, at recovery time, gets a value at least as large as the value of Head that is written back in NVM by the time of the crash. This is ensured by lines 71-80. To find the recovered value of Head, we need to examine the indexes in the unoccupied cells between current Head and Tail. We do so because RECOVERY cannot distinguish if these elements are the result of a *CAS* of line 34 or 41. In the former case, by the way linearization points are assigned, there will be a linearized *deq*, that we have to consider in order to ensure durable linearizability. Scenario 2 provides the details.

Scenario 2. *Let $R = 4$ be the size of the circular array $\mathcal{Q}$. Consider an enqueue, $enq_0(x_0)$, which executes its CAS of line 14 successfully and then becomes slow. Now, a dequeue, $deq_0()$, becomes active, executes its CAS of line 34 successfully removing x_0, and then becomes slow. Next, $enq_0(x_0)$ finishes its execution and*

---
[1] As we explain later, this may occur in two cases. Either because the system, through a system cache invalidation, writes back the triplet, or because an obsolete enqueue operation executes a pwb instruction writing it back.

returns OK. Because of the pair of **pwb-psync** *in line 15,* $Q[0]$ *(which contains the triplet* $(s, 4, \perp)$*) is written back to NVM. The state of* Q *at this point is illustrated in Fig. 1b. Then, a system-wide failure occurs. At crash time, the value of* **Head** *is equal to 1, but it has not been written back.*

Since $enq_0(x_0)$ *has been completed by the time of the crash, durable linearizability requires that it is linearized. The question is whether we will linearize* $deq_0()$*. Since the triplet* $(s, 4, \perp)$ *has been written back, no dequeue can return* x_0 *after recovery. Thus, if we do not linearize* $deq_0()$*, durable linearizability will be violated. If we linearize* $deq_0()$*, the value of* **Head** *must be set to 1 at recovery time. By considering only occupied cells of the array this will not be the case. Thus,* RECOVERY *cannot ignore triplets containing the value* $\perp$.   □

In PERCRQ, RECOVERY traverses all elements from **Head** to **Tail** (line 72) to find the largest index (minus R), max, stored into an unoccupied cell, that is larger than the current value of **Head**. We check if the maximum of these values is larger than or equal to max, where max is initialized to the written back value of **Head** (line 71). If it is, we update max to store the maximum such value (line 74).

RECOVERY also examines the occupied elements between **Head** and **Tail**. We follow a similar approach (as for max) for now finding the minimum index among these elements (lines 76-80). Scenario 3 explains why doing this is necessary.

Scenario 3. *Let* $R = 4$ *be the size of the circular array* Q*. Assume that four enqueue operations,* $enq_0, \ldots, enq_3$ *are executed successfully returning OK. Next,* deq_0 *executes its FAI (line 25) and then becomes slow. There are three more dequeue operations* $deq_1, \ldots, deq_3$ *that execute successfully. These deques return items* $x_1, \ldots, x_3$*, respectively, and write* $\perp_5, \ldots, \perp_7$ *to their respective cells,* $Q[1], \ldots, Q[3]$*. Next, three more enqueue operations* $enq_4, \ldots, enq_6$ *begin their execution. Enqueue* enq_4 *executes its FAI (line 4) and then becomes slow. The other two,* enq_5 *and* enq_6 *insert successfully their items,* x_5 *and* x_6*, and return OK. Now a crash occurs. Figure 1c illustrates the state of* Q *at crash time.*

After the execution of lines 61-68 of RECOVERY*,* **Tail**'s *recovered value is 7. Moreover, because of* deq_3*, the value stored in NVM for* **Head** *is 4 at crash time. After executing lines 71-74,* **Head**'s *value does not change. Thus,* **Head** *points to* $Q[0]$*, which contains element* x_0 *whose index 0 is less than 4. Since* enq_0 *and* $deq_1, \ldots, deq_3$ *have terminated, they have to be linearized. To ensure the FIFO property,* deq_0 *should be linearized as well. Thus, the value of* **Head** *must be higher than 0 at recovery time. To address this problem,* PERCRQ *sets* **Head** *to the smallest index of an occupied cell between the current* **Head** *and* **Tail***. Hence, after lines 69-80 are executed* **Head** $= 5$ *and* **Tail** $= 7$.   □

RECOVERY also initiates all elements of Q outside the range **Head**... **Tail** with the appropriate value for serving subsequent enqueue operations (lines 81-83).

Local Persistence. Experiments show (Fig. 2 of Sect. 5) that having each successful dequeue executing a **pwb** instruction on **Head** is quite costly. The reason for this is that **Head** is a heavily contended variable.

PERCRQ introduces the following technique to reduce the cost of persisting a variable *var* that is highly contended. Every thread p_i maintains a local copy, var_i, of *var*. Thus, there exists an array of n elements, one for each thread, where the thread stores its local copy of the variable. This array is maintained in NVM. Every time the thread performs an operation on *var*, it also updates its local copy with the value read. Then, p_i persists var_i rather than *var*. Since var_i is a single-reader, single-writer variable, persisting it is very cheap [1].

For instance, in PERCRQ, every thread p_i maintains a local copy, Head$_i$, of Head, and persists Head$_i$ rather than Head. This happens every time Head is to be persisted by PERCRQ (lines 35 and 45). The recovery function begins by setting the initial value of Head to be the maximum value found in every element of the array of local copies of the Head values in NVM.

We believe that this technique, which we call *local persistence*, is of general interest, and can be applied to reduce the persistence cost of many other algorithms.

Durable Linearizability. We now discuss how to assign linearization points to ensure durable linearizability. We say that an enqueue (dequeue) operation *has index i* if it reads i into Tail (Head) at its last iteration of the while loop of line 3 (line 24). We use enq_i to denote the enqueue with index i that successfully inserts the triplet $(1, i, x_i)$ in cell $\mathcal{Q}[i \bmod R]$. A dequeue operation *matches enq_i*, if it reads i in Head (line 25); let deq_i be this dequeue.

An enqueue operation is *persisted* in some epoch E_k, if it has executed successfully the *CAS* of line 14 and the write-back of the new value has taken effect by c_k. A persisted enqueue will terminate successfully if the system does not crash. Consider now a dequeue operation *deq* that has read the value i into Head the last time it read it. We say that *deq* is *persisted* in E_k, if either some value of Head $\geq i$ has been written back by c_k or a value $idx \geq i + R$ stored in some element of $\mathcal{Q}$ (by a *CAS* of lines 34, 41) has been written back by c_k.

We assign linearization points to enqueue and dequeue operations of E_k according to the following rules:

Enqueue operations: Consider an enqueue operation, *enq*, for which the *FAI* of the last iteration of the while loop (line 3) before c_k reads i from Tail. We linearize *enq* in the following cases:
(1) If *enq* has been persisted by c_k.
(2) If there exists a dequeue operation, *deq*, that reads i into Head at its last iteration of the while loop, and *deq* has been persisted by c_k.
(3) If *enq* has read 1 in the closed bit of Tail and this value has been written back by c_k.
(4) If *enq* has executed successfully the *TAS* of line 19 and the new value of Tail.cb has been written back by c_k.
In the first three cases, *enq* is linearized at the point that it executed its last *FAI* operation. In the fourth case, *enq* is linearized at the time it executes the *TAS* of line 19.

Algorithm 4. PERCRQ linearization procedure

Execution α and set of epochs $\mathcal{E} := \{E_1, E_2, \ldots\}$ in α
1: LinQ: auxiliary infinite array
2: $head(\text{LinQ})$: head index of LinQ
3: $tail(\text{LinQ})$: tail index of LinQ

4: **for** each epoch $k = 1, 2, \ldots$ **do**
5:     **for** each event $j = 1, 2, \ldots$ in epoch E_k **do**
6:         **if** e_j is a TAS of line 19 by $enq(x)$ and its change has been persisted by c_k **then**
7:             Linearize $enq(x)$ at e_j
8:         **else if** e_j is a FAI of line 4 by $enq(x)$ that reads 1 in the closed bit of Tail **and** the closed bit is persisted by c_k **then**
9:             Linearize $enq(x)$ at e_j
10:        **else if** e_j is a FAI of $enq(x)$ in E_k **then**
11:            **if** e_j is the last FAI of $enq(x)$ and $enq(x)$ is persisted
               or there exists a deq such that:
               a) deq has executed successfully the CAS of line 34,
               b) deq reads i into Head at its last iteration of the while loop
               (line 24), and deq has persisted in E_k **then**
12:                $\text{LinQ}[t] \leftarrow x$
13:                $tail(\text{LinQ}) \leftarrow t + 1$
14:                Linearize $enq(x)$ at e_j
15:            **else if** e_j is a Tail read of line 43 by deq that gives an empty queue **and** deq has been persisted in E_k **then**
16:                Linearize that dequeue operation at e_j
17:            **while** $head(\text{LinQ}) < tail(\text{LinQ})$ **do**
18:                $h \leftarrow \min\{i : \text{LinQ}[i] \neq \bot\}$
19:                Let deq be the dequeue whose FAI at its last iteration returns h
20:                **if** deq not active in $e_1, \ldots, e_j$ of E_k **then**
21:                    **break**
22:                **if** deq has executed successfully the CAS of line 34,
                   deq reads h into Head at its last iteration of the while loop
                   (line 24), and deq has persisted in E_k **then**
23:                    $x \leftarrow \text{LinQ}[h]$
24:                    $\text{LinQ}[h] \leftarrow \bot$
25:                    $head(\text{LinQ}) \leftarrow h + 1$
26:                    Linearize that dequeue operation at e_j

---

Dequeue Operations: Consider a dequeue operation, deq, for which the FAI of the last iteration of the while loop (line 24) before c_k reads i in Head. We linearize deq if it has been persisted by c_k and one of the following two conditions hold:
(1) The matching enqueue operation to deq has successfully executed the CAS of line 14.
(2) deq has read a value $t \leq i + 1$ into Tail (line 43).
In the first case, deq is assigned a linearization point either exactly after the linearization point of the enqueue operation enq that read i in Tail, if deq is active at that point, or at the first point in which it is active and all dequeues that have smaller indices than deq have been linearized. In the second case, deq is assigned a linearization point at the read of Tail in line 43.

Algorithm 4 assigns linearization points to PERCRQ in a formal (algorithmic) way. In [13], we discuss in detail why PERCRQ is durably linerizable.

4.3 Persistent LCRQ

In this section, we describe the persistent version of LCRQ, called PERLCRQ. PERLCRQ implements a lock-free linked list of nodes, each of which represents an instance of PERCRQ. Each PERCRQ instance has its own Head and Tail shared variables that encapsulate the head and the tail of a durably linearizable tantrum queue of finite size, as discussed in Sect. 4.2. PERLCRQ eliminates the disadvantage that the PERCRQ queue is of finite size and transforms it from a tantrum to a FIFO queue. Specifically, when the currently active instance of PERCRQ becomes CLOSED, a new instance is appended in the linked list. Moreover, when the queue of a PERCRQ instance becomes EMPTY, a dequeuer dequeues this instance from the linked list and searches for the item to dequeue in the next element of the linked list. Two pointers, First and Last point to the first and the last element of the linked list, which operates as a queue. These pointers are stored in the NVM.

The pseudocode of PERLCRQ (Algorithm 5 in Appendix) closely follows that of the lock-free queue implementation by Michael and Scott [18]. An ENQUEUE(x) operation starts by reading Last (line 20), to get a reference to an instance, crq, of PERCRQ (of Algorithm 3). If it manages to read the last element of the list (i.e., the next pointer of the node crq pointed by Last is equal to NULL), the enqueue operation calls ENQUEUE(crq, x) method (line 26) to insert x into the queue of crq. If this method returns OK then ENQUEUE(x) is complete and returns also OK. If ENQUEUE(crq, x) returns CLOSED (line 26), then ENQUEUE(x) creates a new node (i.e., a new instance of PERCRQ), containing x (line 17), and tries to append that to the list by executing the *CAS* of line 28. If the *CAS* succeeds, it tries to also change Last to point to the new PERCRQ node (line 30). If the *CAS* fails, another PERCRQ node was inserted. Then, ENQUEUE(x) retries by executing one more loop of its while statement (line 19).

ENQUEUE(x) does not attempt to append its node if Last does not point to the last element of the list (line 22). It rather tries (line 24) to update Last to point to the last element, and starts one more loop (line 19). This scenario might happen if after appending a new node and before updating Last, a thread becomes slow (or fails). Then, Last is *falling behind* (i.e., it does not point to the last element of the list), and other threads will have to help by setting Last to point to the last element of the list, before they try to append their nodes.

A DEQUEUE() reads First to get a pointer to the instance, crq, of PERCRQ that is first in the linked list, and calls DEQUEUE(crq) (lines 8-10) in an effort to dequeue an element from that instance. If DEQUEUE(crq) returns some value, then DEQUEUE is complete and returns also that value (lines 11 and 12). If DEQUEUE(crq) returns EMPTY, then DEQUEUE checks if there is no next PERCRQ node in the list (line 13), and if that is true, then it returns also EMPTY. If there is another PERCRQ node it tries to move the First pointer (line 15) to that node and starts a new loop (line 7).

We now discuss the persistence instructions that we added in LCRQ to make it persistent. We were able to achieve persistence without adding any persistence instruction to the dequeue code. Thus, we focus on enqueue. The pwb-psync pair of line 18 is necessary just in case the enqueue returns on line 31. However, the persistence of nd.next should be performed before the *CAS* of line 28 for

---

Algorithm 5. Persistent LCRQ (PERLCRQ)

```
 1: struct Node {
 2:     Node *next;
 3:     PERCRQ Object crq;
 4: };

 5: Node *First, *Last  // both initially point
    to a Node with next being NULL and crq in
    initial state

 6: function DEQUEUE
 7:     while true do
 8:         f ← First
 9:         crq ← f→crq
10:         v ← DEQUEUE(crq)
11:         if v ≠ EMPTY then
12:             return v
13:         if f→next= NULL then
14:             return EMPTY
15:         CAS(First, f, f→next)

16: function ENQUEUE(x)
17:     Create a new Node nd with next equal to
    NULL, x stored in nd.crq.Q[0], nd.crq.Tail =
    1, and nd.crq.Head = 0
```

```
18:     pwb(nd.next, nd.crq.Q[0], nd.crq.Tail);
        psync();
19:     while true do
20:         ℓ ← Last
21:         crq ← ℓ→crq
22:         if ℓ→next ≠ NULL then
23:             pwb(ℓ→next); psync();
24:             CAS(Last, ℓ, ℓ→next)
25:             continue
26:         if ENQUEUE(crq, x) ≠ CLOSED then
27:             return OK
28:         if CAS(ℓ→next, NULL, nd) then
29:             pwb(ℓ→next); psync();
30:             CAS(Last, ℓ, nd)
31:             return OK

32: function PERLCRQ RECOVERY
33:     while Last→next ≠ NULL do
34:         RECOVERY(Last→crq) // From Algo-
        rithm 3
35:         Last← Last→next
36:         RECOVERY(Last→crq) // From Algo-
        rithm 3
```

---

the following reason. Assume that it occurs after the *CAS* of line 28. Then, if the system evicts the cache line where the next field of the node on which the *CAS* has been executed resides (i.e., ℓ →next), and before the next field of the appended node, **nd**, is persisted, then at recovery time, this field may not be NULL, whereas **nd** is the last node of the recovered list. This jeopardizes the correctness of the algorithm (for instance the condition of the if statement of line 22 may succeed, although **nd** is the last node in the list). Note that **nd.crq.Tail** and **nd.crq**.$Q[0]$ can be persisted after the execution of the *CAS* of line 28. However, since these are fields of a struct of type Node, we place them so that they all reside in the same cache line and can be persisted all together with a single **pwb**. Therefore, we persist them all together to save on the number of **pwb** s the algorithm performs.

We now explain why the **pwb-psync** pair of line 29 is necessary. Assume that this pair is omitted. Consider an enqueue that returns on line 31. If a crash occurs right after the completion of the enqueue, at recovery time, the node appended by the enqueue and its enqueued item will not be included in the state of the linked list. This violates durable linearizability (as the enqueue has been completed, and this it should be linearized).

We next explain why the **pwb-psync** pair of line 23 is necessary. Assume that these persistent instructions are omitted. Assume that an enqueue, enq_1 executes up until ine 28 and then becomes slow. Now assume that another enqueue, enq_2 starts its execution, evaluates the condition of the if statement of line 22 to **true**, executes lines 24-25, and starts a new loop which makes it return on line 27. If a crash occurs at this point, the node appended by enq_1, and the items enqueued by both enqueue operations will not appear in the recovered state of the linked list. This violates durable linearizability (as enq_2 has been completed, and thus it should be linearized).

We now focus on the recovery function. Assume that a crash occurs and the system recovers by calling the recovery function once. Since First and Last are non-volatile variables, at recovery time, they may have any value they had obtained by the crash (as the system may have evicted the cache line they reside and written it back to NVM).

At recovery time, First and Last point either to the first node of the list (if the cache line in which they reside has never been evicted by the system and written back to NVM) or to nodes that has been appended in the linked list by the crash otherwise. In the second case, note that the appended nodes may have been dequeued, by the crash, but they are still connected to the linked list, i.e., if we start from the initial node in the list and follow next pointers we will traverse these nodes).

To recover Last, we start from the value stored for it in NVM (which can be its initial value) and we traverse the list following next pointers of its Nodes, until we reach the last node. Last will be set to point to this Node. On the way, we call RECOVERY on the crq object of each Node we traverse. First pointer never changes at recovery. This has a cost in the performance of dequeues that will be executed after the crash, as they will have to traverse all Nodes until they reach the first such Node whose crq instance contains items to be dequeued.

Durable Linearizability. PERLCRQ is durably linearizable. We next discuss how to assign linearization points to operations.

Dequeue operations: Consider a dequeue operation *deq*. Let crq be the last instance of PERCRQ on which *deq* calls DEQUEUE(crq). *deq* is linearized in the following cases: a) DEQUEUE(crq) is linearized in crq and its response is not EMPTY in the linearization of the corresponding execution of crq. b) DEQUEUE(crq) is linearized in crq, its response is EMPTY in the linearization of the corresponding execution of crq, and *deq* returns EMPTY.

Enqueue operations: Consider an enqueue operation *enq*. Let crq be the last instance of PERCRQ on which *enq* calls ENQUEUE(crq). We linearize an enqueue operation of PERLCRQ if the following three conditions hold

1. If ENQUEUE(crq, x) returns EMPTY (line 26), we assign a linearization point at the point of the corresponding ENQUEUE(crq, x) linearization point.
2. If the change to the next pointer crq.*next* (line 28) is written back to NVM and the new item that was enqueued is also written back by c_k. We assign a linearization point at the time the CAS operation has taken effect.

5   Performance Evaluation

For the evaluation, we used a 48-core machine (96 logical cores) consisting of 2 Intel(R) Xeon(R) Gold 5318Y processors with 24 cores each. Each core executes two threads concurrently. Our machine is equipped with a 128 GB Intel Optane 200 Series Persistent Memory (DCPMM) and the system is configured in AppDirect mode. We use the 1.9.2 version of the *Persistent Memory Development Kit* [20], which provides the `pwb` and `psync` persistency instructions. The operating system is Ubuntu 20.04.6 LTS (kernel Linux 6.6.3-custom x86_64) and we use *gcc* v9.4.0. Threads were bound in all experiments following a scheduling policy which distributes the running threads evenly across the machine's NUMA nodes [6,7].

In the experiments, each run simulates 10^7 atomic operations in total; each thread simulates $10^7/n$ operations (where n is the number of threads). We follow a kind of standard approach [5–7,12,23,24], where each thread performs pairs of ENQUEUE and DEQUEUE starting from an empty queue; this kind of experiment avoids performing unsuccessful (and thus cheap) operations. Experiments where each thread executed random operations (50% of each type) did not illustrate significantly different performance trends.

We compare PERLCRQ with the state-of-the-art persistent queue implementations, namely PBQUEUE and PWFQUEUE [9]. The experimental analysis in [9] compares PBQUEUE and PWFQUEUE with the specialized persistent queue implementation in [12] (FHMP), and those recently published in [24], as well as the persistent queue implementations based on the following general techniques and transformations: a) CAPSULES-NORMAL [2], b) ROMULUS [3] c) ONEFILE [21], d) CX-PUC [4] and CX-PTM [4], and e) REDOOPT [4]. Experiments in [9] showed that PBQUEUE is at least 2x faster than all these algorithms. For this reason, we compare the performance of PERLCRQ only with the performance of PBQUEUE and PWFQUEUE [9].

Results. Figure 2 shows the performance of PERLCRQ against PBQUEUE and PWFQUEUE. We measure throughput, i.e. number of operations executed per time unit (thus, larger is better), as the number of threads increases. We see that PERLCRQ (purple line) is 2x faster than PBQUEUE, which is the best competitor.

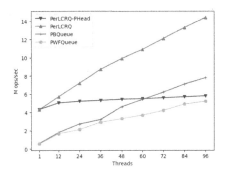
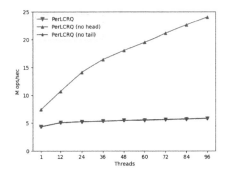
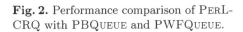

Fig. 2. Performance comparison of PERL-CRQ with PBQUEUE and PWFQUEUE.

Fig. 3. Cost of persisting `Head` and `Tail` in PERLCRQ.

Figure 2 shows also the cost of persisting different instructions. Interestingly, although PERLCRQ executes only one pwb per operation, persisting shared variable Head is too expensive. As we see, the version of PERLCRQ that persists a shared version of Head (PERLCRQ-PHead) instead of local copies of Head maintained by each thread, incurs high performance overhead. The throughput of PERLCRQ-PHead is much lower than that of PERLCRQ, as the number of threads increases, and eventually its performance gets surpassed by PBQUEUE and PWFQUEUE. Figure 3 provides additional evidence that the drop in performance observed in PERLCRQ-PHead, is due to how expensive persisting highly contended shared variables, such as Head is. We denote PERLCRQ in which all pwb instructions on Head have been removed, as PERLCRQ *(no head)* and the PERLCRQ in which all pwb instructions on Tail have been removed, PERL-CRQ *(no tail)*. From Fig. 3, we see that the persistence cost of persisting Tail is negligible. This shows that PERLCRQ persists Tail rarely (only when it closes the current instance of CRQ). Additionally, it provides evidence that the optimization of using the *closedFlag* to avoid persisting Tail more than once in the case a CRQ instance needs to close, works well. By comparing the throughput of PERLCRQ (from Fig. 2) with that of PERLCRQ *(no head)* in Fig. 3, we see the cost of persisting Head even when the algorithm uses a local copy of Head for each thread.

Evaluation of the Recovery Cost. We have developed a custom failure framework to evaluate the recovery cost of algorithms. The framework provides a shared variable called "recovery_steps". All threads" monitor" this variable and each operation periodically (or randomly) lowers the value by 1 step. When it reaches 0, any thread running will cease, effectively simulating a crash of all threads. Afterwards, the recovery function is launched by some thread; this simulates the system running the recovery function.

The above procedure - a standard run where recovery steps are being decreased, a random crash that occurs when these steps become 0, and the recovery function run - is called a *cycle*. Each evaluation test has 10 cycles and we measure only the third part of each cycle, which corresponds to the recovery cost. The average of the 10 cycles is the result shown on the graphs in Figs. 4 and 5.

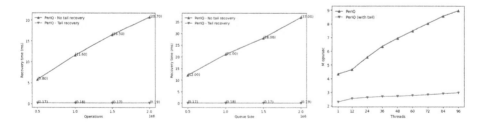

Fig. 4. Recovery time of PERIQ as the number of operations increases.

Fig. 5. Recovery time of PERIQ as the queue size increases.

Fig. 6. Throughput of PERIQ and PERIQ (no Tail).

Algorithm 6. Variant of ENQUEUE of PERIQ, used to illustrate the tradeoff between the persistence and the recovery cost.

---

nOps_i: counter of thread p_i, initially. It counts the the number of enqueue operations p_i has performed.

---

function ENQUEUE(x: Item)
    while true do
        $t \leftarrow$ FETCH&INCREMENT(Tail)
        if $\text{nOps}_i ==$ *Threshold* **then**
            pwb(Tail)
            psync()
            $\text{nOps}_i = 0$
            /* If Threshold = ∞, then persisting Tail is disabled
*/
        if GET&SET($\mathcal{Q}[t], x$) $== \perp$ **then**
            pwb($\mathcal{Q}[t]$)
            psync()
            nOps $\leftarrow$ nOps $+ 1$
            return OK

PERIQ can be designed to illustrate a tradeoff between the persistence cost at normal execution time (where no failures occur) and the recovery cost. The idea is that each thread, periodically, persists Head and Tail, in addition to the values it writes in Q. This may happen e.g. every time the thread has completed the execution of a specific number of operations. Algorithm 6 presents the enhanced pseudocode of PERIQ to accommodate this variant.

Figure 4 compares the time needed to recover the queue after a crash that takes place when a predetermined number of operations have been executed. Figure 5 compares the time (lower is better) needed to recover the queue based on its size. Figure 6 shows the throughput (lower is worse) of the PERIQ compared to the variation of persisting the tail index on each operation. We can see that there is a clear tradeoff between the throughput of the persistence of the tail index and how quick the algorithm's recovery function is. As shown in Figs. 4 to 6, if we do not persist Tail, the recovery function is slower as the queue gets larger, but the throughput of the PERIQ is high. On the other hand, if we persist Tail, the recovery function is very fast, while the throughput of PERIQ is lower.

References

1. Attiya, H., Ben-Baruch, O., Fatourou, P., Hendler, D., Kosmas, E.: Detectable recovery of lock-free data structures. In: Lee, J., Agrawal, K., Spear, M.F. (eds.) PPoPP 2022: 27th ACM SIGPLAN Symposium on Principles and Practice of Parallel Programming, Seoul, Republic of Korea, April 2–6, 2022, pp. 262–277. ACM (2022). https://doi.org/10.1145/3503221.3508444

2. Ben-David, N., Blelloch, G.E., Friedman, M., Wei, Y.: Delay-free concurrency on faulty persistent memory. In: The 31st ACM Symposium on Parallelism in Algorithms and Architectures, pp. 253–264. SPAA 2019, Association for Computing Machinery, New York, NY, USA (2019). https://doi.org/10.1145/3323165.3323187

3. Correia, A., Felber, P., Ramalhete, P.: Romulus: efficient algorithms for persistent transactional memory. In: Proceedings of the 30th on Symposium on Parallelism in Algorithms and Architectures, pp. 271–282. SPAA 2018, Association for Computing Machinery, New York, NY, USA (2018). https://doi.org/10.1145/3210377.3210392

4. Correia, A., Felber, P., Ramalhete, P.: Persistent memory and the rise of universal constructions. In: Proceedings of the Fifteenth European Conference on Computer Systems. EuroSys 2020, Association for Computing Machinery, New York, NY, USA (2020). https://doi.org/10.1145/3342195.3387515

5. Fatourou, P., Kallimanis, N.D.: A highly-efficient wait-free universal construction. In: Proceedings of the Twenty-Third Annual ACM Symposium on Parallelism in Algorithms and Architectures, pp. 325–334. SPAA 2011, Association for Computing Machinery, New York, NY, USA (2011). https://doi.org/10.1145/1989493.1989549

6. Fatourou, P., Kallimanis, N.D.: Revisiting the combining synchronization technique. In: Proceedings of the 17th ACM SIGPLAN Symposium on Principles and Practice of Parallel Programming, pp. 257–626. PPoPP 2012, Association for Computing Machinery, New York, NY, USA (2012). https://doi.org/10.1145/2145816.2145849

7. Fatourou, P., Kallimanis, N.D.: Highly-efficient wait-free synchronization. Theory Comput. Syst. **55**(3), 475–520 (2014)

8. Fatourou, P., Kallimanis, N.D.: Lock oscillation: boosting the performance of concurrent data structures. In: Aspnes, J., Bessani, A., Felber, P., Leitão, J. (eds.) 21st International Conference on Principles of Distributed Systems, OPODIS 2017, Lisbon, Portugal, December 18–20, 2017. LIPIcs, vol. 95, pp. 8:1–8:17. Schloss Dagstuhl - Leibniz-Zentrum für Informatik (2017https://doi.org/10.4230/LIPICS.OPODIS.2017.8

9. Fatourou, P., Kallimanis, N.D., Kosmas, E.: The performance power of software combining in persistence. In: Proceedings of the 27th ACM SIGPLAN Symposium on Principles and Practice of Parallel Programming, pp. 337–352. PPoPP 2022, Association for Computing Machinery, New York, NY, USA (2022). https://doi.org/10.1145/3503221.3508426

10. Fatourou, P., Kallimanis, N.D., Ropars, T.: An efficient wait-free resizable hash table. In: Proceedings of the 30th on Symposium on Parallelism in Algorithms and Architectures, pp. 111–120. SPAA 2018, Association for Computing Machinery, New York, NY, USA (2018).https://doi.org/10.1145/3210377.3210408

11. Friedman, M., Herlihy, M., Marathe, V., Petrank, E.: A persistent lock-free queue for non-volatile memory. SIGPLAN Not. **53**(1), 28–40 (2018). https://doi.org/10.1145/3200691.3178490

12. Friedman, M., Herlihy, M., Marathe, V., Petrank, E.: A persistent lock-free queue for non-volatile memory. ACM SIGPLAN Notices **53**(1), 28–40 (2018)

13. Giachoudis, N., Fatourou, P., Mallis, G.: Highly-efficient persistent fifo queues. Zenodo (2024). https://doi.org/10.5281/zenodo.10699312

14. Hendler, D., Incze, I., Shavit, N., Tzafrir, M.: Flat combining and the synchronization-parallelism tradeoff. In: Proceedings of the Twenty-second Annual ACM Symposium on Parallelism in Algorithms and Architectures, pp. 355–364 (2010)

15. Herlihy, M.P., Wing, J.M.: Linearizability: a correctness condition for concurrent objects. ACM Trans. Program. Lang. Syst. **12**(3), 463–492 (1990). https://doi.org/10.1145/78969.78972

16. Izraelevitz, J., Mendes, H., Scott, M.L.: Linearizability of persistent memory objects under a full-system-crash failure model. In: Gavoille, C., Ilcinkas, D. (eds.) Distributed Computing. LNCS, vol. 9888, pp. 313–327. Springer, Heidelberg (2016). https://doi.org/10.1007/978-3-662-53426-7_23

17. Klaftenegger, D., Sagonas, K., Winblad, K.: Queue delegation locking. IEEE Trans. Parallel Distrib. Syst. **29**(3), 687–704 (2018)

18. Michael, M.M., Scott, M.L.: Simple, fast, and practical non-blocking and blocking concurrent queue algorithms. In: Proceedings of the Fifteenth Annual ACM Symposium on Principles of Distributed Computing, pp. 267–275 (1996)

19. Morrison, A., Afek, Y.: Fast concurrent queues for x86 processors. In: Proceedings of the 18th ACM SIGPLAN Symposium on Principles and Practice of Parallel Programming, pp. 103–112 (2013)

20. PMDK: The persistent memory development kit. https://github.com/pmem/pmdk/, https://github.com/pmem/pmdk/

21. Ramalhete, P., Correia, A., Felber, P., Cohen, N.: Onefile: a wait-free persistent transactional memory. In: 2019 49th Annual IEEE/IFIP International Conference on Dependable Systems and Networks (DSN), pp. 151–163. IEEE (2019)

22. Rusanovsky, M., Attiya, H., Ben-Baruch, O., Gerby, T., Hendler, D., Ramalhete, P.: Flat-combining-based persistent data structures for non-volatile memory. In: Johnen, C., Schiller, E.M., Schmid, S. (eds.) SSS 2021. LNCS, vol. 13046, pp. 505–509. Springer, Heidelberg (2021). https://doi.org/10.1007/978-3-030-91081-5_38

23. Rusanovsky, M., Ben-Baruch, O., Hendler, D., Ramalhete, P.: A flat-combining-based persistent stack for non-volatile memory. CoRR **abs/2012.12868** (2020). (version submitted at 23 December, 2020), https://arxiv.org/abs/2012.12868

24. Sela, G., Petrank, E.: Durable queues: the second amendment. In: Proceedings of the 33rd ACM Symposium on Parallelism in Algorithms and Architectures, pp. 385–397. Association for Computing Machinery, New York, NY, USA (2021). https://doi.org/10.1145/3409964.3461791

Locally Balanced Allocations Under Strong Byzantine Influence

Costas Busch[1] , Paweł Garncarek[2(✉)] , and Dariusz R. Kowalski[1]

[1] School of Computer and Cyber Sciences, Augusta University, Augusta, USA
{kbusch,dkowalski}@augusta.edu
[2] Institute of Computer Science, University of Wrocław, Wrocław, Poland
pawel.garncarek2@uwr.edu.pl

Abstract. The Power of Two Choices (PoTC) is a commonly used technique to balance the incoming load (balls) into available resources (bins) – for each coming ball, two bins are selected uniformly at random and the one with smaller number of balls is chosen as the location of the current ball. We study a generalization of PoTC to a fault-prone setting – faulty bin(s) could present malicious information to enforce allocation decision on any of the two bins. Given m balls and n bins, such that no more than f of the bins are faulty, we show that the maximum loaded honest bin receives a surplus of a logarithmic number of balls with respect to f. Our result generalizes the classic bounds of the Power of Two Choices in the presence of a strong Byzantine adversary. Our solution and methods of analysis can help to efficiently implement and analyze resilient online local decisions made by processes when solving fundamental problems that depend on load balancing under the presence of Byzantine failures.

Keywords: Scheduling and resource allocation · Power of Two Choices · Fault tolerance · Strong Byzantine adversary · Probabilistic analysis

1 Introduction

We study a generalization of the classical Power of Two Choices (PoTC) online principle (in which the ball chooses the bin with a smaller number of balls) to a fault-prone setting – a faulty bin could present the ball with malicious information to enforce the decision on any of the two bins. This generalization could allow using the PoTC paradigm in the design and analysis of distributed protocols, for example, population protocols, multiprocessor job scheduling, and network routing, under the presence of Byzantine failures.

In the traditional balls-in-bins problem, there are m balls and n bins and each ball is thrown into a bin picked uniformly at random [20, 25]. It is well known that for $m = n$ each bin gets at most $\log n / \log \log n$ balls with high probability (i.e., with probability at least $1 - o(1)$) [17]. This improves to at most $\log \log n + \Theta(1)$ balls per bin, if we give two random choices to each ball and the ball goes into

© The Author(s), under exclusive license to Springer Nature Switzerland AG 2024
Y. Emek (Ed.): SIROCCO 2024, LNCS 14662, pp. 262–280, 2024.
https://doi.org/10.1007/978-3-031-60603-8_15

the bin that has the smaller number of balls [5]. The result can be generalized for d random choices per ball to at most $(1+o(1)) \log \log n / \log d + \Theta(m/n)$ balls per bin [5,11].

The balls-in-bins problem has important applications in fundamental decision-making problems [27]. It can be used to solve task load balancing, hash collision, and routing problems. In the task load balancing problem, each ball corresponds to a computing task, where each task has a fixed weight, and the bins are the servers that will execute the tasks. Using the Power of Two Choices,[1] each task picks two random servers and is allocated to the least loaded server. Hence, with n tasks and n servers, each server will be imbalanced by at most an additive $O(\log \log n)$ term. One of the advantages of PoTC is that it is local, i.e., decisions are made by looking into only 2 bins rather than all of them, which is important in distributed settings. A similar approach with PoTC can be used to improve collisions in hash functions [27]. With two perfectly random hash functions giving two distinct table entries, a key is placed to the entry with the smallest chain of keys. The load balancing attribute can also be used in network routing where packets pick two randomly chosen paths and the packet is sent along the path that currently has the lowest congestion [15]. Other applications include load-balancing in virtual machines in fog computing [10], queuing theory analysis [12], and distributed voting [16].

Many of the aforementioned applications for the PoTC problem take place in a distributed computing environment that is prone to failures [1,24]. The failures may take different forms, for example, the computing nodes may crash and communication links may be disrupted. Even worse, nodes and communication links may be compromised and misbehave due to malicious attacks or other reasons. Since typical distributed systems need to continuously operate even under failures, it is important to understand what is the system's resiliency to failures and the impact on the performance of the underlying distributed task.

In this work, we model failures as being generated by a Byzantine adversary that aims to disrupt normal operations and cause the system to misbehave. Byzantine failures model a large range of system failures which include crash failures and malicious attacks. Byzantine adversaries have been considered for classic distributed consensus [9,22]. Here, we examine the impact of Byzantine faults on PoTC.

The Byzantine adversary controls $f < n$ of the bins which are faulty and they may report wrong information about the number of balls in them. The remaining $n - f$ bins are honest and not controlled by the adversary. The Byzantine adversary does not control the scheduler, that is, the scheduler keeps making random picks of two bins out of n for each of the balls. If the scheduler picks two honest bins, the ball will fall into the bin with the least load. However, if the scheduler picks one honest bin together with a faulty bin, then the faulty bin may report a higher number of balls than it actually has, forcing the ball to fall

---

[1] "Power of Two Choices" can denote a principle as well as a greedy algorithm following said principle.

Table 1. Technical contributions. Parameter ϕ can be equal to $\frac{4em^2}{n^2}$ (see Corollary 2) or cem/n for any $c > 2e$ (see Corollary 5 and Lemma 7).

	Number of balls in a bin	m	Additional conditions	Prob.
Theorem 1	$(1+\delta)\frac{2m}{n}$	any m	$\frac{2m\delta^2}{3n} > c\log n$ $\delta \in (0,1)$ any f	$1 - n^{1-c}$
Corollary 1	$\frac{3m}{n}$	$m > 9n\log n$	any f	$1 - n^{-1/2}$
Theorem 2	$\phi + \log\log\frac{n^2}{2emf}$ $+ \log\frac{2f}{\frac{n^2}{2emf}\sqrt{f^2+8\frac{n^2}{m}\log n - f}}+1$	$m =$ $O(n\log n)$	$1 \le f < \frac{n^2}{2em}$	$1 - o(1)$
Corollary 3	$O\left(\log\log\frac{n^2}{mf} + \log\frac{f}{\frac{n^2}{2emf}2e\log n}\right)$	$m = O(n)$	$1 \le f < \frac{n^2}{2em}$	$1 - o(1)$
Corollary 4	$O(\frac{m}{n} + \log\log\frac{n^2}{mf}$ $+ \log\frac{f}{\frac{n^2}{2emf}2e\log n})$	$m \ge en$ $m =$ $O(n\log n)$	$1 \le f < \frac{n^2}{2em}$	$1 - o(1)$
Corollary 5	$O\left(\frac{m}{n} + \log\log\frac{n}{f} + \log f\right)$	$m \ge n$	$1 \le f <$ $\max\{\frac{n}{12e\log n}, \frac{n^2}{2em}\}$	$1 - o(1)$

into the honest bin. Therefore, a Byzantine adversary can create an imbalance in the allocation of balls into the bins, by directing more balls to the honest bins.

We show that PoTC has an inherent tolerance to failures. Given f faulty bins, for the case where $f < n^2/(2em)$ and $m = O(n\log n)$, we show that the maximum loaded honest bin receives only a logarithmic number of balls with respect to f. We also bound the maximum load on an honest bin for the general case. Our results imply that the system can gracefully tolerate failures for certain values of f without significantly affecting the maximum load on any bin. Hence, the affected applications related to load-balancing and collision avoidance have sustained good properties with only an additive logarithmic term on their performance impact due to failures.

Our main technical contributions are summarized in Table 1. All the results occur with high probability. The main results are provided by Theorem 1 and Theorem 2. All the corollaries are derived from these two theorems. Theorem 1 applies to a wide range of number of balls m. A more precise result is shown in Theorem 2, however it applies only to $m = O(n\log n)$ and depends on a parameter ϕ that can be used to provide refined results. Depending on the choice of the problem parameters, and replacing parameter ϕ accordingly, one can derive from Theorem 2 results that can be compared to the results for non-faulty PoTC, e.g., $m/n + O(\log\log n)$ in [11]. A precise description of these results, as well as some additional results and a discussion of why simpler solutions do not work, can be found in Sect. 2.

1.1   Related Work

There are several variants of the PoTC problem. The extension to d choices provides an improvement to the basic $\log\log n$ maximum load bound [5,11]. It is shown that the improvement is inversely proportional to d if there is asymmetry between the d choices [28]. Typically it is assumed that all balls have the same weight, but having different ball weights has also been considered [26], since it represents the load-balancing problem of non-uniform tasks. Variants allowing ball deletions together with insertions have also been studied [7].

Recently, fault-tolerance of PoTC has been studied as well. For example, [23] focuses on load balancing applications and allows incorrect results of comparisons when the compared values differ by less than some parameter.

The classic PoTC problem has been generalized to graphical allocations [6, 21], where the bins are nodes in a graph and for each ball, a uniformly random edge is selected. The ball is allocated to the less loaded of the two bins associated with the edge. The classic problem is a special case when the graph is the clique K_n. It is shown that in a n^ϵ-regular graph (with n nodes) the maximum load is $\log\log n + O(1/\epsilon) + O(1)$. Graphical allocations have also been studied from the perspective of assigning the ball to the lowest loaded bin with probability β and to the other bin with probability $1 - \beta$, also called $(1 + \beta)$-choice process [26]. Notably, $(1 + \beta)$-choice process can be used to model faults caused by an unreliable load reporting mechanism. They show that the gap between the most and least loaded bins is $\Theta(\log n/\beta)$ irrespective of m (w.h.p.). Faults can be modeled with $(1 + \beta)$-choice process for the case where an adversary always gives the ball to the honest node when also a faulty node is picked in the pair. However, a Byzantine adversary can behave in a way that cannot be mimicked by the random $(1+\beta)$-choice process, e.g. deterministically choosing balls that will be put in faulty bins. Additionally, our results are better for some parameters m, n, f than the results for $(1+\beta)$-choice process (see the note after Corollary 5).

PoTC naturally relates to population protocols [3,8], where there are n nodes (similar to the bins) such that a random scheduler each time picks two nodes to participate in an *exchange* (similar to picking two random bins to throw the ball). Population protocols solve basic distributed problems such as leader election and majority consensus using a small set of states per node. A useful primitive in population protocols is the *phase clock* [2], which is a local counter at each node that increments at each exchange modulo a fixed phase duration, such that each phase implements a step of an algorithm. This is similar to the behavior of PoTC as well. There is a relation with faulty population protocols, where some nodes may be malicious or act erratically [4,13,18]. In such settings, a fault-tolerant adaptation of phase clock needs to be developed. Such a phase clock requires bounds on the gap between the most loaded and least loaded bin. Our analysis of the most loaded bin of faulty PoTC is a step towards implementing a fault-tolerant phase clock for fault-resilient population protocols; however, our analysis needs to be adapted to bound the gap for phase clocks.

Outline of the Paper: We continue with the problem specification, description of challenges and our extended results in Sect. 2. The proof of Theorem 2 is given in Sect. 3. The case $m \geq en$ is analyzed in Sect. 4, while the case of large m is analyzed in Sect. 5. Finally, we conclude with a discussion and open problems in Sect. 6.

Algorithm 1: Balls in bins with Power of Two Choices against strong Byzantine adversary

---

Input : n bins and a sequence of m balls $r_1, r_2, \ldots, r_m$; Byzantine adversary controls f bins

Output: Distribution of the m balls to the bins

---

1 Let v_i denote the actual number of balls in bin i;
2 Initially, $v_i \leftarrow 0$, for all $i \in [n]$;

3 Let F, with $|F| = f$, be the set of bins controlled by the adversary;
4 For each $i \in F$, the value v_i is reported by the adversary;
5 For each $i \in [n] \setminus F$, the reported value v_i is the true number of balls in bin i;

6 **for** $k = 1$ *to* m **do**
7   Pick uniformly at random a pair $(i,j) \in [n]^2$, $i \neq j$;
      // Adversary knows the true values of v_i and v_j
      // If $i \in F$ or $j \in F$ then the adversary may report wrong v_i or v_j, resp.
8   Let x be the bin such that $x \in \{i,j\}$ and $v_x = \min(v_i, v_j)$;
9   Place ball r_k into bin x;
10     $v_x \leftarrow v_x + 1$;

---

2 Problem Specification and Our Results

2.1 Problem Description

Consider n bins and m balls. The balls arrive one-by-one in an arbitrary linear order. The goal of an online algorithm is to throw each ball to some bin in such a way that the size of the largest bin (i.e., with the largest number of balls thrown to it) is minimized. f of the n bins are *faulty*, while all other bins are called *honest*.

In this work, we study a fault-tolerant version of an algorithm commonly known as the Power of Two Choices, called *Byzantine-Tolerant Power of Two Choices* or *BT_PoTC* for short. It proceeds as follows. At each time t, upon arrival of ball t, two bins are picked uniformly at random. If both bins are honest, then the ball lands in the bin with fewer balls (ties are solved arbitrarily). If at least one bin is faulty, then a malicious adversary decides which bin receives the ball and is capable to enforce it on the algorithm. See also Algorithm 1.

This is formally done as follows: the adversary controlling the faulty bin could provide the algorithm with an arbitrary number (instead of the actual number of balls in that bin), which is then compared with the number provided by the other bin (potentially honest) and the minimum is selected. We assume that the adversary has the following attributes:

- it is *Byzantine*, i.e., it can provide an arbitrary answer about the number of balls in the selected faulty bin(s),
- it is *computationally unbounded*,
- it *knows the algorithm* BT_PoTC,

– it is *strongly adaptive*, i.e., it knows which bins have been randomly selected so far and which bins will be selected in the future and can use this knowledge to provide a potentially malicious answer to the algorithm about the number of balls in the currently selected faulty bin(s).

We measure the maximum number of balls in honest bins, and the goal is to have as precise asymptotic upper bound on this number as possible. This estimate is supposed to hold *with high probability*, which denotes probability $1-o(1)$. Part of the analysis is done with high probability that is polynomially close to 1, i.e., $1 - n^{-c}$ for some constant $c > 0$. Throughout the paper we use asymptotic notation of $O(\cdot)$ and $o(\cdot)$ to upper bound formulas and (complementary) probabilities: $f(x) = O(g(x))$ means that $f(x) \leq c \cdot g(x)$ for some sufficiently large constant c, while $f(x) = o(g(x))$ means that $f(x)/g(x)$ converges to 0 when x goes to infinity.

Challenges. At first glance it may seem that one can directly apply the results about PoTC process to this problem. E.g., we split the balls into those that choose between 2 honest bins and those that choose between an honest and a faulty bin. The balls that choose between 2 honest bins behave like PoTC process, while the balls that choose between an honest and a faulty bin behave like the standard balls-into-bins process. However, those two processes are dependent on each other – the balls added by the latter (standard balls-into-bins) process affect decisions made during the power-of-two-choices process. This difficulty is not easily worked around. For example, a related problem with similar dependencies required a completely different, non-trivial method of analysis [26].

2.2 Our Results

First we present a result for $m > \frac{3}{2}n \log n$. Note that this result covers any number of faulty bins $f < n$. The proof is given in Sect. 5.

Theorem 1. *For any values of parameters $m, \delta \in (0,1)$ and $c > 1$ such that $\frac{2m\delta^2}{3n} > c \log n$, there is no honest bin with more than $(1 + \delta)\frac{2m}{n}$ balls in it with probability at least $1 - n^{1-c}$.*

E.g., with $m > 9n \log n$, $\delta = 1/2$ and $c = \frac{3}{2}$, we get the following corollary.

Corollary 1. *Let $m > 9n \log n$. Then, there is no honest bin with more than $\frac{3m}{n}$ balls in it with probability at least $1 - n^{-1/2}$.*

Now we present a result complimentary with Theorem 1, i.e., a result that works for $m \leq \frac{3}{2}n \log n$. This is the main result of this article, and its proof is given in Sect. 3.

Theorem 2. *Let $m = O(n \log n)$ and $1 \leq f < \frac{n^2}{2em}$. Let ϕ be such that the number of honest bins with at least ϕ balls is at most $\frac{n^2}{4em}$ with probability at* least $1 - n^{-2}$. Then, there is no honest bin with more than

$$\phi + \log \log \frac{n^2}{2emf} + \log_{\frac{n^2}{2emf}} \frac{2f}{\sqrt{f^2 + 8\frac{n^2}{m} \log n} - f} + 1$$

balls in it, with probability at least $1 - o(1)$.

Next we present corollaries with clearer formulas. We have:

$$O\left(\log_{\frac{n^2}{2emf}} \frac{f}{\sqrt{f^2 + \frac{n^2}{m} \log n} - f}\right) \leq O\left(\log_{\frac{n^2}{2emf}} \frac{f \cdot \left(\sqrt{f^2 + \frac{n^2}{m} \log n} + f\right)}{f^2 + \frac{n^2}{m} \log n - f^2}\right)$$

$$\leq O\left(\log_{\frac{n^2}{2emf}} \frac{f^2 + f\sqrt{\frac{n^2}{m} \log n}}{\frac{n^2}{m} \log n}\right) \leq O\left(\log_{\frac{n^2}{2emf}} \left(\frac{f^2 m}{n^2 \log n} + \sqrt{\frac{f^2 m}{n^2 \log n}}\right)\right)$$

$$\leq O\left(\max\left\{\log_{\frac{n^2}{2emf}} \left(\frac{2emf}{n^2} \cdot \frac{f}{2e \log n}\right), \log_{\frac{n^2}{2emf}} \left(\sqrt{\frac{2emf}{n^2} \cdot \frac{f}{2e \log n}}\right)\right\}\right)$$

$$\leq O\left(\log_{\frac{n^2}{2emf}} \frac{f}{2e \log n}\right). \tag{1}$$

We start with a corollary that is not yet fully optimized, and finish with the one in optimal form.

Corollary 2. *Let* $1 \leq f < \frac{n^2}{2em}$*. Then, there is no honest bin with more than*

$$O\left(\left(\frac{m}{n}\right)^2 + \log \log \frac{n^2}{2emf} + \log_{\frac{n^2}{2emf}} \frac{f}{2e \log n}\right)$$

balls in it, with probability at least $1 - o(1)$*.*

Proof. Let $\phi = \frac{4em^2}{n^2}$. Note that the number of honest bins with at least ϕ balls is at most $m/\phi = \frac{n^2}{4em}$ with probability 1. Therefore, we can use Theorem 2 to obtain the corollary, bounding the last logarithm as in Eq. (1).

Corollary 3. *Let* $1 \leq f < \frac{n^2}{2em}$ *and* $m = O(n)$*. There is no honest bin with more than*

$$O\left(\log \log \frac{n^2}{mf} + \log_{\frac{n^2}{2emf}} \frac{f}{2e \log n}\right)$$

balls in it, with probability at least $1 - o(1)$*.*

Proof. The result follows directly from Corollary 2 and bound $m = O(n)$.

The following corollary uses the fact that the number of honest bins with at least $\phi = \Theta(m/n)$ balls in them is at most $\beta_\phi = \frac{n^2}{4em}$, for $m > en$ and $f < \frac{n^2}{2em}$. This is formally shown in Lemma 7 in Sect. 4.

Corollary 4. *Let $1 \leq f < \frac{n^2}{2em}$, where $m \geq en$ and $m = O(n \log n)$. There is no honest bin with more than*

$$O\left(\frac{m}{n} + \log\log\frac{n^2}{mf} + \log_{\frac{n^2}{2emf}}\frac{f}{2e\log n}\right)$$

balls in it, with probability at least $1 - o(1)$.

Proof. Let $\phi = \frac{cem}{n}$ for some $c > 2e$. According to Lemma 7, the number of honest bins with at least ϕ balls is at most $\beta_\phi = \frac{n^2}{4em}$ (which is denoted by event $\mathcal{E}_\phi$) with probability $1 - n^{-c}$. Therefore, we can use Theorem 2 to obtain the corollary, bounding the last logarithm as in Eq. (1). (The detailed definitions of β_ϕ and $\mathcal{E}_\phi$ are given below in Subsect. 3.1.)

Corollary 5. *Let $1 \leq f < n \cdot \max\{\frac{1}{12e\log n}, \frac{n}{2em}\}$, where $m \geq n$. There is no honest bin with more than*

$$O\left(\frac{m}{n} + \log\log\frac{n^2}{mf} + \log_{\frac{n^2}{2emf}}\frac{f}{2e\log n}\right) \leq O\left(\frac{m}{n} + \log\log\frac{n}{f} + \log f\right)$$

balls in it, with probability at least $1 - o(1)$.

Proof. For $m \geq 6n\log n$, the formula is just $O(m/n)$, by Theorem 1. Consider $en \leq m < 6n\log n$. Observe that $\frac{n}{6f\log n} < \frac{n^2}{mf} \leq \frac{n}{ef}$, therefore we upper bound the formula in Corollary 4 as follows:

$$O\left(\log\log\frac{n^2}{mf} + \log_{\frac{n^2}{2emf}}\frac{f}{2e\log n}\right) \leq O\left(\log\log\frac{n}{f} + \frac{\log\frac{f}{2e\log n}}{\log\frac{n}{f\log n}}\right)$$

$$\leq O\left(\log\log\frac{n}{f} + \log f\right).$$

Note that all the results for $(1 + \beta)$-choice process by Peres et al. [26] are asymptotically at least $\log n$ w.h.p. On the other hand, our analysis provides a tighter bound on the maximum load of a bin for some parameters, e.g., for $m = O(n)$ and $f = o(n)$ we can get from Corollary 5 that the maximum load is $O(\log\log(n/f) + \log f)$ w.h.p.

3   Probabilistic Analysis of Power of Two Choices Under Byzantine Faults—Proof of Theorem 2

3.1   Overview of the Analysis

Our analysis is inspired by [5]. While the general idea of the proof still applies, introduction of faulty bins controlled by a strong Byzantine adversary called for additional more careful analysis to be made. Another challenge lies in finding appropriate values of parameters that take into account malicious influence of faulty bins that may interact with honest bins (when selected together by the random process).

The main effort is in analyzing events $\mathcal{E}_i$, representing that there exist at most β_i honest bins with at least i balls in each, for a range of positive integers i. Values β_i will be defined later.

The starting point of our analysis is for some carefully chosen threshold parameter $i = \phi$ such that the event $\mathcal{E}_\phi$ is guaranteed to happen with high probability. (We prove later in Sect. 4 and Lemma 7 that such $\phi = O(m/n)$, achieving asymptotically perfect allocation threshold, exists and thus could be substituted in the formula of Theorem 2 to obtain an efficient bound in Corollary 4.) As we move to larger parameters $i > \phi$, the events $\mathcal{E}_i$ represent situations where there are fewer and fewer honest bins with more and more balls in them. Eventually, we reach a final event $\mathcal{E}_k$ such that there is less than 1 bin with at least k balls in it, for some value of k to be derived in the analysis; meaning – there are no such bins.

We analyze the probability that each of the events $\mathcal{E}_i$ occurs. The starting point $\mathcal{E}_\phi$ is guaranteed with high probability. We want to show that all the events $\mathcal{E}_i$ occur with high probability until *after* the event we need, $\mathcal{E}_k$. However, the approach we use can only show that $\mathcal{E}_{k-2}$ occurs with high probability. Still, having event $\mathcal{E}_{k-2}$, we can manually show that the number of bins with at least k balls is less than 1.

The proof of Theorem 2 is divided into four steps. Recall that in this analysis we assume $m = O(n \log n)$, while the complementary case is analyzed in Sect. 5 and Theorem 1.

In Step 1, we define critical sequences of parameters β_i and events $\mathcal{E}_i$, taking into account the impact of faulty bins. We obtain the initial estimate of the probability $P(\neg \mathcal{E}_{i+1} \mid \mathcal{E}_i)$. Here $\neg \mathcal{E}_{i+1}$ denotes the event complementary to $\mathcal{E}_{i+1}$, i.e., the event that $\mathcal{E}_{i+1}$ does not hold.

In Step 2, we find that, with an additional assumption for the threshold parameter i, event $\mathcal{E}_i$ occurs with high probability.

In Step 3, we calculate for which values of i the assumption considered in Step 2 actually holds. As long as the assumption holds, the good events $\mathcal{E}_i$ hold too. In this step, the impact of faulty bins on the analysis, and their number f in the formulas, is the most challenging to deal with.

In Step 4, we use the highest event $\mathcal{E}_i$ in the sequence that we could prove to hold with high probability. Instead of continuing to show that further good events in the sequence hold with high probability, we prove directly that the number of bins with many balls is less than 1, which concludes the proof of Theorem 2.

3.2   Worst-Case Adversary

First, we make an observation about the worst case adversary.

Consider a greedy adversary $GREEDY$ that, whenever a ball chooses between a faulty bin and an honest bin, makes the ball land in the honest bin. We claim that such a greedy adversary is the worst-case adversary even among the strongly adaptive adversaries.

Table 2. Table of problem parameters and most important notation in the analysis.

n	The number of bins
m	The total number of balls that will be thrown into the bins
f	The number of faulty bins
e	Euler's number, $e = 2.71828\ldots$
i	Parameter of the considered sequence of events $\{\mathcal{E}_i\}$, in which i corresponds to a threshold on the load of the bins
t	Actually considered time step and ball number
v_i^t	The number of bins that are honest and have at least i balls at time t
h^t	The number of balls in the bin in which the t-th ball has landed at time t (measured after it has landed)
u_i^t	The number of balls in honest bins that have height at least i at time t
β_i	Upper bound on the number of honest bins with at least i balls
$\mathcal{E}_i$	An event such that $v_i^m \leq \beta_i$
ϕ	The starting point for the considered sequence of events $\mathcal{E}_i$. $\beta_\phi = \frac{n^2}{4em}$ and event $\mathcal{E}_\phi$ occurs with high probability
Y_i^t	$Y_i^t = 1$ if the t-th ball landed in an honest bin and $h^t \geq i+1$ and $v_i^{t-1} \leq \beta_i$; otherwise, $Y_i^t = 0$
p_i	Upper bound on probability that a ball landed in a bin with at least i balls in it. We have $p_i = \left(\frac{\beta_i}{n}\right)^2 + \frac{f}{n} \cdot \frac{\beta_i}{n}$
i_1	$\phi + i_1$ is the smallest index of β_i such that $\beta_{\phi+i_1} < f$
i^*	The smallest value such that $mp_{\phi+i^*} < 2\ln n$
$B(m,p)$	Binomial distribution with m trials and probability p

Lemma 1. *GREEDY is the worst-case adversary for maximizing load.*

Proof Idea. Consider any adversary ADV. Consider any bin b and any sequence of exchanges S. Let $l_b^A(S,t)$ be the load of bin b at time t under adversary ADV for the sequence of exchanges S. A simple inductive argument over time t shows that – for every honest bin b, all sequences of exchanges S and at any time t – we have $l_b^{ADV}(S,t) \leq l_b^{GREEDY}(S,t)$.

The full proof is deferred to the full version of the paper.

From now on, we analyze the protocol against the *GREEDY* adversary. According to Lemma 1, the results will also apply to an arbitrary adversary.

3.3 Step 1: Initial Estimate of $P(\neg\mathcal{E}_{i+1} \mid \mathcal{E}_i)$

The goal of the first step of the analysis is to obtain the initial estimate on $P(\neg\mathcal{E}_{i+1} \mid \mathcal{E}_i)$ for any integer $i > 0$.

In our analysis we will use the following two standard facts. Let $B(m,p)$ denote the binomial distribution with m trials and probability of success p.

The following stochastic dominance has been proved and used before in various probabilistic analyses, c.f. [14, in the analysis of Lemma 2].

Fact 1. *Let $X_1, X_2, \ldots, X_m$ be a sequence of random variables with values in an arbitrary domain, and let $Y_1, Y_2, \ldots, Y_m$ be a sequence of binary random variables, with the property that $Y_i = Y_i(X_1, \ldots, X_i)$.*
If $P(Y_i = 1 \mid X_1, \ldots, X_{i-1}) \leq p$ then $P\left(\sum_{i=1}^{m} Y_i \geq k\right) \leq P(B(m, p) \geq k)$.

The proof of the next fact (Chernoff Bounds) can be found, e.g., in [19].

Fact 2 (Chernoff Bounds). *For $a \geq mp$, $P(B(m, p) \geq a) \leq \left(\frac{mp}{a}\right)^a e^{a-mp}$. If $a = (1 + \delta)mp$, for some $\delta \in (0, 1)$, then $P(B(m, p) \geq a) \leq e^{-mp\delta^2/3}$.*

We start the technical analysis by defining three crucial notations: $v_i^t, \mathcal{E}_i, \beta_i$. Here, an integer parameter i denotes the considered threshold on the number of balls in bins, and t stands for the order number of analyzed ball. The list of all important notations and their meaning is given in Table 2. Let v_i^t be the number of bins that are honest and have at least i balls at time t, i.e., after considering and placing t balls. Let $\mathcal{E}_i$ be an event such that $v_i^m \leq \beta_i$, for given parameters β_i that will be determined later. Let index ϕ be such that $\mathcal{E}_\phi$ occurs with probability at least $1 - n^{-2}$.

Our goal is to find parameters β_i such that $\mathcal{E}_i$ holds with high probability and $\beta_i < 1$ for as small parameter i as possible.

Let the *height of ball t*, denoted h^t, be the number of balls in the bin in which the t-th ball has landed (measured after it has landed).

Let Y_i^t be a random variable such that $Y_i^t = 1$ if the t-th ball landed in an honest bin and $h^t \geq i + 1$ and $v_i^{t-1} \leq \beta_i$; otherwise, $Y_i^t = 0$. Intuitively, $Y_i^t = 1$ denotes the event such that a ball landed in a bin with already many balls in it. The additional constraint $v_i^{t-1} \leq \beta_i$ helps to carry on the analysis leading to finding a small value i for which $\beta_i < 1$ and $\mathcal{E}_i$ holds with high probability.

Let ω_j be a random variable equal to the bin number where the j-th ball has landed. We will upper bound the probability $P(Y_i^t = 1 \mid \omega_1, \omega_2, \ldots, \omega_{t-1})$. Note that $Y_i^t = 1$ only if one of the two cases occurs:

- the protocol picks two honest bins with at least i balls in them to choose from, which takes place with probability at most $\left(\frac{\beta_i}{n}\right)^2$, or
- if one chosen bin is faulty and the other is honest, with at least i balls, which happens with probability at most $\frac{f}{n} \cdot \frac{\beta_i}{n}$.

Therefore, we get the following inequality

$$P(Y_i^t = 1 \mid \omega_1, \omega_2, \ldots, \omega_{t-1}) \leq \left(\frac{\beta_i}{n}\right)^2 + \frac{f}{n} \cdot \frac{\beta_i}{n}. \tag{2}$$

Let $p_i = \left(\frac{\beta_i}{n}\right)^2 + \frac{f}{n} \cdot \frac{\beta_i}{n}$.

For parameters i and t, let u_i^t be the number of balls in honest bins that have height at least i at time t. Now we will relate u_{i+1}^m with random variables Y_i^t.

Note that, if $\mathcal{E}_i$ holds, then $v_i^{t-1} \le \beta_i$ for all t. In that case $\sum_{t=1}^{m} Y_i^t = u_{i+1}^m$. Therefore, $P(\sum_{t=1}^{m} Y_i^t \ge k \mid \mathcal{E}_i) = P(u_{i+1}^m \ge k \mid \mathcal{E}_i)$ for any parameter k.

Now we will estimate $P(v_{i+1}^m \ge k \mid \mathcal{E}_i)$ for some values of parameter k. Note that $v_i^t \le u_i^t$ for all i and t, and therefore

$$P(v_{i+1}^m \ge k \mid \mathcal{E}_i) \le P(u_{i+1}^m \ge k \mid \mathcal{E}_i) = P\left(\sum_{t=1}^{m} Y_i^t \ge k \mid \mathcal{E}_i\right) \le \frac{P(B(m, p_i) \ge k)}{P(\mathcal{E}_i)},$$

where the last inequality follows from Fact 1.

For $k = \beta_{i+1}$ we get

$$P(v_{i+1}^m \ge \beta_{i+1} \mid \mathcal{E}_i) \le \frac{P(B(m, p_i) \ge \beta_{i+1})}{P(\mathcal{E}_i)}. \tag{3}$$

We can use Fact 2 with $a = \beta_{i+1}$, where $\beta_{i+1} = emp_i$ and get

$$P(v_{i+1}^m \ge \beta_{i+1} \mid \mathcal{E}_i) \le \left(\frac{1}{e}\right)^{emp_i} e^{(e-1)mp_i} = e^{-mp_i}.$$

Note that the event $(v_{i+1}^m > \beta_{i+1})$ is equivalent to $\neg\mathcal{E}_{i+1}$. Therefore,

$$P(\neg\mathcal{E}_{i+1} \mid \mathcal{E}_i) \le e^{-mp_i}. \tag{4}$$

3.4   Step 2: Analysis of Event $\mathcal{E}_i$ for i Corresponding to $p_i \ge \frac{2\ln n}{m}$

In Step 2 we use the estimate obtained in the previous step to establish that events $\mathcal{E}_i$ hold with high probability (polynomially close to 1) for some parameters i. We will analyze later in Step 3 which values of i hold this property.

Our goal now is to show that $P(\neg\mathcal{E}_{i+1} \mid \mathcal{E}_i) \le 1/n^2$. Assume we have some ϕ such that $\mathcal{E}_\phi$ occurs with high probability for $\beta_\phi = \frac{n^2}{4em}$ – existence of such ϕ will be shown later in Lemma 7. If this is the case, then $\mathcal{E}_i$ holds for all $\phi \le i \le \phi + i^*$ for some i^*, also with high probability.

Suppose first that $mp_i \ge 2\ln n$. Then $e^{-mp_i} \le 1/n^2$ and, by Eq. 4, we have $P(\neg\mathcal{E}_{i+1}|\mathcal{E}_i) \le 1/n^2$. Then, we could follow a simple inductive argument to prove Lemma 2 below. The proof is deferred to the full version of the paper.

Lemma 2. *If $mp_i \ge 2\ln n$ holds for consecutive parameters i starting from ϕ, then $P(\neg\mathcal{E}_\phi \vee \neg\mathcal{E}_{\phi+1} \vee \cdots \vee \neg\mathcal{E}_{\phi+i}) \le (i+1)/n^2$.*

3.5   Step 3: Finding i^* – The Minimum i Satisfying $p_{\phi+i} < \frac{2\ln n}{m}$

The previous step proved that events $\mathcal{E}_i$ hold with high probability for i such that $mp_i \ge 2\ln n$. In this step we answer the question: What are the values of i such that the sought property $mp_i \ge 2\ln n$ holds? To address it, we will now find the minimum i^* such that $mp_{\phi+i^*} < 2\ln n$.

Lemma 3. *The smallest value i^* such that $mp_{\phi+i^*} < 2\ln n$ satisfies*

$$i^* \le \log_2 \log_{\frac{2em\beta_\phi}{n^2}} \frac{2emf}{n^2} + \log_{\frac{n^2}{2emf}} \frac{2f}{-f + \sqrt{f^2 + 8\frac{n^2}{m}\log n}}. \tag{5}$$

In order to prove Lemma 3, we introduce Lemmas 4–6. The first of them, Lemma 4, can be proved by induction on parameter i. The proof of Lemma 4 is deferred to the full version of the paper.

Lemma 4. $\beta_{\phi+i}$ *is monotonically decreasing when parameter i increases.*

Lemma 5. *Let i_1 be the smallest i such that $\beta_{\phi+i} < f$. It holds that $i_1 \le \log_2 \log_{\frac{2em\beta_\phi}{n^2}} \frac{2emf}{n^2}$.*

Proof. Consider values of parameter j such that $\beta_j \ge f$.

We get that $p_i \le 2\left(\frac{\beta_j}{n}\right)^2$ and the following recursive equation $\beta_{j+1} \le em \cdot 2\left(\frac{\beta_j}{n}\right)^2$. Therefore, $\beta_{j+i} \le \frac{n^2}{2em}\left(\frac{2em\beta_j}{n^2}\right)^{2^i}$ for i such that $\beta_j \ge f$, $\beta_{j+1} \ge f$, ..., $\beta_{j+i-1} \ge f$. In particular, we get

$$\beta_{\phi+i} \le \frac{n^2}{2em}\left(\frac{2em\beta_\phi}{n^2}\right)^{2^i} \tag{6}$$

for i such that $\beta_{\phi+i-1} \ge f$ (it follows that $\beta_\phi \ge f$, $\beta_{\phi+1} \ge f$, ..., $\beta_{\phi+i-2} \ge f$ according to Lemma 4).

Now we look at what values of i are such that $\beta_{\phi+i-1} \ge f$. We can look for the smallest i_1 such that the opposite occurs, i.e., $\beta_{\phi+i_1} < f$ (recall that we only consider $f \ge 1$ – see the statement of Theorem 2). Recall our choice $\beta_\phi = \frac{n^2}{4em} < \frac{n^2}{2em}$. Note the base of the power in Inequality 6 is $\frac{2em\beta_\phi}{n^2} < 1$. Therefore, we can find that $\beta_{\phi+i} < f$ for any $i \ge \log_2 \log_{\frac{2em\beta_\phi}{n^2}} \frac{2emf}{n^2}$ (note that $\log_{\frac{2em\beta_\phi}{n^2}} \frac{2emf}{n^2} > 0$, since $\frac{2em\beta_\phi}{n^2} = \frac{1}{2}$ and $\frac{2emf}{n^2} < 1$ due to our assumption that $f < \frac{n^2}{2em}$ in the statement of Theorem 2). It follows that the smallest i_1 such that $\beta_{\phi+i_1} < f$ satisfies: $i_1 \le \log_2 \log_{\frac{2em\beta_\phi}{n^2}} \frac{2emf}{n^2}$.

Note that $\beta_{\phi+i_1}$ is the first β such that $\beta_{\phi+i_1} < f$. $\qquad\square$

Lemma 6. *For $i \ge 0$, $\beta_{\phi+i_1+i} \le \left(\frac{2emf}{n^2}\right)^i \beta_{\phi+i_1}$.*

Proof. According to Lemma 5 and Lemma 4, we have $\beta_j < f$ for all $j \ge \phi + i_1$. We get $\beta_{j+1} < em\left(\frac{f}{n} \cdot \frac{\beta_j}{n} + \frac{f}{n} \cdot \frac{\beta_j}{n}\right) = \frac{2emf}{n^2}\beta_j$. Therefore, $\beta_{j+i} \le \left(\frac{2emf}{n^2}\right)^i \beta_j$ for i, j such that $\beta_j < f$, $\beta_{j+1} < f$, ..., $\beta_{j+i-1} < f$. In particular, for any $i \ge 0$,

$$\beta_{\phi+i_1+i} \le \left(\frac{2emf}{n^2}\right)^i \beta_{\phi+i_1}. \tag{7}$$

Now we are ready to prove Lemma 3.

Proof (Proof of Lemma 3).

First, recall that $\beta_{i+1} = emp_i$ and $p_i = \left(\frac{\beta_i}{n}\right)^2 + \frac{f}{n} \cdot \frac{\beta_i}{n}$ for all i.
We are interested in i such that $mp_{\phi+i} < 2\ln n$, that is to say

$$\left(\frac{\beta_{\phi+i}}{n}\right)^2 + \frac{f}{n} \cdot \frac{\beta_{\phi+i}}{n} < \frac{2\ln n}{m}. \tag{8}$$

We can treat this inequality as a quadratic inequality with variable $\beta_{\phi+i}$ to obtain that $mp_{\phi+i} < 2\ln n$ holds if

$$\beta_{\phi+i} \in \left(\frac{-f - \sqrt{f^2 + 8\frac{n^2}{m}\log n}}{2}, \frac{-f + \sqrt{f^2 + 8\frac{n^2}{m}\log n}}{2}\right). \tag{9}$$

Because of the monotonicity of $\beta_{\phi+i}$, as in Lemma 4, we are interested in the smallest i such that

$$\beta_{\phi+i} < \frac{-f + \sqrt{f^2 + 8\frac{n^2}{m}\log n}}{2}. \tag{10}$$

Based on Lemma 6 and the facts that $\beta_{\phi+i_1} < f$ (see Lemma 5) and $f < \frac{n^2}{2em}$ (assumption in Theorem 2), we get that $\beta_{\phi+i_1+i} < \frac{-f + \sqrt{f^2 + 8\frac{n^2}{m}\log n}}{2}$ holds for any $i \geq \log_{\frac{2emf}{n^2}} \frac{-f + \sqrt{f^2 + 8\frac{n^2}{m}\log n}}{2f}$. It follows that $i^* \leq i_1 + \log_{\frac{n^2}{2emf}} \frac{2f}{-f + \sqrt{f^2 + 8\frac{n^2}{m}\log n}}$.

3.6   Step 4: Finalizing the Analysis - Beyond Threshold Parameter $\phi + i^*$

Now that we have the value of i^* and an analysis of events $\mathcal{E}_i$, for $i \leq i^*$, we can find the probability that there exists an honest bin with more than $\phi + i^*$ balls in it. The result is that there are no honest bins with at least $\phi + i^* + 2$ balls, with high probability of $1 - o(1)$.

Recall that $\beta_{i+1} = emp_i$ for all i. In particular, since $mp_{\phi+i^*} < 2\ln n$ we have:

$$\beta_{\phi+i^*+1} = emp_{\phi+i^*} < 2e\ln n. \tag{11}$$

Consider $v^m_{\phi+i^*+1}$. We get

$$P(v^m_{\phi+i^*+1} \geq 2e\ln n \mid \mathcal{E}_{\phi+i^*}) \leq \frac{P\left(B(m, p_{\phi+i^*}) \geq 2e\ln n\right)}{P(\mathcal{E}_{\phi+i^*})} \tag{12}$$

$$< \frac{P\left(B\left(m, \frac{2\ln n}{m}\right) \geq 2e\ln n\right)}{P(\mathcal{E}_{\phi+i^*})} \leq \frac{1}{n^2 P(\mathcal{E}_{\phi+i^*})}, \tag{13}$$

where the last inequality follows from Fact 2.

Finally, we can bound $P(u^m_{\phi+i^*+2} < 1)$ in the following derivation.

$$P(u^m_{\phi+i^*+2} \geq 1 \mid v^m_{\phi+i^*+1} < 2e\ln n) \leq \frac{P\left(B\left(m, \left(\frac{2e\ln n}{n}\right)^2 + \frac{f}{n}\frac{2e\ln n}{n}\right) \geq 1\right)}{P(v^m_{\phi+i^*+1} < 2e\ln n)} \quad (14)$$

$$\leq \frac{m\left(\left(\frac{2e\ln n}{n}\right)^2 + \frac{f}{n}\frac{2e\ln n}{n}\right)}{P(v^m_{\phi+i^*+1} < 2e\ln n)}. \quad (15)$$

Fact 3 *For any events A, B, we have: $P(A) \leq P(A|B) \cdot P(B) + P(\neg B)$.*

Using Fact 3 twice, we get

$$P(u^m_{\phi+i^*+2} \geq 1) \leq P(u^m_{\phi+i^*+2} \geq 1 \mid u^m_{\phi+i^*+1} < 2e\ln n) \cdot P(u^m_{\phi+i^*+1} < 2e\ln n)$$
$$+ P(u^m_{\phi+i^*+1} \geq 2e\ln n)$$
$$\leq P(u^m_{\phi+i^*+2} \geq 1 \mid u^m_{\phi+i^*+1} < 2e\ln n) \cdot P(u^m_{\phi+i^*+1} < 2e\ln n)$$
$$+ P(u^m_{\phi+i^*+1} \geq 2e\ln n \mid \mathcal{E}_{\phi+i^*}) \cdot P(\mathcal{E}_{\phi+i^*}) + P(\neg\mathcal{E}_{\phi+i^*}).$$

Using Eqs. 13 and 15, we get

$$P(u^m_{\phi+i^*+2} \geq 1) \leq m\left(\left(\frac{2e\ln n}{n}\right)^2 + \frac{f}{n}\frac{2e\ln n}{n}\right) + \frac{1}{n^2} + P(\neg\mathcal{E}_{\phi+i^*})$$
$$\leq m\left(\left(\frac{2e\ln n}{n}\right)^2 + \frac{f}{n}\frac{2e\ln n}{n}\right) + \frac{1}{n^2} + P(\neg\mathcal{E}_{\phi+1} \vee \cdots \vee \neg\mathcal{E}_{\phi+i^*})$$
$$\leq O\left(\frac{\ln^3 n}{n} + \frac{f\ln^2 n}{n}\right) + \frac{1}{n^2} + \frac{i^*+1}{n^2},$$

where the last line follows from Lemma 2 and the assumed bound $m = O(n\log n)$.

For $f = o\left(\frac{n}{\ln^2 n}\right)$, we get

$$P(u^m_{\phi+i^*+2} \geq 1) = o(1). \quad (16)$$

Finally, this means that $u^m_{\phi+i^*+2} < 1$ occurs with probability $1 - o(1)$. Therefore, there are no balls with height at least $\phi + i^* + 2$ with high probability, which also means that there are no honest bins with at least $\phi + i^* + 2$ balls in them which completes the proof of Theorem 2.

4   Case $m \geq en$ – Setting Up Initial Value of ϕ

Lemma 7. *Assume $m \geq en$ and $f < \frac{n^2}{2em}$. Let $\phi = c\frac{em}{n}$ and $\beta_\phi = \frac{n^2}{4em}$, for some constant $c > 2e$ to be defined later in the proof. Then $\mathcal{E}_\phi$ holds with probability at least $1 - n^{-c}$.*

Proof. We first compute an upper bound on the probability of the complementary event, that is, that there is a subset of $\beta_\phi + 1$ of honest bins containing at least ϕ balls each. It is upper bounded by a union of events, parameterized by any subset B of honest bins, that each bin in B has at least ϕ balls. This can be further upper bounded by multiplying the number of such sets B, $\binom{n-f}{\beta_\phi+1}$, by the product of upper bounds on the probability that an y-th bin in B, where $0 \leq y \leq \beta_\phi$ was randomly selected at least ϕ times when allocating the remaining at least $m - y\phi$ balls, $\binom{m-y\phi}{\phi}\left(\frac{2-\frac{1}{n}}{n}\right)^\phi$. This results in the following upper bound formula and its further transformation:

$$\binom{n-f}{\beta_\phi+1} \prod_{y=0}^{\beta_\phi} \left(\binom{m-y\phi}{\phi} \left(\frac{2-\frac{1}{n}}{n} \right)^\phi \right)$$

$$= \left(\frac{n-f}{\frac{n^2}{4em}+1} \right)^{\frac{n^2}{4em}} \prod_{y=0}^{c\frac{em}{n}} \left(\binom{m-yc\frac{em}{n}}{c\frac{em}{n}} \left(\frac{2-\frac{1}{n}}{n} \right)^{c\frac{em}{n}} \right)$$

$$\leq 2^{n-f} \left(\left(\frac{me}{c\frac{em}{n}} \right)^{c\frac{em}{n}} \left(\frac{2}{n} \right)^{c\frac{em}{n}} \right)^{\frac{n^2}{4em}+1} = 2^{n-f} \left(\frac{2}{c} \right)^{c\frac{n}{4}+c\frac{em}{n}} \leq 2^{n-f} \left(\frac{2}{c} \right)^{c\frac{n}{4}},$$

where the first inequality follows from bounds $2 - \frac{1}{n} \leq 2$, $\binom{m-yc\frac{em}{n}}{c\frac{em}{n}} \leq \binom{m}{c\frac{em}{n}}$ for $0 \leq y \leq \beta_\phi$, and $\binom{n}{x} \leq (ne/x)^x$; the second inequality follows from the assumption $c > 2e$ and from the monotonicity of the exponent function. Next, observe that

$$2^{n-f} \left(\frac{2}{c} \right)^{\frac{cn}{4}} = \exp\left((n-f)\ln 2 - \frac{cn}{4}\ln\frac{c}{2} \right) < n^{-c}$$

for any c satisfying $\frac{cn}{4}\ln\frac{c}{2} - c\ln n > (n-f)\ln 2$, for instance, for $c \geq 2e^4$ and any $n \geq 3$,[2] or for any $c > 2e$ and any sufficiently large n.

5   Case $m > \frac{3}{2}n\log n$ – Proof of Theorem 1

In this section we prove Theorem 1. Consider an honest bin j. For any $t \leq m$, let X_t be a random variable equal to 1 if ball t lands in bin j, and equal to 0 otherwise; let Y_t be equal to 1 if bin j is randomly selected at time t, and equal to 0 otherwise. Observe that:

$$\forall_{t \leq m} X_t \leq Y_t \qquad \text{and} \qquad \forall_{t \leq m} P(X_t = 1) \leq P(Y_t = 1) = \frac{2}{n}.$$

---

[2] Observe that the power of two choices makes use of randomness only for $n \geq 3$ bins.

Let $\mu_X = \mathbf{E}\left[\sum_{t \le m} X_t\right]$ and $\mu_Y = \mathbf{E}\left[\sum_{t \le m} Y_t\right] = \frac{2m}{n}$. By Fact 2, for any $\delta \in (0,1)$:

$$P\left(\sum_{t \le m} X_t > (1+\delta)\mu_Y\right) \le P\left(\sum_{t \le m} Y_t > (1+\delta)\mu_Y\right) < e^{\frac{-\mu_Y \delta^2}{3}} = e^{-\frac{2m\delta^2}{3n}}.$$

Assuming that $\frac{2m\delta^2}{3n} > c \log n$, for some $c > 1$, which holds for some $\delta \in (0,1)$ and some $c > 1$ as long as $m > \frac{3}{2}n \log n$, we get $P\left(\sum_{t \le m} X_t > (1+\delta)\frac{2m}{n}\right) < n^{-c}$.

Consequently, by applying the union bound over all bins, the probability that there is a bin with more than $(1+\delta)\frac{2m}{n}$ in it is at most n^{1-c}.

6   Discussion and Open Problems

In this work, we provided an analysis of the efficiency of the popular load balancing rule – the Power of Two Choices – in the system that some bins are controlled by a malicious adversary. For $m = O(n \log n)$, we showed that the maximum load on any honest bin has a logarithmic dependence on the number of faulty nodes f.

There are several open questions to be explored. One open question is related to the tightness of our bounds. It will be interesting to obtain matching lower bounds related to the number of faults f, and also m and n. It will also be interesting to explore the graphical case. In the case without faults, the problem has been studied for regular graphs and the maximum load depends on the node degree. It will be interesting to explore the dependence of the load on the number of faulty graph nodes f and the degree of the regular graph.

Another open problem is the lower bound on the minimum load of a bin as well as the gap between the maximum and minimum loads. These bounds would be used to prove the Fault-Tolerant Phase Clock works in population protocols.

Acknowledgments. This study was partly supported by the National Science Center, Poland (NCN), grant 2020/39/B/ST6/03288, and by the (US) National Science Foundation grant CNS-2131538.

Disclosure of Interests. The authors have no competing interests to declare that are relevant to the content of this article.

References

1. Abdulazeez, M., Garncarek, P., Kowalski, D.R., Wong, P.W.H.: Lightweight robust framework for workload scheduling in clouds. In: IEEE International Conference on Edge Computing, EDGE 2017, pp. 206–209. IEEE Computer Society (2017)

2. Alistarh, D., Aspnes, J., Gelashvili, R.: Space-optimal majority in population protocols. In: Proceedings of the 2018 Annual ACM-SIAM Symposium on Discrete Algorithms (SODA), pp. 2221–2239 (2018)
3. Angluin, D., Aspnes, J., Diamadi, Z., Fischer, M.J., Peralta, R.: Computation in networks of passively mobile finite-state sensors. Distrib. Comput. **18**(4), 235–253 (2006)
4. Angluin, D., Aspnes, J., Eisenstat, D.: A simple population protocol for fast robust approximate majority. Distrib. Comput. **21**(2), 87–102 (2008)
5. Azar, Y., Broder, A.Z., Karlin, A.R., Upfal, E.: Balanced allocations. SIAM J. Comput. **29**(1), 180–200 (1999)
6. Bansal, N., Feldheim, O.N.: The power of two choices in graphical allocation. In: STOC 2022, pp. 52–63. ACM (2022)
7. Bansal, N., Kuszmaul, W.: Balanced allocations: the heavily loaded case with deletions. In: 2022 IEEE 63rd Annual Symposium on Foundations of Computer Science (FOCS), pp. 801–812 (2022)
8. Ben-Nun, S., Kopelowitz, T., Kraus, M., Porat, E.: An $O(\log^{3/2} n)$ parallel time population protocol for majority with $O(\log n)$ states. In: Proceedings of the 39th Symposium on Principles of Distributed Computing, PODC 2020, pp. 191–199. ACM (2020)
9. Ben-Or, M., Pavlov, E., Vaikuntanathan, V.: Byzantine agreement in the full-information model in $O(\log n)$ rounds. In: Proceedings of the 38th Annual ACM Symposium on Theory of Computing, STOC 2006, pp. 179–186. ACM (2006)
10. Beraldi, R., Mattia, G.: Power of random choices made efficient for fog computing. IEEE Trans. Cloud Comput. **10**(02), 1130–1141 (2022)
11. Berenbrink, P., Czumaj, A., Steger, A., Vöcking, B.: Balanced allocations: the heavily loaded case. SIAM J. Comput. **35**(6), 1350–1385 (2006)
12. Bramson, M., Lu, Y., Prabhakar, B.: Asymptotic independence of queues under randomized load balancing. Queueing Syst. Theory Appl. **71**(3), 247–292 (2012)
13. Busch, C., Kowalski, D.R.: Byzantine-resilient population protocols. CoRR abs/2105.07123 (2021)
14. Chrobak, M., Gasieniec, L., Kowalski, D.R.: The wake-up problem in multihop radio networks. SIAM J. Comput. **36**(5), 1453–1471 (2007)
15. Cole, R., et al.: Randomized protocols for low-congestion circuit routing in multistage interconnection networks. In: Proceedings of the 30th Annual ACM Symposium on Theory of Computing, STOC 1998, pp. 378–388 (1998)
16. Cooper, C., Elsässer, R., Radzik, T.: The power of two choices in distributed voting. In: Esparza, J., Fraigniaud, P., Husfeldt, T., Koutsoupias, E. (eds.) ICALP 2014. LNCS, vol. 8573, pp. 435–446. Springer, Heidelberg (2014). https://doi.org/10.1007/978-3-662-43951-7_37
17. Gonnet, G.H.: Expected length of the longest probe sequence in hash code searching. J. ACM **28**(2), 289–304 (1981)
18. Guerraoui, R., Ruppert, E.: Names trump malice: tiny mobile agents can tolerate byzantine failures. In: Albers, S., Marchetti-Spaccamela, A., Matias, Y., Nikoletseas, S., Thomas, W. (eds.) ICALP 2009. LNCS, vol. 5556, pp. 484–495. Springer, Heidelberg (2009). https://doi.org/10.1007/978-3-642-02930-1_40
19. Hagerup, T., Rüb, C.: A guided tour of chernoff bounds. Inf. Process. Lett. **33**(6), 305–308 (1990)
20. Johnson, N.L., Kotz, S.: Urn models and their application: an approach to modern discrete probability theory. Int. Stat. Rev. **46**, 319 (1978)
21. Kenthapadi, K., Panigrahy, R.: Balanced allocation on graphs. In: ACM-SIAM Symposium on Discrete Algorithms (SODA), pp. 434–443 (2006)

22. Lamport, L., Shostak, R., Pease, M.: The byzantine generals problem. ACM Trans. Program. Lang. Syst. **4**(3), 382–401 (1982)
23. Los, D., Sauerwald, T.: Balanced allocations with the choice of noise. In: Proceedings of the ACM Symposium on Principles of Distributed Computing (PODC 2022), pp. 164–175 (2022)
24. Lynch, N.A.: Distributed Algorithms, 1st edn. Morgan Kaufmann (1996)
25. Park, C.J.: Random allocations (valentin f. kolchin, boris a. sevast'yanov and vladimir p. chistyakov). Siam Review **22**, 104–104 (1980)
26. Peres, Y., Talwar, K., Wieder, U.: Graphical balanced allocations and the 1 + β-choice process. Random Struct. Algorithms **47**(4), 760–775 (2015)
27. Richa, A., Mitzenmacher, M., Sitaraman, R.: The power of two random choices: a survey of techniques and results. In: Handbook of Randomized Computing (2001)
28. Vöcking, B.: How asymmetry helps load balancing. In: Proceedings of the 40th Annual IEEE Symposium on Foundations of Computer Science, FOCS 1999, p. 131 (1999)

Distributed Fractional Local Ratio and Independent Set Approximation

Magnús M. Halldórsson[1] and Dror Rawitz[2(✉)]

[1] Department of Computer Science, Reykjavik University, Reykjavik, Iceland
mmh@ru.is
[2] Faulty of Engineering, Bar Ilan University, Ramat Gan, Israel
dror.rawitz@biu.ac.il

Abstract. We consider the MAXIMUM WEIGHT INDEPENDENT SET problem, with a focus on obtaining good approximations for graphs of small maximum degree Δ. We give deterministic local algorithms running in time $\text{poly}(\Delta, \log n)$ that come close to matching the best centralized results known and improve the previous distributed approximations by a factor of about 2. More precisely, we obtain approximations below $\frac{\Delta+1/2}{2}$, and a further improvement to $8/5 + \varepsilon$ when $\Delta = 3$.

Technically, this is achieved by leveraging the *fractional local ratio* technique, for a first application in a distributed setting.

1 Introduction

We consider the fundamental MAXIMUM WEIGHT INDEPENDENT SET problem (MAXWIS): Given a graph $G = (V, E)$ with weights $w : V \to \mathbb{Z}$ on the vertices, find a maximum weight subset of mutually non-adjacent vertices. Independent sets are fundamental to distributed computation, typically representing tasks that can be processed simultaneously.

We seek time- and bandwidth-efficient distributed algorithms to compute independent sets of approximately maximum weight in graphs of low maximum degree Δ. As the performance guarantees are invariably (nearly) linear in Δ, it is natural to focus on the case of small maximum degree Δ (say, constant, or a slowly growing function). The primary focus is on the approximation ratios, with complexity bounds secondary, but we require message size $O(\log n)$ (i.e., the CONGEST model) and round complexity $\text{poly}(\Delta, \log n)$.

Known approximation algorithms in the CONGEST model for MAXWIS as a function of Δ [4,17,23], whether deterministic or randomized, all have performance guarantees at least Δ. The exception is the randomized algorithm of [13] that attains roughly 0.529Δ-ratio, but only in expectation. In contrast, the best performance guarantees known by centralized poly-time algorithms (for small Δ) is $\Delta/2$ [22] for $\Delta \leq 4$ and $(\Delta + 2)/3$ for larger Δ [19]. Those centralized algorithms, however, seem immune to distributed implementation.

This research was done while Dror Rawitz was on sabbatical at Reykjavik University.

All the previously used techniques either have limitations that preclude going below factor Δ or make distributed implementation unlikely. The technique primarily used in distributed MAXWIS algorithms is the *local ratio* method (see related work). With some of the ideas of the current paper, it would be plausible to reduce the performance ratio of this method to $\Delta - 1 + \epsilon$, but going below that would require something quite different. The method of Hochbaum [22] requires both computing a near-exact maximum weight independent set in a bipartite graph and choosing between several possible solutions, both of which are more global problems.

The key idea of the current paper is to bring to the distributed setting a method called *fractional local ratio* [7]. This entails first computing an approximate fractional solution of an LP-relaxation and then using it in local-ratio weight subtractions. The advantage is that fractional values give us bounds on the local contributions to a super-optimal solution, which allows the algorithm to make judicious choices. This is turned into a distributed method by parallelizing the set of reductions performed (in a similar fashion as in previous local ratio-based methods). We explore a range of weight subtractions to further reduce the approximation ratios.

Our Results. We start the paper by describing a centralized fractional local ratio algorithm for MAXWIS whose approximation ratio is $\frac{\Delta+1}{2}$. Then, we improve this algorithm and obtain a $(\frac{\Delta}{2} + \frac{1}{4} - \frac{1}{8\Delta - 12})$-ratio. Finally, we explore in more detail the (perhaps the most interesting) case $\Delta = 3$ and obtain an approximation ratio of $8/5$.

We give a generic technique to turn such methods into a distributed algorithm at the cost of a multiplicative factor of $(1+\varepsilon)$ in the approximation ratio (for any given $\varepsilon > 0$). The run-time of the algorithms is $O(\text{poly}(\Delta, 1/\epsilon) \log n)$. When the weights of the nodes fall in a constant range (a common case of special interest) and the maximum degree is constant, the run-time of the first two algorithms improves to $O(\log^* n)$. All of our algorithms are deterministic.

Technical Introduction. Local ratio algorithms rely on weight subtractions. A new weight function w_1 is constructed in each recursive call and this function is subtracted from the current weight function w. Such a function w_1 is called *r-effective* if the computed solution is r-approximate with respect to w_1. An r-approximation algorithm has to use weight functions and to compute a solution such that all weight functions are r-effective. A drawback of this approach is that the computed solution is compared to several different optima each associated with a different weight function. Bar-Yehuda et al. [7] came up with a more global approach: solutions are compared to a single super-optimal solution $\tilde{x}$ with respect to the constructed weight functions. More specifically, they used an optimal fractional solution to an LP-relaxation. Thus, the fractional local ratio is a hybrid of local ratio and LP rounding.

In this work, we use fractional local ratio in a distributed setting. As a first step, we replace the centralized computation of an optimal fractional solution with a decentralized computation of a $(1 + \varepsilon)$-approximate fractional solution $\tilde{x}$. This is done using the framework by Kuhn, Moscibroda, and Wattenhofer [24].

Our first algorithm is based on a modification of a previously known [3,30] centralized fractional local ratio algorithm for MaxWIS whose approximation ratio is $\frac{\Delta+1}{2}$. The above-mentioned modification makes the algorithm more amenable to a distributed implementation. (The presentation of this algorithm can also serve as a fractional local ratio tutorial.) This algorithm iteratively computes an independent set U, and then constructs a weight function by assigning a weight of $w(u)$ to u and each of its neighbors, for every $u \in U$. The constructed weight function is the total sum of these *star* weight functions. We refer to such a star weight assignment as the *star rule*. We provide implementations of this algorithm that are similar to those given in [4], with crucial differences. First, we compute a fractional solution of an LP-relaxation. Second, the star rule is first used on all vertices whose fractional value is more than half. These differences lower the effectiveness of the star rule from Δ to $\frac{\Delta+1}{2}$.

The integrality gap of the pairwise LP-formulation of MaxWIS is $\frac{\Delta+1}{2}$. To bypass this gap, we consider non-regular graphs, with some nodes of less than maximum degree. We show that by using the star rule carefully, it is possible to lower the effectiveness of the weight functions. Initially, we use the star rule on all vertices whose fractional value is higher than some threshold $\delta > \frac{1}{2}$, and then only consider vertices of degree below Δ. The result is a ratio of $\frac{\Delta}{2} + \frac{1}{4} - \frac{1}{8\Delta-12}$. In the distributed implementation, we first remove a set of low-weight vertices, to ensure progress in all parts of the graph, which increases only slightly the running time and approximation ratio.

Another way to overcome the integrality gap is to strengthen the LP formulation. We do so by adding triangle constraints to the LP, in the context of the case $\Delta = 3$. We also add two additional rules. Intuitively, the *heavy* rule is employed when a degree-2 vertex is sufficiently heavy compared to its neighbors, and the *two siblings* rule is used when the fractional weight of the neighbors is relatively high. As in the previous algorithm, initially we use the star rule on all vertices whose fractional value is higher than some $\delta > \frac{1}{2}$, and then we only consider vertices with degree less than 3. Specifically, we show that given a vertex of degree 2, there is always a rule whose effectiveness is at most $8/5$.

Related Work. Most previous distributed MaxWIS approximation algorithm have used the local ratio technique. Bar-Yehuda et al. [4] presented distributed implementations for a centralized local ratio Δ-approximation algorithm for MaxWIS. The first algorithm runs in $O(T_{\mathrm{MIS}(G)} \cdot \log W)$ rounds in the CONGEST model, where $T_{\mathrm{MIS}(G)}$ is the number of rounds needed to compute a maximal independent set in G and W is the ratio between the largest and smallest weights. They also gave a deterministic implementation with the same approximation ratio that runs in $O(\Delta + \log^* n)$ rounds. Kawarabayashi et al. [23] improved the running time while increasing the approximation ratio by a factor of $1 + \varepsilon$. They presented a deterministic $O(T_{\mathrm{MIS}(n,\Delta)}/\varepsilon)$-round $(1 + \varepsilon)\Delta$-approximation algorithm, where $T_{\mathrm{MIS}(n,\Delta)}$ is the number of rounds needed to compute a maximal independent set in a graph with n vertices and maximum degree Δ. And a randomized $O(\mathrm{poly}(\log\log n)/\varepsilon)$-round $(1 + \varepsilon)\Delta$-approximation algorithm. Faour et al. [17] obtain $(1 + \varepsilon)\Delta$-approximation in time $O(\log^2(\Delta W)(\log 1/\epsilon) + \log^* n)$. Their approach uses a fractional solution (for a slightly different LP), but not as part of the local ratio reductions. All the algorithms mentioned work in the CONGEST model.

Boppana, Halldórsson, and Rawitz [13] gave a one-round algorithm whose expected approximation ratio asymptotically (in Δ) approaches $0.529(\Delta + 1)$.

In the LOCAL model, $n^{\Theta(1/k)}$-approximation is achievable and best possible for LOCAL algorithms running in k rounds [12] (where the upper bound assumes both unlimited bandwidth and computation), and a $(1 + \varepsilon)$-approximation can be achieved in $O(\log n/\varepsilon)$ rounds [18].

Alon [1] gave nearly tight bounds for testing independence properties; his lower bound carries over to distributed algorithms, as shown in [13]. Kuhn et al. [25] showed that achieving any $O(1)$-approximation requires $\Omega(\max(\log \Delta/\log \log \Delta, \sqrt{\log n/\log \log n}))$ rounds of LOCAL. Censor-Hillel, Khoury, and Paz [15] gave a nearly quadratic lower bound on the number of rounds for solving MAXIMUM INDEPENDENT SET exactly of CONGEST.

The local ratio technique was introduced by Bar-Yehuda and Even [6]. Bar-Noy et al. [2] were the first to use it for approximate packing problems. See [3,30] for more details on local ratio. Patt-Shamir, Rawitz and Scalosub [28] were the first to use the local ratio technique in the context of distributed computing. They used it to approximate the CELLULAR COVER problem. Bar-Yehuda, Censor-Hillel, and Schwartzman [5] and Ben-Basat et al. [11] used local ratio to design distributed approximation algorithms for the WEIGHTED VERTEX COVER problem. As mentioned above, Bar-Yehuda et al. [4] and Kawarabayashi et al. [23] used the local ratio technique to design algorithms for MAXWIS.

Bar-Yehuda et al. [7] presented the *fractional local ratio* technique and used it to obtain a $2t$-approximation algorithm for MAXWIS in t-interval graphs. The technique has been further applied to MAXWIS in various other classes of graphs: t-subtree graphs (that generalize the class of t-interval graphs and the class of chordal graphs) [21], a special case of MAXWIS in 2-interval graphs [14], intersection graphs of axis parallel rectangles [26], as well as extensions of some of these to demands [8,29]. For a more detailed description of fractional local ratio the reader is referred to [3,30].

The best sequential approximation ratio known for MAXWIS with large Δ is $O(\Delta \log \log \Delta/\log \Delta)$ [19,20]. The problem is known to be NP-hard to approximate within an $O(\Delta/\log^4 \Delta)$ factor [16], even in the unweighted case.

2   Preliminaries

Definitions and Notation. Let $G = (V, E)$ be a simple graph. The neighborhood of a vertex v is denoted $N(v)$, i.e., $N(v) = \{u : (v, u) \in E\}$. The closed neighborhood of v is defined as $N[v] = N(v) \cup \{v\}$. Given a vertex v, let f_v be a binary vector that indicates whether a vertex is in the closed neighborhood of v, namely define:

$$f_v(u) = \begin{cases} 1 & u \in N[v], \\ 0 & \text{otherwise.} \end{cases}$$

Denote $\deg(v) = |N(v)|$, and let $\Delta = \max_v \deg(v)$ be the maximum degree in G. A graph with $\deg(v) = \Delta$, for every v is called Δ-*regular*. A graph with maximum degree Δ that is not Δ-regular is Δ-*non-regular*.

Denote the weight of a maximum weight independent set in a graph G with a weight function w by $\mathrm{OPT}(G, w)$, or simply OPT.

Recall that W is the ratio between the maximum and the minimum node weight. For simplicity, we use the standard assumptions that W is the largest weight and the weights are integral, and that W is at most polynomial in n, which means that the weight of a node can be transmitted in a single round.

Computation Model. In distributed graph models, each node of the input graph is a processor with communication ports to its neighboring nodes. In synchronous rounds, the nodes can transmit messages to its neighbors and perform arbitrary amount of computation. At the end of the protocol, each node announces whether it is in the independent set solution or not. The time complexity of the protocol is the maximum number of rounds used by a node.

In the LOCAL model, nodes can transmit messages of arbitrary length in each round, while in the CONGEST model, the messages must be $O(\log n)$ bits. By bandwidth-efficiency we mean that the algorithm works in CONGEST.

3   Distributed Fractional Local Ratio

We present a centralized $\frac{\Delta+1}{2}$-approximation algorithm for MAXWIS in graphs with maximum degree Δ which is based on the *fractional local ratio* technique [7]. We then provide a distributed implementation of this algorithm whose approximation ratio is higher by a multiplicative factor of $(1 + \varepsilon)$, for any $\varepsilon > 0$.

3.1   Centralized Algorithm

We start the section by describing a previously known [3,30] centralized fractional local ratio for MAXWIS whose approximation ratio is $\frac{\Delta+1}{2}$. We describe it in a manner that makes it easier to implement in a distributed setting.

MAXWIS can be formulated by the following linear integer program:

$$
\begin{aligned}
\max \; & \textstyle\sum_{v \in V} w(v) \cdot x(v) \\
\text{s.t.} \; & x(v) + x(u) \leq 1 \quad \forall (v, u) \in E \\
& x(v) \in \{0, 1\} \qquad \forall v \in V
\end{aligned}
\tag{1}
$$

An LP-relaxation is obtained by replacing the integrality constraints by: $x(v) \in [0, 1]$, for every $v \in V$. Notice that the integrality gap of (1) is (at least) $\frac{\Delta+1}{2}$. This can be seen by considering a complete graph K_n with unit weights: The integral optimum is 1, while the fractional optimum is $\frac{n}{2} = \frac{\Delta+1}{2}$.

Fractional local ratio compares solutions to a fractional solution $\tilde{x}$ of (1). More formally, given $r \geq 1$, an integral solution x of (1) is said to be r-*approximate relative to* $\tilde{x}$ (with respect to w) if $w \cdot x \geq (w \cdot \tilde{x})/r$, where $\cdot$ stands for dot product.

Fractional local ratio is based on a *fractional* version of the Local Ratio Lemma (the proof is given for completeness).

Lemma 1 (Fractional Local Ratio [7]). *Let $w, w_1, w_2 \in \mathbb{R}^n$ be weight functions such that $w = w_1 + w_2$. Let $\tilde{x}, x \in \mathbb{R}^n$ such that x is r-approximate relative to $\tilde{x}$ with respect to w_1 and with respect to w_2. Then, x is r-approximate relative to $\tilde{x}$ with respect to w as well.*

Proof. $w \cdot x = w_1 \cdot x + w_2 \cdot x \geq (w_1 \cdot \tilde{x})/r + (w_2 \cdot \tilde{x})/r = (w \cdot \tilde{x})/r.$ $\qquad \square$

The first step of a fractional local ratio algorithm is the computation of the super-optimal solution $\tilde{x}$. In the centralized setting, it is typically an optimal fractional solution. The rest of the algorithm is similar to a standard local ratio algorithm. However, in each recursive call (or iteration), the construction of the weight function w_1 is based on the fractional solution $\tilde{x}$. Also, the corresponding analysis compares the weight of the solution returned with the weight of $\tilde{x}$, namely it is based on the Fractional Local Ratio Lemma.

We need the following properties of fractional solutions of (1).

Observation 1. *Let $\tilde{x}$ be a fractional solution of (1). Also, let $U \subseteq V$. Then, $\tilde{x}|_U$ is feasible in $G[U]$.*

Observation 2. *Let $\tilde{x}$ be a fractional solution of (1). Then, $\{v : \tilde{x}(v) > \frac{1}{2}\}$ is an independent set.*

Lemma 2. *Let $\tilde{x}$ be a fractional solution of (1) and let v be a node. If $\sum_{u \in N(v)} \tilde{x}(u) \leq \frac{\Delta}{2}$, then $\sum_{u \in N[v]} \tilde{x}(u) \leq \frac{\Delta+1}{2}$. In particular, this holds if $\tilde{x}(u) \leq \frac{1}{2}$ for every $u \in N(v)$, or when $\tilde{x}(v) \geq \frac{1}{2}$.*

Proof. Let $z = \max_{u \in N(v)} \tilde{x}(u)$ be the largest fractional value among neighbors of v and let $S = \sum_{u \in N(v)} \tilde{x}(u)$ be their sum. Clearly, $z \geq S/\Delta$ and $\tilde{x}(v) \leq 1 - z \leq 1 - S/\Delta$. Thus, $\sum_{u \in N[v]} \tilde{x}(u) \leq (1 - S/\Delta) + S \leq 1 + \frac{\Delta-1}{\Delta} \cdot \Delta/2 = \frac{\Delta+1}{2}$. If $\tilde{x}(v) \geq \frac{1}{2}$, then $z \leq \frac{1}{2}$, and thus $S \leq \Delta/2$. $\qquad \square$

Algorithm **FLR** (Algorithm 1) is recursive and works as follows. First, it discards vertices with non-positive weight. If the remaining graph is empty, then it returns an empty set (recursion base). Otherwise, it computes an independent set U. If there are vertices whose value with respect to $\tilde{x}$ is above half, then U contains these vertices. Otherwise, one may use any non-empty independent set. Next it computes a weight function w_1 such that $w_1(v) = \sum_{u \in U, v \in N[u]} w(u)$, for every v. (Note that $w_1(u) = w(u)$, for every $u \in U$.) Then it recursively solves the problem for $w - w_1$. The vertices in U are added to the solution returned by the recursive call as long as it remains an independent set.

Algorithm **FLR** has at most n recursive calls, and the running time of each such recursive call is polynomial. Thus, the total running time is polynomial.

Let x denote the incidence vector of the computed solution I.

Given an independent set $U = \{u_1, \ldots, u_t\}$, we say that an integral solution S is U-*maximal*, if the following conditions hold for every i: either u_i is in S or

---

Algorithm 1: FLR(V, E, w)

---

1 $V \leftarrow V \setminus \{u : w(u) \leq 0\}$ and $E \leftarrow E \cap (V \times V)$
2 **if** $V = \emptyset$ **then return** $\emptyset$ **if** $\exists v, \tilde{x}(v) > \frac{1}{2}$ **then** $U' \leftarrow \{u : \tilde{x}(u) > \frac{1}{2}\}$ **else**
   $U' \leftarrow \emptyset$ Let U be any non-empty independent set such that $U' \subseteq U$
3 $w_1(v) \leftarrow \sum_{u \in U} w(u) \cdot f_u(v)$, for every v
4 $I \leftarrow \mathbf{FLR}(V, E, w - w_1)$
5 $I \leftarrow I \cup (U \setminus \cup_{v \in I} N[v])$
6 **return** I

---

u_i has a neighbor in S. Formally, $N[u_i] \cap S \neq \emptyset$. The following lemma shows that w_1 is $\frac{\Delta+1}{2}$-effective with relative to $\tilde{x}$, namely that any U-maximal solution is $\frac{1}{2}(\Delta+1)$-approximate relative to $\tilde{x}$ with respect to w_1.

Lemma 3. *Let U be an independent set and let x be a U-maximal solution. Then x is $\frac{1}{2}(\Delta+1)$-approximate relative to $\tilde{x}$ with respect to w_1.*

Proof. By Lemma 2 we have that

$$w_1 \cdot \tilde{x} = \sum_{v \in V} \tilde{x}(v) \sum_{u \in U} w(u) f_u(v) = \sum_{u \in U} w(u) \sum_{v \in N[u]} \tilde{x}(v)$$

$$\leq \sum_{u \in U} w(u) \cdot \frac{\Delta+1}{2} = \frac{\Delta+1}{2} \cdot w(U),$$

On the other hand, since x is U-maximal, it must contain at least one vertex from the closed neighborhood of every $u \in U$. Hence, $w_1 \cdot x \geq w(U)$. $\square$

Lemma 4. *Let (G, w) be an MAXWIS instance of bounded degree $\Delta > 1$. Also, let $\tilde{x}$ be a fractional solution of (1). Then **FLR** computes a $\frac{\Delta+1}{2}$-approximate solution relative to $\tilde{x}$.*

Proof. In order to compare a solution x to $\tilde{x}$, we assume that x and all intermediate weight functions are vectors of size $|V|$. We prove that x is $\frac{\Delta+1}{2}$-approximate relative to $\tilde{x}$ with respect to w by induction on the recursion. If follows that the computed solution is a $\frac{\Delta+1}{2}$-approximate for the original instance.

For the recursive base (Line 2), observe that $\sum_v w(v)x(v) = 0$, while $\sum_v w(v)\tilde{x}(v) \leq 0$. Thus x is $\frac{\Delta+1}{2}$-approximate relative to $\tilde{x}$ with respect to w. As for the inductive step, let x' be the incidence vector that is returned in Line 4. By the induction hypothesis, x' is $\frac{\Delta+1}{2}$-approximate relative to $\tilde{x}$ with respect to $w - w_1$. Since $w_2(u) = w(u) - w_1(u) = 0$, for every $u \in U$, x is also $\frac{\Delta+1}{2}$-approximate relative to $\tilde{x}$ with respect to $w - w_1$. Next, observe that due to Line 5, Algorithm **FLR** computes a U-maximal solution x. By Lemma 3, x is $\frac{\Delta+1}{2}$-approximate relative to $\tilde{x}$ with respect to w_1. Due to the Fractional Local Ratio Lemma, x is r-approximate relative to $\tilde{x}$ with respect to w as well. $\square$

Corollary 1. *Let (G, w) be an MAXWIS instance of bounded degree $\Delta > 1$. Also, let $\tilde{x}$ be an α-approximate solution of (1). Then* ***FLR*** *computes an* $\alpha \cdot \frac{\Delta+1}{2}$- *approximate solution.*

3.2 Distributed Algorithm

In this section we provide a distributed implementation of the fractional local ratio algorithm using a $(1 + \varepsilon)$-approximate solution for (1). This leads to a $(1 + \varepsilon)\frac{\Delta+1}{2}$-approximation algorithm in a distributed setting.

The algorithm consists of two phases. In the first a near-optimal fractional solution is computed. The second phase is a distributed implementation of the fractional local ratio algorithm (Algorithm **FLR**). We present two such implementations that are similar to those given in [4]. However, although the algorithms are reminiscent, they are quite different. The main difference, of course, is the use of a fractional solution. That is, our algorithms start with computing a fractional near-optimal solution, and then the first weight function is constructed around vertices with high fractional values. Also, the analysis of the performance guarantee is quite different since it is based on fractional local ratio.

Computing a Near-Optimal Solution. Kuhn, Moscibroda, and Wattenhofer [24] presented a deterministic distributed algorithm for computing fractional solutions for LPs corresponding to fractional covering and packing problems. Specifically, gave an algorithm that computes a $(1 + \varepsilon)$-approximate fractional solution for LPs corresponding to fractional covering and packing problems in $O(\log(\rho\Delta')/\varepsilon^4)$ rounds in the CONGEST model, where ρ is the ratio between the largest and the smallest non-zero coefficients of the LP, and Δ' stands for the maximum number of times a variable occurs in an inequality of the primal or dual LP. Since $\rho = W$ and $\Delta' = \Delta$ in the case of (1), we have that:

Corollary 2. *There is a MAXWIS algorithm that computes $(1+\varepsilon)$-approximate fractional solutions in $O(\log^2(W\Delta)/\varepsilon^4)$ rounds in the CONGEST model.*

For arbitrary weights one can truncate the computation to complete in $O(\epsilon^{-4}\log^2 n)$ rounds. Note that the above computation is done on a network with a node for each of the primal/dual variables and there is an edge between a primal and a dual variable if the primal variable participates in the primal constraint that corresponds to the dual variable. In the case of (1), dual variables correspond to edges and each edge can be simulated by one of its endpoints.

Using an MIS Algorithm. We first present a distributed implementation of Algorithm 1 that uses a *maximal independent set* (MIS) algorithm.

Given a fractional solution $\tilde{x}$, the algorithm partitions the vertices into classes according to the fractional solution and the current weight $w_v(v)$ of v, namely

$$L_\infty = \left\{v : \tilde{x}(v) > \tfrac{1}{2}\right\}, \qquad L_i = \left\{v : \tilde{x}(v) \le \tfrac{1}{2}, \quad 2^{i-1} < w(v) \le 2^i\right\}.$$

The label $\ell(v)$ of a vertex v represents its class. Note that vertices with high fractional values are in the top level, i.e., $\ell(v) = \infty$, if $\tilde{x}(v) \geq \frac{1}{2}$.

As in Algorithm 1, the weight subtractions are induced by independent sets that are computed by a distributed MIS algorithm. Let $\ell_v(u)$ be the local level of u at a node v. A vertex is *active* if its label is locally high, in which case it participates in the MIS computation. More formally, a vertex v such that $\ell_v(v) \geq \ell_v(u)$, for every $u \in N(v)$, becomes active. However, it joins the MIS computation only if all of its neighbors at the same level are also active. In each of the following phases, we compute an MIS on the set of active nodes. A vertex that belongs to the computed independent set sends a subtract message to its neighbors, reduces its own weight to zero, and becomes a candidate to enter the solution.

We show that the Algorithm terminates after $O(\log W)$ MIS computations. Let $T_{\mathrm{MIS}}(G)$ be the number of rounds of an MIS algorithm on a graph G. Also, let p_{MIS} be its failure probability.

Lemma 5. *The* MIS*-based algorithm terminates after* $O(\log W \cdot T_{\mathrm{MIS}}(G))$ *rounds, with probability* $1 - p_{\mathrm{MIS}}(1 + \log W)$ *in the* CONGEST *model.*

Proof. Consider the first MIS computation. Observe that all nodes which are labeled ∞, namely those in L_∞, constitute an independent set by Observation 2. Hence, it must be that the vertices in L_∞ participate in the MIS computation. Moreover, the first MIS computation cannot include any vertex in $N(L_\infty)$. Hence, all vertices in L_∞ must be contained in the first computed maximal independent set.

Now consider the rest of the MIS computations. Let L_i be the topmost nonempty level, i.e., $i = \arg\max\{j : L_j \neq \emptyset\}$. The next MIS computation must consider all members of L_i. Hence, every $v \in L_i$ is either in the maximal independent set, or has a neighbor from L_i in the maximal independent set. Hence, after the weight subtraction all vertices in L_i are removed, become candidates, or lose at least half of their weight (and the drop to a lower level). It follows that L_i becomes empty. Since there are $1 + \log W$ levels, the algorithm performs $O(\log W)$ MIS computations. The probability of failure is obtained through union bound.

Finally, notice that a vertex that became a candidate after the ith MIS computation, can decide to enter the solution after at most $1 + \log W - i$ rounds. $\square$

The correctness of the algorithm is implied since it is a distributed implementation of Algorithm 1. Hence, by Corollary 2 and Lemma 5 have that:

Theorem 1. *There exists a* MaxWIS *algorithm which computes a* $(1+\varepsilon)\frac{\Delta+1}{2}$*-approximate solutions in* $O(\log^2(W\Delta)/\varepsilon^4 + \log W \cdot T_{\mathrm{MIS}}(G))$ *rounds with probability* $1 - p_{\mathrm{MIS}}(1 + \log W)$ *in the* CONGEST *model.*

Using a Coloring Algorithm. We provide another implementation of the fractional local ratio algorithm which is based on a coloring algorithm. As before the first set consists of vertices of fractional value greater than $\frac{1}{2}$ (color 0). The remaining sets are formed by a $\Delta + 1$-coloring of the remaining graph. Note that the vertices of color 0 do not participate in the coloring algorithm. Starting with color 0, the algorithm uses the vertices in color set i for constructing weight subtractions.

The above algorithm is a distributed implementation of Algorithm 1. If one uses the deterministic $O(\Delta + \log^* n)$-round CONGEST algorithm for $\Delta + 1$-coloring of [9,10], we have the following result.

Theorem 2. *There exists a* MaxWIS *algorithm which computes* $(1 + \varepsilon)\frac{\Delta+1}{2}$-*approximate solutions in* $O(\log^2(W\Delta)/\varepsilon^4 + \Delta + \log^* n)$ *rounds in the* CONGEST *model.*

4   Improving the Approximation Ratio

In this section we improve our fractional local ratio algorithm by considering connected Δ-non regular graphs. This enables us to bypass the $\frac{\Delta+1}{2}$ integrality gap of Eq. (1). The improved fractional local ratio algorithm has an approximation ratio of $\frac{\Delta}{2} + \frac{1}{4} - \frac{1}{8\Delta-12}$. Notice that for $\Delta = 3$, the ratio is $\frac{5}{3}$. Using this algorithm we obtain an algorithm with the same approximation ratio for all graphs of maximum degree Δ. Finally, we provide a distributed implementation.

4.1   Centralized Algorithm

We present an improvement of Algorithm 1 whose approximation ratio is $\frac{\Delta}{2} + \frac{1}{4} - \frac{1}{8\Delta-12}$. Initially we assume that the input graph is connected and Δ-non-regular.

Observation 3. *Let* $G = (V, E)$ *be a connected* Δ-*non-regular graph, and let* $V' \subseteq V$. *Then, every component of* $G[V']$ *is* Δ-*non-regular.*

We describe an algorithm for connected Δ-non-regular graphs (Algorithm 2). We modify Algorithm 1 as follows. Instead of using a threshold of $\frac{1}{2}$, we use a parameter $\delta > \frac{1}{2}$, whose value will be determined later. Namely, the condition in Line 2 is changed to whether there exists a vertex v such that $\tilde{x}(v) > \delta$. Also, in the next line define $U = \{u : \tilde{x}(u) > \delta\}$. Otherwise, the set U contains vertices of degree at most $\Delta - 1$. Observation 3 implies that each connected component has at least one such vertex.

The next two lemmas analyze the approximation ratio of Algorithm 2.

Lemma 6. *Let* $\delta \geq \frac{1}{2}$, *and let* $\tilde{x}$ *be a fractional solution of* (1).

1. *If* $\tilde{x}(v) > \delta$, *then* $\sum_{u \in N[v]} \tilde{x}(u) \leq \Delta - \delta(\Delta - 1)$.
2. *If* $\deg(v) \leq \Delta - 1$ *and* $\tilde{x}(u) \leq \delta, \forall u \in N[v]$, *then* $\sum_{u \in N[v]} \tilde{x}(u) \leq 1 + \delta(\Delta - 2)$.

Proof. Let v be a vertex such that $\tilde{x}(v) > \delta$. Then,

$$\sum_{u \in N[v]} \tilde{x}(u) = \tilde{x}(v) + \sum_{u \in N(v)} (1 - \tilde{x}(v)) \leq 1 + (\Delta - 1) \cdot (1 - \delta) = \Delta - \delta(\Delta - 1).$$

Algorithm 2: Improved-FLR(V, E, w)

1 $V \leftarrow V \setminus \{u : w(u) \leq 0\}$ and $E \leftarrow E \cap (V \times V)$
2 **if** $V = \emptyset$ **then return** $\emptyset$ **if** $\exists v, \tilde{x}(v) > \delta$ **then** $U \leftarrow \{u : \tilde{x}(u) > \delta\}$ **else** Let U
 be a non-empty independent set s.t. $\forall u \in U,\ \deg(u) \leq \Delta - 1$
 $w_1(v) \leftarrow \sum_{u \in U} w(u) \cdot f_u(v)$, for every v
3 $I \leftarrow \mathbf{FLR}(V, E, w - w_1)$
4 $I \leftarrow I \cup (U \setminus \cup_{v \in I} N[v])$
5 **return** I

If $\deg(v) \leq \Delta - 1$ and $\tilde{x}(u) \leq \delta$, for every $u \in N[v]$, then

$$\sum_{u \in N[v]} \tilde{x}(u) \leq \tilde{x}(v) + (1 - \tilde{x}(v)) + (\Delta - 2) \cdot \delta \leq 1 + (\Delta - 2) \cdot \delta.$$

$\square$

Lemma 7. *Let U be an independent set and let x be a U-maximal solution. Then x is $\frac{\Delta^2 - \Delta - 1}{2\Delta - 3}$-approximate relative to $\tilde{x}$ with respect to w_1.*

Proof. By Lemma 6 we have that

$$w_1 \cdot \tilde{x} = \sum_{v \in V} \tilde{x}(v) \sum_{u \in U} w(u) f_u(v) = \sum_{u \in U} w(u) \sum_{v \in N[u]} \tilde{x}(v)$$

$$\leq \max\{\Delta - \delta(\Delta - 1), 1 + \delta(\Delta - 2)\} \cdot w(U).$$

Let $\delta = \frac{\Delta - 1}{2\Delta - 3}$, and we get

$$\Delta - \delta(\Delta - 1) = 1 + \delta(\Delta - 2) = 1 + \frac{(\Delta - 1)(\Delta - 2)}{2\Delta - 3} = \frac{\Delta^2 - \Delta - 1}{2\Delta - 3}.$$

On the other hand, since x is U-maximal, it must contain at least one vertex from the closed neighborhood of every $u \in U$. Hence, $w_1 \cdot x \geq w(U)$. $\square$

Lemma 8. *Let (G, w) be an MaxWIS instance, where G is connected and Δ-non regular. Let $\tilde{x}$ be a fractional solution of (2). Then Algorithm **Improved-FLR** computes a $\left(\frac{\Delta}{2} + \frac{1}{4} - \frac{1}{8\Delta - 12}\right)$-approximate with respect to $\tilde{x}$.*

Proof. The proof is similar to the proof of Lemma 4, where Lemma 7 replaces Lemma 3. $\square$

Theorem 3. *Let (G, w) be an MaxWIS instance, where G is connected and Δ-non regular. Also, let $\tilde{x}$ be an α-approximate solution of (2). Then Algorithm **Improved-FLR** computes $\alpha \cdot \left(\frac{\Delta}{2} + \frac{1}{4} - \frac{1}{8\Delta - 12}\right)$-approximations.*

Corollary 3. *Let (G, w) be an MaxWIS instance of bounded degree $\Delta > 1$. Also, let $\tilde{x}$ be an α-approximate solution of (2). Then there exists an algorithm whose approximation ratio is $\alpha \cdot \left(\frac{\Delta}{2} + \frac{1}{4} - \frac{1}{8\Delta - 12}\right)$.*

Proof. If G is composed of Δ-non regular components, then one may run the improved algorithm of each of them.

Otherwise, consider a connected Δ-regular component C of G. Select a vertex v, and let $G_1 = C[V \setminus \{v\}]$ and $G_2 = C[V \setminus N[v]]$. Notice that both G_1 and G_2 are composed of components that are Δ-non regular. Hence, one may apply Theorem 3 on each of these components. Let I_1 and I_2 be the computed solutions for G_1 and G_2. Since v is either in the optimal solution or not, one may select that the maximum weight solution between I_1 and $I_2 \cup \{v\}$. □

4.2   Distributed Algorithm

The preceding algorithm is local but it is inefficient if there are few nodes of degree less than Δ. We need all nodes to have a such a node nearby. To achieve this, we first remove nodes from all parts of the graph. To ensure that it doesn't affect the performance guarantee too much, we choose them to be both well separated and of relatively low weight.

Recall that, for $t \geq 1$, the power graph $G^t = (V, E^t)$ consists of the set of vertices V with an edge between two nodes if their distance in G is at most t. A subset $T \subseteq V$ *dominates* G if every node is either in T or has a neighbor in T.

We compute a subset S satisfying the following property. Recall that OPT denotes the weight of maximum independent set in G.

Definition 1. *Set $S \subset V$ is an ε-light dominator if $w(S) \leq \varepsilon \cdot \mathrm{OPT}(G)$ and S dominates $G^{\mathrm{poly}(1/\varepsilon)}$.*

It remains to argue that we can compute light dominators efficiently. For this, we need a recent result.

Proposition 1 ([27]). *An (r, r^2)-ruling set is a subset $T \subseteq V$ such that T is independent in G^r and dominates G^{r^2}. There is a $\tilde{O}(r^4 \log^4 n \cdot \log \Delta) = \mathrm{poly}(r, \log n)$ deterministic CONGEST algorithm to compute (r, r^2)-ruling sets.*

Lemma 9. *There exists a deterministic CONGEST algorithm that computes an ε-light dominator S in deterministic $\mathrm{poly}(1/\varepsilon, \log n)$ rounds.*

Proof. We first compute a (r, r^2)-ruling set T for $r = 2r' + 1$, where $r' = 3 \lceil 1/\varepsilon \rceil$. For each $u \in T$, let B_u be the set of nodes within distance r' from u. Each node $u \in T$ finds the node ℓ_u in B_u of least weight. The algorithm outputs $S = \{\ell_u : u \in T\}$. Each node in G has some node in T within distance r^2, by the definition of ruling sets, and each node in T has some node in S within distance r'. Thus, S dominates $G^{r^2 + r'}$. We verify that S is independent in G for the following reason. Nodes in T are of mutual distance more than r, so distinct nodes in S are of distance more than $r - r' - r' = 1$.

We now argue that S is ε-light. Note that B_u and $B_{u'}$ are disjoint, for every $u \neq u'$. Our solution contains exactly one node from each B_u, the lightest ones. Any maximal independent set in G, however, and thus also an optimal solution, contains at least $r'/3 \geq 1/\varepsilon$ nodes from B_u. Hence, $\mathrm{OPT} \geq r'/3 \cdot w(S) \geq w(S)/\epsilon$. □

We explain how we use a light dominator S to further our agenda. We first remove all nodes of S from the graph (or from further consideration), obtaining a graph G'. We then run a distributed version of our independent set Algorithm **Improved-FLR** (Algorithm 2) on G', called **Dist-Improved-FLR**. There are two differences with the implementation of the MIS-based distributed implementation of Algorithm **FLR**. First, the MIS computation ignores the levels, apart from starting with the top level L_∞. Second, only vertices of degree less than Δ participate in the MIS computation. Hence, a vertex of degree Δ that loses a neighbor simply joins the next MIS computation.

Theorem 4. *There exists a deterministic* CONGEST *algorithm that computes* $(1 + \varepsilon')(\frac{\Delta}{2} + \frac{1}{4} - \frac{1}{8\Delta - 12})$*-approximations for* MaxWIS *and runs in* poly$(1/\varepsilon', \Delta, \log n)$ *rounds.*

Proof. We use **Dist-Improved-FLR** with $\epsilon = \varepsilon'/(2(1 + \epsilon'))$ and $\alpha = 1 + \varepsilon'/2$ to obtain an independent set I. Let H be the graph $G[V \setminus S]$, where S is the ε-light dominator given by Lemma 9. By definition, $\text{OPT}(H) \geq (1 - \varepsilon)\text{OPT}$. By Corollary 3,

$$w(I)(\tfrac{\Delta}{2} + \tfrac{1}{4} - \tfrac{1}{8\Delta - 12}) \geq \alpha\text{OPT}(H) \geq (1 - \epsilon)(1 + \epsilon'/2)\text{OPT} = (1 + \epsilon')\text{OPT} .$$

Hence, the performance guarantee claim.

We argue that the number of iterations is $O(\varepsilon^{-2}\Delta)$. Recall that S dominates $G^{O(\epsilon^{-2})}$, so each node v is within distance $k = O(\epsilon^{-2})$ from a node $u \in S$. Denote the path by $u = v_0, v_1, v_2, \ldots, v_k = v$. We claim that v_i is eliminated within $i \cdot \Delta$ iterations, for each i. The proof is by induction. Suppose the claim holds for v_{i-1}. Then, after $(i - 1)\Delta$ rounds, v_{i-1} is eliminated, so v_i has degree less than Δ. Namely, in each iteration, some node u from $N[v_i]$ joins U and gets its degree reduced to zero, in which case it is eliminated. Thus, v_i is eliminated after Δ additional iterations, which yields the claim and the theorem.   □

5   8/5-Approximation Algorithm for Maximum Degree 3

In this section we present a centralized $\frac{8}{5}$-approximation algorithm for MaxWIS in graphs of maximum degree 3. The algorithm is a modified version of the improved fractional local ratio algorithm. This leads to a distributed $(8/5 + \varepsilon)$-approximation for degree-3 graphs.

As a first step we add *triangle* constraints to (1). Let $\mathcal{T}$ be a collection of triplets that induce a triangle in the given graph G, namely $\mathcal{T} = \{\{v_1, v_2, v_3\} : (v_1, v_2), (v_1, v_3), (v_2, v_3) \in E\}$. We add a set of *triangle constraints* to (1), and obtain the following integer program for MaxWIS:

$$
\begin{aligned}
\max \ & \sum_{v \in V} w(v) \cdot x(v) \\
\text{s.t.} \ & x(v) + x(u) \leq 1 && \forall (v, u) \in E \\
& x(v_1) + x(v_2) + x(v_3) \leq 1 \ \forall \{v_1, v_2, v_3\} \in \mathcal{T} \\
& x(v) \in \{0, 1\} && \forall v \in V
\end{aligned}
\tag{2}
$$

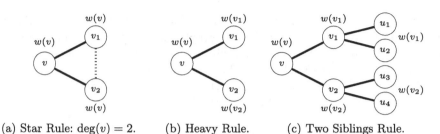

(a) Star Rule: $\deg(v) = 2$.   (b) Heavy Rule.   (c) Two Siblings Rule.

Fig. 1. Rules.

As before, an LP-relaxation is obtained by replacing the integrality constraints by: $x(v) \in [0,1]$, for every $v \in V$.

Our algorithm uses three weight functions or *rules* that are given below:

Star Rule. Similar to the previously described local ratio algorithms, our algorithm relies on a weight function that assigns the weight of a vertex v to its closed neighborhood. Formally, given a vertex v, define $w_\star$ by:

$$w_\star(u) = w(v) \cdot f_v(u).$$

Recall that f_v is a binary vector that indicates whether a vertex is in the closed neighborhood of v. See depiction of the degree 2 case in Fig. 1a.

Heavy Rule. Assume that v weighs a lot compared to the total weight of its neighbors. Intuitively, it would be beneficial to take v into the solution on the expense of its neighbors. The next weight function is an application of this idea. Define:

$$w_h(u) = w(u) \cdot f_v(u).$$

This weight function assigns to each vertex in $N[v]$ its own weight. See depiction in Fig. 1c.

Two Siblings Rule. Now assuming that v is not taken into the solution. In this case, one may activate the star rule on v's neighbors. Hence, we use the following weight function:

$$w_s(u) = \begin{cases} w(v) & u = v, \\ \sum_{v' \in N(v)} f_{v'}(u) \cdot w(v') & \text{otherwise.} \end{cases}$$

This weight function is almost the same as applying the star rule on the vertices of $N(v)$. The only difference is that v is given its own weight. Note that, the intersection of $N(v_1)$ and $N(v_2)$ may contain vertices other than v, where $N(v) = \{v_1, v_2\}$. See depiction in Fig. 1b.

We modify Algorithm **Improved-FLR** (Algorithm 2) as follows. First, we use $\delta = \frac{7}{10}$. As before, we use the $U = \{u : \tilde{x}(u) > \delta\}$ in the first recursive call, if there is a variable whose value is above δ. Otherwise, the set U contains vertices whose degree is at most $\Delta - 1$. Moreover, in this case U is an independent set in G^3. Each vertex $u \in U$ uses the rule which minimizes the approximation ratio. Notice that if a vertex v has degree one, then the star rule is 1-effective. Also, $\deg(v) = 2$ and it is part of a triangle, then the star rule is 1-effective. Hence, one may assume that the heavy and two-sibling rules are used when $\deg(v) = 2$ and it neighbors are non-adjacent. We will show that there is always a rule whose ratio is at most $\frac{8}{5}$.

Another change is the way we augment I. We use the standard approach for the star rule. In the case of the heavy rule, since $w(u) - w_1(u) = 0$, for every $u \in N[v]$, the algorithm can try to add them to the solution in any order. However, in this case we insist on giving precedence to adding v. It follows that v is always taken into the solution, while its neighbors are removed. The Two Siblings rule, also satisfies $w(u) - w_1(u) = 0$, for every $u \in N[v]$. However, in this case we give precedence to $N(v) = \{v_1, v_2\}$.

Algorithm 3: FLR3(V, E, w)

1 $V \leftarrow V \setminus \{u : w(u) \leq 0\}$ and $E \leftarrow E \cap (V \times V)$
2 **if** $V = \emptyset$ **then** **return** $\emptyset$
3 **if** $\exists v, \tilde{x}(v) > \delta$ **then**
4 $\quad$ $U \leftarrow \{u : \tilde{x}(u) > \delta\}$ $w_1 \leftarrow \sum_{u \in U} w_u$
5 **else**
6 $\quad$ Let U be a non-empty independent set in G^3 s.t. $\forall u \in U$, $\deg(u) \leq \Delta - 1$
7 $\quad$ $w_1 \leftarrow \sum_{v \in U} w_{z(v)}$, where $w_z(v)$ is the rule with minimum approx. ratio.
8 $\quad$ Let $U_\star, U_h$, and U_s be a partition of U according to the rules
9 $I \leftarrow$ **FLR**$(V, E, w - w_1)$
10 $I \leftarrow I \cup (U_\star \setminus \cup_{v \in I} N[v])$
11 $I \leftarrow I \cup (U_h \setminus \cup_{v \in I} N[v])$
12 $I \leftarrow I \cup (N(U_s \setminus \cup_{v \in I} N[v]))$
13 **return** I

To analyze the rules we need the following definitions. Given a graph G and a weight function w, we define $\beta_v \triangleq \frac{w(v)}{w(N(v))}$ and $\gamma_v \triangleq \frac{\sum_{u \in N(v)} \tilde{x}(u)w(u)}{w(N(v))}$.

The next lemmas explore the effectiveness of our rules.

Lemma 10 (Star Rule). *Let v be a vertex and let $\tilde{x}$ be a fractional solution of (2). Then, (1) $w_\star \cdot \tilde{x} \leq (\deg(v) - (\deg(v) - 1)\tilde{x}(v))) w(v)$. (2) If $N(v) = \{v_1, v_2\}$, then $w_\star \cdot \tilde{x} \leq (1 + \gamma_v)w(v)$. (3) If $N(v) = \{v_1, v_2\}$ and $\{v, v_1, v_2\} \in T$, then $w_\star \cdot \tilde{x} \leq w(v)$.*

Proof. First, we have that

$$w_\star \cdot \tilde{x} = w(v) \sum_{u \in N[v]} \tilde{x}(u) \leq w(v) (\tilde{x}(v) + \deg(v) \cdot (1 - \tilde{x}(v))).$$

If $N(v) = \{v_1, v_2\}$, then

$$w_v \cdot \tilde{x} = w(v) (\tilde{x}(v) + \tilde{x}(v_1) + \tilde{x}(v_2))$$
$$= w(v) (\tilde{x}(v) + \min\{\tilde{x}(v_1), \tilde{x}(v_2)\} + \max\{\tilde{x}(v_1), \tilde{x}(v_2)\}) \leq w(v) + \gamma_v w(v)$$

Finally, if $\{v, v_1, v_2\} \in \mathcal{T}$ we have that $w_\star \cdot \tilde{x} = w(v)(\tilde{x}(v) + \tilde{x}(v_1) + \tilde{x}(v_2)) \leq w(v)$, since $\tilde{x}$ is a solution of (2). $\qquad\square$

Lemma 11 (Heavy Rule). *Let* $v \in V$. *Then,* $w_h \cdot \tilde{x} \leq \left(1 + \frac{1-\beta_v}{\beta_v}\gamma_v\right) w(v)$.

Proof. We have that

$$w_h \cdot \tilde{x} = w(v)\tilde{x}(v) + \sum_{u \in N(v)} w(u)\tilde{x}(u) = \beta_v w(N(v))\tilde{x}(v) + \sum_{u \in N(v)} w(u)\tilde{x}(u)$$

$$\leq \sum_{u \in N(v)} w(u)[\beta_v(1 - \tilde{x}(u)) + \tilde{x}(u)]$$

$$= \beta_v w(N(v)) + (1 - \beta_v) \sum_{u \in N(v)} w(u)\tilde{x}(u)$$

$$= (\beta_v + (1 - \beta_v)\gamma_v)\, w(N(v))$$

$$= (1 + \gamma_v(1 - \beta_v)/\beta_v)\, w(v).$$

$\qquad\square$

Lemma 12 (Two Siblings Rule). *Let* $N(v) = \{v_1, v_2\}$ *such that* $\{v, v_1, v_2\} \notin \mathcal{T}$. *Also, let* $N(v_1) = \{v, v_{11}, v_{12}\}$ *and* $N(v_1) = \{v, v_{21}, v_{22}\}$. *Then,*

$$\tilde{x} \cdot w_s \leq (1 + (1 + \beta_v)(1 - \gamma_v)) \cdot w(N(v)).$$

Proof. We have that

$$w_s \cdot \tilde{x} = w(v)\tilde{x}(v) + w(v_1)x(v_1) + w(v_2)x(v_2) + \sum_{i=1}^{2}\sum_{j=1}^{2} w(v_{ij})x(v_{ij})$$

$$\leq \beta_v w(N(v))\tilde{x}(v) + \sum_{u \in N(v)} (\tilde{x}(u)w(u) + 2(1 - \tilde{x}(u))w(u))$$

$$\leq \sum_{u \in N(v)} w(u)\left((1 - \tilde{x}(u))\beta_v + 2 - \tilde{x}(u)\right)$$

$$= \sum_{u \in N(v)} w(u)\left(2 + \beta_v - (1 + \beta_v)\tilde{x}(u)\right)$$

$$= (2 + \beta)w(N(v)) - (1 + \beta_v) \sum_{u \in N(v)} x(u)w(u)$$

$$= w(N(v))\left(2 + \beta_v - (1 + \beta_v)\gamma_v\right).$$

$\qquad\square$

The next lemma implies that, given a node with degree 2 or less, one of the above rules gives a ratio of at most $\frac{8}{5}$. The proof is omitted for lack of space.

Lemma 13. *Assume that* $\tilde{x}(u) \leq \frac{7}{10}$, *for every* $u \in V$. *Let* v *be a vertex, such that* $\deg(v) \leq 2$. *Also, let* $w(v) = \beta_v w(N(v))$ *and* $w(v_1)x(v_1) + w(v_2)x(v_2) = \gamma_v w(N(v))$. *Then,* $\min\left\{\gamma_v, \frac{1-\beta_v}{\beta_v}\gamma_v, (1 + \beta_v)(1 - \gamma_v)\right\} \leq \frac{3}{5}$.

Lemma 14. *Let U be an independent set in G^3, and let x be a $U_\star \cup U_h \cup N(U_s)$-maximal solution. Then x is $\frac{8}{5}$-approximate relative to $\tilde{x}$ with respect to w_1.*

Proof. By Lemma 13 we have that

$$w_1 \cdot \tilde{x} = \sum_{v \in V} w_1(v)\tilde{x}(v) = \sum_{v \in U_\star} w_1(v)\tilde{x}(v) + \sum_{v \in U_h} w_1(v)\tilde{x}(v) + \sum_{v \in U_s} w_1(v)\tilde{x}(v)$$

$$\leq \frac{8}{5}\left(w(U_\star \cup U_h) + w(\cup_{v \in U_s} N(v))\right),$$

while $w_1 \cdot x \geq w(U_\star \cup U_h) + w(\cup_{v \in U_s} N(v))$. $\qquad\square$

Corollary 4. *Let (G, w) be an MAXWIS instance of maximum degree $\Delta = 3$. and let $\tilde{x}$ be an α-approximate solution of (2). Then **FLR3** computes an $\alpha \cdot \frac{8}{5}$-approximate solution.*

Distributed Implementation. We can use the same approach for implementing this method distributively as we did in the previous section. The only difference is that we find in each iteration a maximal distance-4 independent set U (or a 4-ruling set), rather than the usual maximal IS. This impacts the complexity by a $\Delta^{O(1)}$, but since $\Delta = 3$ here, this is immaterial.

Another point worth mentioning regarding the computation of a fractional solution using the framework of [24] is that the dual variables that correspond to constraints in (2) are edges and triangles. Thus, each edge can be simulated by one of its endpoints and each triangle by one of its vertices.

Theorem 5. *There exists a distributed CONGEST algorithm that computes $(\frac{8}{5} + \varepsilon)$-approximations for MAXWIS on graphs with maximum degree 3 and runs in $\text{poly}(1/\varepsilon, \log n)$ rounds.*

References

1. Alon, N.: On constant time approximation of parameters of bounded degree graphs. In: Goldreich, O. (ed.) Property Testing. LNCS, vol. 6390, pp. 234–239. Springer, Heidelberg (2010). https://doi.org/10.1007/978-3-642-16367-8_14
2. Bar-Noy, A., Bar-Yehuda, R., Freund, A., Naor, J., Schieber, B.: A unified approach to approximating resource allocation and scheduling. JACM **48**(5), 1069–1090 (2001)
3. Bar-Yehuda, R., Bendel, K., Freund, A., Rawitz, D.: Local ratio: a unified framework for approximation algorithms in memoriam: Shimon Even 1935–2004. ACM Comput. Surv. **36**(4), 422–463 (2004)
4. Bar-Yehuda, R., Censor-Hillel, K., Ghaffari, M., Schwartzman, G.: Distributed approximation of maximum independent set and maximum matching. In: 36th PODC, pp. 165–174 (2017)
5. Bar-Yehuda, R., Censor-Hillel, K., Schwartzman, G.: A distributed $2 + \epsilon$-approximation for vertex cover in $O(\log \Delta/\epsilon \log \log \Delta)$ rounds. JACM **64**(3), 23:1–23:11 (2017)

6. Bar-Yehuda, R., Even, S.: A local-ratio theorem for approximating the weighted vertex cover problem. Ann. Discret. Math. **25**, 27–46 (1985)
7. Bar-Yehuda, R., Halldórsson, M.M., Naor, J., Shachnai, H., Shapira, I.: Scheduling split intervals. SIAM J. Comput. **36**(1), 1–15 (2006)
8. Bar-Yehuda, R., Rawitz, D.: Using fractional primal-dual to schedule split intervals with demands. Discret. Optim. **3**(4), 275–287 (2006)
9. Barenboim, L.: Deterministic $(\Delta + 1)$-coloring in sublinear (in Δ) time in static, dynamic, and faulty networks. JACM **63**(5), 47:1–47:22 (2016)
10. Barenboim, L., Elkin, M., Kuhn, F.: Distributed $(\Delta + 1)$-coloring in linear (Δ) time. SIAM J. Comput. **43**(1), 72–95 (2014)
11. Ben-Basat, R., Even, G., Kawarabayashi, K., Schwartzman, G.: A deterministic distributed 2-approximation for weighted vertex cover in $O(\log n \log \Delta / \log^2 \log \Delta)$ rounds. In: Lotker, Z., Patt-Shamir, B. (eds.) SIROCCO 2018. LNCS, vol. 11085, pp. 226–236. Springer, Cham (2018). https://doi.org/10.1007/978-3-030-01325-7_21
12. Bodlaender, M.H., Halldórsson, M.M., Konrad, C., Kuhn, F.: Brief announcement: local independent set approximation. In: PODC, pp. 377–378. ACM (2016)
13. Boppana, R.B., Halldórsson, M.M., Rawitz, D.: Simple and local independent set approximation. Theoret. Comput. Sci. **846**, 27–37 (2020)
14. Canzar, S., Elbassioni, K.M., Klau, G.W., Mestre, J.: On tree-constrained matchings and generalizations. Algorithmica **71**(1), 98–119 (2015)
15. Censor-Hillel, K., Khoury, S., Paz, A.: Quadratic and near-quadratic lower bounds for the CONGEST model. In: 31st DISC, pp. 10:1–10:16 (2017)
16. Chan, S.O.: Approximation resistance from pairwise-independent subgroups. JACM **63**(3), 27 (2016)
17. Faour, S., Ghaffari, M., Grunau, C., Kuhn, F., Rozhon, V.: Local distributed rounding: generalized to MIS, matching, set cover, and beyond. In: 34th SODA (2023)
18. Ghaffari, M., Kuhn, F., Maus, Y.: On the complexity of local distributed graph problems. In: 49th STOC, pp. 784–797 (2017)
19. Halldórsson, M.M.: Approximations of weighted independent set and hereditary subset problems. Graph Algorithms Appl. **2**, 3–18 (2004)
20. Halperin, E.: Improved approximation algorithms for the vertex cover problem in graphs and hypergraphs. SIAM J. Comput. **31**(5), 1608–1623 (2002)
21. Hermelin, D., Rawitz, D.: Optimization problems in multiple subtree graphs. Disc. Appl. Math. **159**(7), 588–594 (2011)
22. Hochbaum, D.S.: Efficient bounds for the stable set, vertex cover and set packing problems. Discret. Appl. Math. **6**(3), 243–254 (1983)
23. Kawarabayashi, K., Khoury, S., Schild, A., Schwartzman, G.: Improved distributed approximations for maximum independent set. In: 34th DISC, LIPIcs, vol. 179, pp. 35:1–35:16 (2020)
24. Kuhn, F., Moscibroda, T., Wattenhofer, R.: The price of being near-sighted. In: 17th SODA, pp. 980–989 (2006)
25. Kuhn, F., Moscibroda, T., Wattenhofer, R.: Local computation: Lower and upper bounds. JACM **63**(2), 17:1–17:44 (2016)
26. Lewin-Eytan, L., Naor, J., Orda, A.: Admission control in networks with advance reservations. Algorithmica **40**(4), 293–304 (2004)
27. Maus, Y., Peltonen, S., Uitto, J.: Distributed symmetry breaking on power graphs via sparsification. In: 42nd PODC, pp. 157–167 (2023)
28. Patt-Shamir, B., Rawitz, D., Scalosub, G.: Distributed approximation of cellular coverage. J. Parallel Distrib. Comput. **72**(3), 402–408 (2012)

29. Rawitz, D.: Admission control with advance reservations in simple networks. J. Discret. Algorithms **5**(3), 491–500 (2007)
30. Rawitz, D.: Local ratio. In: Gonzalez, T.F. (ed.) Handbook of Approximation Algorithms and Metaheuristics. Methologies and Traditional Applications, 2nd edn, vol. 1, pp. 87–111. Chapman and Hall/CRC (2018)

Towards Singular Optimality
in the Presence of Local Initial Knowledge

Hongyan Ji[ID] and Sriram V. Pemmaraju$^{(\boxtimes)}$[ID]

Department of Computer Science, The University of Iowa, Iowa City, IA, USA
{hongyan-ji,sriram-pemmaraju}@uiowa.edu

Abstract. The *Knowledge Till* ρ (in short, KT-ρ) CONGEST model is a variant of the classical CONGEST model of distributed computing in which each vertex v has initial knowledge of the radius-ρ ball centered at v. The most commonly studied variants of the CONGEST model are KT-0 CONGEST in which nodes initially know nothing about their neighbors and KT-1 CONGEST in which nodes initially know the IDs of all their neighbors. It has been shown that having access to neighbors IDs (as in the KT-1 CONGEST model) can substantially reduce the message complexity of algorithms for fundamental problems such as BROADCAST and MST. For example, King, Kutten, and Thorup (PODC 2015) show how to construct an MST using just $\tilde{O}(n)$ messages in the KT-1 CON-GEST model for an n-node graph, whereas there is an $\Omega(m)$ message lower bound for MST in the KT-0 CONGEST model for m-edge graphs. Building on this result, Gmyr and Pandurangan (DISC 2018) present a family of distributed randomized algorithms for various global problems that exhibit a trade-off between message and round complexity. These algorithms are based on constructing a sparse, spanning subgraph called a *danner*. Specifically, given a graph G and any $\delta \in [0,1]$, their algorithm constructs (with high probability) a danner that has diameter $\tilde{O}(D+n^{1-\delta})$ and $\tilde{O}(\min\{m, n^{1+\delta}\})$ edges in $\tilde{O}(n^{1-\delta})$ rounds while using $\tilde{O}(\min\{m, n^{1+\delta}\})$ messages, where n, m, and D are the number of nodes, edges, and the diameter of G, respectively. In the main result of this paper, we show that if we assume the KT-2 CONGEST model, it is possible to substantially improve the time-message trade-off in constructing a danner. Specifically, we show in the KT-2 CONGEST model, how to construct a danner that has diameter $\tilde{O}(D + n^{1-2\delta})$ and $\tilde{O}(\min\{m, n^{1+\delta}\})$ edges in $\tilde{O}(n^{1-2\delta})$ rounds while using $\tilde{O}(\min\{m, n^{1+\delta}\})$ messages for any $\delta \in [0, \frac{1}{2}]$. This result has immediate consequences for BROADCAST, spanning tree construction, MST, Leader Election, and even local problems such as $(\Delta+1)$-coloring in the KT-2 CONGEST model. For example, we obtain a KT-2 CONGEST algorithm for MST that runs in $\tilde{O}(D+n^{1/2})$ rounds, while using only $\tilde{O}(\min\{m, n^{1+1/4}\})$ messages.

Keywords: Danner · singular optimality · Broadcast · MST · $(\Delta + 1)$-coloring

Partially supported by NSF grant III 1955939. A full version of this paper is available on arXiv [8].

1 Introduction

The CONGEST model is a standard synchronous, message-passing model of distributed computation in which each node can send an $O(\log n)$-bit message along each incident edge, in each round [14]. Algorithms in the CONGEST model are typically measured by their *round complexity* and *message complexity*. The round complexity of an algorithm is the number of rounds it requires to complete, while the message complexity is the total messages exchanged among all the nodes throughout the algorithm's execution. Usually, researchers have focused on studying either the round complexity or the message complexity exclusively. But more recently, researchers have designed *singularly optimal* algorithms in the CONGEST model, which are algorithms that *simultaneously* achieve the best possible *round* and *message* complexity. An excellent example of a singularly optimal algorithm is Elkin's minimum spanning tree (MST) algorithm [5] that runs in $O((D + \sqrt{n}) \log n)$ rounds, using $O(m \log n + n \log n \cdot \log^* n)$ messages. Here n, m, and D are the number of vertices, the number of edges, and the diameter of the underlying graph. Since MST has a $\tilde{\Omega}(D + \sqrt{n})$ round[1] complexity lower bound [15,17] and an $\Omega(m)$ message complexity lower bound [1,12] in the CONGEST model, Elkin's algorithm is singularly optimal, up to logarithmic factors.

The story becomes more nuanced if we consider the *initial knowledge* that nodes have access to. The upper and lower bounds cited above are in the **K**nowledge **T**ill *radius 0* (in short, KT-0) variant of the CONGEST model (aka *clean network* model), in which nodes only have initial knowledge about themselves and have no other knowledge, even about neighbors. In the **K**nowledge **T**ill *radius 1* (in short, KT-1) variant of the CONGEST model, nodes initially possess the IDs of all their neighbors. Round complexity is not sensitive to the distinction between KT-0 CONGEST and KT-1 CONGEST because nodes can spend 1 round to share their IDs with all neighbors. However, message complexity is known to be quite sensitive to this distinction. Specifically, the $\Omega(m)$ message complexity lower bound for MST mentioned above only holds in the KT-0 CONGEST model. In fact, in the KT-1 CONGEST model, King, Kutten, and Thorup (henceforth, KKT) [9] presented an elegant algorithm capable of constructing an MST using just $\tilde{O}(n)$ messages, while running in $\tilde{O}(n)$ rounds. The $\tilde{\Omega}(D + \sqrt{n})$ round complexity lower bound [17] for MST mentioned above holds even in KT-1 CONGEST model. So for an MST algorithm to be singularly optimal in the KT-1 CONGEST model, it would need to run in $\tilde{O}(D + \sqrt{n})$ rounds while using $\tilde{O}(n)$ messages ($\Omega(n)$ is a trivial message lower bound for MST). The KKT algorithm matches the lower bound of message complexity but is far from reaching the lower bound of round complexity. In fact, a fundamental open question in distributed algorithms is whether it is possible to design a singularly optimal algorithm for MST in the KT-1 CONGEST model.

---

[1] We use $\tilde{O}(\cdot)$ to absorb polylog(n) factors in $O(\cdot)$ and $\tilde{\Omega}(\cdot)$ to absorb $1/\text{polylog}(n)$ factors in $\Omega(\cdot)$.

Important progress towards *possible* singular optimality for MST in the KT-1 CONGEST model was made by Ghaffari and Kuhn [6] who presented that MST could be solved in $\tilde{O}(D+\sqrt{n})$ rounds, and uses $\tilde{O}(\min\{m, n^{3/2}\})$ messages. Gymr and Pandurangan [7] who also showed the same results at the same time, where MST could be solved in $\tilde{O}(D + n^{1-\delta})$ rounds using $\tilde{O}(\min\{m, n^{1+\delta}\})$ messages w.h.p.[2] for any $\delta \in [0, \frac{1}{2}]$. Note that by setting $\delta = 1/2$ in this result, one can obtain an algorithm that is round-optimal, i.e., takes $\tilde{O}(D + \sqrt{n})$ rounds, and uses $\tilde{O}(\min\{m, n^{3/2}\})$ messages. And by setting $\delta = 0$, we can recover the KKT result. It is worth emphasizing that singular optimality may not be achievable and instead, we have to settle for a trade-off between messages and rounds. In fact, it is possible that the message-round trade-off shown by Gmyr and Pandurangan is optimal, though showing this is an important open question. The Gmyr-Pandurangan result for MST is based on a randomized algorithm that computes, for any $\delta \in [0, 1]$, a sparse, spanning subgraph that they call a *danner* that has diameter $\tilde{O}(D + n^{1-\delta})$ and $\tilde{O}(\min\{m, n^{1+\delta}\})$ edges. This algorithm runs in $\tilde{O}(n^{1-\delta})$ rounds while using $\tilde{O}(\min\{m, n^{1+\delta}\})$ messages. Once the danner is constructed, it essentially serves as the sparse "backbone" for efficient communication. As a result, Gymr and Pandurangan obtain round-message tradeoffs not just for MST but for a variety of other global problems such as leader election (LE), spanning tree construction (ST), and BROADCAST. They solve all of these problems in the KT-1 CONGEST model in $\tilde{O}(D + n^{1-\delta})$ rounds, using $\tilde{O}(\min\{m, n^{1+\delta}\})$ messages for any $\delta \in [0, 1]$.

An orthogonal direction was previously studied in the seminal paper of Awerbuch, Goldreich, Peleg, and Vainish [1] (henceforth, AGPV). They generalized the notion of initial knowledge and established tradeoffs between the volume of initial knowledge and the message complexity of algorithms. For any integer $\rho \geq 0$, in the *Knowledge Till ρ* (in short, KT-ρ) CONGEST model, each node v is provided initial knowledge of (i) the IDs of all nodes at distance at most ρ from v and (ii) the neighborhood of every vertex at distance at most ρ-1 from v. The KT-0 and KT-1 variants of the CONGEST model can be viewed as the most commonly considered special cases of the KT-ρ CONGEST model, with $\rho = 0$ and $\rho = 1$ respectively. AGPV showed a precise tradeoff between ρ, the radius of initial knowledge, and the message complexity of global problems such as BROADCAST. Specifically, they showed that in the KT-ρ CONGEST model, BROADCAST can be solved using $O(\min\{m, n^{1+c/\rho}\})$ messages, for some constant c. The main drawback of the AGPV message upper bound is that the BROADCAST algorithm that achieves the $O(n^{1+c/\rho})$ message-upper-bound requires $\Omega(n)$ rounds in the worst case. Hence, the algorithm exhibits a round complexity significantly exceeding the optimal $O(D)$ rounds required for BROADCAST.

Main Results. As illustrated by the above discussion, our understanding of singular optimality for global problems in the presence of initial knowledge is severely limited. Motivated specifically by the results of AGPV [1] and those of Gmyr and Pandurangan [7], we consider the design of distributed algorithms

---

[2] We use "w.h.p." as short for "with high probability", representing probability at least $1 - 1/n^c$ for constant $c \geq 1$.

for global problems in KT-2 CONGEST model. Our main contribution is showing that the round-message tradeoff shown by Gmyr and Pandurangan in the KT-1 CONGEST model can be substantially improved in the KT-2 CONGEST model. Specifically, we show the following result.

Main Theorem. There is a danner algorithm in the KT-2 CONGEST model that runs in $\tilde{O}(n^{1-2\delta})$ rounds, using $\tilde{O}(\min\{m, n^{1+\delta}\})$ messages w.h.p. The *danner* constructed by this algorithm has diameter $\tilde{O}(D + n^{1-2\delta})$ and $\tilde{O}(\min\{m, n^{1+\delta}\})$ edges w.h.p.

Like Gmyr and Pandurangan, we obtain implications of this danner construction for various global problems.

- We show that BROADCAST, LE, and ST can be solved in $\tilde{O}(D + n^{1-2\delta})$ rounds, while using $\tilde{O}(\min\{m, n^{1+\delta}\})$ messages for any $\delta \in [0, \frac{1}{2}]$.
- MST can be solved in $\tilde{O}(D + n^{1-2\delta})$ rounds, while using $\tilde{O}(\min\{m, n^{1+\delta}\})$ messages w.h.p., for $\delta \in [0, \frac{1}{4}]$.

Somewhat surprisingly, using recent results of [13], we show that even a local problem such as $(\Delta + 1)$-*coloring* can benefit from a more efficient danner construction. In [13], the authors present a $(\Delta + 1)$-coloring algorithm in the KT-1 CONGEST model that uses $\tilde{O}(\min\{m, n^{1.5}\})$ messages, while running in $\tilde{O}(D + \sqrt{n})$ round. Using our danner construction algorithm in the KT-2 CONGEST model, we show how to solve $(\Delta + 1)$-coloring in $\tilde{O}(D + n^{1-2\delta})$ rounds, while using $\tilde{O}(\min\{m, n^{1+\delta}\})$ messages for any $\delta \in [0, \frac{1}{2}]$.

It is also important to place our result in the context of implications we can obtain using the results of Derbel, Gavoille, Peleg, and Viennot [3]. This paper presents a deterministic distributed algorithm that, given an integer $k \geq 1$, constructs in k rounds a $(2k - 1)$-spanner with $O(k \cdot n^{1+1/k})$ edges for every n-node unweighted graph. This algorithm works in the LOCAL model, which is very similar to the CONGEST model, except that messages in the LOCAL model can be arbitrarily large in size. Now note that a k-round algorithm in the LOCAL model can be executed using 0 rounds and 0 messages if nodes are provided radius-k knowledge initially. This implies that in the KT-2 CONGEST model, a 3-spanner with $O(n^{1+1/2})$ edges can be constructed without communication. One can then use this 3-spanner as a starting point for the various global tasks mentioned above and obtain results that roughly match what we obtain by setting $\delta = 1/2$. However, obtaining any of our results that use fewer messages is impossible, e.g., by using values of $\delta < 1/2$.

1.1 KT-ρ CONGEST Model

We work in the fault-free, message-passing, synchronous distributed computing model, known as the CONGEST model [14]. In this model, the input graph $G = (V, E)$, $n = |V|$, $m = |E|$, also serves as the communication network. Nodes in the graph are processors, and each node has a unique ID drawn from a space whose size is polynomial in n. Edges serve as communication links. Each node can send a possibly distinct $O(\log n)$-bit message per edge per round. We further classify the CONGEST model based on the amount of initial knowledge nodes have. For any integer $\rho \geq 0$, we define the *Knowledge Till ρ* (in short, KT-ρ) CONGEST model as the CONGEST model in which each node v is provided initial knowledge of (i) the IDs of all nodes at distance at most ρ from v and (ii) the neighborhood of every vertex at a distance at most $\rho - 1$ from v. Thus, in the KT-0 CONGEST model, nodes do not know the IDs of neighbors. It is assumed that if a node v has degree d, then the d incident edges are connected to v via "ports" numbered arbitrarily from 1 through d. In the KT-1 CONGEST model, nodes initially know the IDs of neighbors but don't know anything more about their neighbors. In the rest of the paper, we assume that $\rho \leq D$, where D is the diameter of G. If $\rho > D$, then every vertex knows G completely at the start, and all problems become trivial in the KT-ρ CONGEST model.

1.2   Challenges, Approach, and Techniques

Our approach combines ideas from the well-known spanner algorithm of Baswana and Sen [2] with some ideas proposed by Gmyr and Pandurangan [7], which in turn depend on novel techniques proposed by KKT [9]. In the sequential (or centralized) setting, given an edge-weighted graph $G = (V, E)$ and any integer $k \geq 1$, the Baswana-Sen algorithm computes a $(2k-1)$-spanner with $O(k \cdot n^{1+\frac{1}{k}})$ edges in expected $O(k \cdot m)$ time, where m is the number of edges. The algorithm consists of two *phases*. In Phase 1, over a course of $k-1$ *iterations*, clusters are subsampled and then grown. This process establishes disjoint clusters, each resembling a rooted tree with a center. Initially, each vertex is by itself an individual cluster. In Phase 2, clusters are merged; this involves each vertex selecting a minimum-weight edge to each adjacent cluster and incorporating it into the spanner. The natural distributed implementation of the Baswana-Sen algorithm requires $O(k^2)$ rounds and uses $O(k \cdot m)$ messages in the KT-0 CONGEST model. This is clearly too message-inefficient for our purposes. The bottleneck in the Baswana-Sen algorithm is that for each sampled cluster to grow, it needs to inform all its neighbors that it has been sampled. More specifically this challenge appears in two forms.

Too many clusters: In the early iterations (in Phase 1) of the Baswana-Sen algorithm, there are too many clusters. We use too many messages if every cluster tries to inform every neighbor. This issue appears in the first iteration, in which each cluster is an individual node. For example, suppose that we want to produce a spanner with $O(n^{1+\frac{1}{3}})$ edges. Producing such a spanner would imply that downstream tasks such as BROADCAST can be completed using $O(n^{1+\frac{1}{3}})$ messages. So we pick $k = 3$ and the Baswana-Sen algorithm samples clusters with probability $n^{-1/3}$ in the first iteration. This yields $\Theta(n^{2/3})$ clusters w.h.p. and if each cluster sent messages to all neighbors, we could end up using $\Omega(n^{1+\frac{2}{3}})$ messages, well above our target of $O(n^{1+\frac{1}{3}})$ messages.

Redundant messages: Even if we were able to circumvent the above issue, there is a second and even more challenging obstacle. Suppose we have reached a point where the clusters have grown to trees of some constant diameter and the number of clusters is small enough that each cluster is permitted to send n messages informing neighbors. In this setting, if each node in a cluster sent messages to neighbors outside the cluster in an uncoordinated manner, we could end up sending up to $\Omega(n^2)$ messages because each neighbor of the cluster could receive the same message from multiple nodes in the cluster. Removing these redundant messages requires coordination within the cluster before sending messages, but the coordination itself can be quite message-costly.

We overcome these challenges in a variety of ways. First, we design a randomized estimation procedure that clusters can use to estimate if a neighbor w will hear from *other* sampled clusters. There is of course no need to send w a message if it is estimated that someone else will communicate with w. This estimation procedure critically depends on 2-hop initial knowledge. It allows clusters to communicate selectively with neighbors while still guaranteeing that every node w that is a neighbor of a sampled cluster joins one such cluster. To circumvent the challenge of redundant messages, we introduce two new subroutines, for growing a "star" cluster C that is both round and message efficient. Both subroutines critically depend on using 2-hop initial knowledge for their (simultaneous) round and message efficiency. For example, the GROWCLUSTER(C) subroutine (see Sect. 2) takes as input a "star" cluster C with N neighbors and grows the "star" by adding one edge from C to each neighbor. Our implementation requires $O(\sqrt{N})$ rounds while using a total of $O(N)$ messages, which is linear in the size of the constructed cluster. The estimation procedure and subroutines for cluster growing may be of independent interest to anyone designing efficient algorithms in the KT-2 CONGEST model.

Another technique we use is to allow surplus messages in early iterations, which even though not necessary in the early iterations, can improve message complexity in later iterations when combined with estimation procedures.

1.3   Related Work

While the current paper focuses only on synchronous models, we note that there is a growing body of related work in asynchronous models of distributed computation. In [12], a singularly near-optimal randomized leader election algorithm for general synchronous networks in the KT-0 CONGEST model is presented. This result was extended to the asynchronous KT-0 CONGEST model in [10,11]. Even for MST, there has been recent work on singularly optimal randomized MST algorithms in the asynchronous KT-0 CONGEST model [4]. This paper also contains an asynchronous MST algorithm that is sublinear in both time and messages in the KT-1 CONGEST model.

Since a danner is a relaxation of a spanner, it is worth mentioning a recent lower bound result for spanner construction due to Robinson [16]. He considers the KT-1 CONGEST model and shows that any algorithm running in $O(\text{poly}(n))$-time must send at least $\tilde{\Omega}(\frac{1}{t^2}n^{1+1/2t})$ bits to construct a $2t-1$-spanner. It would be interesting to determine if this type of spanner lower bound can be extended to danner construction.

Earlier we mentioned the work of Derbel, Gavoille, Peleg, and Viennot [3]. Another immediate implication of this work is that, for an integer $\rho \geq 1$, it is possible to construct a $(2\rho - 1)$-spanner with $O(\rho \cdot n^{1+1/\rho})$ edges using no communication in the KT-ρ CONGEST model. One can then use this $(2\rho-1)$-spanner as a starting point for various global tasks mentioned earlier (BROADCAST, LE, ST, MST). For example, this implies that for BROADCAST, LE, and ST there are algorithms in the KT-ρ CONGEST model that run in $O(\rho \cdot D)$ rounds, using $O(\rho \cdot n^{1+1/\rho})$ messages. For MST, the corresponding algorithm in the KT-ρ CONGEST model would run in $\tilde{O}(\rho(\cdot D + \sqrt{n}))$ rounds, using $O(\rho \cdot n^{1+1/\rho})$ messages. Setting $\rho = \Theta(\log n)$, this would yield an MST algorithm in the CONGEST model in which nodes have radius-$\Theta(\log n)$ initial knowledge, running in near-optimal $\tilde{O}(D + \sqrt{n})$ rounds, using near-optimal $\tilde{O}(n)$ messages.

1.4   Notation and Definitions

Let $\mathsf{Nbrs}(w)$ denote the set of neighbors of node w and let $\mathsf{Nbrs}_2(w)$ denote the set of 2-hop neighbors of node w. A *cluster* $C = (V(C), E(C))$ is a connected subgraph of graph $G = (V, E)$. All clusters considered in this paper will be constant-diameter trees. Furthermore, every cluster constructed by algorithms in this paper will start as a single node and then grow over the course of the algorithm. For a cluster C, we will use $\mathsf{center}(C)$ to denote the (unique) oldest node in a cluster and we use the ID of $\mathsf{center}(C)$ as the ID of cluster C; we will use the notation ID_C to denote the ID of C. Let $\mathsf{Nbrs}(C)$ denote the set of neighboring vertices of cluster C, i.e., $\mathsf{Nbrs}(C) \cap C = \emptyset$ and every $w \in \mathsf{Nbrs}(C)$ has a neighbor in C.

Note: Due to space limitations, all proofs are omitted from this paper and we refer interested readers to our arXiv version for full proofs [8].

2   Fast Subroutines in the KT-2 CONGEST Model

In this section, we identify 3 key tasks that can be implemented in a round- and message-efficient manner due to access to initial 2-hop knowledge. It is unclear how to execute these tasks efficiently without initial 2-hop knowledge, e.g., in the KT-1 CONGEST model. For each of the 3 tasks, we present subroutines that are round- and message-efficient.

Rank in neighbor's neighborhood: For a given node v and a given neighbor $w \in \mathsf{Nbrs}(v)$, we need to calculate the rank of its identifier (ID_v) within the neighborhood of w. We use $\mathrm{RANK}(v, w)$ to denote the subroutine that completes this task in the KT-2 CONGEST model. It is immediate that $\mathrm{RANK}(v, w)$ completes this task in 0 rounds, using 0 messages because v has all the information it needs within its 2-hop initial knowledge. One might think this task has also been efficiently completed in the KT-1 CONGEST model. In a sense, this is true because in the KT-1 CONGEST model node v can simply ask w to compute the rank of ID_v in w's neighborhood; this would take 2 rounds and 2 messages. Unfortunately, even this is too inefficient for our purposes because the RANK(v, w) subroutine will be used by v as a filter to determine whether v even needs to communicate with w.

Depth-2 BFS tree: Given a node v, our task is to efficiently construct a depth-2 BFS tree rooted at v. We now define a subroutine BUILDD2BFSTREE(v) in the KT-2 CONGEST model that can complete this task in 2 rounds, using $O(K)$ messages, where $K = |\mathsf{Nbrs}_2(v)|$. In other words, our goal is to use constant rounds and bound the number of messages by the size of the depth-2 BFS tree that is constructed.

1. Node v sends a message to each neighbor $w \in \mathsf{Nbrs}(v)$ and the edges $\{v, w\}$ are added to the output tree.
2. Using 2-hop initial knowledge, each node $w \in \mathsf{Nbrs}(v)$ can locally compute the set $\mathsf{Nbrs}(v)$. Then node $w \in \mathsf{Nbrs}(v)$ can use 2-hop initial knowledge to select a subset of neighbors to send messages to. Specifically, node w sends a message to a neighbor x iff ID_w is the lowest ID among the IDs of nodes in $\mathsf{Nbrs}(v) \cap \mathsf{Nbrs}(x)$. Node w can check whether it satisfies this condition using local computation on its initial 2-hop knowledge.

Note that it is possible to construct the second level of the depth-2 BFS tree rooted at v by using a standard "flooding" algorithm in which each node $w \in \mathsf{Nbrs}(v)$ sends a message to each of its neighbors. However, in the worst case, this could take $\Omega(K^2)$ messages. Using 2-hop knowledge allows for a much more message-efficient algorithm, while using constant number of rounds.

Lemma 1. *For any $v \in V$, the subroutine* BUILDD2BFSTREE(v) *runs in $O(1)$ rounds, using $O(|\mathsf{Nbrs}_2(v)|)$ messages.*

Growing a "star" cluster: Consider a cluster C that is "star" graph. In other words, center(C) is some vertex $v \in V$, the rest of the vertices satisfy $V(C) \setminus \{\mathrm{center}(C)\} \subseteq \mathsf{Nbrs}(v)$, and there are $|V(C)| - 1$ edges, from center(C) to each node in $V(C) \setminus \{\mathrm{center}(C)\}$. Note that $\mathsf{Nbrs}(C) \subseteq \mathsf{Nbrs}_2(\mathrm{center}(C))$ and it is possible for $|\mathsf{Nbrs}(C)|$ to be much smaller than $|\mathsf{Nbrs}_2(\mathrm{center}(C))|$. Let $N = |\mathsf{Nbrs}(C)|$. We need to complete this task efficiently: grow the cluster C by adding an edge from C to each of its N neighbors.

We now define an efficient subroutine GROWCLUSTER(C) for this task that uses $O(\sqrt{N})$ rounds and $O(N)$ messages. As with the previous subroutines, 2-hop initial knowledge plays a critical role in achieving these round and message complexities. Note that if $N \approx |\mathsf{Nbrs}_2(\mathrm{center}(C))|$, we can simply call the BUILDD2BFSTREE$(\mathrm{center}(C))$ subroutine defined above to complete this task in $O(1)$ rounds and $O(N)$ messages. So the challenge is in designing an efficient algorithm (in terms of N) even when $N \ll |\mathsf{Nbrs}_2(\mathrm{center}(C))|$.

Given a rooted tree T and a node u in T, we use $ch_T(u)$ to denote the set of children of u in T. We now describe our algorithm for GROWCLUSTER(C).

1. (**Local computation.**) center(C) uses 2-hop initial knowledge and knowledge of $V(C)$ to locally construct a tree T obtained by adding edges to C, where each added edge is from a node $w \in$ Nbrs(C) to a neighbor of w in C with minimum ID. Viewing T as a tree rooted at itself, center(C) classifies each of its children u as a *low-degree* node if $|ch_T(u)| \leq \sqrt{N}$; the rest of its children are classified as *high-degree* nodes.
   Notation: We use LDC to denote the set of low-degree children of center(C) and similarly HDC as the set of high-degree children of center(C).
2. To each node $u \in LDC$, center(C) sends the IDs of all nodes in $ch_T(u)$, one ID at a time. To each node $u \in HDC$, center(C) sends IDs of all nodes in HDC, again one ID at a time.
3. Each node $u \in LDC$ sends message Msg_1 to each node $v \in ch_T(u)$.
4. Each node $u \in HDC$ sends message Msg_2 to each node $v \in$ Nbrs(u), if u is the node with the smallest ID in $HDC \cap$ Nbrs(v).
5. Each node $v \notin V(C)$ adds edge $\{u,v\}$ to the output, where u is the node with smallest ID from which it has received a message.

Lemma 2. *Let T be the tree locally constructed by center(C) in Step 1 of subroutine* GROWCLUSTER(C). *The subroutine* GROWCLUSTER(C) *runs in $O(\sqrt{N})$ rounds and uses $O(N)$ messages and at the end of the subroutine edges in T that are not already in C are added to the output.*

3   Distributed Danner Construction in the KT-2 CONGEST Model

As mentioned earlier, our danner algorithm is inspired by the celebrated Baswana-Sen spanner algorithm [2]. For any integer $k \geq 1$, this algorithm constructs a $(2k-1)$-spanner by subsampling and growing clusters for $k-1$ iterations and then merging them. We consider this algorithm for $k = 3$ and implement the 2 iterations of the Baswana-Sen algorithm in a round- and message-efficient manner by leveraging 2-hop initial knowledge in a fundamental way. These correspond to the two *cluster growing* phases (described in Algorithm 1 and Algorithm 2(a), 2(b)). We use two versions of Phase 2 of the cluster growing algorithm, one for high δ ($\delta \in (1/3, 1/2]$) and one for low δ ($\delta \in [0, 1/3]$). It is unclear how to implement the merging step of the Baswana-Sen algorithm in a round- and message-efficient manner. So instead, we use ideas similar to those used by Gmyr and Pandurangan [7] for merging the clusters. This *cluster merging* phase is described in Algorithm 3. Since the cluster merging phase uses ideas similar to those used by Gmyr and Pandurangan [7], we sketch this phase and refer the reader to [7] for more details.

3.1 Cluster Growing: Phase 1

Algorithm 1 starts with a set $\mathcal{C}_0$ of initial clusters created with each node by itself being a cluster. We then sample each cluster with probability $n^{-\delta}$ and create a set $\mathcal{C}_1$ of sampled clusters. Let U denote the set of nodes not in sampled clusters, i.e., $U := \{v \in V \mid \{v\} \notin \mathcal{C}_1\}$. The rest of the algorithm aims to "grow" these sampled clusters by having unsampled nodes join neighboring sampled clusters. Since there are $\Theta(n^{1-\delta})$ sampled clusters w.h.p., it is not message efficient for each cluster $C \in \mathcal{C}_1$ (which is just a node at this point) to communicate with all neighbors $w \in \mathsf{Nbrs}(C)$. Instead, cluster $C \in \mathcal{C}_1$ uses the rank computation subroutine (see Sect. 2) to reduce the message complexity of this step. Specifically, cluster C first checks (in Step 4) if its ID belongs to the smallest $\lceil n^{2\delta} \rceil$ IDs of *neighbors* of w^3. In Step 5, the sampled cluster $C \in \mathcal{C}_1$ sends a message $\mathsf{Msg}_1(\mathsf{ID}_C)$ just to those neighbors w who pass this check. For $w \in V$, let $M_1(w)$ denote a set of tuples, including the IDs of clusters in $\mathcal{C}_1$ that sent w a message, and the edge which the cluster uses to send a message to w, i.e., $M_1(w) := \{(\mathsf{ID}_C, e) \mid w \text{ receives } \mathsf{Msg}_1(\mathsf{ID}_C) \text{ along edge } e\}$. The choice of the $\lceil n^{2\delta} \rceil$-sized "bucket" of smallest ID neighbors of w is critical in ensuring two properties we need: (i) every node w receives $\tilde{O}(n^\delta)$ messages (Lemma 3) and (ii) every node w that does not receive a message has low, i.e., $\tilde{O}(n^\delta)$, degree (Lemma 4). Subsequently, every node w that does not belong to a sampled cluster, can take one of two actions. If w receives a message from a neighboring sampled cluster, it joins the sampled cluster S_w with minimum ID among all sampled clusters from which it receives a message (Steps 9–11). When w joins a cluster, the edge connecting w to the cluster is added to the cluster and the danner H. For nodes w that do not receive any message from a sampled cluster, all incident edges of w are added to the danner (Step 13). The fact that such nodes are guaranteed to have a low degree is critical to ensuring this step is message-efficient.

We now prove bounds on the round and message complexity of Algorithm 1 and then prove properties of the output produced by the algorithm.

Lemma 3. *In Algorithm 1, for any $w \in U$, $|M_1(w)| = O(n^\delta)$ w.h.p.*

Lemma 4. *In Algorithm 1, for any $w \in U$, if $|M_1(w)| = \emptyset$ then w.h.p. $|\mathsf{Nbrs}(w)| = \tilde{O}(n^\delta)$.*

Lemma 5. *Algorithm 1 runs in $O(1)$ rounds and uses $\tilde{O}(n^{1+\delta})$ messages.*

Lemma 6. *After Algorithm 1 completes (a) $\mathcal{C}_1$ contains $\Theta(n^{1-\delta})$ clusters w.h.p. and every cluster is a star graph, (b) H is a spanning subgraph of G containing all cluster edges and all edges incident on nodes not in clusters, and (c) H contains $\tilde{O}(n^{1+\delta})$ edges.*

---

[3] Note that this step requires 2-hop knowledge; all steps in our algorithms which assume 2-hop knowledge are highlighted in gray.

Algorithm 1. Cluster Growing: Phase 1

Input: $G = (V, E)$, $\mathcal{C}_0 = \{\{v\} \mid v \in V\}$, $H = (V, \emptyset)$
Output: a set $\mathcal{C}_1$ of clusters, partially constructed danner H
1: Independently sample each cluster $C \in \mathcal{C}_0$ with probability $n^{-\delta}$;
   Notation: $\mathcal{C}_1 \subseteq \mathcal{C}_0$ denotes the set of sampled clusters; for each cluster $C = \{v\} \in \mathcal{C}_1$, v is the center of C, denoted by center(C); $U := \{v \in V \mid \{v\} \notin \mathcal{C}_1\}$ is the set of nodes not in sampled clusters.
2: **for** $C \in \mathcal{C}_1$ **do**   ▷ Actions by sampled clusters
3:     **for** $w \in$ Nbrs(C) **do**
4:         **if** RANK(center$(C), w) \leq \lceil n^{2\delta} \rceil$ **then**
5:             center(C) sends a message Msg$_1$(ID$_C$) to w.
6: **for** $w \in V$ **do**
7:     $M_1(w) := \{(\text{ID}_S, e) \mid w$ receives Msg$_1$(ID$_S$) along edge $e\}$
8:     **if** $w \in U$ **then**   ▷ Actions by nodes not in sampled clusters
9:         **if** $M_1(w) \neq \emptyset$ **then**   ▷ Actions by nodes that hear from a sampled cluster
10:            $S_w :=$ cluster S with minimum ID in $M_1(w)$.
11:            w joins the cluster S_w, the edge $\{w, \text{center}(S_w)\}$ is added to cluster S_w
    and to the danner H.
12:        **else**   ▷ Actions by nodes that don't hear from a sampled cluster
13:            w adds all incident edges to H by sending messages along incident edges.
           Note: w becomes inactive and does not participate any further in the
    algorithm.

3.2   Cluster Growing: Phase 2

In Algorithm Cluster Growing: Phase 2 (refer to pseudocode in Algorithm 2(a) and 2(b)), the clusters constructed in Algorithm 1 are further subsampled and grown. The details of this algorithm are more complicated than Algorithm 1, so we first describe it at a high level. At the start of the algorithm, each cluster constructed in Algorithm 1 is sampled with probability $n^{-\delta}$. This produces a collection $\mathcal{C}_2$ of $\Theta(n^{1-2\delta})$ clusters. We show that the clusters that are not sampled can be partitioned into two groups: (i) *high-degree* clusters, which are guaranteed to have a sampled cluster in their neighborhood, and (ii) *low-degree* clusters, which (as the name suggests) are guaranteed to have a small neighborhood, i.e., $\tilde{O}(n^{2\delta})$ nodes in their neighborhood. Each high-degree cluster C connects to a sampled cluster C' in its neighborhood, thus leading to the growth of cluster C'. For each low-degree cluster C and each neighbor $w \in$ Nbrs(C), we add an edge from C to w to the danner. We can afford to do this because such clusters have low degrees. We now explain the algorithm in more detail. In fact, we have two separate algorithms, one for high δ, i.e., $\delta \in (\frac{1}{3}, \frac{1}{2}]$ (Algorithm 2(a)), and one for low δ, i.e., $\delta \in [0, \frac{1}{3}]$ (Algorithm 2(b)). The high δ algorithm is easier and we explain it first.

High δ Case: In this case, each sampled cluster $C \in \mathcal{C}_2$ can (at least in theory) communicate the fact that it has been sampled to all its neighboring nodes. This is because there are $\Theta(n^{1-2\delta})$ sampled clusters in $\mathcal{C}_2$ w.h.p. and for $\delta > 1/3$,

$\Theta(n^{1-2\delta}) \times n = \Theta(n^{2-2\delta}) = O(n^{1+\delta})$. The actual communication is implemented by center(C) using the depth-2 BFS tree subroutine described in Sect. 2 to build a depth-2 BFS tree and broadcast via this tree to its 2-hop neighborhood (see Step 3). As established in the description of the depth-2 BFS tree subroutine, all of this takes $O(1)$ rounds and $O(n)$ messages. If a cluster C is not sampled, and some node $w \in C$ receives a message from a sampled cluster, then C is identified as a high-degree cluster (Step 8). Every high-degree cluster identified in this manner connects to the sampled cluster C' with the lowest ID that it hears from (Step 10). As a result, the (non-sampled) cluster C joins sampled cluster C' and an edge via which C heard about C' is added to cluster C' as well as the danner H. If a cluster C is not sampled and it does not hear from a sampled cluster, it is identified as a *low-degree cluster*. As the name suggests, we show in Lemma 7 that w.h.p. the center of every such low-degree cluster C has only $O(n^{2\delta})$ nodes in its 2-hop neighborhood. Since the total number of clusters is $O(n^{1-\delta})$ w.h.p., each low-degree cluster can afford to communicate with all nodes in its 2-hop neighborhood and add one edge connecting w to C, for each $w \in \mathsf{Nbrs}(C)$, to the danner H (Step 12). Again, the actual implementation of this step uses the depth-2 BFS tree subroutine.

Algorithm 2(a). Cluster Growing: Phase 2 [High Delta]

Input: $\delta \in (\frac{1}{3}, \frac{1}{2}]$; $G = (V, E)$, $\mathcal{C}_1$ and H are the set of clusters and partial danner output by Algorithm 1.

Output: a set $\mathcal{C}_2$ of clusters, partially constructed danner H

1: Independently sample each cluster $C \in \mathcal{C}_1$ with probability $n^{-\delta}$.
   Notation: $\mathcal{C}_2 \subseteq \mathcal{C}_1$ denotes the set of sampled clusters; for each cluster $C \in \mathcal{C}_2$, each node $w \in V(C) \setminus \{\mathsf{center}(C)\}$ is a child of $\mathsf{center}(C)$.

2: **for** $C \in \mathcal{C}_2$ **do**   ▷ C is sampled.

3:    center(C) uses BUILDD2BFSTREE(center(C)) to broadcast $\mathsf{Msg}_2(\mathsf{ID}_C)$ to $w \in \mathsf{Nbrs}_2(\mathsf{center}(C))$.

4: **for** $C \in \mathcal{C}_1 \setminus \mathcal{C}_2$ **do**   ▷ C is not sampled.

5:    Each node $w \in V(C)$ computes $M_2(w) := \{(\mathsf{ID}_S, e) \mid w \text{ receives } \mathsf{Msg}_2(\mathsf{ID}_S)$ along edge $e\}$

6:    Each child w in C with $M_2(w) \neq \emptyset$ sends $\mathsf{Msg}_3(\min M_2(w))$ to center(C).

7:    center(C) computes $M_3 := \{(\mathsf{ID}_S, e) \mid \mathsf{center}(C) \text{ receives } \mathsf{Msg}_3(\mathsf{ID}_S, e)\}$

8:    **if** $M_2(\mathsf{center}(C)) \cup M_3 \neq \emptyset$ **then**   ▷ C is a **high-degree cluster**

9:       center(C) computes $(\mathsf{ID}_{C'}, e) = \min(M_2(\mathsf{center}(C)) \cup M_3)$

10:      center(C) connects to cluster C' via edge e; edge e is added to cluster C' and to H.

11:   **else**   ▷ C is a **low-degree cluster**

12:      center(C) uses BUILDD2BFSTREE(center(C)) and adds edges of the tree to H.

Lemma 7. *In Algorithm 2(a), for every low-degree cluster $C \in \mathcal{C}_1 \setminus \mathcal{C}_2$, $|\mathsf{Nbrs}_2(\mathsf{center}(C))| = O(n^{2\delta})$, w.h.p.*

Lemma 8. *Algorithm 2(a) takes $O(1)$ rounds and uses $\tilde{O}(n^{1+\delta})$ messages.*

Low δ Case: When $\delta \in [0, \frac{1}{3}]$, it is message-inefficient for the center of a sampled cluster C to inform its 2-hop neighbors that C has been sampled. Specifically, Step 3 in Algorithm 2(a) uses $O(n^{2-2\delta})$ messages, which is bounded above by $O(n^{1+\delta})$ *only for large δ*, i.e., $\delta \geq 1/3$. To obtain the $O(n^{1+\delta})$ message complexity even for small δ, unsampled clusters have to learn if there is a neighboring sampled cluster in a more message-frugal manner. This challenge is overcome in Algorithm 2(b). Towards this goal, each unsampled cluster $C \in \mathcal{C}_1 \setminus \mathcal{C}_2$ considers all messages received in Algorithm 1, from sampled clusters in $\mathcal{C}_1$. More specifically, recall that we use $M_1(w)$ to denote the set of IDs of clusters in $\mathcal{C}_1$ that sent w a message in Algorithm 1 (see Steps 5 and 7 in Algorithm 1). Each center of a cluster $C \in \mathcal{C}_1 \setminus \mathcal{C}_2$ (i.e., an unsampled cluster) then gathers the IDs of all clusters in $\mathcal{C}_1$ that sent a message to some node $w \in C$ in Algorithm 1. This is done in Step 5 of Algorithm 2(b) by each node $w \in C$, $w \neq \mathsf{center}(C)$, simply sending the IDs in $M_1(w)$ one-by-one to $\mathsf{center}(C)$. This is still round-efficient because w.h.p. $|M_1(w)| = O(n^\delta) = O(n^{1-2\delta})$, with the latter equality being true for $\delta \leq 1/3$. This is also message-efficient, again because $|M_1(w)| = O(n^\delta)$, and so $O(n^{1+\delta})$ messages are sent in this step. $\mathsf{center}(C)$ computes the set $M_1(C)$ of IDs of clusters sampled in Algorithm 1 that sent the cluster C a message. If this set is large, i.e., at least $\frac{1}{2}n^\delta \ln n$ in size, then the cluster C is classified as a *high-degree cluster*. Such a cluster can be confident that w.h.p. at least one of the clusters with ID in $M_1(C)$ is sampled in Algorithm 2(b) and belongs to $\mathcal{C}_2$ (Lemma 10). Cluster C can then find and connect to one such cluster C' (Steps 8–10). On the other hand, if $|M_1(C)|$ is small, then we show that it must be the case that the total number of neighbors of cluster C is small (Lemma 9). In this case (as in Algorithm 2(a)), we want to grow cluster C, i.e., add edges connecting cluster C to each of its neighbors w, to the danner H. For this purpose we use the subroutine GROWCLUSTER(C) defined earlier. This subroutine uses $O(\sqrt{D})$ rounds and $O(|V(C)| + D)$ messages, where D is the size of the neighborhood of C. In Lemma 9 we show that $|\mathsf{Nbrs}(C)| = \tilde{O}(n^{2\delta})$, w.h.p. Combining this with the round and message complexity of GROWCLUSTER(C), we see that each low-degree cluster C can connect to all its neighbors in $\tilde{O}(n^\delta)$ rounds and $\tilde{O}(|V(C)| + n^{2\delta})$ messages. Since $\delta \in [0, 1/3]$, this yields a round complexity of $\tilde{O}(n^{1-2\delta})$ rounds. To get a bound on the overall message complexity, we sum over all clusters $C \in \mathcal{C}_1$ and get an $\tilde{O}(n^{1+\delta})$ bound on the message complexity using the fact that the $|\mathcal{C}_1| = O(n^{1-\delta})$ w.h.p.

Lemma 9. *In Algorithm 2(b), if C is a low-degree cluster then $|\mathsf{Nbrs}(C)| = O(n^{2\delta} \ln n)$, w.h.p.*

Lemma 10. *At the end of Algorithm 2(a), each cluster $C \in \mathcal{C}_1 \setminus \mathcal{C}_2$ (i.e., a non-sampled cluster) that is designated a high-degree cluster will connect to a cluster $C' \in \mathcal{C}_2$ (i.e., a sampled cluster), w.h.p.*

Lemma 11. *Algorithm 2(b) takes $\tilde{O}(n^{1-2\delta})$ rounds and uses $\tilde{O}(n^{1+\delta})$ messages.*

Algorithm 2(b). Cluster Growing: Phase 2 [Low Delta]

Input: $\delta \in [0, \frac{1}{3}]$; $G = (V, E)$, $\mathcal{C}_1$ and H are the set of clusters and partial danner output by Algorithm 1.

Output: a set $\mathcal{C}_2$ of clusters, partially constructed danner H

1: Independently sample each cluster $C \in \mathcal{C}_1$ with probability $n^{-\delta}$.
   Notation: $\mathcal{C}_2 \subseteq \mathcal{C}_1$ denotes the set of sampled clusters; for each cluster $C \in \mathcal{C}_2$, each node $w \in V(C) \setminus \{\text{center}(C)\}$ is a child of $\text{center}(C)$.

2: **for** $C \in \mathcal{C}_1$ **do**
3:    $\text{center}(C)$ broadcasts information on whether C is sampled to all its children.

4: **for** $C \in \mathcal{C}_1 \setminus \mathcal{C}_2$ **do**   ▷ C is not sampled
5:    Each node $w \in V(C) \setminus \{\text{center}(C)\}$ transmits *all* the elements in $M_1(w)$ to $\text{center}(C)$ along edge $\{w, \text{center}(C)\}$.
6:    $\text{center}(C)$ computes $M_1(C)$, a maximal subset of $\cup_w M_1(w)$ with unique IDs.
7:    **if** $|M_1(C)| \geq \frac{1}{2}n^\delta \ln n$ **then**   ▷ C is a **high-degree cluster**
8:       $\text{center}(C)$ chooses $X \subseteq M_1(C)$, consisting of the smallest $\lfloor \frac{1}{2}n^\delta \ln n \rfloor$ IDs from $M_1(C)$.
9:          for each $(\text{ID}_{C'}, e) \in X$, $\text{center}(C)$ sends a message along edge e to check if $C' \in \mathcal{C}_2$.
10:          On finding $C' \in \mathcal{C}_2$, $\text{center}(C)$ connects to cluster C' along edge e and adds edge e to H.
11:    **else**   ▷ C is a **low-degree cluster**
12:       $\text{center}(C)$ calls the subroutine GROWCLUSTER(C) ▷ Refer to 2
13:       Edges returned by this subroutine are added to H.

Lemma 12. *After Algorithm Cluster Growing: Phase 2 (refer to Algorithm 2(a), 2(b)) completes (a) there are $\Theta(n^{1-2\delta})$ clusters w.h.p., (b) every cluster is a tree with $O(1)$ diameter and all cluster edges belong to H, and (c) for every cluster $C \in \mathcal{C}_1 \setminus \mathcal{C}_2$ that is designated low-degree, there is one edge in H connecting cluster C to each of its neighbors, and (d) H contains $\tilde{O}(n^{1+\delta})$ edges w.h.p.*

3.3   Cluster Merging

The cluster merging algorithm in this subsection is similar to the corresponding steps in the Gmyr-Pandurangan KT-1 CONGEST danner algorithm [7], with some key differences in the analysis, which we point out.

Recall that $\mathcal{C}_2$ is the set of clusters returned by Phase 2 of the Cluster Growing algorithm (Algorithms 2(a) and 2(b)). In Algorithm 3, let $V(\mathcal{C}_2)$ denote the set of vertices belonging to clusters in $\mathcal{C}_2$, i.e., $V(\mathcal{C}_2) = \cup_{C \in \mathcal{C}_2} V(C)$. The steps of Algorithm 3 are performed on two induced subgraphs, $G[V(\mathcal{C}_2)]$, which we denote by $\hat{G}$ and $H[V(\mathcal{C}_2)]$, which we denote by $\hat{H}$. The algorithm executes a distributed Borůvka-style merging of the connected components of $\hat{H}$ using edges from the underlying graph $\hat{G}$. To do this in a manner that is both round and message efficient, we employ the FINDANY algorithm of KKT [9]. During each merging phase, FINDANY is employed by each connected component to efficiently locate an outgoing edge, which is then added to the danner H. Specific properties of the FINDANY algorithm are described in Theorem 1 below. The process of finding an outgoing edge is coordinated by a leader, elected within each component. For this purpose, we use the leader election algorithm from [12] that is both round and message efficient. Theorem 2 below specifies the properties of this leader election algorithm. The entire process requires only $\log n$ iterations to merge all fragments into a set of maximally connected components, i.e., reach a stage where no further merging is possible. This takes only $\log n$ iterations because in each iteration, each connected component with an outgoing edge merges with at least one other connected component. See [9] for further details of how FINDANY works and how it is used to merge connected components. The analysis below shows that the diameter of every connected component before every iteration is bounded above by $\tilde{O}(n^{1-2\delta})$ w.h.p. Thus, this is an upper bound on the number of rounds it takes for a leader to coordinate the process of finding an outgoing edge. So waiting for $\tilde{O}(n^{1-2\delta})$ rounds in each iteration ensures that all the iterations proceed in lock-step.

Algorithm 3. Cluster Merging

Input: $G = (V, E)$, partially constructed danner H, set $\mathcal{C}_2$ of clusters returned by Algorithms 2(a) or 2(b)
Output: fully constructed danner H
1: **for** $i = 1$ to $\log n$ **do**   ▷ Do the following steps in parallel in each connected component K of $\hat{H}$.
2:     Elect a leader using the algorithm from Theorem 2.
3:     Using the algorithm FINDANY from Theorem 1 to find an edge in $\hat{G}$ leaving K. The leader elected in Step 2 coordinates this process. If such an edge exists, add it to H and $\hat{H}$.
4:     Wait until $\tilde{O}(n^{1-2\delta})$ rounds have passed in this iteration before starting the next iteration.   ▷ To synchronize the execution between the connected components.

Analysis. The Cluster Merging algorithm (Algorithm 3) relies on two well-known previously designed algorithms. The first algorithm, FINDANY is the core of the KKT MST algorithm [9]. As shown in the theorem below, given a connected component H within a graph G, it efficiently identifies an outgoing edge from H, if such an edge exists. The natural algorithm for this task would be for each node v in H to scan its neighborhood and identify a neighbor outside H. Then, all nodes v in H can upcast one identified edge each to the leader of H. Finally, the leader can pick one edge from among these. The problem with this algorithm is that the step that requires v to scan its neighbors is extremely message-inefficient and could require $\Omega(m)$ edges. KKT overcame this issue by cleverly using random hash functions with certain specific properties.

Theorem 1 (KKT [9]). *Consider a connected subgraph H of a graph G. An algorithm* FINDANY *in the KT-1* CONGEST *model exists that w.h.p. outputs an arbitrary edge* ID *in G leaving H if such an edge exists and $\emptyset$ if no such edge exists. This algorithm takes $\tilde{O}(D(H))$ rounds and $\tilde{O}(E(H))$ messages.*

We also need an efficient leader election algorithm because we need each connected component H to have a leader that can coordinate the process of finding an outgoing edge. We use the following theorem stated in Gmyr and Pandurangan [7], which in turn is a reformulation of Corollary 4.2 in the paper by Kutten, Pandurangan, Peleg, Robinson, and Trehan [12].

Theorem 2 ([12]). *There exists an algorithm in the KT-0* CONGEST *model that, for any graph G, elects a leader in $O(D(G))$ rounds and utilizes $\tilde{O}(E(G))$ messages, w.h.p.*

The Gmyr-Pandurangan analysis requires two key properties to hold *before* the Cluster Merging algorithm: (i) there is a set $\mathcal{C}$ of clusters, each with constant diameter and (ii) the partially constructed danner H contains all the edges belonging to the clusters along with *all* edges incident on nodes not in clusters. If these two properties hold, then they can show that the following crucial property holds *after* the Cluster Merging algorithm:

> Before each iteration of the algorithm and after the algorithm ends, the sum of the diameters of all the connected components in $\hat{H}$ is $O(|\mathcal{C}|)$.

After Phase 2 of our Cluster Growing algorithm ends, we do have a set $\mathcal{C}_2$ of clusters, with each cluster $C \in \mathcal{C}_2$ having constant diameter. However, we do not have the second property. This is because some nodes not in clusters in $\mathcal{C}_2$ belong to low-degree clusters not sampled in Phase 2 of the Cluster Growing algorithm. Specifically, consider a cluster $C \in \mathcal{C}_1 \setminus \mathcal{C}_2$ such that cluster C is designated as a low-degree cluster in Phase 2 of the Cluster Growing algorithm. Here, we refer to both versions of our algorithm, i.e., Algorithm 2(a) and 2(b). Nodes in C may have only a small number of incident edges belonging to C and therefore to H. So Property (ii) above, which is required by the analysis of the Gmyr-Pandurangan algorithm, may not hold. However, we know that for every such cluster C, we add edges connecting C to each of its neighbors $w \in \mathsf{Nbrs}(C)$ to the danner. This means we can treat each low-degree cluster $C \in \mathcal{C}_1 \setminus \mathcal{C}_2$ as a *super node* and contract it. Each super node now has the property that all incident edges are in H. Given that after Phase 1 of our Cluster Growing algorithm, we have set aside a set of nodes (see Step 11 in Algorithm 1) and added all incident edges to H, we now have both properties needed by the Gmyr-Pandurangan analysis. As a result, we obtain the following lemma. It is worth highlighting that since $|\mathcal{C}_2| = \tilde{O}(n^{1-2\delta})$, the sum of the diameters of the connected components in $\hat{H}$ is also $\tilde{O}(n^{1-2\delta})$. This is in contrast with the corresponding Gmyr-Pandurangan lemma that obtains a weaker $\tilde{O}(n^{1-\delta})$ because that is the number of clusters they have before starting the Cluster Merging algorithm.

Lemma 13. *Let $K_1, \ldots, K_r$ be the connected components of $\hat{H}$ before any iteration of the loop in Algorithm 3 or after the final iteration. It holds $\sum_{i=1}^{r} diam(K_i) = \tilde{O}(n^{1-2\delta})$, w.h.p.*

The rest of the analysis is identical to that of Gmyr and Pandurangan [7] and we obtain the following lemmas.

Lemma 14. *Algorithm 3 computes a danner in $\tilde{O}(n^{1-2\delta})$ rounds and using $\tilde{O}(min\{m, n^{1+\delta}\})$ messages w.h.p.*

Lemma 15. *Algorithm 3 computes a danner with $D(H) \leq D(G) + \tilde{O}(n^{1-2\delta})$ and with $\tilde{O}(min\{m, n^{1+\delta}\})$ edges, w.h.p.*

With Lemmas 5, 8, 11, and 14 in place, it is easy to see that the Danner Algorithm (including Algorithm 1, 2(a), 2(b), and 3) takes $\tilde{O}(n^{1-2\delta})$ rounds and sends $\tilde{O}(n^{1+\delta})$ messages w.h.p. With Lemmas 6, 12, and 15 in place, after the algorithm terminates it holds that H has $\tilde{O}(n^{1+\delta})$ edges and $\tilde{O}(D+n^{1-2\delta})$ diameter, where $\delta \in [0, \frac{1}{2}]$. As a result, we obtain the following theorem directly.

Theorem 3. *The Danner Algorithm (including Algorithm 1, 2(a), 2(b), and 3) takes $\tilde{O}(n^{1-2\delta})$ rounds and sends $\tilde{O}(n^{1+\delta})$ messages w.h.p. After the algorithm terminates, it holds that H has $\tilde{O}(min\{m, n^{1+\delta}\})$ edges and $\tilde{O}(D+n^{1-2\delta})$ diameter w.h.p., where $\delta \in [0, \frac{1}{2}]$.*

4   Conclusion

Our main contribution is showing that the round-message tradeoff shown by Gmyr and Pandurangan in the KT-1 CONGEST model can be substantially improved in the KT-2 CONGEST model. Specifically, we show that there is a danner algorithm in the KT-2 CONGEST model that runs in $\tilde{O}(n^{1-2\delta})$ rounds, using $\tilde{O}(min\{m, n^{1+\delta}\})$ messages w.h.p. The *danner* constructed by this algorithm has diameter $\tilde{O}(D + n^{1-2\delta})$ and $\tilde{O}(min\{m, n^{1+\delta}\})$ edges w.h.p. Similar to Gmyr and Pandurangan, we obtain implications of this danner construction for a variety of global problems, namely BROADCAST, LE, ST, and MST, as well as the $(\Delta + 1)$-coloring problem (which is a local problem).

Since we don't show lower bounds, it is not clear if the round-message tradeoff we show is optimal. This is open in the KT-1 CONGEST model as well because we don't know if the tradeoff shown by Gmyr and Pandurangan is optimal. One possible way to improve the tradeoff we show is to construct a constant-spanner, rather than a danner, which imposes a large additive factor $n^{1-2\delta}$ on the diameter of the subgraph. However, it is not clear if a constant-spanner can be constructed in the KT-2 CONGEST model in $\tilde{O}(n^{1-2\delta})$ rounds, using $\tilde{O}(min\{m, n^{1+\delta}\})$ messages. This problem would be a natural follow-up to our current work.

References

1. Awerbuch, B., Goldreich, O., Vainish, R., Peleg, D.: A trade-off between information and communication in broadcast protocols. J. ACM (JACM) **37**(2), 238–256 (1990)
2. Baswana, S., Sen, S.: A simple and linear time randomized algorithm for computing sparse spanners in weighted graphs. Random Struct. Algorithms **30**(4), 532–563 (2007)
3. Derbel, B., Gavoille, C., Peleg, D., Viennot, L.: On the locality of distributed sparse spanner construction. In: Proceedings of the Twenty-Seventh ACM Symposium on Principles of Distributed Computing, pp. 273–282 (2008)
4. Dufoulon, F., Kutten, S., Moses, W.K., Jr., Pandurangan, G., Peleg, D.: An almost singularly optimal asynchronous distributed MST algorithm. arXiv preprint arXiv:2210.01173 (2022)
5. Elkin, M.: A simple deterministic distributed mst algorithm with near-optimal time and message complexities. J. ACM (JACM) **67**(2), 1–15 (2020)
6. Ghaffari, M., Kuhn, F.: Distributed MST and broadcast with fewer messages, and faster gossiping. In: 32nd International Symposium on Distributed Computing (DISC 2018), vol. 121, pp. 30:1–30:12. Schloss Dagstuhl-Leibniz-Zentrum fuer Informatik (2018)
7. Gmyr, R., Pandurangan, G.: Time-message trade-offs in distributed algorithms. In: 32nd International Symposium on Distributed Computing, DISC 2018, pp. 32:1–32:18 (2018)
8. Ji, H., Pemmaraju, S.V.: Towards singular optimality in the presence of local initial knowledge (2024). https://arxiv.org/abs/2402.14221
9. King, V., Kutten, S., Thorup, M.: Construction and impromptu repair of an MST in a distributed network with o (m) communication. In: Proceedings of the 2015 ACM Symposium on Principles of Distributed Computing, pp. 71–80 (2015)
10. Kutten, S., Moses, W.K., Jr., Pandurangan, G., Peleg, D.: Singularly optimal randomized leader election. arXiv preprint arXiv:2008.02782 (2020)
11. Kutten, S., Moses, W.K., Jr., Pandurangan, G., Peleg, D.: Singularly near optimal leader election in asynchronous networks. arXiv preprint arXiv:2108.02197 (2021)
12. Kutten, S., Pandurangan, G., Peleg, D., Robinson, P., Trehan, A.: On the complexity of universal leader election. J. ACM (JACM) **62**(1), 1–27 (2015)
13. Pai, S., Pandurangan, G., Pemmaraju, S.V., Robinson, P.: Can we break symmetry with o(m) communication? In: Proceedings of the 2021 ACM Symposium on Principles of Distributed Computing, pp. 247–257 (2021)
14. Peleg, D.: Distributed Computing: A Locality-Sensitive Approach. Society for Industrial and Applied Mathematics, Philadelphia (2000)
15. Peleg, D., Rubinovich, V.: A near-tight lower bound on the time complexity of distributed minimum-weight spanning tree construction. SIAM J. Comput. **30**(5), 1427–1442 (2000)
16. Robinson, P.: Being fast means being chatty: the local information cost of graph spanners. In: Proceedings of the Thirty-Second Annual ACM-SIAM Symposium on Discrete Algorithms, SODA 2021, pp. 2105–2120. Society for Industrial and Applied Mathematics, USA (2021)
17. Sarma, A.D., et al.: Distributed verification and hardness of distributed approximation. SIAM J. Comput. **41**(5), 1235–1265 (2012)

Efficient Wait-Free Linearizable Implementations of Approximate Bounded Counters Using Read-Write Registers

Colette Johnen[1], Adnane Khattabi[2], Alessia Milani[3], and Jennifer L. Welch[4]([✉])

[1] LaBRI, Université de Bordeaux, Bordeaux, France
johnen@labri.fr
[2] ESI Group, Lyon, France
adnane.khattabi@esi-group.com
[3] LIS, Aix-Marseille Université, Marseille, France
alessia.milani@univ-amu.fr
[4] Texas A&M University, College Station, TX 77843, USA
welch@cse.tamu.edu

Abstract. Relaxing the sequential specification of a shared object is a way to obtain an implementation with better performance compared to implementing the original specification. We apply this approach to the *Counter* object, under the assumption that the number of times the Counter is incremented in any execution is at most a known bound m. We consider the *k-multiplicative-accurate* Counter object, where each read operation returns an approximate value that is within a multiplicative factor k of the accurate value. More specifically, a read is allowed to return an approximate value x of the number v of increments previously applied to the counter such that $v/k \leq x \leq vk$. We present three algorithms to implement this object in a wait-free linearizable manner in the shared memory model using read-write registers. All the algorithms have read operations whose worst-case step complexity improves exponentially on that for an *exact* m-bounded counter (which in turn improves exponentially on that for an exact *unbounded* counter). Two of the algorithms have read step complexity that is asymptotically optimal. The algorithms differ in their requirements on k, step complexity of the increment operation, and space complexity.

Keywords: Bounded counters · Approximate counters · Linearizable implementations · Wait-freedom · Read-write registers · Complexity

1 Introduction

Finding efficient ways to implement linearizable shared objects out of other shared objects in crash-prone asynchronous distributed systems is central to

This work has been partially supported by the ANR projects SKYDATA (ANR-22-CE25-0008) and TEMPOGRAL (ANR-22-CE48-0001).

concurrent programming. In this paper, we focus on implementing the *Counter* object, which provides an Increment operation that increases the value of the counter by one, and a Read operation that returns the current value of the counter. Counters are a fundamental data structure for many applications, ranging from multicore architectures to web-based e-commerce.

A general result by Jayanti, Tan and Toueg [13] implies the discouraging result that any implementation of a Counter using "historyless" objects—those whose modifying operations over-write each other—has an execution in which some operation takes $\Omega(n)$ steps on the building block objects, where n is the number of processes in the system. As noted in [2], this lower bound is tight, as a Counter can be implemented with an atomic snapshot object, which in turn can be implemented using read-write registers with linear step complexity [12].

There are several ways one could attempt to circumvent this linear lower bound on the worst-case step complexity. The proof in [13] constructs a slow execution in which a very large number of operations are performed on the implemented object. A natural restriction is to consider the case when the number of operations, especially Increments, is *bounded*, say at most m. This approach is taken by Aspnes, Attiya and Censor-Hillel [2], resulting in an algorithm for an exact counter whose worst-case step complexity is $O(\log m)$ for the Read operation and $O(\log n \cdot \log m)$ for the Increment operation[1]. The step complexity for Read is tight, as shown by an $\Omega(\log m)$ lower bound in [2,4].

Another approach is to consider the *amortized* step complexity instead of worst case. It could be that in any execution, most of the operations are fast, while only a few are slow. Baig, Hendler, Milani and Travers [6] present an exact unbounded Counter implementation that has $O(\log^2 n)$ amortized step complexity. This performance is close to tight, thanks to an $\Omega(\log n)$ lower bound in [6], which is based on a result in [5].

Finally, the semantics of the Counter object being implemented could be relaxed so that the value returned by a Read is not necessarily exactly the number of preceding Increments. Approximate counting has many applications (e.g., [1,4]); approximate *probabilistic* counting has been studied extensively both in the sequential setting (e.g., [8,16]) and the concurrent setting (e.g., [3,7]). We focus on the *deterministic* situation. A Counter is said to be *k-multiplicative-accurate* if, informally speaking, each Read returns a value that is within a factor of k of the Counter value. This approximation is exploited by Hendler, Khattabi, Milani and Travers [9] to improve the *amortized* step complexity of an unbounded counter. They achieve $O(1)$ amortized step complexity in any execution if $k \geq n$, while for certain long executions the constant amortized step complexity is achieved for $k \geq \sqrt{n}$ [15]; if $k < \sqrt{n}$, they show a lower bound of $\Omega(\log \frac{n}{k^2})$ on the amortized step complexity. The constant step complexity

---

[1] $\log n$ means $\log_2 n$; any other base of a logarithm is explicitly given. Since Read and Increment can be implemented in $O(n)$ steps, technically each asymptotic bound of the form $f(k, m, n)$ for some function f should be $\min\{n, f(k, m, n)\}$. To simplify the expressions, we implicitly assume that $f(k, m, n)$ is $o(n)$ so that we can drop the "minimum of n" part of the expression.

upper bound does not contradict the $\Omega(\log n)$ lower bound in [6] since the lower bound is for exact counters, indicating the performance benefit resulting from the approximation.

In this paper, we present three implementations of a *k-multiplicative-accurate m-bounded* Counter and analyze their *worst-case* step complexities. All of our algorithms use only read-write registers. All our algorithms have $O(\log \log m)$ or smaller Read complexity, which is an exponential improvement on the $O(\log m)$ bound in [2], thanks to the approximation. Our results are incomparable to those in [6,9] since we consider *worst-case* complexity of *bounded* Counters instead of *amortized* complexity of *unbounded* Counters.

We first present a simple Counter implementation in which both Read and Increment have $O(\log \log m)$ step complexity, as long as k is a real number with $k \geq \sqrt{2n}$. (See Sect. 3.) The algorithm uses a shared max-register object that is bounded by $\lceil \log m \rceil$ in which processes store the logarithm of the number of Increments that they know about so far. When a process executes an Increment, it keeps track in a local variable of the number of Increments invoked at it; every time the number has doubled, it writes a new value to the max-register, which is one larger than the previous value it wrote. The Read operation reads the max-register and returns k times 2^r, where r is the value read from the max-register. By using the max-register implementation in [2], we obtain the claimed step complexity. The idea of waiting to expose the number of Increments until a power has been reached and then writing the logarithm of the number is taken from an algorithm in [9]; however, that algorithm waits for powers of k and does not have good complexity in executions with few increments. Our innovations are to wait for powers of 2 instead of k, and to carefully control the number of increments exposed together, as well as the evolution of this value.

Our second result is a Counter implementation in which the step complexity of Read is $O(\log \log_k m)$ and that of Increment is $O(\max\{\log n \cdot \log(\frac{kn}{k-1}), \log \log_k m\})$, for any real number k with $k > 1$. (See Sect. 4.) The Read complexity is asymptotically optimal due to a lower bound of $\Omega(\log \log_k m)$ in [9]. Like the first algorithm, this one uses a shared max-register. To track the number of Increments more accurately while keeping the fast Read operation, each Increment could increment an exact counter, then read the exact counter and write the logarithm of the value read to the max-register. However, the exact counter is bounded by m, and when implemented with registers using the algorithm in [2] results in an Increment step complexity of $\Omega(\log n \cdot \log m)$. To reduce the Increment step complexity in the common case when m is much larger than n, our algorithm uses an array of smaller exact counter objects, which we call "buckets". There are $\lceil (k-1)\frac{m}{n} \rceil$ buckets, with the maximum value stored in a bucket being $\lceil \left(\frac{k}{k-1}\right)n \rceil$. Using buckets reduces the step complexity, with the tradeoff that a process cannot always decide if an increment in a bucket has an effect (i.e., was stored). By using the exact bounded counter and max-register implementations in [2], we obtain the claimed step complexity.

Our third algorithm works for any integer k with $k \geq 2$. It combines the techniques of the first two algorithms: exposing increments in batches and using buckets. It has better space complexity than, and the same Read step complexity as, the second algorithm, but its Increment step complexity is worse when k is super-constant with respect to n. (See Sect. 5.) In more detail, the Increment step complexity is $O(\max\{\log n \cdot \log n, \log \log_k m\})$, the number of buckets is $\lceil \log_k \frac{m}{n} \rceil$, and the maximum value stored in a bucket is $4n$.

A key challenge for both our second and third algorithms was to prove linearizability. The definitions of the linearizations are subtle and take into account interactions between multiple operations.

It is easy to see that all our algorithms are wait-free as they have no loops or waiting statements. Our results are summarized in Table 1.

Table 1. Comparison of our wait-free linearizable k-multiplicative-accurate m-bounded Counter implementations using read-write registers

	Algorithm 1	Algorithm 2	Algorithm 3
k	$k \geq \sqrt{2n}$	$k > 1$	$k \geq 2$
Read step complexity	$O(\log \log m)$	$O(\log \log_k m)$ optimal	$O(\log \log_k m)$ optimal
Increment step complexity	$O(\log \log m)$	$O(\max\{\log n \cdot \log \frac{k}{k-1} n, \log \log_k m\})$	$O(\max\{\log^2 n, \log \log_k m\})$
space complexity	1 max-register	1 max-register; $\lceil \frac{(k-1)m}{n} \rceil$ exact counters, each $\lceil \frac{kn}{k-1} \rceil$-bounded	1 max-register; $\lceil \log_k \frac{m}{n} \rceil$ exact counters, each $4n$-bounded

2   Preliminaries

Overview. We consider an asynchronous shared memory system in which a set $\mathcal{P}$ of n crash-prone processes communicate by applying operations on shared objects. The objects are *linearizable*, which means that each operation appears to occur instantaneously at some point between its invocation and response and it conforms to the sequential specification of the object [11]. Our ultimate goal is for the system to implement a linearizable k-multiplicative-accurate counter by communicating using linearizable read-write registers. However, for convenience, our algorithms are described using linearizable max-register objects and linearizable exact counter objects, which in turn can be implemented using linearizable read-write registers. When proving the correctness of our algorithms, we assume that the operations on the max-registers and exact counters are instantaneous, but when analyzing the step complexity of our algorithms, we take into account the specific algorithms used to implement the max-registers and exact counters out of registers.

Sequential Specifications of Objects. We next give the sequential specifications of the objects under consideration.

– A *read/write register* has operations Read and Write; in every sequence of operations, each Read returns the value of the latest preceding Write (or the initial value if there is none).
– A *max-register* has operations MaxRead and MaxWrite; in every sequence of operations, each MaxRead returns the largest value among all the preceding MaxWrites (or 0 if there is none). A max-register is *h-bounded* if attention is restricted to sequences of operations in which the largest value of any MaxWrite is h.
– An *exact Counter* has operations Read and Increment; in every sequence of operations, each Read returns the number of preceding Increments.
– A *k-multiplicative-accurate Counter* has operations Read and Increment; in every sequence of operations, each Read returns a value x such that $v/k \leq x \leq kv$, where v is the number of preceding Increments.
– A Counter (either exact or multiplicative-accurate) is *h-bounded* if attention is restricted to sequences of operations in which at most h Increments occur.

Executions of Implementations. An *implementation* of a shared object provides a specific data representation for the object from base objects, each of which is assigned an initial value. The implementation also provides sequential algorithms for each process in $\mathcal{P}$ that are executed when operations on the implemented object (a k-multiplicative-accurate m-bounded Counter in our case) is invoked; these algorithms involve local computation and operations on the base objects.

Each *step* of a process contains at most one invocation of an operation of the implemented object, at most one operation on a base object, and at most one response for an operation of the implemented object. A step can also contain local computation by the process.

An *execution* of an implementation of a shared object is a possibly-infinite sequence of process steps such that the subsequence of the execution consisting of all the steps by a single process is *well-formed*, meaning that the first step is an invocation, invocation and responses alternate, and the steps between an invocation and its following response are defined by the algorithm provided by the implementation. Since we put no constraints on the number of steps between consecutive steps of a process or between steps of different processes, we have modeled an *asynchronous* system.

Wait-Freedom. We desire algorithms that can tolerate any number of crash failures. This property is captured by the notion of *wait-freedom* [10]: in every execution, if a process takes an infinite number of steps, then the process executes an infinite number of operations on the implemented object. In other words, each process completes an operation if it performs a sufficiently large number of steps, regardless of how the other processes' steps are scheduled.

The rest of this section is devoted to a lemma, whose proof is in the full version [14], showing that a generic way of ordering the operations in an execution ensures the relative order of non-overlapping operations. The lemma is independent of the semantics of the object being implemented. This behavior is part of what is needed to prove linearizability. The other part, showing that the sequential specification is respected, of course depends on the specific object being implemented and is not addressed by this lemma. The purpose of extracting this observation as a stand-alone lemma is that essentially the same argument is used in the linearizability proofs for Algorithms 2 and 3.

Partition all the complete, and any subset of the incomplete, operations in a (concurrent) execution E into two sets, A and B. Create a total order L of the operations as follows: Choose any point inside each operation in A and order these operations in L according to the chosen points. Consider the operations in B in increasing order of when they start in E. Let op be the next operation under consideration. Let op' be the earliest operation already in L that begins in E after op ends in E. Place op immediately before op' in L.

Lemma 1. *L respects the order of non-overlapping operations in E.*

3   Algorithm for $k \geq \sqrt{2n}$

In this section, we present a wait-free linearizable m-bounded k-multiplicative-accurate Counter, implemented using a shared bounded max-register object, assuming that k is a real number with $k \geq \sqrt{2n}$. The worst-case step complexity is $O(\log \log m)$ for both the Read and Increment operations. Pseudocode is given in Algorithm 1.

3.1   Algorithm Description

Each process MaxWrites to a shared max-register *logNumIncrems* after it has experienced a certain number of Increment invocations. The value MaxWritten to *logNumIncrems* is stored in a local variable *nextVal*, which starts at 0 and is incremented by 1 every time *logNumIncrems* is written by the process. (Of course, since *logNumIncrems* is a max-register, if another process has already MaxWritten a larger value, this MaxWrite will have no effect.) The rule for MaxWriting to *logNumIncrems* is that a local variable *lcounter* has reached a given value, stored in a local variable *threshold*. Variable *threshold* starts at 1 and is doubled every time *logNumIncrems* is MaxWritten (with the exception that the threshold remains 1 after the first MaxWrite). Variable *lcounter* starts at 0, is incremented by one every time an Increment is invoked, and is reset to 0 when *logNumIncrems* is written. This procedure ensures that the value MaxWritten to *logNumIncrems* by the process is approximately the logarithm (base 2) of the number of Increments invoked at the process. An exponential amount of time (and space) is saved by storing the logarithm of the number instead of the number itself, although some accuracy is lost. Since at most m Increments are assumed to occur, the max-register can be bounded by $\lceil \log m \rceil$.

During an instance of Read, a process simply MaxReads the value r of *logNumIncrems*, and if $r \geq 0$, then it returns $k \cdot 2^r$. Otherwise, the process returns 0. Multiplying by k, which is at least $\sqrt{n}$, takes care of the uncertainty caused by the possibility of concurrent Increments. We show that the return value falls within the approximation range defined by the sequential specification of the k-multiplicative-accurate counter.

Algorithm 1: Implementation of a k-multiplicative m-bounded counter with $k \geq \sqrt{2n}$.

Shared variable
- $logNumIncrems$: $\lceil \log m \rceil$-bounded max register object that stores the logarithm of the number of increments exposed to the readers, initialized to -1.

Local persistent variables
- $lcounter$: counts the number of increments invoked locally, initially 0.
- $threshold$: stores the current required number of locally-invoked increments to update $logNumIncrems$, initially 1.
- $nextVal$: stores the next value to MaxWrite into $logNumIncrems$, initially 0.

1 **Function** $Increment()$
2 | $lcounter + +$
3 | **if** $lcounter == threshold$ **then**
4 | | $logNumIncrems.MaxWrite(nextVal)$
5 | | $nextVal + +$
6 | | $lcounter \leftarrow 0$
7 | | **if** $nextVal \geq 2$ **then** $threshold \leftarrow 2 \times threshold$

8 **Function** $Read()$
9 | $r \leftarrow logNumIncrems.MaxRead()$
10 | **if** $r \geq 0$ **then** **return** $k \cdot 2^r$
11 | **return** 0

3.2 Proof of Linearizability

Let E be an execution of the k-multiplicative-accurate m-bounded counter implemented in Algorithm 1. We construct a linearization L of E by removing some specific instances of the Increment and Read operations, then ordering the remaining operations in E.

Let op be an incomplete Increment operation in E. We remove op from E in all but the following scenario: op executes a MaxWrite on $logNumIncrems$ during E. We also remove from E, any incomplete Read operation.

From the remaining operations in E, we denote by OP_w the set of Increment operations that do a MaxWrite on $logNumIncrems$, and OP_l the set of remaining Increment operations. And let OP_r denote the set of Read operations in E. We construct L by first identifying linearization points in E for operations in $OP_w \bigcup OP_r$ using Rules 1 and 2 below and then putting those operations in L according to the order in which the linearization points occur in E. Rule 3 below describes a procedure for completing the construction of L by inserting the operations in OP_l at appropriate places.

1. Each Increment operation in OP_w is linearized at its MaxWrite on $logNumIncrems$ at line 4 of Algorithm 1.

2. Each Read operation in OP_r is linearized at its MaxRead of $logNumIncrems$ at line 9 of Algorithm 1.
3. Consider the operations in OP_l in increasing order of when they begin in E. Let op be the next operation in OP_l to be placed. Let op' be the earliest element already in L such that op ends before op' begins and insert op immediately before op' in L. If op' does not exist, then put op at the end of L.

Rules 1 and 2 ensure that each operation in $OP_w \cup OP_r$ is linearized at a point in the interval of its execution. Thanks to the definition of Rule 1, Lemma 1 shows that the relative order of non-overlapping operations in the execution is preserved in the linearization:

Lemma 2. *Let op_1 and op_2 be two operations in E such that op_1 ends before op_2 is invoked. We have that op_1 precedes op_2 in L.*

We show in Lemma 4 that the implementation respects the sequential specification of the k-multiplicative-accurate counter. The proof relies on the following lemma, whose proof is in the full version [14].

Lemma 3. *Each new value of $logNumIncrems$ during E is an increment by 1 of the previous value of $logNumIncrems$.*

Lemma 4. *Let op denote an instance of the Read operation that returns x in execution E, and let v be the number of Increment operations before op in L. We have $v/k \le x \le k \cdot v$ for $k \ge \sqrt{2n}$.*

Proof. Let r denote the value of $logNumIncrems$ MaxRead during op at line 9 of Algorithm 1 in E. Thus $x = k \cdot 2^r$. From Lemma 3, the values MaxWritten to $logNumIncrems$ in E before op MaxReads the value r are increments of 1, starting from -1 to r. Therefore, the minimum number of Increment operations necessary to reach this value of $logNumIncrems$ is $v_{min} = 1 + \sum_{j=1}^{r} 2^{j-1} = 2^r$. Indeed, $v + 1$ Increment instances by a process are required for that process to execute MaxWrite(v) on $logNumIncrems$ with $v \le 1$. Subsequently, the number of invocations required is multiplied by a factor of 2 each time the threshold is reached. Thus in E, at least v_{min} Increment instances start before op's linearization point, and therefore in L, at least v_{min} Increments occur before op.

Furthermore, the maximum number of Increment operations invoked in E before op is $v_{max} = n(1 + \sum_{i=1}^{r} 2^{i-1}) + n(2^r - 1) = n(2^{r+1} - 1)$. Each process has set $nextVal$ to r (i.e., it executes v_{min} Increment operations) plus an additional $2^r - 1$ instances, which is the maximum number a process can count locally after setting $nextVal$ to r. Thus in L at most v_{max} Increments occur before op.

Let v be the actual number of Increments that precede op in L. To show that op's return value x is at most $k \cdot v$, observe that $x = k \cdot 2^r$, which equals $k \cdot v_{min}$ by the argument above, which is at most $k \cdot v$. To show that x is at least v/k, note that $v \le v_{max}$, which equals $n(2^{r+1} - 1)$ by the argument above. This expression is less than $n \cdot 2^{r+1}$, which is at most $k^2 \cdot 2^r = k \cdot x$ by the assumption that $k \ge \sqrt{2n}$. $\qquad\square$

3.3   Complexity Analysis

We analyze the step complexity of Algorithm 1 when $logNumIncrems$ is implemented with the h-bounded max-register algorithm given by Aspnes et al. [2] which uses 1-bit read-write registers and has a step complexity of $O(\log h)$ for both MaxWrite and MaxRead operations.

Lemma 5. *A process executes $O(\log \log m)$ steps during a call to the Read or Increment operation.*

Proof. At most, each Read (resp., Increment) performs one MaxRead (resp., MaxWrite) on $logNumIncrems$. The maximum number of calls to Increment is m, by assumption, and thus the largest argument to MaxWrite on $logNumIncrems$ is $\lceil \log_2 m \rceil$. Since we use the h-bounded max register implementation from [2] with $O(\log h)$ step complexity, substituting $h = O(\log m)$ gives the claim.   □

4   Algorithm for $k > 1$

In this section, we present a wait-free linearizable k-multiplicative-accurate m-bounded Counter working properly for any real number k with $k > 1$ and any positive integer m. The step complexity of a Read operation is $O(\log \log_k m)$. The step complexity of an Increment operation is $O(\max\{\log n \cdot \log(\frac{k}{k-1}n), \log \log_k m\})$. Pseudocode is given in Algorithm 2.

4.1   Algorithm Description

As in Algorithm 1, each Increment MaxWrites the logarithm of the number of Increments into a shared max-register $logNumIncrems$, and each Read returns an exponential function of the value it MaxReads from $logNumIncrems$. Recall that this approach saves an exponential amount of time (and space) at the cost of the loss of some accuracy. In order to efficiently accommodate an approximation factor of any real $k > 1$, we incorporate several new ideas.

First, the base of the logarithm is k, not 2, and the value returned by a Read is k^{r+1}, where r is the value MaxRead from $logNumIncrems$. As in Algorithm 1, the extra factor of k accommodates the uncertainty caused by the possibility of concurrent Increments. The max-register is bounded by $\lceil \log_k m \rceil$.

Second, in order to estimate the number of Increments more closely, processes communicate more information through additional shared objects. In particular, processes communicate the number of increments they have experienced through an array *Bucket* of "buckets". Each bucket is an exact counter, bounded by $\lceil \frac{kn}{k-1} \rceil$ and the number of buckets is $\lceil \frac{(k-1)m}{n} \rceil$, as explained below. To ensure that every increment in a bucket has an effect (i.e., changes the stored value), a process stops using a bucket if the stored value is larger than or equal to $\lceil \frac{n}{k-1} \rceil$.

Each process keeps a local variable *index*, starting at 0 and incremented by 1, that indicates the current bucket to be used. Each Increment increments the current bucket by one and then reads a value from that bucket. If the bucket is "full", i.e., the value read is at least $n/(k-1)$, then the process moves on to the next bucket by incrementing $index$. However, because of the possibility of up to n concurrent Increments, the maximum value of each bucket is approximately $n/(k-1) + n = kn/(k-1)$. Since there are at most m Increments, each one causes one bucket to be incremented by one, and each bucket is incremented at least $n/(k-1)$ times, the maximum number of buckets needed is approximately $m/(n/(k-1)) = (k-1)m/n$.

Unlike in Algorithm 1, the value MaxWritten to $logNumIncrems$ in Algorithm 2 reflects information about the number of Increments by other processes that is learned through the buckets. Since each bucket holds about $n/(k-1)$ Increments, the total number of Increments is approximately $n/(k-1)$ times the number of full buckets (stored in $index$) plus the number of Increments read from the current, non-full, bucket. The value written to $logNumIncrems$ is the logarithm, base k, of this quantity.

Algorithm 2: Implementation of a k-multiplicative m-bounded counter with $k > 1$.

Constants: $X = \frac{1}{(k-1)}$

Shared variables:
- $Bucket[\lceil \frac{m}{Xn} \rceil]$: array of $\lceil (X+1)n \rceil$-bounded exact counter objects indexed starting at 0, initialized to all 0's.
- $logNumIncrems$: $\lceil \log_k m \rceil$-bounded max register object that stores the logarithm (base k) of the number of increments exposed to the readers, initialized to -1.

Local persistent variables:
- $index$: stores the current $Bucket$ index, initially 0.

1 **Function** $Increment()$
2     $Bucket[index].Increment()$
3     $val \leftarrow Bucket[index].Read()$
4   **if** $val < X \cdot n$ **then**
5 | $logNumIncrems.MaxWrite(\lfloor \log_k(val + index \cdot X \cdot n) \rfloor)$
6   **else**
7        $index{+}{+}$
8        $logNumIncrems.MaxWrite(\lfloor \log_k(index \cdot X \cdot n) \rfloor)$

9 **Function** $Read()$
10     $r \leftarrow logNumIncrems.MaxRead()$
11   **if** $r \geq 0$ **then return** k^{r+1}
12   **return** 0

4.2   Proof of Linearizability

Let E be an execution of the k-multiplicative-accurate m-bounded counter implemented in Algorithm 2. We construct a linearization L of E by removing some specific instances of the CounterIncrement and Read operations, then ordering the remaining operations in E.

We remove from E any incomplete Read operation. We also remove from E any incomplete Increment that increments $Bucket[i]$ for some i but such that there is no subsequent Read of $Bucket[i]$ in an operation that contains a MaxWrite to $logNumIncrems$.

From the remaining operations in E, we denote by S_{CR}: the set of (completed) Reads. We also denote by S_{CI}^{inc}: the set of Increments that increment $Bucket[i]$ for any $i \geq 0$; an operation in this set might or might not perform a MaxWrite on $logNumIncrems$. Note that if an operation op in this set is incomplete, then its bucket Increment must be followed by a Read of that bucket inside an operation that performs a MaxWrite; otherwise op would have been excluded from consideration by the second bullet given above.

We construct L by identifying linearization points in E for operations in $S_{CR} \bigcup S_{CI}^{inc}$ using the following Rule 1 and Rule 2. These rules ensure that each operation in $S_{CR} \bigcup S_{CI}^{inc}$ is linearized at a point in its execution interval. L is built according the following directive: when an operation op has its linearization point before the linearization point of op', op precedes op' in L.

1. The linearization point of each operation $op \in S_{CR}$ is the enclosed MaxRead of $logNumIncrems$ done by op.
2. The linearization point of each operation $op \in S_{CI}^{inc}$ is the earlier of (i) the return of op and (ii) the time of the earliest MaxWrite to $logNumIncrems$ in an operation op' such that op' reads $Bucket[i]$ after op increments $Bucket[i]$. It is possible for op' to be op. If op finishes, then at least (i) exists. If op is incomplete, then at least (ii) exists by the criteria for dropping incomplete operations. Operations that have the same linearization point due to this rule appear in L in any order.

Thanks to Rules 1 and 2, the relative order of non-overlapping operations in the execution is preserved in the linearization:

Lemma 6. *Let op_1 and op_2 be two operations in E such that op_1 ends before op_2 is invoked. We have that op_1 precedes op_2 in L.*

We call a MaxWrite to $logNumIncrems$ *effective* if it changes the value of $logNumIncrems$. We define op_r as the Increment containing the effective MaxWrite of r to $logNumIncrems$, for $r \geq 0$. The sequence of values taken on by $logNumIncrems$ in execution E (i) begins at -1, and (ii) is increasing. op_r is well-defined in that there is at most one such operation for each r. For each $r \geq 0$, if op_r exists then let t_r be the time when the effective MaxWrite of r to $logNumIncrems$ occurs in E. The proof of the next lemma is in the full version [14].

Lemma 7. *For each $r \geq 0$, if op_r exists then the linearization point of op_r is at or before t_r.*

We next show that L satisfies the sequential specification of a k-multiplicative-accurate m-bounded counter. We start by showing that the number of Increments linearized before a Read cannot be too small: specifically, if the Read returns x, then the number of preceding Increments is at least x/k; see Lemma 10. We need two preliminary lemmas, whose proofs are in [14].

Lemma 8. *For any process p, the sequence of values taken by its index variable in E (i) begins at 0, (ii) is increasing, and if $i \geq 1$ appears in the sequence, then so does $i - 1$.*

Lemma 9. *Let op be a Read operation that returns 0. Then no Increment operation is linearized before op in L.*

Lemma 10. *Let op be a Read in L that returns x and let y be the number of Increments that precede op in L. Then $y \geq x/k$.*

Proof. Case 1: $x = 0$. By Lemma 9, $y = 0$ and the claim follows.

Case 2: $x > 0$. Then op MaxReads some value $r \geq 0$ from $logNumIncrems$, where $x = k^{r+1}$. By the definition of L and Lemma 7, op_r precedes op in L. Let p be the process that executes op_r. By the code, p increments $Bucket[i]$, then reads val, which is at least 1, from $Bucket[i]$, and finally MaxWrites r to $logNumIncrems$. Then $Bucket[i] \geq val$ immediately before p MaxWrites to $logNumIncrems$. If $val < Xn$, then $r = \lfloor \log_k(val + iXn) \rfloor$, otherwise $r = \lfloor \log_k((i+1)Xn) \rfloor$.

According to Lemma 8, for each j, $0 \leq j \leq i - 1$, there is an Increment instance op_j performed by p such that p reads a value greater than or equal to Xn from $Bucket[j]$ in the execution of op_j. By the linearization Rule 2, the linearization point of each Increment instance that applies one of the first Xn increments to $Bucket[j]$ is before or at the time op_j does its MaxWrite. By Lemma 6, L respects the real-time order, and thus for each j, $0 \leq j \leq i - 1$, op_j precedes op_r and also op in L. Thus the number of Increments linearized before op is at least $val + iXn$.

Note that $val + iXn = k^{\log_k(val+iXn)} \geq k^{\lfloor \log_k(val+iXn) \rfloor}$. If $val < Xn$, then the exponent on k is $\lfloor \log_k(val + iXn) \rfloor$. If $val \geq Xn$, then the exponent on k is at least $\lfloor \log_k((i+1)Xn) \rfloor$. In both cases, the value of the exponent on k is equal to the value of r for the corresponding case. Thus at least k^r Increments appear in L before op, including op_r.

We have $y \geq k^r = k^{r+1}/k = x/k$. The claim follows.   □

We now show that the number of Increments linearized before a Read cannot be too big: specifically, if the Read returns x, then the number of preceding Increments is at most kx. This is proved in Lemma 13. We need some preliminary definitions and lemmas.

For every $i \geq 0$, let V_i be the set of all Increment operations performed in E such that the value of the local variable $index$ at the beginning of the operation is i.

Lemma 11. *If* $op \in V_i$, *then* op *appears in* L *after* op_r *where* $r = \lfloor \log_k(iXn) \rfloor$ *for all* $i \geq 1$.

Proof. We first show that op begins after t_r in E. Let p be the process executing op. Since op is in V_i, p's *index* variable equals i at the beginning of op with $i > 0$. By Lemma 8, the previous value of *index* was $i - 1$ and when p incremented *index* to i, it MaxWrote $\lfloor \log_k(iXn) \rfloor$ to *logNumIncrems* (cf. Lines 7–8).

Thus the first MaxWrite of $r = \lfloor \log_k(iXn) \rfloor$ to *logNumIncrems*, which occurs at t_r by definition, precedes the beginning of op in E. The linearization point of op occurs during its interval and thus follows t_r, which by Lemma 7 is at or after the linearization point of op_r. Hence, in L, op_r precedes op.   □

Lemma 12. $|V_i| < n(X + 1)$ *for all* $i \geq 0$.

Proof. We show that *Bucket*[i] is incremented at most $Xn + n - 1$ times, for all $i \geq 0$. After the first $Xn - 1$ Increments that increment *Bucket*[i], each subsequent Increment by a process p reads a value greater than or equal to Xn from *Bucket*[i]. Thus, p moves to the next bucket by incrementing its *index* variable. No subsequent Increment by p contributes to V_i. Thus each of the n processes does at most one Increment of *Bucket*[i] after the first $Xn - 1$ Increments of *Bucket*[i].

Every Increment of *Bucket*[i] by that process corresponds to one Increment (of the approximate Counter). Since *Bucket*[i] is incremented at most $Xn + n - 1$ times, $|V_i| < n(X + 1)$.   □

Lemma 13. *Let* op *be a Read operation that returns* x, *and let* y *be the number of Increment instances that precede* op *in* L. *Then* $y \leq kx$.

Proof. Case 1: $x = 0$. By Lemma 9, no Increment operation is linearized before op in L. Then, $y = 0$ and the claim follows.

Case 2: $x > 0$. Then op MaxReads some value $r \geq 0$ from *logNumIncrems*, where $x = k^{r+1}$. Note that the MaxWrite to *logNumIncrems* in op_r precedes op's MaxRead of *logNumIncrems* and no MaxWrite of a value larger than r precedes op's MaxRead of *logNumIncrems*.

Also, every process has *index* 0 as long as *logNumIncrems* is less than $\lfloor \log_k(Xn) \rfloor$. The reason is that when a process sets its *index* to a value greater than 0, it also MaxWrites $\lfloor \log_k(Xn) \rfloor$ (or larger) to *logNumIncrems* (cf. Lines 7–8). Then, we analyze the two possible cases:

Case 2.1: $r < \lfloor \log_k(Xn) \rfloor$. Since op_r MaxWrites $r = \lfloor \log_k v \rfloor$ to *logNumIncrems* where v is the value op_r read from *Bucket*[0], we have $r \leq \log_k v < r + 1$. Then, $k^r \leq v < k^{r+1}$. This means that at most $k^{r+1} - 1$ Increments have been applied to *Bucket*[0] before the Read of op_r.

In the following we prove that the number y of Increment operation instances linearized before op is at most $\lfloor k^{r+1} \rfloor - 1$, which is less than $kx = k^{r+2}$.

Let W be the set of Increment operation instances that enclose the first $\lfloor k^{r+1} \rfloor - 1$ Increments of *Bucket*[0]. Since Increments of *Bucket*[0] are instantaneous, W is well-defined. We prove that no Increment other than those in W can appear before op in L. Suppose by contradiction that some $op' \in S_{CI}^{inc} \setminus W$

by a process p' appears in L before op. Since op' is not in W, and because in the execution of an Increment operation, a process first increments the $Bucket$ and then reads its value, p' reads a value $v' \geq \lfloor k^{r+1} \rfloor$ from $Bucket[0]$.

Thus, the corresponding MaxWrite to $logNumIncrems$ MaxWrites the value $r' = \lfloor \log_k v' \rfloor \geq r + 1 > r$ to $logNumIncrems$.

- Suppose op' is linearized when it ends. Before it ends, the MaxWrite of $r' > r$ is applied to $logNumIncrems$. But then op would read r' instead of r from $logNumIncrems$, a contradiction.
- Suppose op' is linearized at the time of the MaxWrite to $logNumIncrems$ by some operation op'' that reads $Bucket[0]$ after op' increments $Bucket[0]$. Since $op' \notin W$, the value of $Bucket[0]$ is larger than or equal to $\lfloor k^{r+1} \rfloor$. Then the value of op'''s effective MaxWrite is some value $r'' \geq r' > r$. Since this MaxWrite occurs in E before op's MaxRead of $logNumIncrem$, op would return r'' instead of r, a contradiction.

Case 2.2: $r \geq \lfloor \log_k(Xn) \rfloor$. Since op precedes op_{r+1} in L (if op_{r+1} exists), Lemma 11 implies that every Increment in V_i with $i \geq j$ appears after op_{r+1}, and thus after op, where $r + 1 = \lfloor \log_k(jXn) \rfloor$. Solving for j, we get $j \geq k^{r+1}/Xn$. Thus y is at most the number of operations in $V_0 \cup \ldots \cup V_{\lfloor k^{r+1}/Xn \rfloor - 1}$. By Lemma 12,

$$\left| \bigcup_{j=0}^{\lfloor k^{r+1}/Xn \rfloor - 1} V_j \right| \leq n(X+1)\lfloor k^{r+1}/Xn \rfloor$$

$$\leq (X+1)/X \cdot Xn\lfloor k^{r+1}/Xn \rfloor$$
$$\leq (X+1)/X \cdot k^{r+1}$$
$$\leq k \cdot k^{r+1}$$
$$\leq kx.$$

$\square$

4.3 Complexity Analysis

We analyze the step complexity of Algorithm 2 when $logNumIncrems$ is implemented with the h-bounded max register algorithm given by Aspnes et al. [2] which uses 1-bit read-write registers and has a step complexity of $O(\log h)$ for both MaxWrite and MaxRead operations. Also we consider the l-bounded exact counter algorithm given by Aspnes et al. [2] which has a step complexity of $O(\log l)$ for the Read operation and $O(\log n \cdot \log l)$ for the Increment.

Lemma 14. *A process executes $O(\log \log_k m)$ steps during a call to the Read, and $O(\max\{\log n \cdot \log(\frac{k}{k-1}n), \log \log_k m\})$ steps during a call to the Increment, for any $k > 1$.*

Proof. Since we use the h-bounded max register implementation from [2] with $O(\log h)$ step complexity, substituting $h = O(\log_k m)$ gives the worst case step complexity for the MaxRead and MaxWrite operations applied to the max-register *logNumIncrems*.

Thus, since an instance of Read does one MaxRead on *logNumIncrems* and then computes the return value, the claim follows for the Read operation.

An instance of Increment does one Increment and one Read on a bucket, which is a low-level h-bounded exact counter where $h = \lceil (X+1)n \rceil$ and $X = \frac{1}{k-1}$. It also does one MaxWrite on *logNumIncrems* whose worst-case step complexity is in $O(\log \log_k m)$.

The worst-case step complexities for the Increment and Read on the h-bounded exact counter are in $O(\log n \log(\lceil (X+1) \rceil)n)$ and $O(\log \lceil (X+1) \rceil n)$ with $X = \frac{1}{k-1}$, respectively. Thus, the worst case step complexity for the Increment operation is in $O(\max\{\log n \log(\frac{k}{k-1}n), \log \log_k m)\})$.   □

5   Algorithm for $k \geq 2$

In this section, we present a wait-free linearizable implementation of a k-multiplicative-accurate m-bounded counter for any positive integer m and any integer $k \geq 2$. The algorithm is a variation of that in Sect. 4; it works for a more restricted range of k but the space complexity is better, as discussed below. The step complexity of a Read operation is $O(\log \log_k m)$ and the step complexity of an Increment operation is $O(\max\{\log^2 n, \log \log_k m\})$. The pseudocode appears in Algorithm 3.

5.1   Algorithm Description

When k is at least 2, we can reduce the number of buckets needed exponentially, from about $(k-1)m/n$ to about $\log_k(m/n)$, while the size of each bucket at most quadruples. The key idea is to store some of the Increments locally instead of incrementing a bucket for each one.

To deal with values of n that are not powers of k, we use the notation *cln* for $\lceil \log_k n \rceil$ and let N be k^{cln}, i.e., N is the smallest power of k that is at least as large as n.

Each bucket is an exact counter, bounded by $2N$ as explained below, and the number of buckets is approximately $\log_k(m/n)$. Note that $\log_k(m/n) + 1$ buckets are enough to store m Increments as the number of Increments stored in the first $i + 1$ buckets is k times the number stored in the first i buckets, and the first bucket stores N Increments.

Each process keeps a local variable, *index*, starting at 0 and incremented by 1. Depending on the value of *index*, a certain number of Increments need to be invoked by a process before the process posts them by incrementing *Bucket[index]*. This number is stored in a local variable *threshold* and it increases as *index* increases. Each process keeps track of the progress toward reaching *threshold* using its local variable *lcounter*.

When an Increment is invoked, *lcounter* is incremented. If *threshold* has been reached, then *lcounter* is reset to 0, the process increments the appropriate bucket and then reads that bucket. If the value read is less than N, then no further action

---

Algorithm 3: Implementation of a k-multiplicative m-bounded counter, $k \geq 2$.

Constants: $cln = \lceil \log_k n \rceil$ and $N = k^{cln}$

Shared variables:
- $Bucket[\max\{1, \lceil \log_k \lceil \frac{m}{n} \rceil \rceil + 1\}]$: array of $2N$-bounded exact counter objects indexed starting at 0, initialized to all 0's.
- $logNumIncrems$: $\lceil \log_k m \rceil$-bounded max register object that stores the logarithm (base k) of the number of increments exposed to the readers, initialized to -1.

Local persistent variables:
- $lcounter$: counts the number of increments invoked locally, initially 0.
- $index$: stores the index of the current entry in $Bucket$, initially 0.
- $threshold$: stores the current number of locally-invoked increments required to update the current entry in $Bucket$, initially 1.

```
1  Function Increment()
2  │  lcounter + +
3  │  if lcounter == threshold then
4  │  │  Bucket[index].Increment()
5  │  │  lcounter ← 0
6  │  │  val ← Bucket[index].Read()
7  │  │  if (index == 0) and (val < N) then
8  │  │  │  logNumIncrems.MaxWrite(⌊log_k val⌋)
9  │  │  if val ≥ N then
10 │  │  │  logNumIncrems.MaxWrite(cln + index)
11 │  │  │  index + +
12 │  │  │  if index > 1 then  threshold ← k · threshold
13 │  │  │  if index == 1 then  threshold ← k − 1

14 Function Read()
15 │  r ← logNumIncrems.MaxRead()
16 │  if r ≥ 0 then  return k^{r+1}
17 │  return 0
```

---

is taken except in the corner case when $index$ is 0, causing the logarithm (base k) of the value to be MaxWritten to $logNumIncrems$. If the value read is at least N, i.e., the bucket is "full", then $cln + index$ is MaxWritten to $LogNumIncrems$, $index$ is incremented, and $threshold$ is updated. If $index$ is 1, then the threshold is set to $k - 1$, otherwise it is set to k times its previous value. The method for updating $threshold$ is key to the correct working of the algorithm.

The reason that each bucket is bounded by $2N$ instead of N is that all the processes can have the same value i for their $index$ variables when $Bucket[i]$ equals $N - 1$, and then have each process start an Increment, causing N additional increments to be done on $Bucket[i]$.

The Read operation is the same as in Algorithm 2.

5.2   Proof of Linearizability

The proof of linearizability of Algorithm 3 uses some of the same ideas as in the proof of linearizability for Algorithm 2, although the definition of the linearization is more involved. Due to space limitations, its proof is deferred to [14].

5.3   Complexity Analysis

Let m be the maximum number of Increment operations invoked in any execution of Algorithm 3. Then the largest value ever MaxWritten to $logNumIncrems$ is $\lceil \log_k m \rceil$. We implement $logNumIncrems$ with the algorithm in [2] for an h-bounded max-register, where $h = \lceil \log_k m \rceil$. Thus the number of steps for both MaxRead and MaxWrite on $logNumIncrems$ is $O(\log h) = O(\log \log_k m)$.

Each bucket is incremented at most $2N$ times. We implement each bucket using the algorithm in [2] for an exact $2N$-bounded counter. The number of steps for incrementing a bucket is $O(\log n \cdot \log(2N))$, which is $O(\log^2 n)$. The number of steps for reading a bucket is $O(\log(2N)) = O(\log n)$.

Thus the number of steps required for a Read, which performs a MaxRead on $logNumIncrems$, is $O(\log \log_k m)$. Each Increment consists of at most one Increment of a bucket, at most one Read of a bucket, and at most one MaxWrite of $logNumIncrems$, requiring $O(\max\{\log^2 n, \log \log_k m\})$ steps.

Algorithm 3, which requires $k \geq 2$, and Algorithm 2, which works for any $k > 1$, both use a shared max-register that is bounded by $\lceil \log_k m \rceil$. They both also use an array of shared buckets. In Algorithm 3, the number of buckets is, roughly, $\log_k(m/n)$, while each bucket is, roughly, $(n+n)$-bounded. In contrast, Algorithm 2 uses $(k-1)m/n$ buckets, each bounded by $n + n/(k-1)$. When $k \geq 2$, the number of buckets used by Algorithm 3 is exponentially smaller than that used by Algorithm 3 and the bound on each bucket in Algorithm 3 is less than four times that needed in Algorithm 2.

6   Conclusion

We have presented three algorithms for implementing a wait-free linearizable k-multiplicative-accurate m-bounded counter using read-write registers. By combining the assumption that the number of increments is bounded by m and relaxing the semantics of the object to allow a multiplicative error of k, our algorithms achieve improved worst-case step complexity for the operations over prior algorithms for exact or unbounded counters. The three algorithms provide different tradeoffs between the step complexity, space complexity, and range of values for k, as depicted in Table 1.

A natural open question is to find the optimal worst-case step complexity of *CounterIncrement* for such implementations and see if it can be achieved simultaneously with an optimal *CounterRead*. Similarly, finding a tight bound on the space complexity would be interesting. Would allowing the implementation to use more powerful primitives, such as the historyless ones considered in [13], help? Finally, we are not aware of any *worst-case* results for *approximate* counters in the *unbounded* case or of any *amortized* results for *exact* counters in the *bounded* case; such results would flesh out our general understanding of counter implementations.

References

1. Afek, Y., Kaplan, H., Korenfeld, B., Morrison, A., Tarjan, R.E.: The CB tree: a practical concurrent self-adjusting search tree. Distrib. Comput. **27**(6), 393–417 (2014)
2. Aspnes, J., Attiya, H., Censor-Hillel, K.: Polylogarithmic concurrent data structures from monotone circuits. J. ACM **59**(1), 2:1–2:24 (2012)
3. Aspnes, J., Censor, K.: Approximate shared-memory counting despite a strong adversary. ACM Trans. Algorithms **6**(2), 25:1–25:23 (2010)
4. Aspnes, J., Censor-Hillel, K., Attiya, H., Hendler, D.: Lower bounds for restricted-use objects. SIAM J. Comput. **45**(3), 734–762 (2016)
5. Attiya, H., Hendler, D.: Time and space lower bounds for implementations using k-CAS. IEEE Trans. Parallel Distrib. Syst. **21**(2), 162–173 (2010)
6. Baig, M.A., Hendler, D., Milani, A., Travers, C.: Long-lived counters with polylogarithmic amortized step complexity. Distrib. Comput. **36**(1), 29–43 (2023)
7. Bender, M.A., Gilbert, S.: Mutual exclusion with O(log² log n) amortized work. In: 52nd IEEE Annual Symposium on Foundations of Computer Science, pp. 728–737 (2011)
8. Flajolet, P.: Approximate counting: a detailed analysis. BIT Comput. Sci. Sect. **25**(1), 113–134 (1985)
9. Hendler, D., Khattabi, A., Milani, A., Travers, C.: Upper and lower bounds for deterministic approximate objects. In: 41st IEEE International Conference on Distributed Computing Systems (ICDCS), pp. 438–448 (2021)
10. Herlihy, M.: Wait-free synchronization. ACM Trans. Program. Lang. Syst. (TOPLAS) **13**(1), 124–149 (1991)
11. Herlihy, M.P., Wing, J.M.: Linearizability: a correctness condition for concurrent objects. ACM Trans. Program. Lang. Syst. (TOPLAS) **12**(3), 463–492 (1990)
12. Inoue, M., Masuzawa, T., Chen, W., Tokura, N.: Linear-time snapshot using multi-writer multi-reader registers. In: Tel, G., Vitányi, P. (eds.) WDAG 1994. LNCS, vol. 857, pp. 130–140. Springer, Heidelberg (1994). https://doi.org/10.1007/BFb0020429
13. Jayanti, P., Tan, K., Toueg, S.: Time and space lower bounds for nonblocking implementations. SIAM J. Comput. **30**(2), 438–456 (2000)
14. Johnen, C., Khattabi, A., Milani, A., Welch, J.L.: Efficient wait-free linearizable implementations of approximate bounded counters using read-write registers (2024). arXiv:2402.14120
15. Khattabi Riffi, A.: On improving complexity of linearizable and wait-free implementations of concurrent objects by relaxing their specifications. Theses, Université de Bordeaux, March 2023. https://theses.hal.science/tel-04147099
16. Morris, R.H.: Counting large numbers of events in small registers. Commun. ACM **21**(10), 840–842 (1978)

On Distributed Computation
of the Minimum Triangle Edge
Transversal

Keren Censor-Hillel[ID] and Majd Khoury[(✉)][ID]

Technion, Haifa, Israel
{ckeren,majd-khoury}@cs.technion.ac.il

Abstract. The distance of a graph from being triangle-free is a fundamental graph parameter, counting the number of edges that need to be
removed from a graph in order for it to become triangle-free. Its corresponding computational problem is the classic *minimum triangle edge
transversal* problem, and its normalized value is the baseline for triangle-
freeness testing algorithms. While triangle-freeness testing has been successfully studied in the distributed setting, computing the distance itself
in a distributed setting is unknown, to the best of our knowledge, despite
being well-studied in the centralized setting.

This work addresses the computation of the minimum triangle edge
transversal in distributed networks. We show with a simple warm-up
construction that this is a global task, requiring $\Omega(D)$ rounds even in
the LOCAL model with unbounded messages, where D is the diameter
of the network. However, we show that approximating this value can be
done much faster. A $(1 + \epsilon)$-approximation can be obtained in poly $\log n$
rounds, where n is the size of the network graph. Moreover, faster approximations can be obtained, at the cost of increasing the approximation
factor to roughly 3, by a reduction to the minimum hypergraph vertex
cover problem. With a time overhead of the maximum degree Δ, this can
also be applied to the CONGEST model, in which messages are bounded.

Our key technical contribution is proving that computing an exact
solution is "as hard as it gets" in CONGEST, requiring a near-quadratic
number of rounds. Because this problem is an *edge selection* problem, as
opposed to previous lower bounds that were for *node selection* problems,
major challenges arise in constructing the lower bound, requiring us to
develop novel ingredients.

Keywords: Distributed graph algorithms · Distance from
triangle-freeness · Lower bounds

This paper takes a reduced form of the original paper. A full version is available at
https://arxiv.org/abs/2402.13985 and referred in [12]. The section numbering is the
same in both versions. Note that the numbering of some claims may be different in the
full version.

© The Author(s), under exclusive license to Springer Nature Switzerland AG 2024
Y. Emek (Ed.): SIROCCO 2024, LNCS 14662, pp. 336–358, 2024.
https://doi.org/10.1007/978-3-031-60603-8_19

1 Introduction

Subgraph-freeness, and specifically triangle-freeness, are cornerstone graph properties that have been widely studied in many computational settings. Moreover, in distributed computing, triangle-free networks are known to exhibit faster algorithms for finding large cuts [28] and girth [10], and for coloring [41].

For a graph that is not triangle-free, a central computational question is to compute how far it is from being so. The distance of a graph from satisfying a property has been extensively studied in the area of property testing, which asks whether the removal of any ϵ-fraction of edges from the graph makes it satisfy the property [27]. In particular, distributed triangle-freeness testing has been studied in [9] and in [22,24], where the latter two showed solutions within $O(1/\epsilon)$ rounds of communication in the CONGEST model [39], where n nodes of a network communicate in synchronous rounds with $O(\log n)$-bit messages to their neighbors in the underlying graph.

A fundamental computational question is what is the *exact* number of edges that need to be removed from a graph to make it triangle-free. This is called the *minimum triangle edge transversal (MTET) problem*, and has received attention in centralized settings [21,33,34] but its complexity in distributed networks has not been studied.

In this paper, we study MTET in the aforementioned classic distributed setting of CONGEST, as well as in the LOCAL model [36], which is similar but without the restriction on the message size. In a distributed solution for triangle edge transversal, each node of the graph marks some of its adjacent edges, such that the set of marked edges cover all triangles in the graph, where we say that an edge covers a triangle if it is a part of that triangle. A distributed solution for MTET thus outputs a cover of minimum size.

1.1 Our Contribution

As a warm-up for studying the problem, we first show that the minimum triangle edge transversal problem is *global*, in the sense that solving it necessitates communication throughout the network. This translates to a tight complexity $\Theta(D)$ in the LOCAL model, where D is the diameter of the graph, as can be seen by a simple graph construction.

Theorem 1. *Any distributed algorithm in the LOCAL model for computing a minimum triangle edge transversal requires $\Omega(D)$ rounds.*

Despite the problem being a global one, we show that approximate solutions can be obtained fast. Using the ball-carving technique of [26], we provide a $(1 + \epsilon)$-approximation algorithm in the LOCAL model.

Theorem 2. *There is a distributed algorithm in the LOCAL model that obtains a $(1+\epsilon)$-approximation to the minimum triangle edge transversal in $poly(\log n, 1/\epsilon)$ rounds.*

Moreover, we show that we can approximate the task even faster, at the expense of a slightly larger approximation factor, as follows.

Theorem 3. *There are distributed algorithms in the LOCAL model that obtain a 3-approximation and a $(3 + \epsilon)$-approximation to the minimum triangle edge transversal in $O(\log n)$ and $O(\frac{\log n}{\log \log n})$ rounds, respectively. The same approximation guarantees can be obtained in the CONGEST model with a multiplicative overhead of Δ in the round complexity.*

To obtain the above, we prove a reduction from minimum triangle edge transversal to the *minimum hypergraph vertex cover* problem (MHVC), in which one needs to find the smallest set of vertices that cover all hyperedges. The high-level outline of the reduction is that we consider each edge in the input graph G as a node in a 3-uniform hypergraph H_G and each triangle in G is a hyperedge in it. To allow the reduction to go through, we show that we can simulate H_G over G without a significant overhead. Finally, we plug the MHVC algorithm of [6] into our reduction to obtain the stated round complexity. We note that all of our approximation algorithms apply also to the weighted version of the minimum triangle edge transversal.

Finally, our key technical contribution is in showing that finding an exact solution for this problem is essentially "as hard as it gets" in the CONGEST model, requiring a near-quadratic number of rounds.

Theorem 4. *Any distributed algorithm in the CONGEST model for computing a minimum triangle edge transversal or deciding whether there is a triangle edge transversal of size M requires $\Omega(n^2/(\log^2 n))$ rounds.*

We emphasize that the fact that MTET is NP-hard is insufficient for deducing that it is hard for CONGEST. As explained in [3], there are NP-hard problems that are solvable in $o(n^2)$ rounds in CONGEST (and there are problems in P that require $\tilde{\Omega}(n^2)$ rounds). To prove Theorem 4, we follow the successful approach of reducing the 2-party set disjointness problem to the distributed minimum triangle edge transversal problem. We stress that our construction requires novel insights, as we overview in what follows.

1.2   Technical Overview of the CONGEST Lower Bound

We consider the framework for reducing a 2-party communication problem to a problem P in the CONGEST model, as attributed first to the work of [40]. Our presentation follows the one of [3]. The goal is to construct a family of graphs, who differ only based on the inputs to the two parties, and show that a solution to the graph problem P determines the output of the joint function f that they have to compute. Then, the parties, Alice and Bob, simulate a distributed algorithm for solving P to derive their output. The round complexity of the distributed algorithm is then bounded from below by the number of bits that must be exchanged between the two players for solving f over their joint inputs, divided by the number of bits that can be exchanged between them per each round of the simulation. The latter is strongly affected by the size of the *cut* between the two node sets that the players are responsible for simulating. Specifically, for the lower bound to be high, we need the cut between the two sets of nodes that the players simulate to be as small as possible.

To obtain a lower bound for MTET in the CONGEST model, we need to find a 2-party function f and to construct a family of lower bound graphs so that the value of MTET for a graph determines the value of f on the joint inputs that the graph represents. The challenges for a near-quadratic lower bound are: (i) we need the communication complexity of f to be linear in the input size, (ii) we need the input size of f to be quadratic in the number of nodes of the lower bound graphs, and (iii) we need the cut to have at most a polylogarithmic size.

Many of the CONGEST problems for which lower bounds were obtained by this framework use the well-known 2-party set disjointness problem, in which the input to each of the two players is a binary string, and the output of the joint function is 1 if and only if there exists an index in which the bit in both strings is 1.

The Challenges . A crucial difference between MTET and all of these problems is that *MTET is an edge selection problem, rather than a node selection one.* This characteristic appears to make lower bounds harder to obtain, and we extract the central challenges below.

To illustrate the difficulties, we consider the lower bound construction for the minimum vertex cover problem [3]. Even without the need yet to delve into the specifics of the construction, a reasonable attempt for converting it to MTET would be to replace each node in the construction with an edge, and each edge with a triangle, so that edges cover triangles in the new construction, just as nodes cover edges in the one for minimum vertex cover.

The first glaring problem with this is that the original construction has $\Theta(n^2)$ edges, so the new one will have this number of nodes. This violates requirement (ii) above, and is a major issue since any lower bound for MTET that one could obtain this way would be at most $\tilde{\Omega}(n)$. Yet, we stress that even this is insufficient, as attaching the edges into triangles in such a conversion imposes significant challenges. We focus below on two major ones.

Challenge 1: One obstacle is that an essential property of the lower bound graphs for the minimum vertex cover problem is their symmetry, as follows. There are four sets of cliques in the construction (A_1, A_2, B_1, B_2 in Fig. 1), two for each player, and the input to each player corresponds to a bipartite graph between its two cliques, by adding an edge if and only if its corresponding index in the input is a 0 bit (Dashed edges in Fig. 1. There are additional fixed nodes and edges which are not yet needed for the intuition here). An optimal solution for the fixed part of the construction does not *prefer* any node of a clique over the others, in the sense that for each node in each clique, there is an optimal solution not containing it.

This *impartiality* feature is vital because we want to be able to choose any single one of the clique nodes to be omitted from the vertex cover, once we take

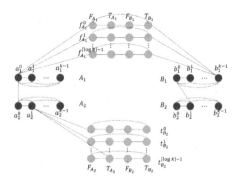

Fig. 1. The family of lower bound graphs for minimum vertex cover [3].

into account also the need to cover the edges between the two cliques which are added according to the set disjointness input. In other words, it allows to not select a node from each of the two cliques such that this pair of nodes *do not have an edge between them*, thus capturing an index in the input with a 1 bit.

Once we try this approach for MTET, we need to switch to *edge selection*. To this end, we replace each clique with an independent set of the same size and connect all nodes of the two independent sets of a certain side to a central node on that side (node a for Alice and node b for Bob). This implies that each input edge corresponds to a triangle between its two endpoints and the central node of its side. There will be more edges and triangles in the construction, such that we use the edges to the central node in order to cover them. Our hope is that a missing input edge (due to an input bit of 1) will allow us to avoid taking the corresponding edge from this pair of nodes to the central node into an MTET solution. However, we arrive at the following issue. For MTET, the only graphs that have this flavor of symmetry in optimal solutions, in the sense that they do not prefer certain edges over others, are either cliques or empty-edge graphs. We thus must attach a clique to the set of edges for which we want all but one of them to be in an optimal solution.

This obtains the property of symmetry, but doing this in a naïve manner prevents an optimal solution from having the ability to capture a missing input edge. To cope with this caveat, we need to configure the fixed part of the lower bound graph in a way that it prefers choosing all but one edge in certain edge sets over any other number of edges from it, or otherwise the size of the solution grows by 1 (due to the need to then take other edges in the construction that touch this set). Therefore, the solution favors choosing all but one edge from the set, *but without any preference as to which edge it leaves out.* Then, for the final graph, the edge chosen to be left out corresponds to not needing to cover a missing triangle due to a missing input edge, as we wanted. Our construction that eventually overcomes the above consists of various cliques that we very carefully attach to the nodes among which we insert input edges.

By using the set disjointness function, we satisfy requirement (i) above, of having the complexity of the input function be linear in its input size. Due to our above approach for the general structure of the lower bound graphs, our construction satisfies requirement (ii) of having an input size that is quadratic in the number of nodes.[1]

Challenge 2: The above discussion provided intuition about how to construct a lower bound graph that relates its MTET *for each of the two players* to their set disjointness *input*. A crucial aspect that remains to be handled is how to relate *the overall size* of the MTET in the entire graph to the set disjointness *output*. Specifically, in our MTET solution, we need to be able to omit the same pair of edges on both sides of the construction in case the inputs to the 2-party task are not disjoint.

For the vertex cover problem, a similar property was shown using the general notion of bit-gadgets [3]. Roughly speaking, each of the two main cliques on each side of the construction (to which input edges were added) had a logarithmic number of corresponding nodes to which they were attached (a clique $S \in \{A_1, A_2, B_1, B_2\}$ has the bit-gadgets F_S and T_S, see Fig. 1). These bit-gadget nodes were connected to the clique nodes according to the binary representation of their order. Each pair of bit-gadget nodes on one side are connected to the corresponding pair on the other side with a 4-cycle. This allows properly indexing the input, in the sense that if the 2-party inputs are not disjoint then an optimal solution omits the same pair of clique nodes on both sides of the construction.

When we move from nodes that cover edges to edges that cover triangles, we must incorporate a new mechanism. To overcome this, we develop a gadget that we call a *ring-of-triangles*, that replaces the simple cycles that are used for the vertex cover construction. We stress that attaching this gadget to the bit-gadget is not straightforward, and must be done in a careful manner. Eventually, because we make sure to still use only a logarithmic number of nodes for the cut, our construction also satisfies requirement (iii) of having a polylogarithmic cut size.

1.3 Related Work

Distributed triangle finding has been extensively studied, as well as finding other subgraphs (see [8] for a survey). In [18], a tight $O(n^{1/3})$-round algorithm is given for listing all triangles in the Congested Clique model [37], which resembles the CONGEST model but allows all-to-all communication. A matching lower bound was later established in [31,38]. Triangle listing in sparse graphs was addressed in [13,38]. In the CONGEST model, triangle listing was shown to be solvable

---

[1] It was suggested to us to replace this part of the construction with a different underlying graph with $\Theta(n)$ nodes, which is known to have $\Theta(n^2/2^{\sqrt{\log n}})$ edge-disjoint triangles [43]. Such a graph would give a smaller lower bound, but an even more severe issue is that it is not clear how to incorporate it in a construction that will allow the MTET size to correspond to the solution to the set disjointness input.

342   K. Censor-Hillel and M. Khoury

within the same complexity in [15], by developing a fast distributed expander decomposition and using its clusters for fast triangle listing. Recently, [14] showed that this complexity can also be obtained in a deterministic manner, up to sub-polynomial factors, based on the prior work of deterministic expander decomposition and routing of [16]. In [30], a listing algorithm in terms of the maximum degree Δ is given.

Triangle detection can be done in the **Congested Clique** model in $O(n^{0.158})$ rounds [11], but no faster-than-listing solutions are known for it in **CONGEST**. In terms of lower bounds, it is only known that a single round does not suffice [2], even with randomization [23]. Any polynomial lower bound would imply groundbreaking results in circuit complexity [20].

Distributed testing algorithms were first studied in [7]. Triangle-freeness testing in $O(1/\epsilon^2)$ rounds was given in [9], and improved to $O(1/\epsilon)$ rounds by [22,24].

In the centralized setting, [34] present an LP-based 2-approximation algorithm for MTET, while the integrality graph of the underlying LP was posed as an open problem. In [33] a reduction is given from MTET to the minimum vertex cover problem for $(2 - \epsilon)$-approximations, proving that approximating MTET to a factor less 2 is NP-hard. We stress that it is not clear how to make this reduction work in the **CONGEST** model.

Near-quadratic lower bounds for **CONGEST** have been obtained before, e.g., for computing a minimum vertex cover, a minimum dominating set, a 3-coloring, and weighted max-cut [3,4].

2   Hardness of Exact Computation of MTET

In this section, we show that computing the exact MTET is a global problem for both the **LOCAL** and **CONGEST** models, in the sense that it takes $\Omega(D)$ rounds in the **LOCAL** model, and that in the **CONGEST** model its complexity is near-quadratic in n. We first provide some preliminaries which we use for proving the difficulty of solving the problem in both models.

2.1   Preliminaries

We proceed by showing properties of MTET in two special graphs, as follows.

Definition 1 (A t-line of triangles). *Let $G = (V, E)$ be an undirected graph with $t + 2$ nodes, where $t \geq 2$ is even. Denote by $V = \{v_0, \ldots, v_{t/2}\} \cup \{u_0, \ldots, u_{t/2}\}$ the nodes of the graph. Denote the edges by $E = \{\{v_i, v_{i+1}\} : 0 \leq i \leq t/2 - 1\} \cup \{\{u_i, u_{i+1}\} : 0 \leq i \leq t/2 - 1\} \cup \{\{v_i, u_i\} : 0 \leq i \leq t/2\} \cup \{\{v_i, u_{i+1}\} : 0 \leq i \leq t/2 - 1\}$. Then G is called a t-line of triangles (see Fig. 2).*

It is immediate to see that a t-line of triangles has exactly t triangles. Moreover, Claim 1 easily shows that it has a single MTET, and its size is $t/2$.

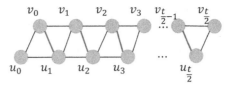

Fig. 2. An illustration of a line of triangles. The bold red edges are the only optimal solution of triangle edge transversal. (Color figure online)

Claim 1. *There exists only one MTET in a t-line of triangles, and its size is $t/2$.*

Definition 2 (A t-ring of triangles). *Let $G = (V, E)$ be an undirected graph with t triangles, where t is even. Then G is called a t-ring of triangles if the triangles can be ordered $T_0, \ldots, T_{t-1}$, such that every two consecutive triangles, T_i and $T_{(i+1) \bmod t}, 0 \leq i \leq t-1$, share an edge, and any other pair of triangles are edge disjoint (see Fig. 3).*

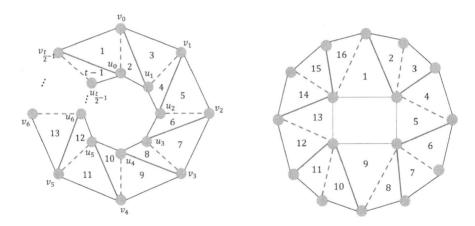

Fig. 3. Two illustrations of t-rings of triangles. The numbers inside the triangles indicates their order. The bold red edges show an optimal solution to triangle edge transversal, and the dashed blue edges show the other optimal solution. (Color figure online)

Claim 2. *There exist exactly two MTETs in a t-ring of triangles, and their sizes are $t/2$.*

The proofs of both claims are given in the full version of the paper [12].

2.2 A Linear Lower Bound in the **LOCAL** Model

We prove that computing an exact MTET is a global problem, requiring $\Omega(D)$ rounds, where D is the diameter of the graph.

Theorem 1. *Any distributed algorithm in the* LOCAL *model for computing a minimum triangle edge transversal requires* $\Omega(D)$ *rounds.*

The proof of this theorem is given in the full version of the paper [12]. The distinguish-ability argument is the premise of the proof, as the following: If the algorithm finishes in less than a linear number (in D) of rounds, then certain nodes cannot distinguish between two graphs that they should output different outputs for, and therefore output the same answer for both graphs.

2.3   A Near-Quadratic Lower Bound in the CONGEST Model

In this section, we focus on proving the following theorem, which concludes a near-quadratic lower bound on the exact MTET in the CONGEST model.

Theorem 4. *Any distributed algorithm in the* CONGEST *model for computing a minimum triangle edge transversal or deciding whether there is a triangle edge transversal of size* M *requires* $\Omega(n^2/(\log^2 n))$ *rounds.*

Note that a lower bound on deciding whether there is a triangle edge transversal of size M implies a lower bound on computing a minimum triangle edge transversal, since counting the edges of a given triangle edge transversal set can be done in $O(D)$ via standard techniques (pipelining through a BFS tree). To prove the lower bound on MTET, we take the approach of reducing problems from 2-party communication along the lines pioneered in the framework of [40], and specifically, we define a family of lower bound graphs [3, Definition 1] for MTET. The family of lower bound graphs we define bears some similarity to the family of the lower bound class that was defined for the minimum vertex cover problem in [3]. However, in order to handle triangles, we need to develop much additional machinery. We start by formalizing the required background.

Triangle Edge Transversal Properties in Cliques. Let $\mathcal{ET}(G)$ be the set of MTETs of the graph G. Let $N(v)$ be the neighborhood of a node v. Let $\mu(G, v)$ denote the maximum number of edges that touch v and are in the same MTET, that is, $\mu(G, v) = \max_{S \in \mathcal{ET}} |\{\{u, v\} : u \in N(v)\} \cap S|$.

Definition 3 (Edge mapping under node mapping). *Let* $G = (V, E)$ *be an undirected graph, and let* $S \subseteq E$ *be a subset of the edges. Let* $g : V \to V$ *be a bijective node mapping. Then the edge mapping* φ_g *of* S *under* g *is defined as:* $\varphi_g(S) = \{\{g(u), g(v)\} : \{u, v\} \in S\}$. *If* $\varphi_g(S) \subseteq E$, *then we call* φ_g *a well-defined edge mapping, as it can be seen as a function* $\varphi_g : S \to E$.

We start with the following simple observation, which holds since every pair of nodes in a clique has an edge.

Observation 1. *Let* $K_n = (V, E)$ *be a clique with* n *nodes, let* $S \subseteq E$ *be a subset of the edges, and let* $g : V \to V$ *be a bijective node mapping. Then* $\varphi_g(S)$ *and* $\varphi_{g^{-1}}(S)$ *are well-defined edge mappings.*

Claim 3 (Triangle edge transversal preserved under node mapping in cliques). *Let $K_n = (V, E)$ be a clique with n nodes. Let S be a triangle edge transversal of K_n, and let $g : V \rightarrow V$ a bijective node mapping function. Then $\varphi_g(S)$ is a triangle edge transversal.*

The proof of Claim 3 is given in the full version of the paper [12]. We thus obtain the following.

Corollary 1. *For a clique K_n with n nodes, the following hold:*

1. *For every two nodes u, v, it holds that $\mu(K_n, u) = \mu(K_n, v)$. Thus, we can denote this value by μ_n (where $\mu_n = \mu(K_n, v)$ for all v).*
2. *For any node v and μ_n nodes $u_1, u_2, \ldots, u_{\mu_n}$, there exists a MTET, which we denote S, such that the set of edges incident to v that are in S is exactly $\{\{v, u_i\}, 1 \le i \le \mu_n\}$.*

Next, we show that the maximum number of incident edges to a node in a clique with n nodes, that all are in the same MTET cannot be small, nor can it be equal to $n - 1$.

Claim 4. *Let K_n be a clique with n nodes, then $\mu_n \ge (n-1)/6$ and $\mu_n < n-1$.*

The proof of Claim 4 is given in the full version of the paper [12]. We later incorporate the above properties of K_n in our construction of the lower bound graphs.

2-Party Communication Complexity. In the 2-party communication complexity setting [35], there are two players, Alice and Bob, who are given two inputs $x, y \in \{0, 1\}^K$, respectively, and wish to evaluate a function $f : \{0, 1\}^K \times \{0, 1\}^K \rightarrow \{\texttt{true}, \texttt{false}\}$ on their joint input. The communication between Alice and Bob is carried out according to some fixed protocol π (which only depends on the function f, known to both players). The protocol consists of the players sending bits to each other until the value of f is determined by either of them. The maximum number of bits that need to be sent in order to compute f over all possible inputs x, y, is called *the communication complexity* of the protocol π, and is denoted by $CC(\pi)$. The *deterministic communication complexity* for computing f, denoted $CC(f)$, is the minimum over $CC(\pi)$ taken over all deterministic protocols π that compute f. In a randomized protocol, Alice and Bob have access to (private) random strings r_A, r_B, respectively, of some arbitrary length, chosen independently according to some probability distribution. The *randomized communication complexity* for computing f, denote $CC^R(f)$, is the minimum over $CC(\pi)$ taken over all randomized protocols π that compute f with success probability at least $2/3$.

Set Disjointness. The *Set Disjointness* function, denoted by $f(x, y) = DISJ_K(x, y)$, whose inputs are of size K, returns $\texttt{false}$ if the inputs represent sets that are not disjoint, i.e., if there exists an index $i \in \{0, \ldots, K-1\}$ such that $x_i = y_i = 1$, and returns $\texttt{true}$ otherwise. The deterministic and randomized communication complexities of $DISJ_K$ are $\Omega(K)$ [5,32,42].

Lower Bound Graphs. We quote the following formalization of the reduction of optimization problems from 2-party communication complexity to distributed problems in CONGEST.

Definition 4. *Fix an integer K, a function $f : \{0,1\}^K \times \{0,1\}^K \to \{true, false\}$, and a predicate P for graphs. The family of graphs $\{G_{x,y} = (V, E_{x,y}) : x, y \in \{0,1\}^K\}$, is said to be a family of lower bound graphs w.r.t f and P if the following properties hold:*

1. *The set of nodes V is the same for all graphs, and $V = V_A \cup V_B$ is a fixed partition of it;*
2. *Only the existence or the weight of edges in $V_A \times V_A$ may depend on x;*
3. *Only the existence or the weight of edges in $V_B \times V_B$ may depend on y;*
4. *$G_{x,y}$ satisfies the predicate P iff $f(x, y) = true$.*

The following theorem reduces 2-party communication complexity problems to CONGEST problems (see e.g [3,17,19,25,29]).

Theorem 5. *Fix a function $f : \{0,1\}^K \times \{0,1\}^K \to \{true, false\}$ and a predicate P. Let $\{G_{x,y}\}$ be a family of lower bound graphs w.r.t f and P and denote $C = E(V_A, V_B)$. Then any deterministic algorithm for deciding P in the CONGEST model requires $\Omega(CC(f)/(|C|\log n))$ rounds, and any randomized algorithm for deciding P in the CONGEST model requires $\Omega(CC^R(f)/(|C|\log n))$ rounds.*

The Lower Bound. We begin by constructing the fixed graph on which we proceed by adding edges corresponding to the input strings x and y of the 2-party set disjointness problem. Then we show that having a solution for MTET of size K implies the disjointness of the input strings and vice versa, hence the complexity of the set disjointness problem provides a lower bound to the complexity of the MTET problem, according to Theorem 5. The construction is illustrated by Figs. 4, 5, 6 and 7.

The Fixed Graph Construction – The Node Set. Let k be a natural number. The fixed graph has two center nodes a and b, and four sets of nodes $A_i = \{a_i^\ell : 0 \leq \ell \leq \mu_{k+1}\}, B_i = \{b_i^\ell : 0 \leq \ell \leq \mu_{k+1}\}, i \in \{1,2\}$, which we call *bit nodes*. For each set $S \in \{A_1, A_2, B_1, B_2\}$, the graph also has four additional corresponding sets of nodes, as follows. Two such sets are *bit-gadgets* (as defined in [1] and explained below) of size $\lceil \log(\mu_{k+1} + 1) \rceil$, $T_S = \{t_S^\ell : 0 \leq \ell \leq \lceil \log(\mu_{k+1} + 1) \rceil - 1\}$ and $F_S = \{f_S^\ell : 0 \leq \ell \leq \lceil \log(\mu_{k+1} + 1) \rceil - 1\}$, another set is a clique of size k, $C_S = \{c_S^\ell : 0 \leq \ell \leq k - 1\}$, and another set $H_S = \{h_S^\ell : 0 \leq \ell \leq \mu_{k+1}\}$ called *connectors* is of size $\mu_{k+1}+1$ (this set connects between the bit nodes to the corresponding cliques with a line of 2-triangles as will be explained later and illustrated in Fig. 6). In addition, we add *ring-auxiliary nodes* $M_i^\ell = \{m_{i,\ell}^j, 0 \leq j \leq 13\}$, for $i \in 1, 2$ and $0 \leq \ell \leq \lceil \log(\mu_{k+1} + 1) \rceil - 1$, that connect the edges

$\{f_{A_i}^\ell, a\}, \{t_{A_i}^\ell, a\}, \{f_{B_i}^\ell, b\}, \{t_{B_i}^\ell, b\}$ as part of a 20-ring of triangles, as defined in Definition 2 (see Fig. 7, where the bold edges indicate the four previously mentioned edges for M_1^0, that is, for $i = 1$ and $\ell = 0$).

The Fixed Graph Construction – The Edge Set. For clarity, we divide the edge set to five sets and explain the contents of each set.

Central edges: The center node a is connected to all the nodes in the bit-gadgets $F_{A_1}, T_{A_1}, F_{A_2}, T_{A_2}$, the cliques C_{A_1}, C_{A_2} and the connector nodes H_{A_1}, H_{A_2}. The center node b is connected to all the nodes in the bit-gadgets $F_{B_1}, T_{B_1}, F_{B_2}, T_{B_2}$, the cliques C_{B_1}, C_{B_2}, and the connector nodes H_{B_1}, H_{B_2} (see Figs. 4 and 6).

Clique edges: Each of the sets $C_{A_1}, C_{A_2}, C_{B_1}, C_{B_2}$ is a clique, i.e., is isomorphic to the graph K_k. Therefore, every two nodes in the same set C_S where $S \in \{A_1, A_2, B_1, B_2\}$ are connected (see Fig. 4).

Bit edges: The sets T_S, F_S behave as bit-gadgets, meaning that the connections between the bit-gadgets and their corresponding bit nodes in S match their binary representation. Formally, for $(s, i) \in \{(a, 1), (a, 2), (b, 1), (b, 2)\}$ and $0 \le \ell \le \mu_{k+1} + 1$, let s_i^ℓ be a node in the corresponding set of bit nodes (i.e, in $\{A_1, A_2, B_1, B_2\}$), and let $(\ell)_j$ be the bit value of jth bit in the binary representation of ℓ. Then the *corresponding binary representation nodes* of s_i^ℓ, denoted $\mathtt{bin}(s_i^\ell)$, are $\mathtt{bin}(s_i^\ell) = \left\{ f_S^j | (\ell)_j = 0 \right\} \cup \left\{ t_S^j | (\ell)_j = 1 \right\}$ for the corresponding set $S \in \{A_1, A_2, B_1, B_2\}$ that the node s_i^ℓ is a part of. Therefore, each bit node s_i^ℓ is connected to $\mathtt{bin}(s_i^\ell)$ (see Fig. 5).

Connector edges: For any $(s, i, S) \in \{(a, 1, A_1), (a, 2, A_2), (b, 1, B_1), (b, 2, B_2)\}$, each node h_S^ℓ in H_S is connected to the nodes with the corresponding index in S and the clique C_S, i.e., to the nodes s_i^ℓ and c_S^ℓ (see Fig. 6). Notice that the connections from H_S to S and C_S are possible because the size of S equals the size of H_S, and because due to the inequality $\mu_{k+1} < k$ that is satisfied according to Claim 4, it holds that the number of nodes in C_S is sufficiently large to connect all the nodes in H_S to C_S.

Ring edges: For $i \in \{1, 2\}$, the edges between a and F_{A_i}, T_{A_i} (that is, the edges $\{a, f_{A_i}^\ell\}, \{a, t_{A_i}^\ell\}$) participate in a 20-ring of triangles with the edges between b and F_{B_i}, T_{B_i} (that is, the edges $\{b, f_{B_i}^\ell\}, \{b, t_{B_i}^\ell\}$) with the ring-auxiliary nodes acting as connectors. To be specific, let $i \in \{1, 2\}$ and $0 \le \ell \le \lceil \log(\mu_{k+1} + 1) \rceil - 1$. Then the following edges are part of the set of ring edges: $\{\{a, m_{i,\ell}^j\} \mid j \in \{0, 1, 12\}\}, \{\{b, m_{i,\ell}^j\} \mid j \in \{5, 6, 7\}\}, \{\{t_{A_i}^\ell, m_{i,\ell}^j\} \mid j \in \{1, 2, 3\}\}, \{\{f_{B_i}^\ell, m_{i,\ell}^j\} \mid j \in \{3, 4, 5\}\}, \{\{t_{B_i}^\ell, m_{i,\ell}^j\} \mid j \in \{6, 7, 8\}\}, \{\{f_{A_i}^\ell, m_{i,\ell}^j\} \mid j \in \{0, 12, 13\}\}$, as well as the edges $\{\{m_{i,\ell}^j, m_{i,\ell}^p\}\}$ for any (j, p) that are $(0, 1), (2, 3), (2, 4), (3, 4), (5, 6), (7, 8), (7, 9), (8, 9), (8, 10), (9, 10), (9, 11), (10, 11), (10, 13), (11, 12), (11, 13)$, or $(12, 13)$ (see Fig. 7). Note that the choice of having exactly 20 triangles per ring is arbitrary and any even number greater than 12 works.

The important part is to have an odd number of triangles (at least 3) between the four pairs of edges $(\{f^\ell_{A_j}, a\}, \{t^\ell_{A_j}, a\})$, $(\{t^\ell_{A_j}, a\}, \{f^\ell_{B_j}, b\})$, $(\{f^\ell_{B_j}, b\}, \{t^\ell_{B_j}, b\})$, and $(\{t^\ell_{B_j}, b\}, \{f^\ell_{A_j}, a\})$.

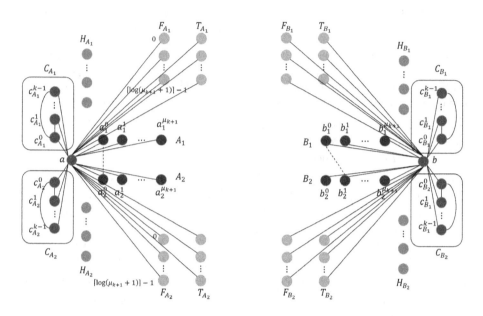

Fig. 4. The basic structure of the family of lower bound graphs for deciding the size of the minimum triangle edge transversal, with many edges and nodes omitted for clarity. See the additional figures for more detailed illustrations.

This concludes the construction of the fixed graph. We next analyze its required properties, describe the edges that are added according to the set disjointness inputs strings x, y, and complete the lower bound proof. We denote the size of the MTET of a graph by $\tau(G)$.

Claim 5. *Any triangle edge transversal of the fixed graph* G *must contain* $\tau(K_{k+1})$ *edges of each clique, 10 edges of each 20-ring of triangles (Fig. 7), and an edge from each triangle* $\left\{a^j_i, h^j_{A_i}, a\right\}$ *and* $\left\{b^j_i, h^j_{B_i}, b\right\}$*, for* $i \in \{1, 2\}, 0 \leq j \leq \mu_{k+1}$.

Instead of proving the claim directly, we prove a more generalized claim, which implies it.

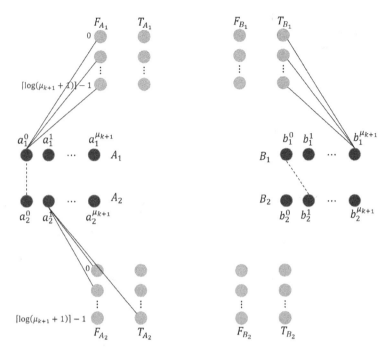

Fig. 5. The bit-gadget layer of connections, each bit node is connected to the corresponding nodes in the bit gadget with respect to the binary representation of that node.

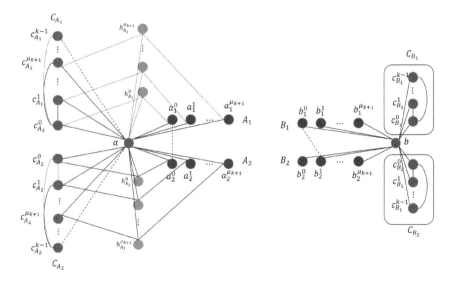

Fig. 6. The clique layer of connections (Connector edges), each edge between a bit node and a center node (e.g., $\{a, a_1^i\}$) and the edge between a clique node with the same index and the center node (i.e., $\{a, c_{A_1}^i\}$) are connected via a line of 2-triangles ($\{a, h_{A_1}^i, c_{A_1}^i\}$ and $\{a, h_{A_1}^i, a_1^i\}$).

Claim 6 (Sub-optimality of parts of a triangle edge transversal). *Let $E' \subseteq E$ be a subset of edges of G. Let $H = (V, E')$ be the subgraph of G with only the edges from E'. Then every triangle edge transversal S of G contains at least $\tau(H)$ edges of E'.*

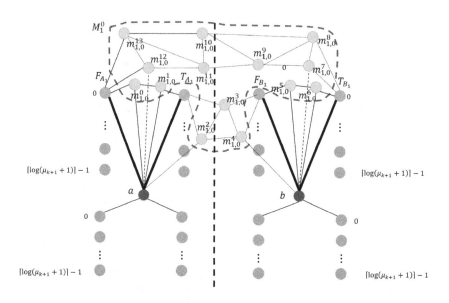

Fig. 7. This figure shows the connection between the bit-gadgets. Each tuple of the four edges $S_i^j = \left(\left\{ a, f_{A_j}^i \right\}, \left\{ a, t_{A_j}^i \right\}, \left\{ b, f_{B_j}^i \right\}, \left\{ b, t_{B_j}^i \right\} \right)$ for $0 \leq i \leq \lceil \log(\mu_{k+1} + 1) \rceil + 1$ and $j \in \{1, 2\}$ are connected with each other via a 20-ring of triangles, such that between each two consecutive edges of S_i^j (cyclic) there is an odd number of triangles. For example, between $\{a, f_{A_j}^i\}$ and $\{a, t_{A_j}^i\}$ there are three triangles and between $\{b, t_{B_j}^i\}$ and $\{a, f_{A_j}^i\}$ there are 9 triangles.

Proof. The set $S \cap E'$ is a triangle edge transversal of H, since all the triangles that are fully contained in H can be covered using only edges of $S \cap E'$. Thus, $|S \cap E'| \geq \tau(H)$. Note that $S \cap E' \subseteq E'$ and $S \cap E' \subseteq S$, thus, S contains at least $|S \cap E'| \geq \tau(H)$ edges of E'.

Having multiple edge-disjoint subsets of the edges gives a stronger lower bound on the size of a triangle edge transversal.

Claim 7. *Let $E_1, \ldots, E_m$ be disjoint subsets of E, and for every $1 \leq i \leq m$ let $H_i = (V, E_i)$ be the subgraph of G with only the edges of E_i. Then every triangle edge transversal S contains at least $\tau(H_i)$ edges of E_i and hence $|S| \geq \sum_{i=1}^{m} |\tau(H_i)|$.*

Proof. The first part of the claim follows directly from Claim 6. The second part follows from the disjointness of the subsets.

Corollary 2 (Corollary of Claims 5 and 7). *The size of any triangle edge transversal of the fixed graph G is at least $4\tau(K_{k+1}) + 4(\mu_{k+1} + 1) + 2 \cdot 10\lceil\log(\mu_{k+1} + 1)\rceil$.*

The proof of Corollary 2 and the proof of the following claim are given in the full version of the paper [12].

Claim 8. *If $S \subseteq E$ is a triangle edge transversal of G of size $4\tau(K_{k+1}) + 4(\mu_{k+1} + 1) + 2 \cdot 10\lceil\log(\mu_{k+1} + 1)\rceil$, then there are two indices $i, j \in \{0, \ldots, \mu_{k+1}\}$ such that the edges $\{a, a_1^i\}, \{a, a_2^j\}, \{b, b_1^i\}, \{b, b_2^j\}$ are not in S.*

Adding Edges Corresponding the Strings x and y. Given two strings $x, y \in \{0, 1\}^{(\mu_{k+1}+1)^2}$, we modify the graph G accordingly to obtain a graph $G_{x,y}$. Assuming that the strings are indexed by a pair of two indices of the form $(i, j), 0 \leq i, j \leq \mu_{k+1}$, we update the graph G by adding the edges $\{a_1^i, a_2^j\}$ if $x_{ij} = 0$ and the edges $\{b_1^i, b_2^j\}$ if $y_{ij} = 0$.

Before proving the next lemma, implying equivalence between Set Disjointness and MTET in our lower bound graphs, we first provide some intuition. We partition the nodes to two parts, one on Alice's side, informally all sets of nodes with A_1 or A_2 in their name, and the other nodes on Bob's side (see Fig. 4). The structure of the graph forces Alice to choose $\lceil\log(\mu_{k+1} + 1)\rceil$ bit-edges, and thus forces Bob to choose the same mirrored bit-edges on his side due to the 20-ring of triangles connection between those bit-edges (see Fig. 7). The chosen bit-edges correspond to the binary representations of the nodes $a_1^i, a_2^j, b_1^i, b_2^j$, therefore omitting the unnecessary edges $\{a, a_1^i\}, \{a, a_2^j\}, \{b, b_1^i\}, \{b, b_2^j\}$ from the MTET iff the edges $\{a_1^i, a_2^j\}, \{b_1^i, b_2^j\}$ don't exist, which implies the disjointness of the inputs. If Alice and Bob chose bit edges with corresponding binary representations that do not match, then the solution size will increase. The connections to the other cliques $C_{A_1}, C_{A_2}, C_{B_1}, C_{B_2}$ (See Fig. 6) are necessary to prevent other bit edges from being chosen (they correspond to exactly one binary representation).

Lemma 1. *The graph $G_{x,y}$ has a triangle edges-transversal of cardinality $M = 4\tau(K_{k+1}) + 4(\mu_{k+1} + 1) + 2 \cdot 10\lceil\log(\mu_{k+1} + 1)\rceil$ iff $DISJ(x, y) = \mathtt{false}$.*

Proof. If $DISJ(x, y) = \mathtt{false}$, then there exists a pair of indices $(i, j) \in \{0, \ldots, \mu_{k+1}\}^2$ such that $x_{ij} = y_{ij} = 1$. Consider the following sets of edges:

- A subset of the edges between the center node a and the clique C_{A_1}:
  $E_{a,C_{A_1}} = \{\{a, c_{A_1}^\ell\} : 0 \leq \ell \leq \mu_{k+1}, \ell \neq i\}$.
- A subset of the edges between the center node a and the clique C_{A_2}:
  $E_{a,C_{A_2}} = \{\{a, c_{A_2}^\ell\} : 0 \leq \ell \leq \mu_{k+1}, \ell \neq j\}$.
- A subset of the edges between the center node b and the clique C_{B_1}:
  $E_{b,C_{B_1}} = \{\{b, c_{B_1}^\ell\} : 0 \leq \ell \leq \mu_{k+1}, \ell \neq i\}$.
- A subset of the edges between the center node b and the clique C_{B_2}:
  $E_{b,C_{B_2}} = \{\{b, c_{B_2}^\ell\} : 0 \leq \ell \leq \mu_{k+1}, \ell \neq j\}$.

Let $S_{C_{A_1}}, S_{C_{A_2}}, S_{C_{B_1}}, S_{C_{B_2}}$ be the optimal triangle edge transversals of each of the cliques $C_{A_1} \cup \{a\}, C_{A_2} \cup \{a\}, C_{B_1} \cup \{b\}, C_{B_2} \cup \{b\}$ respectively, such that $E_{a,C_{A_1}} \subseteq S_{C_{A_1}}, E_{a,C_{A_2}} \subseteq S_{C_{A_2}}, E_{b,C_{B_1}} \subseteq S_{C_{B_1}}, E_{b,C_{B_2}} \subseteq S_{C_{B_2}}$, which exist due to Corollary 1.

Let us denote the union of these triangle edge transversals by $S_{cliques} = S_{C_{A_1}} \cup S_{C_{A_2}} \cup S_{C_{B_1}} \cup S_{C_{B_2}}$. Consider the optimal solutions of each ring of triangles (Fig. 7) that include the edges $\{\{a, u\}, u \in \mathtt{bin}(a_1^i)\}$ and the edges $\{\{a, v\}, v \in \mathtt{bin}(a_2^j)\}$ (these solutions also include the edges $\{\{b, u\}, u \in \mathtt{bin}(b_1^i)\}, \{\{b, v\}, v \in \mathtt{bin}(b_2^j)\}$). Let us denote the union of the solutions of each ring of triangles above by S_{rings}. Let us denote $S_H = \{\{a, h_{A_1}^i\}, \{a, h_{A_2}^j\}, \{b, h_{B_1}^i\}, \{b, h_{B_2}^j\}\}$ and $S_{bits} = \{\{a, a_1^\ell\} : 0 \le \ell \le \mu_{k+1}, \ell \ne i\} \cup \{\{a, a_2^\ell\} : 0 \le \ell \le \mu_{k+1}, \ell \ne j\} \cup \{\{b, b_1^\ell\} : 0 \le \ell \le \mu_{k+1}, \ell \ne i\} \cup \{\{b, b_2^\ell\} : 0 \le \ell \le \mu_{k+1}, \ell \ne j\}$. Thus, we claim that the subset $S = S_{cliques} \cup S_{rings} \cup S_{bits} \cup S_H$ is a triangle edge transversal of $G_{x,y}$ of the required cardinality.

It is easy to validate that the cardinality of S is equal to the required cardinality in the statement of the lemma. We must show that S is a triangle edge transversal. The triangles in each clique C_R, for $R \in \{A_1, A_2, B_1, B_2\}$, are covered by S_{C_R}. The triangles between the bit-nodes (A_1, A_2, B_1, B_2), the center node (a, b), the connector nodes H_R, for $R \in \{A_1, A_2, B_1, B_2\}$, and the cliques, are covered by either $S_{cliques}$, S_H or S_{bits} (see the triangles in Fig. 6). The triangles between the bit nodes, the bit-gadgets and the center nodes are covered by either S_{rings} (specifically, for the triangles that contain the edge $\{a, a_{A_1}^i\}, \{a, a_{A_2}^j\}, \{b, b_{B_1}^i\}$ or $\{b, b_{B_2}^j\}$) or S_{bits}. The triangles in the 20-rings of triangles are covered by S_{rings}. Lastly, all the other triangles in the graph $G_{x,y}$ are created after adding the edges corresponding to the strings x, y, and they are of the form $\{a, a_{A_1}^{i'}, a_{A_2}^{j'}\}$ or $\{b, b_{B_1}^{i'}, b_{B_2}^{j'}\}$. Those triangles are covered by S_{bits}, which includes at least one edge from every possible triplet of that form except for the two triplets $\{a, a_{A_1}^i, a_{A_2}^j\}$ and $\{b, b_{B_1}^i, b_{B_2}^j\}$. However, these triplets are not triangles in the graph, as it is given that $x_{ij} = 1$ (i.e., the edges $\{a_{A_1}^i, a_{A_2}^j\}, \{b_{B_1}^i, b_{B_2}^j\}$ do not exist).

On the other hand, if the triangle edge transversal S is of the given size, then by Claim 8, there exist two indices i, j such that none of the edges $\{a, a_{A_1}^i\}, \{a, a_{A_2}^j\}, \{b, b_{B_1}^i\}, \{b, b_{B_2}^j\}$ are in S. Thus, the edges $\{a_{A_1}^i, a_{A_2}^j\}, \{b_{B_1}^i, b_{B_2}^j\}$ do not exist, as otherwise they must be in S and the size of the cover is larger than given. The non-existence of the edges $\{a_{A_1}^i, a_{A_2}^j\}, \{b_{B_1}^i, b_{B_2}^j\}$ corresponds to the strings x, y having a bit 1 in the index (i, j). Thus, the inputs are not disjoint.

We are now finally ready to prove our main theorem.

Proof (Proof of Theorem 4). Let $MA_i^\ell \subseteq M_i^\ell$ be the ring-auxiliary nodes $MA_i^\ell = \{m_{i,\ell}^j \mid j \in \{0, 1, 2, 10, 11, 12, 13\}\}$ and let $MB_i^\ell \subseteq M_i^\ell$ be the ring-auxiliary nodes $MB_i^\ell = \{m_{i,\ell}^j \mid j \in \{3, 4, 5, 6, 7, 8, 9\}\}$ for $i \in 1, 2$ and $0 \le \ell \le \lceil \log (\mu_{k+1} + 1) \rceil - 1$. Let MA be the union of the sets MA_i^ℓ and let MB be the union of the sets MB_i^ℓ. We divide the nodes of the graph G (and $G_{x,y}$) into two sets. One set is

$V_A = A_1 \cup A_2 \cup F_{A_1} \cup F_{A_2} \cup H_{A_1} \cup H_{A_2} \cup C_{A_1} \cup C_{A_2} \cup MA$ and the other set is $V_B = B_1 \cup B_2 \cup F_{B_1} \cup F_{B_2} \cup H_{B_1} \cup H_{B_2} \cup C_{B_1} \cup C_{B_2} \cup MB$.

Note that the number of nodes n is $\Theta(k)$, as Claim 4 gives that $\mu_{k+1} = \Theta(k)$. In particular, $|A_1| = |A_2| = |H_{A_1}| = |H_{A_2}| = \Theta(k)$. In addition, the size of the inputs x and y, which is equal to μ_{k+1}^2, is $K = \Theta(k^2) = \Theta(n^2)$. Furthermore, the cut between the two parts V_A, V_B consists only of the edges of the rings of triangles. Precisely, each ring contributes a constant number of edges to the cut. Hence, the cut size is $\Theta(\log \mu_{k+1}) = \Theta(\log n)$. Since Lemma 1 proves that $G_{x,y}$ is a family of lower bound graphs for triangle edge transversal, and since the communication complexity of set disjointness is linear in the input size, then by applying Theorem 5 on the partition $\{V_A, V_B\}$, we deduce that any algorithm in the CONGEST model for deciding whether a given graph has a triangle edge transversal of size $M = 4\tau(K_{k+1}) + 4(\mu_{k+1} + 1) + 2 \cdot 10\lceil \log(\mu_{k+1} + 1)\rceil$ requires at least a near quadratic number of rounds $\Omega(K/\log^2(n)) = \Omega(n^2/\log^2(n))$.

3   A $(1 + \epsilon)$-Approximation for MTET in the LOCAL Model

Despite MTET being a global problem, we use the approach of [26] to show that it can be well approximated very efficiently.

Theorem 2. *There is a distributed algorithm in the LOCAL model that obtains a $(1+\epsilon)$-approximation to the minimum triangle edge transversal in poly$(\log n, 1/\epsilon)$ rounds.*

The proof of Theorem 2 is given in the full version of the paper [12]. The outline of the proof follows the ball-carving technique of [26] for various tasks such as the problem of approximating the minimum dominating set. The idea is to carve non-intersecting balls (neighborhoods) around multiple nodes up to radius $O(\log n)$ and compute the optimal solution in each one, by iterating over the radius length (increasingly) until the ratio between the new and previous optimal solutions is capped by $(1+\epsilon)$. The procedure carves the inner ball while pays for the optimal solution for the outer ball (thus covering the solution of the carved-out ball) and repeats the process on another node. The cap on the ratio ensures the $(1 + \epsilon)$ approximation factor. The process of parallelizing the above technique is done via network-decomposition, a partition of the nodes to clusters. Then the technique is applied on each cluster concurrently.

4   Faster Approximations for the Minimum Triangle Edge Transversal

In this section, we provide a reduction from the MTET problem to the MHVC problem and use it to show how the $(3 + \epsilon)$-approximation algorithm for MHVC introduced in [6, Section 3] can be simulated in the LOCAL and CONGEST models to solve MTET. With a slight increase in the time complexity, the algorithm can also approximate MHVC by a factor of 3. The reduction and the adjusted algorithm are also applicable for the weighted case.

The faster approximation we obtain for MTET in the LOCAL and CONGEST models is summarized in the following.

Theorem 3. *There are distributed algorithms in the* LOCAL *model that obtain a 3-approximation and a $(3 + \epsilon)$-approximation to the minimum triangle edge transversal in $O(\log n)$ and $O(\frac{\log n}{\log \log n})$ rounds, respectively. The same approximation guarantees can be obtained in the* CONGEST *model with a multiplicative overhead of Δ in the round complexity.*

We denote the input graph for the MTET problem by $G = (V, E)$. An MHVC algorithm considers a communication network on a hypergraph H, where every node v_H and hyperedge e_H have their own computation units, and v_H can communicate with e_H if and only if $v_H \in e_H$, i.e., every hyperedge can communicate with its nodes and every node can communicate with all the hyperedges it is a part of. The complexity in this model is measured by rounds. The degree of each node is bounded by Δ and the cardinality of every hyperedge is bounded by f (the rank of the hypergraph).

Reduction of MTET to MHVC. We construct the hypergraph $H_G = (V_{H_G}, E_{H_G})$ as the following. Each edge in the original graph G becomes a node in H_G, i.e., $V_{H_G} = E$. Every triangle $t = \{e_1, e_2, e_3\}$ in the original graph becomes a hyperedge in H_G, i.e., $E_{H_G} = \{\{e_1, e_2, e_3\} : e_1, e_2, e_3 \in E \text{ form a triangle}\}$. For the weighted case, the weight of every node in H_G is the same as the weight of the corresponding edge in E.

We call the obtained hypergraph *the reduced hypergraph of G* and denote it by H_G. It is easy to see that the reduced hypergraph H_G is 3-uniform. Specifically, the rank of the graph is $f = 3$. We denote the node in the reduced hypergraph H_G corresponding to an edge e in G by $v_{H_G}^e$, and the hyperedge corresponding to a triangle t in G by $e_{H_G}^t$. We also denote the set of nodes of the edge e and the triangle t by V_e and V_t, respectively.

In our case, each node in the reduced hypergraph $v_{H_G}^e$ corresponds to the edge $e = \{u, v\}$ in the original graph. Therefore, the messages of this node will be simulated by the two endpoints of the corresponding edge, i.e., V_e. The hyperedges corresponding to the triangles in the original graph will be simulated by the three nodes that forms the triangle, i.e., V_t.

We claim that any algorithm for MHVC that runs on the reduced hypegraph in LOCAL can be simulated with the same complexity in the original graph. We further claim that any algorithm for MHVC that runs on H_G in CONGEST can be simulated with a multiplicative overhead of Δ over the original complexity.

Claim 9. *Let $G = (V, E)$ be an undirected graph and H_G be the reduced hypergraph of G. Then:*

1. *Any algorithm that solves MHVC in H_G in the* LOCAL *model that runs in $O(r)$ rounds can be simulated in the graph G in $O(r)$ rounds.*

2. *Any algorithm that solves MHVC in H_G in the* CONGEST *model that runs in $O(r)$ rounds can be simulated in the graph G in $O(\Delta \cdot r)$ rounds.*

The proof of Claim 9 is given in the full version of the paper [12].

We now plug the MHVC algorithm of [6, Section 3] into our reduction. The following claim summarizes its round complexity.

Claim 10 ([6, **Corollaries 4.10, 4.12 and Appendix B**]). *There are distributed algorithms for MHVC that:*

1. *compute an f-approximation in $O(f \log n)$ rounds in the* LOCAL *model,*
2. *for $f = O(1)$ and $\epsilon = 2^{-O\left((\log \Delta)^{0.99}\right)}$, compute an $(f + \epsilon)$-approximation in $O\left(\frac{\log \Delta}{\log \log \Delta}\right)$ rounds in the* LOCAL *model, and*
3. *can be adapted to the* CONGEST *model without affecting the round complexity.*

Proof. [Proof of Theorem 3] The proof combines Claims 9 and 10. By Claim 9, the execution of the MHVC algorithm of Claim 10 on the reduced hypergraph can be simulated on the original graph in the LOCAL model in the same round complexity and in the CONGEST model up to a factor Δ of the round complexity on the hypergraph. Then, for its solution to MTET, each node v marks the edges in $\{e = \{u, v\} \mid u \in N(v)\}$ that correspond to the nodes in the solution for MHVC (of H_G), which it knows about since it simulates the nodes $v_{H_G}^e$.

This yields a 3-approximation algorithm that runs in $O(f \log |V_{H_G}|) = O(3 \log |E_G|) = O(\log n^2) = O(\log n)$ and $O(\Delta \cdot f \log |V_{H_G}|) = O(\Delta \log n)$ rounds in the LOCAL and CONGEST models, respectively. For the $(3 + \epsilon)$-approximation, we obtain round complexities of $O(\frac{\log \Delta_H}{\log \log \Delta_H}) = O(\frac{\log n}{\log \log n})$ and $O(\frac{\Delta \log \Delta_H}{\log \log \Delta_H}) = O(\Delta \frac{\log n}{\log \log n})$, respectively. The correctness of the algorithm (feasibility and approximation ratio) are directly derived from the correctness of the MHVC algorithm (see [6, Section 4.1]).

Acknowledgements. This project has received funding from the European Union's Horizon 2020 research and innovation programme under grant agreement no. 755839, and from ISF grant 529/23. The authors thank Seri Khoury for many useful discussions.

References

1. Abboud, A., Censor-Hillel, K., Khoury, S.: Near-linear lower bounds for distributed distance computations, even in sparse networks. In: Gavoille, C., Ilcinkas, D. (eds.) Distributed Computing - 30th International Symposium, DISC 2016, Paris, France, 27–29 September 2016. Proceedings. Lecture Notes in Computer Science, vol. 9888, pp. 29–42. Springer, Heidelberg (2016). https://doi.org/10.1007/978-3-662-53426-7_3

2. Abboud, A., Censor-Hillel, K., Khoury, S., Lenzen, C.: Fooling views: a new lower bound technique for distributed computations under congestion. Distrib. Comput. **33**, 545–559 (2020). https://doi.org/10.1007/s00446-020-00373-4

3. Abboud, A., Censor-Hillel, K., Khoury, S., Paz, A.: Smaller cuts, higher lower bounds. ACM Trans. Algorithms **17**(4), 30:1–30:40 (2021). https://doi.org/10.1145/3469834, https://doi.org/10.1145/3469834

4. Bacrach, N., Censor-Hillel, K., Dory, M., Efron, Y., Leitersdorf, D., Paz, A.: Hardness of distributed optimization. In: Proceedings of the 2019 ACM Symposium on Principles of Distributed Computing, pp. 238–247 (2019)

5. Bar-Yossef, Z., Jayram, T.S., Kumar, R., Sivakumar, D.: An information statistics approach to data stream and communication complexity. J. Comput. Syst. Sci. **68**(4), 702–732 (2004). https://doi.org/10.1016/j.jcss.2003.11.006

6. Ben-Basat, R., Even, G., Kawarabayashi, K., Schwartzman, G.: Optimal distributed covering algorithms. In: Robinson, P., Ellen, F. (eds.) Proceedings of the 2019 ACM Symposium on Principles of Distributed Computing, PODC 2019, Toronto, ON, Canada, 29 July–2 August 2019, pp. 104–106. ACM (2019). https://doi.org/10.1145/3293611.3331577

7. Brakerski, Z., Patt-Shamir, B.: Distributed discovery of large near-cliques. Distrib. Comput. **24**(2), 79–89 (2011). https://doi.org/10.1007/s00446-011-0132-x

8. Censor-Hillel, K.: Distributed subgraph finding: progress and challenges. CoRR **abs/2203.06597** (2022). https://doi.org/10.48550/arXiv.2203.06597

9. Censor-Hillel, K., Fischer, E., Schwartzman, G., Vasudev, Y.: Fast distributed algorithms for testing graph properties. Distrib. Comput. **32**(1), 41–57 (2019). https://doi.org/10.1007/S00446-018-0324-8

10. Censor-Hillel, K., Fischer, O., Gonen, T., Gall, F.L., Leitersdorf, D., Oshman, R.: Fast distributed algorithms for girth, cycles and small subgraphs. CoRR **abs/2101.07590** (2021). https://arxiv.org/abs/2101.07590

11. Censor-Hillel, K., Kaski, P., Korhonen, J.H., Lenzen, C., Paz, A., Suomela, J.: Algebraic methods in the congested clique. Distrib. Comput. **32**(6), 461–478 (2019). https://doi.org/10.1007/s00446-016-0270-2

12. Censor-Hillel, K., Khoury, M.: On distributed computation of the minimum triangle edge transversal (2024). https://arxiv.org/abs/2402.13985

13. Censor-Hillel, K., Leitersdorf, D., Turner, E.: Sparse matrix multiplication and triangle listing in the congested clique model. Theor. Comput. Sci. **809**, 45–60 (2020). https://doi.org/10.1016/j.tcs.2019.11.006

14. Censor-Hillel, K., Leitersdorf, D., Vulakh, D.: Deterministic near-optimal distributed listing of cliques. In: Proceedings of the Symposium on Principles of Distributed Computation (PODC), pp. 271–280 (2022)

15. Chang, Y.J., Pettie, S., Saranurak, T., Zhang, H.: Near-optimal distributed triangle enumeration via expander decompositions. J. ACM **68**(3), 1–36 (2021)

16. Chang, Y., Saranurak, T.: Deterministic distributed expander decomposition and routing with applications in distributed derandomization. In: Proceedings of the 61st IEEE Annual Symposium on Foundations of Computer Science (FOCS), pp. 377–388 (2020). https://doi.org/10.1109/FOCS46700.2020.00043

17. Das Sarma, A., et al.: Distributed verification and hardness of distributed approximation. In: Proceedings of the Forty-Third Annual ACM Symposium on Theory of Computing, STOC 2011, pp. 363-372. Association for Computing Machinery, New York (2011). https://doi.org/10.1145/1993636.1993686

18. Dolev, D., Lenzen, C., Peled, S.: "tri, tri again": finding triangles and small subgraphs in a distributed setting - (extended abstract). In: Aguilera, M.K. (ed.) Distributed Computing - 26th International Symposium, DISC 2012, Salvador, Brazil, 16–18 October 2012. Proceedings. Lecture Notes in Computer Science, vol. 7611, pp. 195–209. Springer, Heidelberg (2012). https://doi.org/10.1007/978-3-642-33651-5_14

19. Drucker, A., Kuhn, F., Oshman, R.: On the power of the congested clique model. In: Halldórsson, M.M., Dolev, S. (eds.) ACM Symposium on Principles of Distributed Computing, PODC 2014, Paris, France, 15–18 July 2014, pp. 367–376. ACM (2014). https://doi.org/10.1145/2611462.2611493

20. Eden, T., Fiat, N., Fischer, O., Kuhn, F., Oshman, R.: Sublinear-time distributed algorithms for detecting small cliques and even cycles. Distrib. Comput. **35**(3), 207–234 (2022). https://doi.org/10.1007/s00446-021-00409-3

21. Erdös, P., Gallai, T., Tuza, Z.: Covering and independence in triangle structures. Discret. Math. **150**(1–3), 89–101 (1996). https://doi.org/10.1016/0012-365X(95)00178-Y

22. Even, G., et al.: Three notes on distributed property testing. In: Richa, A.W. (ed.) 31st International Symposium on Distributed Computing, DISC 2017, Vienna, Austria, 16–20 October 2017. LIPIcs, vol. 91, pp. 15:1–15:30. Schloss Dagstuhl - Leibniz-Zentrum für Informatik (2017). https://doi.org/10.4230/LIPIcs.DISC.2017.15

23. Fischer, O., Gonen, T., Kuhn, F., Oshman, R.: Possibilities and impossibilities for distributed subgraph detection. In: Proceedings of the 30th Symposium on Parallelism in Algorithms and Architectures (SPAA), pp. 153–162 (2018). https://doi.org/10.1145/3210377.3210401

24. Fraigniaud, P., Olivetti, D.: Distributed detection of cycles. In: Proceedings of the 29th ACM Symposium on Parallelism in Algorithms and Architectures (SPAA), pp. 153–162 (2017)

25. Frischknecht, S., Holzer, S., Wattenhofer, R.: Networks cannot compute their diameter in sublinear time. In: Rabani, Y. (ed.) Proceedings of the Twenty-Third Annual ACM-SIAM Symposium on Discrete Algorithms, SODA 2012, Kyoto, Japan, 17–19 January 2012, pp. 1150–1162. SIAM (2012). https://doi.org/10.1137/1.9781611973099.91

26. Ghaffari, M., Kuhn, F., Maus, Y.: On the complexity of local distributed graph problems. In: Hatami, H., McKenzie, P., King, V. (eds.) Proceedings of the 49th Annual ACM SIGACT Symposium on Theory of Computing, STOC 2017, Montreal, QC, Canada, 19–23 June 2017, pp. 784–797. ACM (2017). https://doi.org/10.1145/3055399.3055471

27. Goldreich, O., Goldwasser, S., Ron, D.: Property testing and its connection to learning and approximation. J. ACM **45**(4), 653–750 (1998). https://doi.org/10.1145/285055.285060

28. Hirvonen, J., Rybicki, J., Schmid, S., Suomela, J.: Large cuts with local algorithms on triangle-free graphs. Electron. J. Comb. **24**(4), P4.21 (2017). http://www.combinatorics.org/ojs/index.php/eljc/article/view/v24i4p21

29. Holzer, S., Pinsker, N.: Approximation of distances and shortest paths in the broadcast congest clique. In: Anceaume, E., Cachin, C., Potop-Butucaru, M.G. (eds.) 19th International Conference on Principles of Distributed Systems, OPODIS 2015, Rennes, France, 14–17 December 2015. LIPIcs, vol. 46, pp. 6:1–6:16. Schloss Dagstuhl - Leibniz-Zentrum für Informatik (2015). https://doi.org/10.4230/LIPIcs.OPODIS.2015.6

30. Huang, D., Pettie, S., Zhang, Y., Zhang, Z.: The communication complexity of set intersection and multiple equality testing. In: Proceedings of the 2020 ACM-SIAM Symposium on Discrete Algorithms (SODA), pp. 1715–1732 (2020). https://doi.org/10.1137/1.9781611975994.105

31. Izumi, T., Gall, F.L.: Triangle finding and listing in CONGEST networks. In: Schiller, E.M., Schwarzmann, A.A. (eds.) Proceedings of the ACM Symposium on

Principles of Distributed Computing, PODC 2017, Washington, DC, USA, 25–27 July 2017, pp. 381–389. ACM (2017).https://doi.org/10.1145/3087801.3087811

32. Kalyanasundaram, B., Schnitger, G.: The probabilistic communication complexity of set intersection. SIAM J. Discret. Math. **5**(4), 545–557 (1992). https://doi.org/10.1137/0405044

33. Kortsarz, G., Langberg, M., Nutov, Z.: Approximating maximum subgraphs without short cycles. SIAM J. Discret. Math. **24**(1), 255–269 (2010). https://doi.org/10.1137/09074944X

34. Krivelevich, M.: On a conjecture of tuza about packing and covering of triangles. Discret. Math. **142**(1-3), 281–286 (1995). https://doi.org/10.1016/0012-365X(93)00228-W

35. Kushilevitz, E., Nisan, N.: Communication Complexity. Cambridge University Press, Cambridge (1997)

36. Linial, N.: Locality in distributed graph algorithms. SIAM J. Comput. **21**(1), 193–201 (1992). https://doi.org/10.1137/0221015

37. Lotker, Z., Patt-Shamir, B., Pavlov, E., Peleg, D.: Minimum-weight spanning tree construction in O(log log n) communication rounds. SIAM J. Comput. **35**(1), 120–131 (2005). https://doi.org/10.1137/S0097539704441848

38. Pandurangan, G., Robinson, P., Scquizzato, M.: On the distributed complexity of large-scale graph computations. ACM Trans. Parallel Comput. **8**(2), 7:1–7:28 (2021). https://doi.org/10.1145/3460900

39. Peleg, D.: Distributed Computing: A Locality-Sensitive Approach. Society for Industrial and Applied Mathematics, Philadelphia (2000)

40. Peleg, D., Rubinovich, V.: A near-tight lower bound on the time complexity of distributed minimum-weight spanning tree construction. SIAM J. Comput. **30**(5), 1427–1442 (2000). https://doi.org/10.1137/S0097539700369740

41. Pettie, S., Su, H.: Fast distributed coloring algorithms for triangle-free graphs. In: Proceedings of the 40th International Colloquium on Automata, Languages, and Programming (ICALP), pp. 681–693 (2013)

42. Razborov, A.A.: On the distributional complexity of disjointness. In: Proceedings of the 17th International Colloquium on Automata, Languages and Programming (ICALP), pp. 249–253 (1990)

43. Ruzsa, I.Z., Szemerédi, E.: Triple systems with no six points carrying three triangles. Combinatorics (Keszthely, 1976), Coll. Math. Soc. J. Bolyai **18**(939-945), 2 (1978)

Stability of P2P Networks Under Greedy Peering

Lucianna Kiffer[1]($\boxtimes$) and Rajmohan Rajaraman[2]

[1] ETH Zurich, Zürich, Switzerland
lkiffer@ethz.ch
[2] Northeastern University, Boston, USA
r.rajaraman@northeastern.edu

Abstract. Historically, major cryptocurrency networks have relied on random peering choice rules for making connections in their peer-to-peer networks. Generally, these choices have good properties, particularly for open, permissionless networks. Random peering choices however do not take into account that some actors may choose to optimize who they connect to such that they are quicker to hear about information being propagated in the network. In this paper, we explore the dynamics of such greedy strategies. We study a model in which nodes select peers with the objective of minimizing their average distance to a designated subset of nodes in the network, and consider the impact of several factors including the peer selection process, degree constraints, and the size of the designated subset. The latter is particularly interesting in the context of blockchain networks as generally only a subset of nodes are miners/content producers (i.e., the propagation source for content), or are running specialized hardware that would make them higher performing.

We first analyze an idealized version of the game where each node has full knowledge of the current network and aims to select the d best connections, and prove the existence of equilibria under various model assumptions. Since in reality nodes only have local knowledge based on their peers' behavior, we also study a greedy protocol which runs in rounds, with each node replacing its worst-performing edge with a new random edge. We exactly characterize stability properties of networks that evolve with this peering rule and derive regimes (based on number of nodes and degree) where stability is possible and even inevitable. We also run extensive simulations with this peering rule examining both how the network evolves and how different network parameters affect the stability properties of the network. Our findings generally show that the only stable networks that arise from greedy peering choices are low-diameter and result in disparate performance for nodes in the network.

1 Introduction

The use of cryptocurrency networks has exploded in recent years, with billions of dollars being transferred daily across various platforms [6]. As these networks grow, it becomes increasingly important to understand the factors that

© The Author(s), under exclusive license to Springer Nature Switzerland AG 2024
Y. Emek (Ed.): SIROCCO 2024, LNCS 14662, pp. 359–383, 2024.
https://doi.org/10.1007/978-3-031-60603-8_20

impact their structure and stability. Traditionally, networks like Bitcoin and Ethereum[1] have implemented peer selection strategies which aim at randomizing peer choices to avoid attacks [13,14,24], and favor properties such as low diameter [5,10] for fast information propagation. Recently in Ethereum's new Proof-of-Stake network, the Beacon network, some clients have taken a new approach of greedily choosing peers based on performance [21] where nodes drop some fraction of worst-behaving peers. In this work, we analyze a simplified model of such greedy peering strategies.

This paper explores the dynamics of greedy peering strategies, specifically the case where nodes aim to optimize their distance to all nodes in the network or to a special subset of high-value nodes (which we term the *miners*). We are interested in characterizing how heterogeneous peering preferences affect the evolution and stability of the network. Our study is driven by the following questions:

1. Do greedy peering strategy games have stable networks and can we characterize the conditions for stability?
2. What are the properties of stable networks? When stability is reachable, how do networks converge to a stable state?
3. Do protocol parameters (e.g., degree constraints, number of miners) alleviate or exacerbate stability dynamics?

1.1 Our Results

We consider networks of n nodes where $m \leq n$ nodes are special and are labeled as *miners*; these special nodes represent miners in a cryptocurrency network or content generators in a gossip-based dissemination network. The primary objective for each node is to be "close" to the miners in the network; we assign a score for each node as the average hop-distance to the miners in the *undirected network* formed by all the edges. The P2P network topologies that evolve in the process depend on a number of factors, including the fraction of the network that are miners, and nodes' out- and in-degree constraints.

We begin by evaluating an idealized version of a non-cooperative game where nodes choose their optimal peers to minimize their own score (i.e., average distance to the miners).

- **Existence of stable networks:** We show that *there always exist pure Nash equilibria (or stable networks) when nodes have unbounded in-degree*. We also prove that *Nash equilibria always exist for a non-trivial infinite family of networks with bounded in-degree*; the general case with bounded in-degrees is open. These results are presented in Sect. 4.1.

While the idealized game provides a sense of stable network topologies that can arise, the game is hard to implement in practice since nodes do not have global knowledge of the network and cannot make arbitrary peering choices.

---

[1] Historically the two largest by market cap [6].

Our main results concern a natural peering protocol, which we call *Single-Exploratory-Greedy*, where nodes have knowledge of only their neighbors' scores and can add an exploratory random link to replace an existing link. For this protocol, we consider a notion of stability in the network with respect to stable edges (edges that will never be dropped).

- **Conditions for stability:** We find that under the Single-Exploratory-Greedy protocol, *often the only stable networks involve a centralized, well-connected core of nodes (the miners), with all other nodes directly connecting to this core.* We also show that *when nodes are faced with ties on which connection to drop, the Tie Breaker Rule is a determining factor on whether any stable topologies exist;* for instance, we show the impossibility of stability with two such rules. Our detailed analysis of Single-Exploratory-Greedy is in Sect. 4.2 with a summary of results in Table 1.
- **Reaching stability:** We also show that when the number of miners is capped (to at most the node out-degree capacity), if a stable network exists, networks starting from any initial state will *always eventually* reach a stable state.

The bulk of our theoretical results are for networks with unbounded in-degree. To explore how and if bounded networks differ in their evolution to stable topologies, we run simulations of the Single-Exploratory-Greedy protocol in diverse scenarios (see Sect. 5).

- **Connectivity:** Our simulations largely support that in networks where a subset of the nodes are miners, even with bounded in-degree, the miners become more connected to one another and more centralized in the network over time (i.e., lower miner diameter and eccentricity), and that this behavior is quicker in networks where miners have a higher probability of connecting (e.g., smaller networks, larger inbound capacity, more than one exploratory edge).
- **Fairness:** Additionally, across the board, our simulations show that though Single-Exploratory-Greedy lowers network properties like average distance to miners, diameter, and average eccentricity, this often comes at the price of fairness, where some nodes are worse off in the network (e.g., have a higher average distance to miners than in the random network) and that the network-level average advantage is likely due to a few nodes being much better off and skewing the average.

2 Related Work

We aim to understand the topological impact of a greedy peering protocol on a peer-to-peer network with a subset of preferential nodes everyone is trying to get close to. Most major existing P2P networks of this kind, namely cryptocurrencies, implement peering protocols that approximate low-diameter random networks [5,10]. Topological analysis of these networks, however, is hard as they tend to obfuscate peer connections to avoid attacks [13,14,24], and existing measurement techniques are prohibitively expensive or invasive [8,20], thus they tend to be run on test networks as a proof of concept. In [8], Delgado et al. map the topology of the Bitcoin Ropsten test network and compared it with simulated random networks of the same size and similar degrees. They find the Ropsten network to have some behavior similar to networks in our simulations (e.g., a small number of concentrated central nodes).

An inspiration for the Single-Exploratory-Greedy of this paper is the Perigee protocol of Mao et al. [23], which also implements a greedy peering protocol. The authors consider a network of all miners in an Euclidean space and show theoretically that nodes choosing neighbors geographically close to them results in network distance that is close to the optimum distance (shortest path is a constant away from the Euclidean distance). They show via simulations that the Perigee protocol approximates Euclidean distances between all nodes. In their simulations, they show that as processing time at each node increases, the performance of Perigee and the random network converge. In our work, this is the regime we are interested in studying: negligible link latency in comparison to node processing latency such that the properties of the overlay topology (and hop distance) are more important than the underlying communication link latency. We also focus on networks where a subset of the nodes are miners, as this is often the case [18,26]. Node processing times and congestion have also been a recent focus in security analyses of blockchain consensus protocols [16,27].

There is a growing body of work on peering protocols in cryptocurrency networks. Recent work has shown that variations of the Perigee protocol run by a single entity can work well as a strategy for reducing a node's own latency to the source of transactions in a network [3,33]. In [30], Park et al. use their own measurements of Bitcoin to propose an IP-based distance peer selection protocol to reduce round trip time of peers. Though this is in the same spirit of Perigee-like greedy protocols, using IP location is notoriously unreliable [12,31], and tricky as a basis for routing as the Internet does not satisfy the triangle inequality [22]. Other proposals for overlay networks include randomly replacing edges to maintain approximate sparse random graphs [36], and more structured proposals such as one based on the DHT Kademlia [32], and a hypercubic overlay network [2], as well as others [34].

Our game-theoretic formulation falls under the category of network creation games that have been extensively studied by researchers in computer science, game theory, and economics. One of the earliest works in this space is due to Jackson and Wolinsky who introduced a model in which there is an underlying cost for each link, and studied tradeoffs between stability and efficiency [15]. Demaine et al. studied the price of anarchy in network creation, quantifying the overhead of the worst stable network, when compared with the social optimum [9]. Many different models based on cost, direction of links, degree constraints, and initial conditions have been introduced, with the aim of capturing diverse scenarios including social networks and the Internet [4,7,19,25]. Our game-theoretic model is different from models studied in past work in the following way. Previous work has considered either directed networks with the node actions being given by directed edges (e.g., [19]) or undirected networks with the node actions being given by undirected edges (e.g., [25]). In contrast, P2P networks underlying blockchains are better modeled by processes in which nodes select directed out-going edges to peers under some capacity constraints, while the latency incurred in the resulting network is independent of the edge direction. Previous models in network creation games do not consider this important distinction on how links are created and their impact on delays. As a result, the stability and convergence properties of past models differ from the results we derive for our model. Furthermore, past work on network creation games is primarily on the nature of Nash equilibria; while we also study equilibria for our model, our focus is on the analysis of a greedy peering process that captures the peering approaches of real P2P networks.

3   Model

We consider a network with n nodes, the first m of which are miners, with each node having d out-going edges and at most $d_{in} \leq n$ incoming edges. Let G denote the graph consisting of the nodes and the directed edges. We use (i, j) to denote the directed edge from i to j. For convenience, we also use $e_{i,j}$ as the indicator variable for whether either of edges (i, j) or (j, i) exist; so, $e_{i,j}$ is 0 (resp., 1) if neither (i, j) nor (j, i) exist (resp., if (i, j) exists or (j, i) exists) in G. Thus, $e_{i,j} = e_{j,i}$. We use the terms peer and neighbor interchangeably, and edge and connection interchangeably. We also refer to d_{in} as the inbound connection capacity.

The **Single-Exploratory-Greedy** protocol operation proceeds in rounds, each of which consists of the following steps:

1. In an arbitrary order, each node i selects at random a peer j s.t. currently $e_{i,j} = 0$ (thus preventing bi-directional edges) and j has not reached it's inbound capacity, and adds (i, j) to G (thus setting $e_{i,j} = 1$). If i has an edge to or from every node, it adds no new edges.
2. Every node i updates its score $score_G(i)$ as follows. If $dist_G(i, j)$ denotes the shortest distance between i and j in the *undirected* version of G, then

$$score_G(i) = \sum_{j \in miners} \frac{dist_G(i, j)}{m},$$

which is the average distance to all miners. A miner's distance to themselves is 0, thus the minimum score for a miner is $\frac{m-1}{m}$ and the minimum score of a non-miner is 1.
3. Each node i with d out-going edges after the above steps, drops an out-going edge (i, j) (i.e., sets $e_{i,j} = 0$), where j is the out-neighbor of i with the maximum score; if i does not have d out-going edges then it does not drop a peer. If two or more peers have the highest score, then a Tie Breaker Rule as defined below is used to pick one peer.

Definition 1 (Tie Breaker Rule). *We define the Tie Breaker Rule as the rule used for deciding ties when a node has multiple peers sharing the highest score in step (3). We consider nodes deciding ties according to the following* possible rules: uniformly at random (Random), First-In-First-Out (FIFO), Last-In-First-Out (LIFO), or a global ordering where all nodes are ranked in some order that does not change during the protocol and the node with highest rank wins (Global-Ordering).

Definition 2 (Stability). *We define an edge as* **stable** *at some round r, if its value at the end of the round (after step 3), does not change for all future rounds. Formally, let $e_{i,j}^r$ be the value of $e_{i,j}$ at the end of round r. Edge $e_{i,j}$ is stable at round r, if $\forall r' \geq r$, $e_{i,j}^{r'} = e_{i,j}^r$.*

A **network-stable topology** *is one where all nodes have $d-1$ stable outgoing edges. Formally, G is network-stable at round r if $\forall i, j \leq n, r' \geq r$ $e_{i,j}^{r'} = e_{i,j}^r$. A network-stable topology is therefore an equilibrium since each node has one outgoing edge they keep replacing with a random node, but the choice of this edge will never cause a stable edge to be dropped.*

A **miner-stable topology** *is one where all edges between miners are stable, and therefore no further edge between miners will arise outside of the d-th edge. Formally, graph is miner-stable at round r if $\forall i, j \leq m, r' \geq r$ $e_{i,j}^{r'} = e_{i,j}^r$.*

Note that for a topology to be network-stable or miner-stable, we require all nodes to have at least $d - 1$ outgoing edges since a node will never drop an edge if they have less than d. Additionally, the following event arises often in our arguments related to stability.

Definition 3 (Special Event). *We define Minimum-Score-Event$_i$ as the event where at step (1) of some round, all miners who were not yet connected to some node i, become connected to i. At step (2) of the protocol, node i thus has its minimal score ($\frac{m-1}{m}$ for miners, and 1 for non-miners). Assuming nodes have no inbound connection capacity, there is always some probability this event happens for a given i in the next round.*

This event goes to show that it is always possible for any node to come to have its minimal score in any round, thus, most of our proofs involve showing that stable connections must be to minimal scoring peers so that they are not replaced.

Model Motivation. In our model (and in practice), nodes do not know if a peer is a miner or not, all they know is a peer's relative performance. Thus, both miners and non-miners are trying to minimize their distance to all miners via the score function. While for simplicity, in our model, we use hop distance to score peers, in practice, network propagation delays can be used locally to score peers against each other. We also adopt a graph model where the set of connections are made in a directed fashion, while their performance is evaluated over the undirected edges. This mirrors the behavior of real P2P networks, where nodes generally have a set inbound and outbound connection cap, and will accept connections as long as they have slots (allowing new nodes who join the system to find peers) but will be more selective with their out-going edges (e.g., to prevent DoS). When it comes to information propagation however (e.g., gossiping of blocks and transactions), all peers are treated the same and thus message delay is dependent only on undirected hops. Additionally, we omit a peer discovery process and instead assume nodes can sample connections at random from all nodes in the network[2]. We also consider a greedy peering process that proceeds in rounds, though within each step there is no coordination between nodes (except the assumption that in step 1 no two nodes choose each other). These simplifying assumptions allow for an initial clean theoretical evaluation.

4   Theoretical Analyses

Our goal is to analyze Single-Exploratory-Greedy and the underlying network formation process. To begin, we consider an idealized game-theoretic version of Single-Exploratory-Greedy where nodes have full knowledge of the graph and can choose all peers to optimize their score, and explore the equilibria of this game. We then analyze the Single-Exploratory-Greedy protocol, studying the stability of the networks that arise during the protocol.

4.1   An Idealized Game-Theoretic Model

We consider first an idealized game-theoretic model in which each node wants to select their outgoing peers so as to minimize the average distance to a miner. A natural approach is via the concept of Nash equilibrium [29]. Let d denote the out-degree of each node. For a given node v, an action is a set of at most d nodes of the network representing the out-neighbors that v selects. Note that in this idealized game, nodes can choose *all* new outgoing edges. Recall that while we distinguish between in- and out-edges, the distances among nodes and the average diameter are based on the graph that is obtained by viewing each edge as undirected.

Let $\mathcal{G}(n, m, d)$ (resp., $\mathcal{G}(n, m, d, d_{in})$) denote the *uncapped* (resp., *capped*) *game* with n nodes, $m \leq n$ miners, out-degree d, and unbounded in-degree (resp., in-degree at most d_{in}). We first show that the uncapped game always has a pure Nash equilibrium, a stable topology in which no node wants to change its out-neighbors.

Lemma 1 (Every Uncapped Game has a Stable Network).   *For all* n, m, d, $\mathcal{G}(n, m, d)$ *has a Nash equilibrium.*

Proof. Consider any graph G with the following properties: (i) there exists a node r such that every node in G has an edge to r; (ii) every edge is directed to a miner; (iii) there does not exist any pair of nodes u and v such that (u, v) and (v, u) are both present in G; and (iv) if a node v has less than d outgoing edges, then it has an edge to or from every miner.

We claim that G is a Nash equilibrium. We compute the score for each node in G. The score of miner r equals $(m-1)/m$ since every other miner is at distance

---

[2] In reality, information of new nodes in the network needs to propagate and, though this generally happens quickly [11], it is possible that propagation of information on new nodes impacts how nodes cluster in the network.

1; this is the lowest score possible and no action of r can improve it. Let $u \neq r$ denote any other miner, and let i_u denote the number of miners that have edges to u. Since r is adjacent to every other node, u is at distance at most 2 from any other miner. Given the choices of all other nodes, a best-response action of u is to set all its out-edges to miners that do not have an edge to u; the best score achievable equals $\max\{(d + i_u) + 2(m - 1 - i_u - d), (m - 1)\}/m$. This is precisely what is achieved in G given the conditions: by (ii), all edges are directed to a miner; by (iii) u does not put any out-edges to a miner that already has an edge to u; finally, by (iv), if u does not use all of its d edges, then it already has a score of $(m - 1)/m$.

We finally consider a non-miner v. By the conditions, all of the edges of a non-miner are to miners. Its distance to $\min\{d, m\}$ miners is 1 and its distance to every other miner is 2, leading to an optimal possible score of $\max\{(d + 2(m - d))/m, 1\}$. Thus, v has no incentive to change its out-edges. We have thus shown that G is a pure Nash equilibrium.   □

We next consider the network formation game with bounded in-degree. Since the endpoints of the out-edges are constrained by this bound, reasoning about stable networks is much more challenging. We first show that the simplest network formation games in the capped setting, the family $\mathcal{G}(n, m, 1, d_{in})$, have pure Nash equilibrium for all d_{in} and infinitely many choices of n and m. We sketch the graphs constructed in the following proofs in Fig. 1.

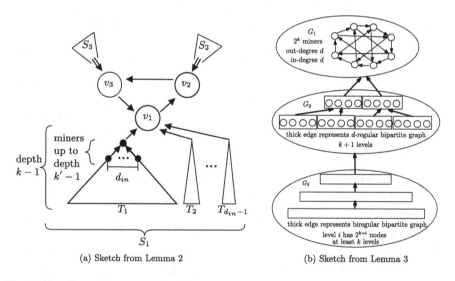

(a) Sketch from Lemma 2   (b) Sketch from Lemma 3

Fig. 1. For the existence proofs of Lemmas 2 and 3 we construct the above network families. Both networks are essentially layered, and the crux of the proof is that the miners are positioned in the "top layer(s)" with the non-miners occupying the remaining layers, such that all nodes except the non-miners in the last layer have full in-capacities. The nodes in the last layer have the highest score and no node higher up in the layers would give up their position to connect to a node in the bottom layer.

Lemma 2 (Stable Networks in Capped Games with Unit Out-Degree).
For any $n = 3 \cdot d_{in}^k$ and $m = 3 \cdot d_{in}^{k'}$ for any integers $k, k', k \geq k' \geq 0$, $\mathcal{G}(n, m, 1, d_{in})$ has a pure Nash equilibrium.

Proof. Note that n and m are at least 3. Let C be a directed triangle over three miners v_1, v_2, and v_3. Partition the remaining nodes into three equal-sized sets S_1, S_2, and S_3. By the assumption on n, each S_i has $d_{in}^k - 1$ nodes and $d_{in}^{k'} - 1$ miners. Split each S_i into $d_{in} - 1$ balanced complete d_{in}-ary directed trees, with every edge directed to the parent. Note that a balanced complete d_{in}-ary tree of depth $h - 1$ has $\frac{d_{in}^h - 1}{d_{in} - 1}$ nodes. Furthermore, assume that all nodes in each d_{in}-ary tree T_j up to depth $k' - 1$ are miners (there are exactly $\frac{d_{in}^{k'} - 1}{d_{in} - 1}$ such nodes in each T_j). Let $r_{i,j}$ for $1 \leq j \leq d_{in} - 1$ be the root of each tree T_j in S_i. Consider the network G consisting of the union of C, S_1, S_2, S_3, and the edges $\{(r_{i,j}, v_i) : 1 \leq i \leq 3\}$. We show that G is a pure Nash equilibrium.

For each v_i, consider the directed tree Tv_i made up of the trees in S_i and the edges from their roots to v_i. By symmetry, the scores of all nodes at a given depth in Tv_i are the same. Let δ_1 and δ_2 be integers such that $k \geq \delta_1 > \delta_2$. Consider two nodes n_1 and n_2 in S_i at depths δ_1 and δ_2, respectively. The score of n_1 is strictly greater than that of n_2. This is because a node v at some depth δ has a lower score than any of the nodes in the sub-tree below it since there are twice as many miners not in Tv_i as are in Tv_i, and v is in the only path of its sub-tree to those miners. This also implies that for any leaf ℓ and any non-leaf node v, the score of ℓ is at least the score of v. Given the symmetry of the leaves, this applies for all n_1, n_2 at depths δ_1 and δ_2 in different Tv_i trees.

To complete the proof, we consider each node and verify that their out-edge in G gives the node the best score, given the choices of other nodes. By construction, every non-leaf node in G has a full in-degree of d. Hence, the action of any node v in the game is to either have their current out-edge in G or to direct the out-edge to one of the leaf nodes. If v selects a leaf node ℓ, note that by symmetry this node has score s at least that of v. If v is not a miner, then its distance to all miners will be one more than that of ℓ, so the new score of v is $s + 1$.

If v is a miner at depth h, we consider the change in its score as new_score(v)−old_score(v) and prove this change is positive. Note first that v's distance to itself and all miners in the sub-trees that point to v remains the same, so a difference of 0. WLOG, assume v was in Tv_1 and chose to connect to a leaf in Tv_2. Thus v's distance to miners in Tv_3 has only increased by $k + 1 - h$. Next, let v' be the new ancestor of v in Tv_2 at height h. Since v and v' were at the same height in G, by the symmetric property of G, they had the same cumulative distance to miners in Tv_1 and Tv_2 who are not in the sub-trees of v or v'. Thus v's score to these miners has increased by its distance to v' which is again $k + 1 - h$.

The only remaining miners to consider in the change in score is from v to the miners in the sub-tree of v' which may now be closer (i.e., take away from the change in score). In the original structure, v was a maximum $2h + 1$ from v' ($2h$

if v and v' are in the same tree), thus these miners are now at most $k + h + 1$ from v. If we consider only the sub-trees of G whose roots are at depth h, let X be the number of miners in each of these sub-trees. There are $(d - 1) * d^{h-1}$ such sub-trees in each S_i. Then new_score(v)−old_score(v) is at least the score difference for these sub-trees (since we are just under-counting miners who are now further away from v), we get that

$$\text{new\_score}(v) - \text{old\_score}(v) \geq (k+1-h) \cdot \left[3 \cdot (d-1) \cdot d^{h-1} - 2\right] \cdot X - (k+h+1) \cdot X.$$

Let's assume this score difference is non-positive. Remembering that $d \geq 2$, we can simplify the above inequality to be

$$2^{h-1} \leq (d-1)d^{h-1} \leq \frac{3k + h - 3}{3k - 3h + 3}$$

which is not satisfiable. Thus the score difference is strictly positive, meaning v would not choose to change its out-edge. □

We next extend Lemma 2 to capped games with arbitrary out-degrees, establishing the existence of stable networks in much more general settings in Lemma 3. At a very high level, the proof establishes stability of networks with a certain structure: a strongly connected core G_1 consisting of the miners, a hierarchical layered network G_2 with nodes in its top layer having directed edges to G_1, and a more loosely structured layered network G_3 with nodes in its top layer having directed edges to G_2 and nodes in its bottom layer having no incoming edges. The nodes in the bottom layer of G_3 are farthest from the miners while the in-degree of every other node in the network equals the in-capacity, which ensures that the network is stable. Figure 1(b) illustrates this proof.

Lemma 3 (Stable Networks in Capped Games with Arbitrary Out-Degree). *For integers $k \geq 0$, $d \geq 2$, $m \leq 2^k$, and $n \geq 2^{3k}$, $\mathcal{G}(n, m, d, 2d)$ has a pure Nash equilibrium.*

Proof. We consider a network G that has three components. The first component G_1 is a graph with its vertex set being all the miners, each node with an out-degree of d to other miners, leaving $d2^k$ incoming capacity.

The second component G_2 consists of $k + 1$ levels of nodes, with level $i \geq 0$ consisting of set L_i of 2^{k+i} nodes. For each i, we partition the 2^{k+i} nodes into 2^i groups of 2^k nodes each; we label these groups $S_{i,0}, \ldots, S_{i,2^{i+1}-1}$. For any non-negative integer $j_{k-1} < 2^k$, the node groups S_{i,j_i}, $j_i = \lfloor j_{i+1}/2 \rfloor$, $-1 \leq i < k$, include a butterfly network. That is, the edges between S_{i,j_i} and $S_{i+1,j_{i+1}}$, with $j_i = \lfloor j_{i+1}/2 \rfloor$ include the ith level of an 2^k-input butterfly network. This can be set up since $d \geq 2$. This ensures that every node in L_k has a path of length k to every node in $S_{1,0}$. (Instead of using a butterfly network, we could also have random biregular bipartite graphs between consecutive levels, achieving the same distance property between L_k and $S_{1,0}$ with high probability.) We can then place all remaining edges from each L_i arbitrarily to nodes in L_{i-1}.

We then have all edges from $S_{1,0}$ go into G_1, thus all nodes at level L_k have the same score as all their paths to the miners converge at L_0.

The final component G_3 consists of at least k levels of nodes, which we number from k. For level $i \geq k$, the set L_i consists of 2^{k+i} nodes, with the last level ℓ possibly having fewer than $m2^\ell$ nodes. There is an arbitrary bipartite graph connecting L_i to L_{i-1} with d outgoing edges for each node in level i and $2d$ incoming edges for each node L_{i-1}.

We now argue that G is a stable network. We make three observations about distances, which are critical in establishing that no node benefits by changing any of its outgoing edges.

First, for any $0 \leq i < k$, for any node u in level i and miner v, the distance between u and v is at most $2k - i + \delta_v$, where δ_v is the minimum, over all nodes in L_0, of the distance between the node and v. This follows from the fact that any node in L_k has a shortest path to every node in L_0; consequently, any node in G_2 has a path to v by concatenating a subpath of length $k - i$ to the closest node in L_k, followed by a subpath of length k to the node in L_0 that is closest to v, followed by a subpath of length δ_v to v.

Second, for any $i \geq k$, for any node u in level i and miner v, the distance between u and v is exactly equal to $i + \delta_v$. This holds since any node in L_i has a path to v by concatenating a subpath of length $i - k$ to an arbitrary node in L_k, followed by a subpath of length k to the node in L_0 that is closest to v followed by a subpath of length δ_v to v.

Finally, we note that for any leaf node u and any miner v, the distance between u and v is at least $\ell + \delta_v$. To complete the proof, we observe that for any u in G and any miner v, the distance between u and v is at most $\max\{i, 2k - i\} + \delta_v \leq \ell + \delta_v$; consequently, replacing any out-edge with an out-edge to a leaf will not decrease its cost. Therefore, G is a stable network. $\quad\square$

Lemma 3 can be further extended to allow for more general relationships between the in-degree and the out-degree (e.g., when $d_{in} = cd$ for any integer constant $c \geq 3$). We do not know, however, if stable networks exist for all choices of n, m, d, and d_{in}. We also note that the proofs of Lemmas 1 through 3 are all existence proofs in the sense that they present *specific* families of stable networks. There are, however, many other families of stable networks, as suggested by the proofs. In particular, the proof of Lemma 3 presents a layered structure for stable networks that accommodates a variety of different topologies.

4.2   Single-Exploratory-Greedy

Our results in Sect. 4.1 indicate that there exist Nash equilibria in the full-information setting, where every node knows the current network topology at all times. In practice, however, nodes operate with information only about their own peers. We consider now the Single-Exploratory-Greedy protocol defined in Sect. 3 and analyze networks that evolve in this process. Instead of equilibria, we consider miner and network stability as defined in Sect. 3. We focus on two questions: does there always exist a stable topology? what properties do stable topologies have? Table 1 summarizes our results and the impact of the Tie Breaker Rule on the ability of the network to stabilize. We begin by defining conditions for miner stability.

Table 1. Summary of our results of Sect. 4.2 for the existence and impossibility of miner-stable and network-stable topologies in the unbounded Single-Exploratory-Greedy game. Note for network-stability, we assume $n \gg d$, as discussed in their proofs.

	$m < 2d$		$m \geq 2d$	
	miner-stable	network-stable	miner-stable	network-stable
Global-Ordering/LIFO	✓ Proposition 1	✓ Lemma 5	✓ Lemma 4	✗ Lemma 6
Random/FIFO	✓ Proposition 1	✗ Lemma 7	✗ Proposition 2	✗ Lemma 7

Proposition 1 (Clique Miner Stability for Small m). *If $m < 2d$ there exists a miner-stable topology for any Tie Breaker Rule.*

Proof. Since $m < 2d$, there are enough edges between miners such that they can connect in a clique with each miner having at most $d - 1$ outgoing edges to any other miner. Each miner i has a distance 0 to itself and a distance 1 to each of the other miners, thus their weighted distance to miners in the clique is $w_i = \frac{m-1}{m}$. Any node i drops a connection to the peer with the largest score among the peers of i. Any non-miner node h must have a distance of at least 1 to each miner, thus $score(h) \geq 1 > \frac{m-1}{m}$, thus a non-miner score will not tie with a miner score, therefore miners will only ever drop a peer who is not a miner. Since all miners are already connected, a miner will never establish a new connection to another miner.   □

Lemma 4 (Global-Ordering and LIFO Miner Stability). *For all m and d, there exists a miner-stable topology for the cases where the Tie Breaker Rule is Global-Ordering or LIFO.*

Proof. For $m < 2d$, by Proposition 1 a clique is miner-stable for any tie-breaking rule. For all other values of m and d, we first examine the Global-Ordering rule. Suppose the first $d - 1$ miners (by global ordering) connect in a clique and all other miners have their $d - 1$ edges to the first $d - 1$ miners. The first $d - 1$ miners are therefore connected to all miners and have a minimum score $\frac{m-1}{m}$. In a round, these $d - 1$ miners will not drop one another since they have the lowest score and priority ranking, and all other miners will not drop these first $d - 1$ miners for the same reason. Since the first $d - 1$ miners are connected to all miners, they will never develop a new connection to a miner. In a round, two miners i, j could connect to each other via their d-th outgoing edge. Since either miner could at best have the same score, but worst ranking as the first $d - 1$ miners, this new edge will not replace the existing $d - 1$ out-going edges to the first $d - 1$ miners.

Next, consider the LIFO Tie Breaker Rule. Consider the graph where some set of $d-1$ miners connect in a clique, and all other miners have outgoing edges to these miners. Let the non-clique miners have their d-th edge to some non-miner. Note that the clique does not have any outgoing edges to the non-clique miners and the non-clique miners have no edges among themselves. Since all edges between miners are going into the clique, for any new edge to replace one of these edges, it would need to have an equal score but would be the newest edge, by LIFO it would be dropped. The clique is connected to all miners so it can't make a connection to any other miner. Miners not in the clique thus have $d-1$ stable edges.   □

We note that the networks presented in the proof of Lemma 4 are not stable under FIFO (or Random) as there is always a chance all miners connect to some non-clique miner and this miner would always (or with some probability) replace an existing miner in the clique. We extend this logic to the following claim.

Proposition 2 (Impossibility of Miner-Stability under FIFO or Random for Large m). *If $2d \leq m$, and ties are broken by FIFO or randomly, there is no miner-stable topology.*

Proof. Assume there is some miner-stable topology. This means all edges between miners are either stable or will always be dropped at the end of the round (i.e., are the maximum-score edge after the Tie Breaker Rule). Since $2d \leq m$, miners cannot form a *stable* clique, thus there is some miner m_i that is not connected to all miners. There is some probability that in the next round, all miners connect to this miner with at least one miner m_j having a new out-going edge to m_i. Miner m_i now has score $\frac{m-1}{m}$; so, m_j will not drop the edge to m_i since it is either smaller than some existing edge or ties with all edges and by FIFO (or random) Tie Breaker Rule replaces some edge (has a probability of replacing some edge).   □

We now derive conditions under which stable networks exist and conditions where stability is impossible. For the following cases, we generally assume $n \gg d$, and $n > 2d^2$ for proofs of instability[3]. Recall that since all nodes have at least $d-1$ outgoing edges at the start of the protocol, nodes always have at least $d-1$ outgoing edges during any portion of the protocol.

Lemma 5 (Stable Topology for Small m). *For all $m < 2d$ with Tie Breaker Rules LIFO or global ordering, there exists a stable topology.*

Proof. For $m < d$, the following topology is stable: Miners connect in a clique thus all have score $\frac{m-1}{m}$. All non-miners connect to all miners, either via incoming or outgoing edges. With global ordering, we have that in the miner clique, a miner with k outgoing edges to other miners has $d-1-k$ outgoing edges to the first $d-1-k$ non-miners by the global order. Every non-miner connects to all k

---

[3] Note that this is true of Bitcoin (where $d = 8$ and $n > 10000$) and Ethereum ($d \approx 16$ and $n > 2000$).

miners it is not connected to and similarly has its remaining $d - 1 - k$ outgoing edges to the first $d - 1 - k$ (or if no global ordering, to any) of the remaining non-miners. Thus all non-miners have the same score, and no non-miner will drop an edge to a miner. We then set everyone's d-th edge, note miners never have a d-th edge since they are already connected to everyone. For non-miners, any new edge will be to a node not in the first $d - 1 - k$ non-miners defined above (or any non-miner for LIFO), and therefore will never win the tie-breaker.

For $d \leq m < 2d$, we present a stable topology with all miners connecting in a clique using $(m - 1)/2$ out-going edges and all remaining outgoing edges from the miners going to some (top-ranked) n' non-miners who have a score of 1 and all outgoing edges into the clique. Thus the miners have a miner-stable topology and the n' non-miners will keep their edges to the miners as long as the miners keep their edges to these non-miners and vice versa. All non-miners that are not part of the n' non-miners connect to the top $d - 1$ miners, thus the top $d - 1$ miners have no d-th edge and the rest have an edge to some non-miner that will at best have a score of 1 but be higher-ranked than the n' non-miners. The m miners use $(m - 1)/2$ outgoing edges each to connect to the clique, leaving $d - 1 - (m - 1)/2$ to connect to the n' non-miners. The n' non-miners have a cumulative $n'(d - 1)$ edges, that need to be enough to connect them to all m miners, thus $n'm = n'(d - 1) + d - 1 - (m - 1)/2$, yielding $n' = \frac{(2d-1)-m}{2}$. Note that when $m = 2d - 1$ the miners use all their outgoing edges to connect in the clique so $n' = 0$. □

Next, we show how all remaining cases cannot reach stability. We provide a proof sketch, referring the reader to the full version [17] for the full proofs.

Lemma 6 (LIFO and Global Ordering Tie-Breaking Instability for Large m). *For all $m \geq 2d$, with Tie Breaker Rules LIFO or global ordering, there exists no stable topology for $n \gg d$.*

Proof. We essentially show that in a stable topology, 1.) there is a subset of miners who have the minimum score and *all nodes* must have all of their outgoing edges to this subset for the edge to be stable, and 2.) this subset does not have enough miners to ensure that all their outgoing edges are to one another, so some

Lemma 7 (FIFO and Random Tie-Breaking Instability for Large n). *For all m, d, and $n > 2d^2$ with Tie Breaker Rules FIFO or random, there exists no stable topology.*

Proof. Where miner-stable topologies exist, we show that even if all non-miners connect to all miners (i.e., have the same minimal score), any edge between themselves cannot be stable as the Tie Breaker Rule allows the new edge to replace an existing edge with the same score. □

All examples given in the affirmative proofs of Table 1 rely on specific topologies involving a highly connected sub-network of miners, and generally low-diameter networks. Next we show that such properties are in fact necessary for any stable network.

Properties of Stable Topologies. We can extend the above analysis to derive the following central properties that *any* miner-stable or network-stable topologies must have. Principally, these state that stable networks must necessarily have a highly-connected miner component.

Proposition 3. *If $m < 2d$, and ties are broken by FIFO or randomly, the only miner-stable topology is a clique.*

Proof. The proof follows similarly to that of Proposition 2. Assume there is some non-clique miner-stable topology. Thus there is some miner m_i who is not connected to all miners. There is some probability that in the next round, all miners connect to this miner with at least one miner, m_j having a new out-going edge to m_i. Miner m_i now has score $\frac{m-1}{m}$, thus m_j will not drop the edge to m_i since it is either smaller than some existing edge or ties with all edges and by FIFO (or random) Tie Breaker Rule replaces some edge (has a probability of replacing some edge). □

Proposition 4. *For all m, d and a global order (or LIFO) Tie Breaker Rule, in any miner-stable topology, all miners must connect to the first (or some) $d - 1$ miners.*

Proof. (Case 1: global ordering) Assume there is some miner-stable topology where there is a miner m_i who does not connect to some miner m_j where m_j is in the top $d - 1$ miners. In any round, there is a chance that all miners connect to miner m_j with miner m_i forming an outgoing edge to m_j. Thus in this round m_j has the minimum score and won't be dropped by m_i since it either has a lower score than some outgoing edge of m_i or is ranked higher.

(Case 2: LIFO) Assume that there are less than $d - 1$ miners who all miners connect to in a miner stable topology. There is some miner m_i with score $> \frac{m-1}{m}$ which some miner, m_j, is not connected to. There is a chance that in the next round all miners who are not yet connected to m_i, connected to it, including m_j with an outgoing edge to m_i. In this round, m_i has the minimum score of $\frac{m-1}{m}$, and since less than $d - 1$ miners have such a score, m_i will now be in the top $d - 1$ edges of m_j. □

Proposition 5. *Any miner-stable topology (regardless of m, d) must have miner diameter at most 2 and will lead to a network diameter of at most 3.*

Proof. This follows directly from requirements of stability. Assume there is some miner-stable topology where no miner connects to all miners. There is always some possibility that in a round all miners connect to some miner m_i with some miner m_j having a new outgoing edge to m_i. As has been argued, this new edge contradicts that the initial topology was miner-stable. Thus any stable topology has at least one miner who connects to all other miners. Since the lowest-scoring node is a miner, eventually all non-miners will connect to at least one miner and any future state has all non-miners connecting to at least one miner. Note there is no guarantee of stability of these edges, but we get that the distance between any two non-miners is at most 3. □

Recall from Lemma 1 that the Nash Equilibrium we present for the idealized game also has a miner diameter of 2 and network diameter of 3.

Inbound Connection Cap. We briefly explore the impact of inbound connection caps on the stability properties we explored above. Without loss of generality, we assume that the inbound cap $d_{in} > d$, otherwise nodes would not have the inbound capacity to form new connections.

Lemma 8. *For $m < 2d$ and any Tie Breaker Rule, there exists a miner-stable topology. For tie-breaking rules LIFO and global ordering, miner-stable topologies exist for $m \leq max(2d - 1, d_{in} + d/2)$.*

Proof. For $m < 2d$, there are enough edges between miners for cliques to form, thus they are still stable under Proposition 1. For $m \geq 2d$, consider a topology where the first $d - 1$ miners form a clique, call them set M, then the rest of the miners have their outgoing edges to all miners in M. Thus all miners in M have the minimum score, and won't be dropped by any miner; all miners not in M have their outbound capacity filled by these stable edges so they will not connect to each other. We now calculate the maximum miners not in M. The maximum in-bound capacity for all miners in M is for each miner to have equal incoming/outgoing edges $= \frac{d-2}{2}$. Thus they each have $d_{in} - \frac{d-2}{2}$ capacity left for miners not in M, this happens for $m \leq d - 1 + d_{in} - \frac{d-2}{2} = d_{in} + d/2$. □

It remains an open problem if there exists a miner-stable topology for other regimes.

4.3   Reachable Stability

In the previous section, we characterized the exact conditions for when miner-stable and network-stable topologies exist for the unbounded Single-Exploratory-Greedy protocol under different Tie Breaker Rules. We next consider *if* these stable topologies are necessarily reachable. We show that *for $m \leq d$, all Tie Breaker Rules will lead to miner-stable topologies*, in particular a miner-clique. Additionally, for $m < d$, we show that *if there exists a stable network, then the network will stabilize*. For larger values of m, our proofs of stability in the previous subsection lead us to believe that the stable networks that do exist are hard to reach, we leave this for future work.

We consider the state transition model of the Single-Exploratory-Greedy game. We define a state of the protocol as the graph G of the network at some round r, and the set of reachable states as the states possible after the execution of one round of the protocol.

Proposition 6 (Eventual Miner-Stable Clique). *If the number of miners $m \leq d$, the network will reach a miner-stable topology.*

Proof. To prove this we show that some miner-stable topology is reachable from any state (network topology and round). To do this we present the following *clique process* which defines an exact path from *any* state to some miner clique which we've shown in the previous section is the only miner-stable topology for $m \leq d$. For convenience, we define some arbitrary order over all nodes such that the miners are the first m nodes.

Def. *clique-process*:

For each round r, take $\min_i$ s.t. $\text{score}(m_i) > \frac{m-1}{m}$:

1. $\forall j \leq m$, if $e_{i,j} = 0$, then m_j chooses m_i as their new random outgoing edge.
2. $\forall j \leq i$ $\text{score}(m_j) = \frac{m-1}{m}$
3. Note: $\forall j \leq i$ no edges *to* m_j are dropped

For the remainder of the proof, we re-define an edge as stable in some state, if the value of the edge does not change for all future states on the path defined by the clique-process. Equally, a miner is miner-stable if all of its edges to other miners are stable in the set of future states defined by the clique-process path. Note that once all miners are miner-stable on the clique-process path, then the network is miner-stable for any path as this only happens when a clique is formed by the miners.

We now prove the following statement: Starting from any state, the path defined by the clique-process will result in the first j miners being miner-stable, for $j \leq m$.

Base case: (Miner m_1 will miner-stabilize). Assume that m_1 never stabilizes. This means $\exists j \leq m$ s.t. $e_{1,j}$ does not stabilize. By definition of the clique-process, $i \geq 1$ for all rounds, thus at step (2.) of each round $\text{score}(m_1) = \frac{m-1}{m}$, thus $e_{1,j} = 1$. For this edge to not be stable, $\exists r' \geq r$ where $e_{1,j} = 0$ at step (3.) which only happens if this edge was an outgoing edge from m_1, and m_1 drops the edge at round r'. Then, at round $r' + 1$ we have $i = 1$, so miner m_j will form a *stable* outgoing edge to m_1 (note step 3 of the clique-process).

Inductive Step: Assume that $\forall j' < j \leq m$ miner $m_{j'}$ is miner stable at some round r. It must be that $\forall j' < j$ $\text{score}(m_{j'}) = \frac{m-1}{m}$. Note that this means for each round $r' \geq r$, $\text{score}(m_j) = \frac{m-1}{m}$ at step (2.), even if m_j is not stable, as it must be that $i \geq j$. The proof that m_j stabilizes follows similarly to the base case.

Assume m_j never stabilizes, this means $\exists h \leq m$ s.t. $e_{j,h}$ does not stabilize. By definition of the clique-process, $i \geq j$ for all rounds, thus at step (2.) of each round $\text{score}(m_j) = \frac{m-1}{m}$, thus $e_{j,h} = 1$. For this edge to not be stable, $\exists r' \geq r$ where $e_{j,h} = 0$ at step (3.) which only happens if this edge was an outgoing edge from m_j, and m_j drops the edge at round r'. If this happens, at round $r' + 1$ we have that $i = j$, so miner m_h will form a *stable* outgoing edge to m_j (note step 3 of the clique-process). □

We now extend our results to include network stability guarantees. Recall that only Tie Breaker Rule rules LIFO and Global-ordering have stable networks and only for $m < 2d$. Note that the following proposition is for m strictly less than d as the stable networks we used in our proof of Lemma 5 for $d \leq m < 2d$ were quite contrived and may not be reachable.

Proposition 7 (Eventual Network Stability). *If the number of miners $m < d$, given Tie Breaker Rule of Global-ordering or LIFO, the network will arrive at a stable topology.*

Proof. We give a proof sketch. By Proposition 6, the miners will eventually form a miner-stable topology which is a miner clique. Given this, we first argue that all edges involving miners will stabilize, and lastly that all edges between non-miners will stabilize. To do so, we define similar stabilizing processes as in the proof above. For details, see [17].

Corollary 1. *For $m < d$, and any Tie Breaker Rule, if there exists a miner-stable or network-stable topology, then the network will arrive at some such topology.*

5   Simulations

In the previous section, we prove the conditions for stability in a network implementing Single-Exploratory-Greedy and some conditions for when a network running Single-Exploratory-Greedy will stabilize. In this section, we simulate Single-Exploratory-Greedy to explore how the networks evolve, possible properties that stabilize for different network sizes and in-bound connection capacities, and the impact of having more exploratory edges in the protocol.

As in the theory, we assume each node can uniformly pick a random new peer from the set of all nodes in the network, and knows the score of their peers at step (3) of each round. In simulations, at step (1) of the protocol, each node adds a new edge to a peer they are not yet connected to. The order of the peer selection is randomized in each round, and with capped d_{in}, if the new peer they try to connect to has full in-edge capacity, then the node tries another random node until they succeed, or run out of peers (e.g., if all nodes that are not currently peers have full in-connections)[4]. In the previous section, we prove that of the Tie Breaker Rules we consider, only LIFO and a global ordering have the property that for all m, n there exists a miner-stable topology. Since a global ordering over all nodes is not intuitive in a decentralized network, we use LIFO as the Tie Breaker Rule in our simulations. We run each of the simulations below 20 times and plot the averages of the runs.

Main Results of the Simulations

1. Our simulations suggest that the effect of the d_{in} cap is smooth (as the cap is raised, the behavior shifts smoothly to the no cap model).

---

[4] We emphasize that all other steps of each round are done in parallel for all nodes. We explored simulations where nodes completed each full round one at a time and found no differences in the simulation results.

2. Though theoretical results show the existence of stable topologies, in particular for $m \leq d$ that cliques between miners are stable and inevitable, even for low n (e.g., 400 and 900), arriving at such a topology is hard/improbable and our simulations do not reach a miner clique in the first 256 rounds.
3. An extension of Single-Exploratory-Greedy where nodes drop the k-worst outgoing edges each round suggests the additional randomness can lead to the network converging quicker to the same values, but that this advantage is maximized at low values of k (e.g., 3).

Our analysis indicates that, consistent with the theoretical results, a hierarchical structure emerges within these networks with respect to node scores, degree, and centrality, which are highly correlated. In general, these networks converge to overall lower properties, but often with some nodes (primarily the miners) being at a clear advantage and skewing the average.

Starting Topology. For most of our simulations, we begin with a random graph where each node chooses d initial random peers it is not yet connected to (see Fig. 2(a).). As we discuss below, the Single-Exploratory-Greedy protocol tends to converge to a structure like Fig. 2(b) where few nodes in the middle connect to most others on the periphery. We also run simulations starting with a small-world graph (using the Watts-Strogratz [35] model with probability 0.5) and with a scale-free graph (using the Barabasi-Albert [1] model with an initial connected component of size 20). For a full analysis of these, see [17]. We observe that the initial graph does have a small impact on the properties the *capped* simulations are converging to (e.g., scale-free network converges to higher average distance to miners).

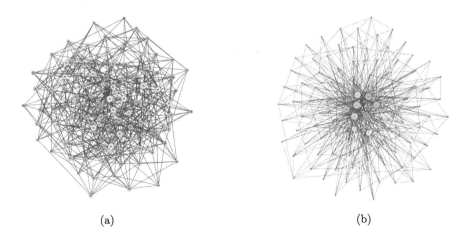

(a)   (b)

Fig. 2. (a) the random graph created from all nodes choosing d random nodes to connect to. (b) the graph we seem to converge to from running Single-Exploratory-Greedy with all nodes being homogeneous miners, size scaled by degree. This structure appears after ∼20 rounds.

5.1   Effect of In-Degree Caps

To complement our limited theoretical analysis with bounded d_{in}, we explore the impact the inbound cap has on networks of all miners, and networks with a subset of 10 nodes being miners. Figure 3 shows the average distance over time for different inbound caps. We see that if the network is all miners, we need a substantial inbound cap for the average distance between all nodes to be better than the random graph within the span of our simulations. We see, however, in Fig. 4 that this lowered average distance is skewed by a few nodes being better off than the majority of the network. When only some of the nodes are miners, the network has the potential to converge to one with a better average distance to the miners; higher the inbound cap, lower the average distance to miners. Figure 5 supports this, showing the network diameter approaching 2. Additionally, the diameter between just the miners is approaching 1 meaning that even with bounded capacity, the miners are forming a clique. *Our main observation here is that the value of the bound has a smooth effect on network properties.*

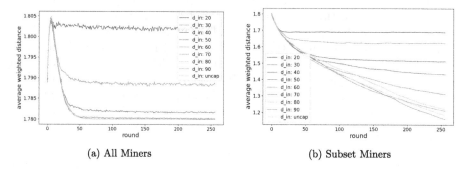

(a) All Miners   (b) Subset Miners

Fig. 3. Average distance to miners per round for an $n = 100$ node network with all miners and $m = 10$ miners, for different in-degree caps.

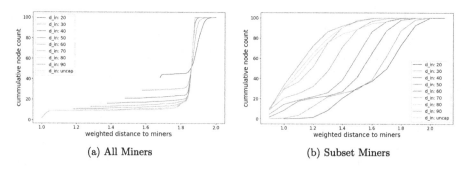

(a) All Miners   (b) Subset Miners

Fig. 4. Distribution of distance to miners(node scores) at round 256 for an $n = 100$ node network with all miners and $m = 10$ miners, for different in-degree caps.

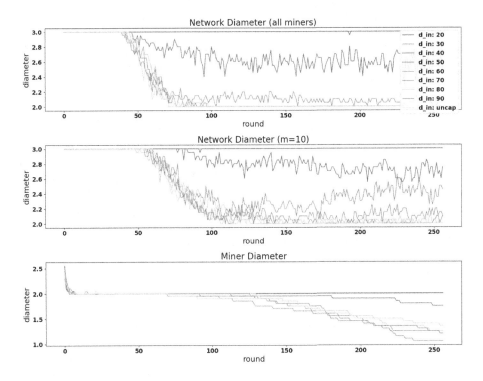

Fig. 5. Diameter of a 100 node network with all miners and a subset of 10 miners, and the diameter between the 10 miners.

5.2 Network Size

We also explore the impact of network size on our simulations when a subset of the network is miners. Again, we consider $m = 10$ and $d = 10$, with $d_{in} = 20$ or uncapped. Our theoretical results show that for $m \leq d$, the miners connecting in a clique is a miner-stable topology, regardless of what the rest of the network is doing (for uncapped d_{in}, this is the only miner-stable topology). We run simulations for $n = 100, 400$ and 900 and look at the network/miner diameter and eccentricities in [17]. *Our main take away is that, though miners are becoming better connected over time, with more non-miners, it is harder for the miners to form a clique.* This is likely due a decreasing probability of another miner being randomly selected by a miner as a neighbor as the number of nodes increases. The non-miners also maintain connections to the miners (seen in the eccentricity of the miners in [17]), it is thus possible the non-miners will occupy all the incoming edges of the miners before they form a clique. This seems to be the case for even $n = 100$ as the capped miner diameter is stabilizing at 2.

5.3   K-Worst Edges

So far, in both the theory and our simulations, we've considered dropping only the single worst outgoing edge. We also explore what happens when nodes drop the $k < d$ worst outgoing edges. Note that $k = d$ is essentially a random graph. Conceptually, more dropped edges leads to more randomness in the network, as more nodes have more incoming connections (see Fig. 7). When all nodes are miners, there is not a big difference in network behavior (Fig. 6 a. and b. show some difference in behavior but the scale is quite small). For most cases (low enough k) in the subset miner simulations, miners are finding each other over time, but the additional randomness in the network appears to lead to worse average distances (Fig. 6 d.). The main advantage that higher values of k sometimes give is a faster convergence to the same values, we show this in [17]. *Generally, dropping more outgoing edges does not change the behavior of the network much, and where it does, the optimal k seems to be up to 3.*

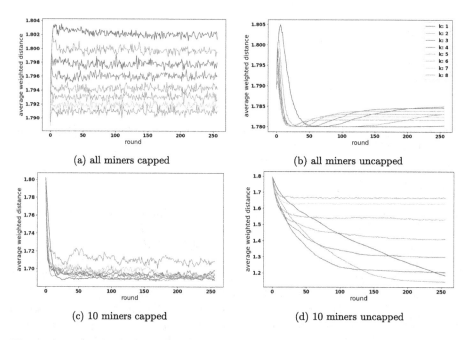

Fig. 6. Average distance to miners for different k-worst values, for networks of $n = 100$, incoming cap of 20 or unlimited, and all miners or subset miners.

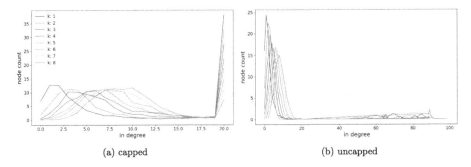

(a) capped   (b) uncapped

Fig. 7. In-degree distribution for different k-worst values for sub-network of miners. The case when all nodes are miners looks the same. For the capped case, the larger the k, the closer the network behaves to the initial random graph.

6   Discussion

In this work, we study the impact of greedy peering choices in a P2P network. In our models, we consider some nodes (either all or a subset of nodes) as "miners" as the source of messages that all nodes are trying to optimize their distance to. We mostly consider the case of homogeneous miners[5], so it is often the case that a nodes' peers will tie in their scores. In fact, in our theoretical results, the choice of tie-breaking rule is a determinant in whether the network will stabilize. Our choice of edges having a weight of 1 symbolizes the delay in propagation each hop adds, normalized over some period of time. In reality, this weight would be an average of delays, and therefore ties between peers would be quite rare. We consider our model of lowering hop distance (as opposed to link-latency distance) for a network where link latency has less of an impact than the processing latency added by each node. It seems that the global-ordering rule fits best as a stand-in for this, assuming the processing latency of any two nodes don't differ too much. Node heterogeneity of this kind and in default parameters is left for future work.

Both our theoretical results and simulations look at the conditions for stability and how the network evolves into a possibly stable state. In both cases, we see a two-tier system for the miners in a stable topology and an additional third tier of all other nodes. An interesting direction for future work is what happens when new nodes (with or without the preferential property) join these stable networks: Does the network (re)stabilize and how long does it take?

In both our simulations and theoretical analysis, stability properties are observed in the regime of a small m relative to n, we note that cryptocurrency networks historically fall under this regime with a minority of network participants being miners [18]. In our simulations, we explore how networks stabilize, particularly for network conditions that are difficult to theorize about. While we

---

[5] We explore miners having different weights (representing different mining powers) in [17] and find that the more disparate the weights, the more the network behaves like the subset miner network.

explore several parameter choices, one caveat of our simulations is that we stay within relatively small network sizes (compared to the major existing cryptocurrencies, e.g., Bitcoin and Ethereum) so that the computation is tractable and we are still able to observe network dynamics[6]. In the same spirit, we consider a single d value while changing the network size and d_{in} caps to explore the proportional effect of the out-degree. Our theoretical results, however, deal with more general parameters.

References

1. Albert, R., Barabási, A.-L.: Statistical mechanics of complex networks. Rev. Mod. Phys. **74**(1), 47 (2002)
2. Aradhya, V., Gilbert, S., Hobor, A.: Overchain: building a robust overlay with a blockchain. arXiv preprint arXiv:2201.12809 (2022)
3. Babel, K., Baker, L.: Strategic peer selection using transaction value and latency. In: Proceedings of the 2022 ACM CCS Workshop on Decentralized Finance and Security, pp. 9–14 (2022)
4. Bala, V., Goyal, S.: A noncooperative model of network formation. Econometrica **68**(5), 1181–1229 (2000)
5. Bitcoin core 24.0.1 (2023). https://github.com/bitcoin/bitcoin
6. Today's cryptocurrency prices by market cap (2023). https://coinmarketcap.com/
7. Corbo, J., Parkes, D.C.: The price of selfish behavior in bilateral network formation. In: Proceedings of PODC 2005, pp. 99–107 (2005)
8. Delgado-Segura, S., et al.: Txprobe: discovering bitcoin's network topology using orphan transactions. In: Financial Cryptography and Data Security: 23rd International Conference, FC 2019, Frigate Bay, St. Kitts and Nevis, 18–22 February 2019, Revised Selected Papers 23, pp. 550–566. Springer, Heidelberg (2019). https://doi.org/10.1007/978-3-030-32101-7_32
9. Demaine, E.D., Hajiaghayi, M., Mahini, H., Zadimoghaddam, M.: The price of anarchy in network creation games. ACM Trans. Algor. **8**(2), 1–13 (2012)
10. devp2p (2023). https://github.com/ethereum/devp2p
11. Dotan, M., Pignolet, Y.-A., Schmid, S., Tochner, S., Zohar, A.: Survey on blockchain networking: context, state-of-the-art, challenges. ACM Comput. Surv. (CSUR) **54**(5), 1–34 (2021)
12. Gouel, M., Vermeulen, K., Fourmaux, O., Friedman, T., Beverly, R.: IP geolocation database stability and implications for network research. In: Network Traffic Measurement and Analysis Conference (2021)
13. Heilman, E., Kendler, A., Zohar, A., Goldberg, S.: Eclipse attacks on Bitcoin's peer-to-peer network. In: 24th {USENIX} Security Symposium ({USENIX} Security 15), pp. 129–144 (2015)
14. Henningsen, S., Teunis, D., Florian, M., Scheuermann, B.: Eclipsing Ethereum peers with false friends. arXiv preprint arXiv:1908.10141 (2019)
15. Jackson, M., Wolinsky, A.: A strategic model of social and economic networks. J. Econ. Theory **71**, 44–74 (1996)

---

[6] We note, however, that our simulations are within the realm of many smaller cryptocurrency networks such as the Ethereum Classic Network with around 500 observed nodes [28].

16. Kiffer, L., Neu, J., Sridhar, S., Zohar, A., Tse, D.: Security of blockchains at capacity. arXiv preprint arXiv:2303.09113 (2023)
17. Kiffer, L., Rajaraman, R.: Stability of p2p networks under greedy peering (full version). http://arxiv.org/abs/2402.14666 (2024)
18. Kiffer, L., Salman, A., Levin, D., Mislove, A., Nita-Rotaru, C.: Under the hood of the Ethereum gossip protocol. In: International Conference on Financial Cryptography and Data Security, pp. 437–456. Springer, Heidelberg (2021). https://doi.org/10.1007/978-3-662-64331-0_23
19. Laoutaris, N., Poplawski, L., Rajaraman, R., Sundaram, R., Teng, S.-H.: Bounded budget connection (BBC) games or how to make friends and influence people, on a budget. J. Comput. Syst. Sci. 80(7), 1266–1284 (2014)
20. Li, K., Tang, Y., Chen, J., Wang, Y., Liu, X.: Toposhot: uncovering Ethereum's network topology leveraging replacement transactions. In: Proceedings of the 21st ACM Internet Measurement Conference, pp. 302–319 (2021)
21. Lighthouse book: Advanced networking. https://lighthouse-book.sigmaprime.io/advanced_networking.html
22. Lumezanu, C., Baden, R., Spring, N., Bhattacharjee, B.: Triangle inequality variations in the internet. In: Proceedings of the 9th ACM SIGCOMM Conference on Internet Measurement, pp. 177–183 (2009)
23. Mao, Y., Deb, S., Venkatakrishnan, S.B., Kannan, S., Srinivasan, K.: Perigee: efficient peer-to-peer network design for blockchains. In: Proceedings of the 39th Symposium on Principles of Distributed Computing, pp. 428–437 (2020)
24. Marcus, Y., Heilman, E., Goldberg, S.: Low-resource eclipse attacks on Ethereum's peer-to-peer network. Cryptology ePrint Archive (2018)
25. Meirom, E.A., Mannor, S., Orda, A.: Network formation games with heterogeneous players and the internet structure. In: Proceedings of the Fifteenth ACM Conference on Economics and Computation, EC 2014, pp. 735–752. Association for Computing Machinery, New York (2014)
26. Miller, A., et al.: Discovering Bitcoin's public topology and influential nodes (2015)
27. Neu, J., Sridhar, S., Yang, L., Tse, D., Alizadeh, M.: Longest chain consensus under bandwidth constraint. In: AFT. ACM (2022)
28. Etc node explorer. https://etcnodes.org/. Accessed 08 Aug 2023
29. Osborne, M.J., Rubinstein, A.: A Course in Game Theory. MIT Press, Cambridge (1994)
30. Park, S., Im, S., Seol, Y., Paek, J.: Nodes in the Bitcoin network: comparative measurement study and survey. IEEE Access 7, 57009–57022 (2019)
31. Poese, I., Uhlig, S., Kaafar, M.A., Donnet, B., Gueye, B.: IP geolocation databases: Unreliable? ACM SIGCOMM Comput. Commun. Rev. 41(2), 53–56 (2011)
32. Rohrer, E., Tschorsch, F.: Kadcast: a structured approach to broadcast in blockchain networks. In: Proceedings of the 1st ACM Conference on Advances in Financial Technologies, pp. 199–213 (2019)
33. Tang, W., Kiffer, L., Fanti, G., Juels, A.: Strategic latency reduction in blockchain peer-to-peer networks. Proc. ACM Meas. Anal. Comput. Syst. 7(2), 1–33 (2023)
34. Toshniwal, B., Kataoka, K.: Comparative performance analysis of underlying network topologies for blockchain. In: 2021 International Conference on Information Networking (ICOIN), pp. 367–372. IEEE (2021)
35. Watts, D.J., Strogatz, S.H.: Collective dynamics of 'small-world' networks. Nature 393(6684), 440–442 (1998)
36. Zich, J., Kohayakawa, Y., Rödl, V., Sunderam, V.: Jumpnet: improving connectivity and robustness in unstructured P2P networks by randomness. Internet Math. 5(3), 227–250 (2008)

Universal Coating by 3D Hybrid Programmable Matter

Irina Kostitsyna[1], David Liedtke[2]([⊠]), and Christian Scheideler[2]

[1] KBR at NASA Ames Research Center, Moffett Field, USA
irina.kostitsyna@nasa.gov
[2] Paderborn University, Paderborn, Germany
liedtke@mail.upb.de, scheideler@upb.de

Abstract. Motivated by the prospect of nano-robots that assist human physiological functions at the nanoscale, we investigate the coating problem in the three-dimensional model for hybrid programmable matter. In this model, a single agent with strictly limited viewing range and the computational capability of a deterministic finite automaton can act on passive tiles by picking up a tile, moving, and placing it at some spot. The goal of the coating problem is to fill each node of some surface graph of size n with a tile. We first solve the problem on a restricted class of graphs with a single tile type, and then use constantly many tile types to encode this graph in certain surface graphs capturing the surface of 3D objects. Our algorithm requires $\mathcal{O}(n^2)$ steps, which is worst-case optimal compared to an agent with global knowledge and no memory restrictions.

Keywords: Programmable Matter · Coating · Finite Automaton · 3D

1 Introduction

Recent advances in the field of molecular engineering gave rise to a series of computing DNA robots that are capable of performing simple tasks on the nano scale, including the transportation of cargo, communication, movement on the surface of membranes, and pathfinding [1,3,23,29]. These results foreshadow future technologies in which a collective of computing particles cooperatively act as programmable matter - a homogenous material that changes its shape and physical properties in a programmable fashion. Robots may be deployed in the human body as part of a medical treatment: they may repair tissues by covering wounds with proteins or apply layers of lipids to isolate pathogens. The common thread uniting these applications is the *coating problem*, in which a thin layer of some specific substance is applied to the surface of a given object.

In the past decades, a variety of models for programmable matter has been proposed, primarily distinguished between passive and active systems. In passive systems, particles move and bond to each other solely by external stimuli,

This work was supported by the DFG Project SCHE 1592/10-1.
Due to space constraints, the proof of Lemma 5 is deferred to the full version [22].

e.g., current or light, or by their structural properties, e.g., specific glues on the sides of the particle. Prominent examples are the DNA tile assembly models aTAM, kTAM and 2HAM (see survey in [25]). In contrast, particles in active systems solve tasks by performing computation and movement on their own. Noteworthy examples include the Amoebot model, modular self-reconfigurable robots and swarm robotics [7,26,31,32]. While computing DNA robots are difficult to manufacture, passive tiles can be folded from DNA strands efficiently in large quantities [18]. A trade-off between feasibility and utility is offered by the hybrid model for programmable matter [16,17,19], in which a single active agent that acts as a deterministic finite automaton operates on a large set of passive particles (called tiles) serving as building blocks. A key advantage of the hybrid approach lies in the reusability of the agent upon completing a task. While agents in purely active systems are expended as they become part of the formed structure, hybrid agents can be deployed again. This property is beneficial in scenarios requiring the coating of numerous objects, such as isolating malicious cells within the human body. Coating multiple objects concurrently, with each being individually coated by a single agent, allows for efficient pipelining.

In the 3D hybrid model, we consider tiles of the shape of rhombic dodecahedra, i.e., polyhedra with 12 congruent rhombic faces, positioned at nodes of the adjacency graph of face-centered cubic (FCC) stacked spheres (see Figs. 1 and 2). In contrast to rectangular graphs (e.g., [12]), this allows the agent to fully revolve around tiles without losing connectivity, which prevents the agent and tiles from drifting apart, e.g., in liquid or low gravity environments. In this paper, we investigate the coating problem in the 3D hybrid model, in which the goal is to completely cover the surface of some impassable object with tiles, where tiles can be gathered from a material depot somewhere on the object's surface.

1.1   Our Results

We present a generalized algorithm that solves the coating problem assuming that the agent operates on a graph $G_\triangle$ of size n and degree $\Delta \leq 6$ that is a triangulation of a closed 3D surface. We assume a fixed embedding of $G_\triangle$ in $\mathbb{R}^3$ in which edges have constantly many possible orientations, and that the boundary of each node in $G_\triangle$ is a chordless cycle. Our algorithm requires only a single type of passive tiles and solves the coating problem in $\mathcal{O}(n^2)$ steps, which is worst-case optimal compared to an algorithm for an agent with global knowledge and no restriction on its memory or the number of tile types. In the 3D hybrid model, the surface graph arises as the subgraph induced by nodes adjacent to a given object (some subset of nodes) where we assume holes in the object to be sufficiently large. While that subgraph is not necessarily a triangulation with the properties described above, we show that our algorithm can be emulated on a restricted class of objects with a single type of tiles. To realize the algorithm outside of that class, we construct a virtual surface graph on which our algorithm can be emulated in $\mathcal{O}(\Delta^2 n^2)$ steps using $2^{2\Delta}$ types of passive tiles. Notably, Δ is a constant in the 3D hybrid model.

1.2 Related Work

In recent years much work on the 2D version of the hybrid model has been carried out, yet the only publication that considers the 3D variant is a workshop paper from EuroCG 2020 in which an arbitrary configuration of n tiles is rearranged into a line in $\mathcal{O}(n^3)$ steps [19]. 2D shape formation was studied in [17]; the authors provide algorithms that build an equilateral triangle in $\mathcal{O}(nD)$ steps where D is the diameter of the initial configuration. The problem of recognizing parallelograms of a specific height to length ratio was studied in [16]. The most recent publication [24] solves the problem of maintaining a line of tiles in presence of multiple agents and dynamic failures of the tiles.

Closely related to the hybrid model is the well established Amoebot model, in which computing particles move on the infinite triangular lattice via a series of expansions and contractions. In this model, a variety of problems was researched in the last years, including convex hull formation [5], shape formation [9,11], and leader election [6]. A recent extension considers additional circuits on top of the Amoebot structure which results in a significant speedup for fundamental problems [13]. In [8,10], the authors solve the coating problem in the 2D Amoebot model; in their variant, the objective is to apply multiple layers of coating to the object. In [30], the authors solve the coating problem in the 3D Amoebot model. In their approach, the object's surface is greedily flooded by amoebots that remain connected in a tree structure. The process terminates for each amoebot when there are no more surrounding empty positions to move to. Notably, given our agent's requirement to retrieve a tile after each placement, their approach cannot be applied or transferred to the hybrid model.

Deterministic finite automata that navigate the infinite 2D grid graph $\mathbb{Z}^2$ are considered in [4,21]. The authors of [4] address the challenge of building a compact structure, termed a nest, using available tiles. They present a solution with a time complexity of $O(s^n)$, where s denotes the initial span, and n denotes the number of tiles. In [21], the authors focus on constructing a fort, a shape with minimal span and maximum covered area, in $\mathcal{O}(n^2)$ time.

In the field of modular reconfigurable robots, coating is often part of the shape formation problem. In the 3D Catom model, a module of robots first assembles into a scaffolding [28] that is then coated by another module of robots [27]. The robots have spherical shape and reside in the FCC lattice; in contrast to the hybrid model, they assume more powerful computation, sensing and communication capabilities. The problem of leader election and local identifier assignment by generic agents in the FCC lattice is considered in [15]. Coating is approached differently in the field of swarm robotics where robots form a non-uniform spatial distribution around objects that are too heavy to be lifted alone [32].

2 Model and Problem Statement

In the *3D hybrid model*, we consider a single active agent with limited sensing and computational power that operates on a finite set of passive *tiles* positioned at nodes of some underlying graph G that has a fixed embedding in $\mathbb{R}^3$.

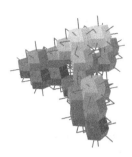

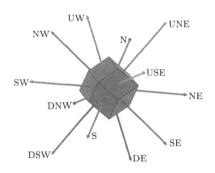

Fig. 1. Tiled nodes of the underlying graph G and their incident edges to empty nodes.

Fig. 2. A passive tile (rhombic dodecahedron) and the twelve compass directions.

2.1 Underlying Graph

Consider the close packing of equally sized spheres at each point of the infinite face-centered cubic lattice. Let $G = (V, E)$ be the adjacency graph of those spheres (see Fig. 1). This graph can be embedded in $\mathbb{R}^3$ such that nodes have alternating cubic coordinates, i.e., each node v has a coordinate $\vec{v} = (x, y, z)$ with $x, y, z \in \mathbb{Z}$ and $x + y + z$ is even. Each node has twelve neighbors whose relative positions are described by the directions UNE, UW, USE, N, NW, SW, S, SE, NE, DNW, DSW and DE, which correspond to the following vectors:

$$\overrightarrow{\text{UNE}} = (1,1,0), \quad \overrightarrow{\text{UW}} = (0,1,1), \quad \overrightarrow{\text{USE}} = (1,0,1), \quad \overrightarrow{\text{N}} = (0,1,-1),$$
$$\overrightarrow{\text{NW}} = (-1,1,0), \quad \overrightarrow{\text{SW}} = (-1,0,1), \quad \overrightarrow{\text{S}} = (0,-1,1), \quad \overrightarrow{\text{SE}} = (1,-1,0),$$
$$\overrightarrow{\text{NE}} = (1,0,-1), \quad \overrightarrow{\text{DNW}} = (-1,0,-1), \quad \overrightarrow{\text{DSW}} = (-1,-1,0), \quad \overrightarrow{\text{DE}} = (0,-1,-1).$$

Cells in the dual graph of G w.r.t. the above embedding have the shape of rhombic dodecahedra, i.e., polyhedra with 12 congruent rhombic faces (see Fig. 2). This is also the shape of every cell in the Voronoi tessellation of G, i.e., that shape completely tessellates 3D space. Consistent to the embedding, we denote by $v + \text{X}$ the node w that is neighboring v in some direction X, i.e., $\vec{w} = \vec{v} + \vec{\text{X}}$. Consider a finite set of tiles of k distinguishable types (until Sect. 4 we only consider $k = 1$). Tiles have the shape of rhombic dodecahedra and are passive, in the sense that they cannot perform computation or movement on their own. A node $v \in V$ is either *tiled*, if there is a tile positioned at v, part of the object (see Sect. 2.3), or *empty*, otherwise. Except for the material depot (see Sect. 2.3), nodes can hold at most one tile at a time. We denote by $\mathcal{T}$ the set of tiled nodes, and by $\mathcal{E}$ the (infinite) set of empty nodes.

2.2 Agent Model

The agent r is the only active entity in this model. It can place and remove tiles of any type at nodes of G and loses and gains a unit of material in the process. We assume that it initially carries no material and that it can carry at most one unit of material at any time. The agent has the computational capabilities of a deterministic finite automaton performing *Look-Compute-Move* cycles. In the *look*-phase, it observes tiles at its current position p and the twelve neighbors of p, and if there are tiles, it observes their types as well. The agent is equipped with a compass that allows it to distinguish the relative positioning of its neighbors. Its initial rotation and chirality can be arbitrary, but we assume that it remains consistent throughout the execution. In the *compute*-phase the agent determines its next state transition according to the finite automaton. In the *move*-phase, the agent executes an *action* that corresponds to that state transition. It either (i) moves to an empty or tiled node adjacent to p, (ii) places a tile (of any type) at p, if $p \notin T$ and r carries material (we call that *tiling* node p), (iii) removes a tile from p, if $p \in T$ and r carries no material, (iv) changes the tile type at p, or (v) terminates. During (ii) and (iii), the agent loses and gains one unit of material, respectively. While the agent is technically a finite automaton, we describe algorithms from a higher level of abstraction textually and through pseudocode using a constant number of variables of constant size domain.

2.3   Problem Statement

Denote by $G(W)$ the subgraph of G induced by some set of nodes $W \subseteq V$, by $d(v, w)$ the distance (length of the shortest path) between nodes $v, w \in V$ w.r.t. G, and by $d_W(v, w)$ the distance w.r.t. $G(W)$. Consider a connected subset $\theta \subset V$ of impassable and static nodes, called *object*. Any node is either an object node, empty or tiled such that θ and the sets of empty nodes $\mathcal{E}$ and tiled nodes T are pairwise disjoint. A *configuration* is the tuple $C = (T, \theta, p)$ containing the set of tiled nodes, object nodes, and the agent's position p. A configuration C is *valid*, if $G(T \cup \theta \cup \{p\})$ is connected. We assume that holes in the object have width larger than one, i.e., $d_\theta(v, w) \leq 2$ for any $v, w \in \theta$ with $d(v, w) \leq 2$.

Let $C^0 = (T^0, \theta, p^0)$ be a valid initial configuration with $T^0 = \{p^0\}$. Superscripts generally refer to step numbers and may be omitted if they are clear from the context. Define the *coating layer* as the maximum subset $L \subset V \backslash \theta$ such that for each node $v \in L$ there is an object node $w \in \theta$ with $d(v, w) = 1$ and $d_L(v, p^0) < \infty$ (w.r.t. $G(L)$). The latter condition excludes unreachable nodes that are separated by the object, e.g., the inner surface of a hollow sphere. We assume a *material depot* at the agent's initial position p^0, that is p^0 is a node with the special property of holding at least $|L|$ units of material. An algorithm solves the *coating problem*, if its execution results in a sequence of valid configurations $C^0, \ldots, C^{t^*}$ such that $T^{t^*} = L$, C^t results from C^{t-1} for $1 \leq t \leq t^*$ by performing any action (i)–(iv) at p^{t-1}, and the agent terminates (v) in step t^*.

3   The Coating Algorithm

In this section, we give a generalized coating algorithm for a surface graph $G_\triangle$ whose node set is the coating layer L. The agent operates only in $G_\triangle$ and all notation in this section is exclusively w.r.t. $G_\triangle$. We assume that (1) $G_\triangle$ is a triangulation of a closed 3D surface (e.g., Fig. 3a) with an embedding in $\mathbb{R}^3$ in which edges

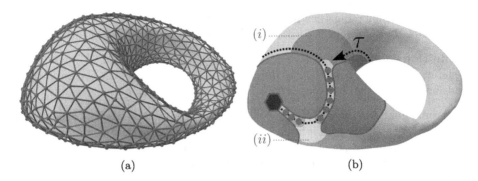

Fig. 3. (a) Example of a triangulation $G_\triangle$. (b) The simple path τ that starts at the material depot (hexagon) along the boundary of tiled nodes (opaque surface area). Links are depicted as circles. The blue area (i) is the *range* explored by the agent. In the yellow area (ii), τ is 'exposed' to nodes that are not tiled and not links. The agent explores the blue area to find an overlap with the yellow area, in which case a link can be tiled while preserving connectivity of $G_\triangle(\mathcal{E})$. (Color figure online)

have constantly many possible orientations. Define the *i-neighborhood* $N_i(W)$ of W as the set of nodes $v \in L$ with $d(v, w) \leq i$ for some node $w \in W$, and the *boundary* as $B(W) := N_1(W)\backslash W$. We write $N_i(w)$ and $B(w)$ for $W = \{w\}$, and use subscripts to denote subsets of only empty or only tiled nodes, e.g., $B_\mathcal{E}(w) = B(w) \cap \mathcal{E}$. Further, we assume that (2) $G_\triangle(B(v))$ contains precisely one simple cycle for any $v \in L$ (we say that $B(v)$ is *chordless*). Notably, this assumption does not weaken our results, since there are surface graphs in which properties (1) and (2) naturally arise, and otherwise we can emulate the properties using additional tile types (see Sect. 4).

3.1   High-Level Description and Preliminaries

The main challenge is the exploration of $G_\triangle$ to find an empty node that can be tiled while maintaining a path back to the material depot, as it is the only source of additional tiles. The exploration of all graphs of diameter D and degree Δ requires $\Omega(D\log(\Delta))$ memory bits [14]. With constant memory already the exploration of plane labyrinths (grid graphs in $\mathbb{Z}^2$) requires two pebbles (placable markers) [2,20]. Our agent has only constant memory and while it is not provided with pebbles, we can use tiles to aid the exploration of $G_\triangle$. Since at some point all nodes in $G_\triangle$ must be tiled, and our algorithm uses only a single type of tiles, i.e., tiles are indistinguishable, we cannot directly use tiles as markers. Instead, our strategy involves strategically placing tiles to create a 'narrow tunnel' of empty nodes leading to the tile depot. We then employ the left- and right-hand-rule (LHR, RHR) commonly used in labyrinth traversal to navigate through the resulting tile structure. In this process, the agent consistently maintains contact with the set of tiled nodes, either on its left or right. Essentially, we 'mark' empty nodes by disconnecting tiles in their boundary such that they become part of the tunnel. This concept is formalized through the introduction of *links*:

Definition 1. *A node* $v \in \mathcal{E}$ *is a* link, *if* $B_{\mathcal{E}}(v)$ *is disconnected. A node* $v \in \mathcal{E}$ *generates (a link at) node* w, *if tiling node* v *turns* w *into a link. Similarly,* v *consumes (a link at) node* w, *if by tiling* v, w *is no longer a link.*

From a high level perspective, the agent traverses the boundary of tiled nodes by following the LHR until it finds a node to place its carried tile at. It then moves back to the material depot following the RHR, gathers material and repeats the process. Following the LHR and RHR, the agent cannot leave the connected component of $G_{\triangle}(\mathcal{E})$ that contains its position. Hence, it is crucial that each tile placement preserves connectivity of $G_{\triangle}(\mathcal{E})$, as otherwise the agent might never find its way back to the material depot, or it might terminate without tiling some node in L. A naive approach to maintain connectivity would be to never place a tile at a link. However, that strategy only works if the surface that is captured by $G_{\triangle}$ is simply connected, i.e., it does not contain any hole. In fact, the link property is necessary for some node $v \in \mathcal{E}$ to be a cut node w.r.t. $G_{\triangle}(\mathcal{E})$ but it is not sufficient. If the surface contains holes, then the naive approach would converge $G_{\triangle}(\mathcal{E})$ to a cyclic graph containing only links, which again may be impossible to explore. Hence, links must be tiled eventually.

In our algorithm, we build a simple path τ of empty nodes that contains all links and is completely contained in the boundary $B(\mathcal{T})$ of tiled nodes (see Fig. 3b). We cannot ensure connectivity of links on τ, i.e., there can be multiple sections at which τ is 'exposed' to empty nodes that are not links. However, we can ensure that links on τ are 'sufficiently close' to each other which we will formalize below by the notion of *segments*. In each traversal of τ, the agent visits all links precisely once by following the LHR until there are no more links in the segment at the agent's position. Afterwards, it explores a constant size neighborhood called *range*, where we define ranges such that they contain all 'close' links (all segments in some local neighborhood). The presence of a link in that range indicates that we have found one of the above mentioned 'exposures' of τ that was already visited before, i.e., there exists a link in some cycle of $G_{\triangle}(\mathcal{E})$ that can safely be tiled. Note that the agent does not necessarily traverse τ fully. As an example, the subpath of τ that follows the topmost link v in Fig. 3b is not traversed, since the next segment following node v contains no link.

Additional Terminology. The agent (at node p) maintains a direction pointer to some tiled node $a(p) \in B(p)$, called *anchor* of p. Using the analogy of wall-following in a labyrinth, $a(p)$ is the node at which the agent's 'hand' is currently placed. Initially, the agent leaves the tile depot p^0 to an arbitrary neighbor and sets its anchor to p^0. Afterwards, it can follow the LHR (RHR) by moving to the first empty node v in a clockwise (counter-clockwise) order of $B(p)$ starting at $a(p)$ and sets its anchor to the last tiled node between $a(p)$ and v in that order (see Fig. 4a). Note that the agent's anchor does not necessarily change after each move. Subsequently, L and R denote the direction to the next node according to the LHR and RHR, respectively.

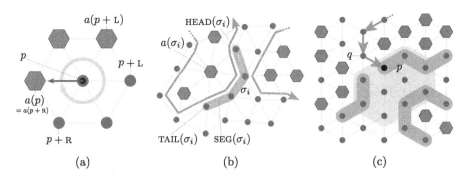

Fig. 4. Tiled nodes are depicted as hexagons, empty nodes as disks, links are depicted in red (node σ_i in (b), q in (c) and the uppermost central node in (c)). (a) Example of the updates made to the agent's anchor $a(p)$ while following the LHR and RHR. (b) $\text{SEG}(\sigma_i)$, $\text{TAIL}(\sigma_i)$ and $\text{HEAD}(\sigma_i)$ for some node $\sigma_i \in \sigma$ in an example configuration. (c) Example of the i-range $R_i(p,q)$ for $i = 2$ (orange, opaque) together with $N_2(p)$ w.r.t. $G_\triangle(\mathcal{E} \backslash \{q\})$ (green, transparent) (Color figure online).

Let $s^0 \in B(p^0)$ be a dedicated starting node chosen arbitrarily in the initialization. Define $\sigma = (\sigma_0, \ldots, \sigma_m)$ as the path along nodes of $B(\mathcal{T})$ according to a LHR traversal that starts and ends at $s^0 = \sigma_0 = \sigma_m$. Note that σ is not necessarily a simple cycle. In fact, in a full traversal of σ, the agent can visit a link multiple times with its anchor set to different tiled nodes on each visit (e.g., σ_i in Fig. 4b). To avoid ambiguity, $a(\sigma_i)$ refers to the anchor *after* moving from σ_{i-1} to σ_i, and $a(v)$ refers to the anchor at the first occurence of v on σ.

The *segment* $\text{SEG}(\sigma_i)$ of a node σ_i (see Fig. 4b) contains all nodes that can be reached from σ_i by following the LHR or RHR while keeping the anchor fixed at $a(\sigma_i)$. Simply put, $\text{SEG}(\sigma_i)$ is the set of nodes that touch the anchor $a(\sigma_i)$ from the same 'side' as σ_i. Any node $\sigma_j \in \text{SEG}(\sigma_i)$ is a *successor* of σ_i, if $j > i$, or a *predecessor* of σ_i, if $j < i$. As an example, node σ_i in Fig. 4b has two predecessors and one successor. Denote by $\text{HEAD}(\sigma_i)$ the node without a successor in $\text{SEG}(\sigma_i)$, and $\text{TAIL}(\sigma_i)$ the node without a predecessor in $\text{SEG}(\sigma_i)$.

We can now formally introduce the above mentioned path τ. Define $\tau = (\tau_0, \ldots, \tau_l)$ as a maximal simple sub-path of σ that starts at s^0, i.e., $\tau_i = \sigma_i$ for any $0 \le i \le l$. The definition of $\text{SEG}(\cdot), \text{HEAD}(\cdot)$ and $\text{TAIL}(\cdot)$ directly carry over from σ. For simplicity, we write $v \in \tau$ or $v \in \sigma$ if τ or σ contains node v.

Finally, we introduce the aforementioned *range* that is explored by the agent before each tile placement. Consider the agent to be positioned at some empty node p, such that the node $q = p + \text{R}$ is a link. The *i-range* $R_i(p,q)$ (see Fig. 4c) is a specific neighborhood of empty nodes that is defined as if node q were tiled. Consider a node v that can be reached from p in at most i steps without moving through q or any tiled node. If the segment $\text{SEG}(v)$ does not contain q, we add it to $R_i(p,q)$. Otherwise, that segment is separated at q and only the part that contains v is added. The formal definition is as follows:

Definition 2. *The i-range of p w.r.t. q is the set of nodes*

$$R_i(p, q) := \bigcup_{v \in N_i(p)} \bigcup_{w \in B_T(v)} \text{SEG}(v)$$

where $N_i(p)$ and $\text{SEG}(v)$ are w.r.t. $G_\triangle(\mathcal{E} \backslash \{q\})$ and anchor w.

3.2   Algorithm Details and Pseudocode

The coating algorithm (see pseudocode in Algorithm 1) consists of an initialization INIT (lines 1–3) and two phases COAT (lines 4–10), dedicated to the tile placement, and FETCH (lines 11–16), dedicated to the gathering of material, the traversal of τ, and the termination. The agent switches between phases COAT and FETCH after each tile placement. In the pseudocode, LI and GEN denote the set of all links and generators (see Definition 1).

In phase INIT (lines 1–3), the agent gathers material and moves to an arbitrary node $s^0 \in B(p^0)$. It stores the direction of p^0 w.r.t. s^0 such that it can recognize s^0 by the adjacent material depot at a later visit, and it initializes $a(s^0) \leftarrow p^0$ and SKIP $\leftarrow$ *false*. Note that s^0 is the first node of the paths σ and τ, $a(p)$ is the agent's anchor, and SKIP is a flag that indicates that the search for links in the 3-range is skipped in the next execution of phase COAT (see line 4). The flag is necessary to maintain a crucial invariant which we elaborate in the proof of Lemma 5. Afterwards, the agent moves L and enters phase COAT.

Phase COAT is always entered such that $p \notin$ LI $\cup \{s^0\}$ and $p + $ R is the last node $v \in \tau$ (i.e., with maximum index) for which $v \in$ LI $\cup \{s^0\}$. In each execution

---

Algorithm 1: Coating Algorithm

Phase INIT:

1   gather material from p^0; move to an arbitrary $s^0 \in B(p^0)$

2   store the direction of p^0 w.r.t. s^0; $a(s^0) \leftarrow p^0$; SKIP $\leftarrow$ *false*

3   move L; enter phase COAT   ▷ $p \leftarrow s^0 + $ L

Phase COAT:

4   **if** SKIP or $R_3(p, p + $ R$) \cap ($LI $\cup \{s^0\}) = \emptyset$ **then**

5 | place a tile at p; SKIP $\leftarrow$ *false*; enter phase FETCH

6   **else**

7 | move R   ▷ $p \leftarrow p + $ R

8 | **if** $p \in$ GEN **then** move R;   ▷ $p \leftarrow p + $ R

9 | **if** $p \in$ GEN and $p + $ R $\notin$ LI $\cup \{s^0\}$ **then** SKIP $\leftarrow$ *true*

10 | place a tile at p; enter phase FETCH

Phase FETCH:

11   **while** $p \neq s^0$ **do** move R;   ▷ $p \leftarrow p + $ R

12   move to p^0; gather material from p^0; move to s^0

13   **if** $B_\mathcal{E}(s^0) = \emptyset$ **then** place a tile at s^0; terminate

14   **while** $p \in$ LI $\cup \{s^0\}$ or $v \in$ LI for a successor v of p in $\text{SEG}(p)$ **do**

15 | move L   ▷ $p \leftarrow p + $ L

16   enter phase COAT

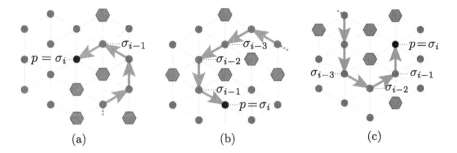

Fig. 5. Examples in which the next tile is placed at some link (red disk) in phase COAT: σ_{i-1} is tiled in (a); σ_{i-2} is tiled in (b) and (c). Only in (c), SKIP is set to *true* since $\sigma_{i-3} \notin \text{LI} \cup \{s^0\}$. As a result, σ_{i-3} is tiled on the next visit. (Color figure online)

of phase COAT the tile that is carried by the agent is either placed directly at p (lines 4–5), or at some link that can be reached by following the RHR for at most two steps (lines 7–10). In any case, the agent enters phase FETCH afterwards. The position of the next tile depends on the following criteria: If the flag SKIP is set to *true*, then the tile is placed at p and SKIP is set to *false* afterwards. The tile is also placed at p, if $R_3(p, p + \text{R})$ (see Definition 2) does not contain any node of $\text{LI} \cup \{s^0\}$. Otherwise, the agent must have found some link or the starting node, which implies that $R_3(p, p + \text{R})$ contains a node $v \in \tau$ with smaller index than $p \in \tau$, i.e., the agent has detected a cycle in $G_\triangle(\mathcal{E})$ in which it can safely place a tile at some link. It is crucial that no link is generated on the 'wrong side' of the newly placed tile as this link may later lead to a false detection. Let the agent's position w.r.t. σ be $p = \sigma_i$, i.e., $\sigma_{i-1}, \sigma_{i-2}$ and σ_{i-3} are the next three nodes visited by following the RHR (see Fig. 5). If $\sigma_{i-1} \notin \text{GEN}$, then node σ_{i-1} is tiled, otherwise node σ_{i-2} is tiled. In the latter case, SKIP is set to *true* whenever $\sigma_{i-2} \in \text{GEN}$ and $\sigma_{i-3} \notin \text{LI}$ (line 9). The flag SKIP ensures that node σ_{i-3} is tiled next such that the agent again enters phase COAT L of the last node $v \in \tau$ with $v \in \text{LI} \cup \{s^0\}$.

In phase FETCH, the agent moves R until it is positioned at s^0 (line 11), which it detects by the adjacent material depot and the direction stored in phase INIT. It moves to p^0, gathers material and returns to s^0 (line 12). If s^0 has no empty neighbors, it places a tile at s^0 and terminates (line 13). Otherwise, the agent moves L as long as $p \in \text{LI} \cup s^0$ or whenever a successor of p in SEG(p) is a link, and switches to phase COAT afterwards (lines 14–16). In that step, the agent implicitly explores SEG(p), since p can have multiple successors.

3.3   Analysis

Consider an initial configuration $C^0 = (T^0, \theta, p^0)$ with $p^0 \in T^0 \subseteq L$, and a material depot of size at least $|L| - |T^0| + 1$ at p^0. In the problem statement we assume that p^0 is the only initially tiled node, but now we allow T^0 to contain multiple tiled nodes besides p^0. This will later become useful in Sect. 4 where we construct a virtual graph in which some nodes are tiled initially. We analyze Algorithm 1 given that C^0 satisfies the following definition:

Definition 3. *A configuration $C^0 = (T^0, \theta, p^0)$ is coatable w.r.t. $G_\triangle$, if $|B(v)| \leq 6$ for any $v \in \mathcal{E}^0$, $B_{\mathcal{E}^0}(p^0) \neq \emptyset$, $\mathrm{LI}^0 = \emptyset$, and $\mathcal{E}^0$ is connected.*

We aim to maintain five properties as invariants: *P1*: Links may only occur on the simple path τ, i.e., $\mathrm{LI} \subseteq \tau$. *P2*: All links are connected by a sequence of overlapping segments to the starting node, i.e., $\mathrm{TAIL}(v) \in \mathrm{LI} \cup \{s^0\}$ for any $v \in \mathrm{LI}$. *P3*: The subpath of τ from s^0 to the last link on τ induces no cycle in $G_\triangle$, i.e., for any $i < j \leq k$ with $\tau_k \in \mathrm{LI}$: if $d(\tau_i, \tau_j) = 1$, then $j = i + 1$. *P4*: The boundary of any link contains precisely two connected components of empty nodes, i.e., $||B_{\mathcal{E}}(v)|| = 2$ for any $v \in \mathrm{LI}$ where $||\cdot||$ denotes the number of connected components. *P5*: There exists a node of τ at which the agent enters phase COAT, i.e., either $\mathrm{LI} = \emptyset$ or there is an i such that $\tau_i \notin \mathrm{LI} \cup \{s^0\}$ and $\mathrm{HEAD}(\tau_i) = \tau_i$.

Observe that all properties hold initially by $\mathrm{LI}^0 = \emptyset$. The structure of our proof is as follows: We prove termination given that $P5$ is maintained, and that $\mathcal{E}$ never disconnects given that $P1$–$P4$ are maintained. Since the agent always finds a node to place the next tile at by $P5$, there must eventually be a step in which $B_{\mathcal{E}}(s^0) = \emptyset$. Since $\mathcal{E}$ remains connected until that step, $L = T$ holds after the last tile is placed at s^0. Finally, we show that $P1$–$P5$ are maintained as invariants. We start with the termination of the algorithm:

Lemma 1. *If $P5$ holds in step t in which the agent gathers material at p^0, then there is a step $t^+ > t$ in which the agent enters p^0 again or terminates.*

Proof. Assume by contradiction that the agent does not terminate or enter p^0 again in any step $t' > t$. There are two cases: the agent places a tile in phase COAT and moves to a connected component of $\mathcal{E}$ that does not contain s^0, or the agent never enters phase COAT, i.e., it moves indefinitely L in phase FETCH. If the agent traverses a simple path from s^0 to some node v by moving L, places a tile at v and moves R afterwards, it must enter the connected component of $\mathcal{E}$ that contains s^0. Hence, in the first case, the agent places a tile at v after visiting v at least twice, i.e., it has fully traversed τ, which contradicts the existence of node τ_i specified by $P5$. In the second case, the agent never reaches τ_i, as it would otherwise enter phase COAT and place a tile. This implies a cycle $(\tau_j, ..., \tau_k)$ with $\tau_{k+1} = \tau_j$ and $j < i$, which contradicts that τ is a simple path. Hence, there is a step t^+ in which p^0 is entered again or $B_{\mathcal{E}}(s^0) = \emptyset$ and the agent terminates. $\square$

Subsequently, our notation refers to some step t in which the agent gathers material at p^0. With a slight abuse of notation we will use a superscript $^+$ $(^-)$ to denote the next (previous) step t^+ (t^-) in which the agent gathers material at p^0 or terminates, e.g., $\mathcal{E}^+$ denotes the set of empty nodes in step t^+.

Lemma 2. *If $P2$ holds in step t and phase COAT is entered at some τ_i between step t and t^+, then τ_{i-1} is the last node v on τ with $v \in \mathrm{LI} \cup \{s^0\}$.*

Proof. As long as $p \in \text{LI} \cup \{s^0\}$, the agent moves L in phase FETCH which implies $\tau_{i-1} \in \text{LI} \cup \{s^0\}$. It also moves L if a successor of p in SEG(p) is a link. Hence, no successor v of τ_{i-1} in SEG(τ_{i-1}) is a link. Assume by contradiction that τ_{i-1} is not the last node on τ that is contained in LI $\cup \{s^0\}$. Let τ_k be the first link after τ_{i-1}, i.e., $\tau_k \in \text{LI}$ and $k > i - 1$. Since no successor of τ_{i-1} in SEG(τ_{i-1}) is a link, τ_k cannot be contained in SEG(τ_{i-1}). Let $S = (v_0, ..., v_m)$ be the sequence of nodes with $v_0 = \tau_k$, $v_m = s^0$ and $v_j = \text{TAIL}(v_{j-1})$ for any $0 < j \leq m$. The agent's anchor changes only if there is no further successor in its current segment, i.e., *after* moving L at some node v of τ with HEAD$(v) = v$. This implies HEAD(TAIL(v)) = TAIL(v) for all $v \in \tau$. It follows that S must contain some node v_j with $0 < j < m$ for which $v_j = \text{HEAD}(\tau_i)$. Since HEAD$(\tau_i) \notin \text{LI}$, there must exists some $v_{j'}$ with $j' < j$ for which TAIL$(v_{j'}) \notin \text{LI} \cup \{s^0\}$ which contradicts $P2$ and concludes the lemma. $\qquad\square$

Lemma 3. *If $\mathcal{E}$ is connected and P1–P4 hold in step t, then $\mathcal{E}^+$ is connected.*

Proof. Tiling a node $v \notin \text{LI}$ cannot disconnect $\mathcal{E}$ by Definition 1. This covers the tile placement at the start of phase COAT, especially the case SKIP = *true*, and the last placement before termination in phase FETCH. Thus, we must only consider cases in which a tile is placed at some link $v \in \text{LI}$. By Lemma 2, the agent enters phase COAT at τ_i such that τ_{i-1} is the last node $w \in \tau$ with $w \in \text{LI} \cup \{s^0\}$. Together with $P1$ follows that whenever it detects some node $w \in R_3(\tau_i, \tau_{i-1})$ with $w \in \text{LI} \cup \{s^0\}$, then w was visited in the previous execution of phase FETCH. Let τ_j be the last node of τ that is contained in $R_3(\tau_i, \tau_{i-1})$ with $j < i - 1$, and P be the shortest path from τ_i to τ_j in $G_\triangle(R_3(\tau_i, \tau_{i-1}))$. Then $C = P \circ (\tau_{j+1}, \tau_{j+2}, ..., \tau_{i-2}, \tau_{i-1})$ is a simple cycle in $G_\triangle$, where $\circ$ is the concatenation of paths. Placing a tile at τ_{i-1} or τ_{i-2} cannot disconnect C since it is a cycle, and it cannot disconnect $\mathcal{E}$ since C contains nodes of all connected components of $B_\mathcal{E}(\tau_{i-1})$ (and $B_\mathcal{E}(\tau_{i-2})$, if $\tau_{i-2} \in \text{LI}$) by $P4$. $\qquad\square$

Lemma 4. *If $B_T(v) \cap B_T(w) \neq \emptyset$, then tiling v cannot increase $||B_\mathcal{E}(w)||$.*

Proof. The lemma follows trivially if $v \notin B(w)$. Consider arbitrary $v, w \in L$ with $v \in B(w)$. Since $G_\triangle$ is a triangulation, the edge $\{v, w\}$ is contained in precisely two triangular faces, each of which contains another node u_1 and u_2, respectively. Since $B(v)$ and $B(w)$ are chordless, these are the only nodes adjacent to both v and w, which implies that $B(v) \cap B(w) = \{u_1, u_2\}$ and that $\{u_1, v, u_2\}$ is connected in $B(w)$. If any u_i is tiled, then $||B_T(w) \cup \{v\}|| \leq ||B_T(w)||$. Thereby, $||B_\mathcal{E}(w) \backslash \{v\}|| = ||B_T(w) \cup \{v\}|| \leq ||B_T(w)|| = ||B_\mathcal{E}(w)||$. $\qquad\square$

We can now deduce the precise neighborhood of v for the case where v is both a link and a generator, i.e., $v \in \text{LI} \cap \text{GEN}$. By Definition 3, $B(v)$ contains at most six nodes, at least two of which must be tiled, as otherwise v cannot be a link. Hence, at most four empty nodes in at least two connected components of $B_\mathcal{E}(v)$ remain. If each connected component has size at most two, then all nodes in $B_\mathcal{E}(v)$ share a tiled neighbor with v, which contradicts that v is a generator by the previous lemma. Hence, as a corollary we obtain the following:

Corollary 1. *For any $v \in$ LI $\cap$ GEN: $B_{\mathcal{T}}(v)$ contains two connected components of size one, and $B_{\mathcal{E}}(v)$ contains two connected components of size one and three.*

Lemma 5. *If P1–P5 hold and a node is tiled in step t, then either P1–P5 hold in step t^+ or the agent sets SKIP $=$ true and P1–P5 hold in step t^{++}.*

The proof of Lemma 5 is deferred to the full version of the paper [22]. Essentially, we distinguish the type of node v that is tiled in step t, i.e., whether v is not a link, a link but no generator, or a link and a generator, and finally whether $v + \mathbb{R} \in$ LI $\cup \{s^0\}$ (see line 9). In all but the last case, we can show that P1–P5 immediately hold in step t^+, and in the last case, we know by the algorithm that SKIP is set to true after tiling v in step t^+. Here we can show that only property P2 is violated in step t^+, and only within the neighborhood of the previously placed tile. Using Corollary 1, we can precisely determine at which node the next tile is placed (with SKIP $=$ true), and that this tile placement restores P2 in step t^{++}. In the other cases the properties mostly follow from Lemma 4.

All properties hold in an initial configuration, and they are maintained as invariants by Lemma 5. By Lemma 2, the agent eventually terminates in some step t^*, and by Lemma 3 $\mathcal{E}$ never disconnects. Hence, $\mathcal{T}^{t^*} = L$ holds after termination which concludes the following:

Theorem 1. *Following Algorithm 1, a finite-state agent solves the coating problem on $G_\triangle$, given a configuration $C^0 = (\mathcal{T}^0, \theta, p^0)$ that is coatable w.r.t. $G_\triangle$.*

3.4   Runtime Analysis

Since the agent does not sense any node outside of $N_1(p)$ in its look-phase, exploring $R_i(p, q)$ requires additional steps. From $R_i(p, q) \subset N_{i+2}(p)$ it follows that the number of steps is upper bounded by $2 \cdot |N_{i+2}(p)| = \mathcal{O}(|N(p)|^i)$. Since i is a constant and $G_\triangle$ has constant degree, each execution of phase COAT takes $\mathcal{O}(1)$ steps. Each execution of phase FETCH takes $\mathcal{O}(|\tau|)$ steps as the agent traverses a sub-path of τ twice. Since τ is simple, it follows that $|\tau| = \mathcal{O}(n)$, where $n = |L|$. The agent can place at most n tiles until $L = \mathcal{T}$, thereby performs at most n executions of COAT and FETCH, which results in $\mathcal{O}(n^2)$ steps in total.

An agent $\tilde{r}$ with unlimited memory and global vision can reach any node via a shortest path instead of sticking to the boundary of tiled nodes. Except for the last placed tile it must always return to the material depot which implies that the last tile is placed at a node w with maximum distance to p^0. In the worst case, the surface graph is the triangulation of an object resembling a straight line such that $d_L(p^0, w) = \Theta(n)$. Each node on the shortest path P from p^0 to w must be tiled. Hence, $\tilde{r}$ takes at least $2 \left(\sum_{u \in P} d_L(p^0, u) \right) - d_L(p^0, w) = \left(\sum_{i=1}^{\Theta(n)} 2 \cdot i \right) - \Theta(n) = \Theta(n^2)$ steps which implies worst-case optimality of Algorithm 1.

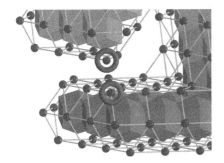
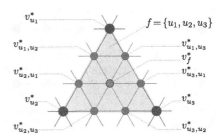

Fig. 6. A snapshot of $G_\diamondsuit$: the circled nodes are adjacent in $G(L)$.

Fig. 7. A triangular face f of $G_\triangle$ and its corresponding virtual edges and nodes in $G_\triangle^*$.

4 Coating in the 3D Hybrid Model

In this section, we apply our coating algorithm to the 3D hybrid model. We first define a triangulation on nodes of L with degree $\Delta \leq 8$, and afterwards construct a virtual graph on which we emulate our algorithm using $2^{2\Delta}$ types of tiles.

Recall the definition of graph $G = (V, E)$ and its embedding in $\mathbb{R}^3$ from Sect. 2. Define $G_\diamondsuit = (L, E')$ as the subgraph of $G(L)$ that contains only those edges $\{v, w\}$ for which v and w share adjacent object neighbors (see Fig. 6), i.e., $E' = \{\{v, w\} \mid d_\theta(N_1(v), N_1(w)) \leq 1\}$. We can view $G_\diamondsuit$ as embedded on the surface of our 3D object. That embedding contains triangular and tetragonal faces (see Fig. 6) where tetragonal faces can occur in one of three orientations: (1) $v, v + \mathrm{NE}, v + \mathrm{NE} + \mathrm{USE}, v + \mathrm{USE}$, (2) $v, v + \mathrm{NW}, v + \mathrm{NW} + \mathrm{UNE}, v + \mathrm{UNE}$, and (3) $v, v + \mathrm{N}, v + \mathrm{N} + \mathrm{UW}, v + \mathrm{UW}$. Apart from rotation, Fig. 8 shows all possible arrangements of faces within $G_\diamondsuit$. We define the class of *smooth objects* $\mathcal{S}$ as all objects for which $G_\diamondsuit$ contains only the cases (a)–(f) from Fig. 8. Let $G_\triangle$ be the triangulation of $G_\diamondsuit$ in which the same diagonal edge is added for each tetragonal face of the same orientation (1)–(3) (since we want the agent to be able to deduce

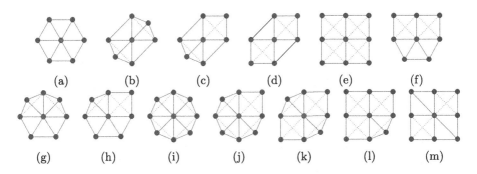

Fig. 8. All possible arrangements of faces in $G_\diamondsuit$ apart from rotation. Dashed edges indicate that the distance between its endpoints is precisely two w.r.t. G.

the triangulation). Since $d_L(v, w) = 1$ w.r.t. $G_\triangle$ implies $d_L(v, w) \leq 2$ w.r.t. $G_\lozenge$, the agent can emulate moving on $G_\triangle$ with a multiplicative time and memory overhead of at most two. It is easy to see that $B(v)$ is chordless and v has degree at most six for all $v \in L$ within the class $\mathcal{S}$. Together with Theorem 1 follows:

Theorem 2. *A finite-state agent with a single tile type solves the coating problem on any object $\theta \in \mathcal{S}$ with coating layer L in $\mathcal{O}(n^2)$ steps, where $n = |L|$.*

4.1   Emulation of Coatable Surface Graphs

Consider an arbitrary triangulation $G_\triangle = (L, E)$ of constant degree Δ and an initially valid configuration C_0. We construct a virtual graph $G_\triangle^* = (L^*, E^*)$ with virtual initial configuration C^{0*} such that $G_\triangle^*$ is coatable w.r.t. C^{0*}. During that construction, we define a partial surjective function $\mathcal{R} : L^* \to L$ that maps virtual nodes to real nodes. We show that an agent r operating on $G_\triangle$ w.r.t. C_0 with $2^{2\Delta}$ tile types can emulate an agent r^* that executes Algorithm 1 on $G_\triangle^*$ w.r.t. C^{0*} such that throughout the emulation $\mathcal{R}(p^*) = p$.

Virtual Graph Construction. The virtual graph $G_\triangle^*$ is the result of subdividing each face of $G_\triangle$ into nine triangular faces (see Fig. 7). The node set L^* contains a virtual node v_u^* for each node $u \in L$, two virtual nodes $v_{u,w}^*$ and $v_{w,u}^*$ for each edge $\{u, w\} \in E$, and a virtual node v_f^* for each triangular face f of $G_\triangle$. For each edge $\{u, w\} \in E$ the edge set E^* contains three virtual edges $\{v_u^*, v_{u,w}^*\}$, $\{v_{u,w}^*, v_{w,u}^*\}$ and $\{v_{w,u}^*, v_w^*\}$. For each triangular face $f = \{u_1, u_2, u_3\}$ of $G_\triangle$, E^* contains six virtual edges $\{v_f^*, v_{u_i,u_j}^*\}$ and three virtual edges $\{v_{u_i,u_j}^*, v_{u_i,u_k}^*\}$, where $u_i, u_j, u_k \in f$ are pairwise distinct. We define $\mathcal{R}(v_{u,w}^*) = u$ for any virtual node $v_{u,w}^* \in L^*$. Consider an arbitrary but fixed order on the vectors $\overrightarrow{x_1}, ..., \overrightarrow{x_m}$ that correspond to edges in the embedding of $G_\triangle$. Let π represent that order, i.e., $\pi(\overrightarrow{x_i}) = i$. For some face $f = \{u_1, u_2, u_3\}$ of $G_\triangle$, we define $\mathcal{R}(v_f^*) = u_i$, where u_i is the node minimizing $\pi(\overrightarrow{u_i} - \overrightarrow{u_j})$ for any $u_i, u_j \in f$ with $i \neq j$. We define the virtual initial configuration C^{0*} such that all v_u^* are tiled, i.e., $T^{0*} = \cup_{u \in L} v_u^*$, $p^{0*} = v_{p^0}^*$ and assume a material depot of size at least $|L^*| - |L|$ at $v_{p^0}^*$.

Lemma 6. *C^{0*} is coatable w.r.t. $G_\triangle^*$.*

Proof. Each face of $G_\triangle$ is triangular, and two virtual nodes are added for each edge of $G_\triangle$. Hence, $|B(v_f^*)| = 6$ w.r.t. $G_\triangle^*$ for any face f of $G_\triangle$. Any $v_{u,w}^*$ is adjacent to $v_{f_1}^*$ and $v_{f_2}^*$, where f_1, f_2 are the two faces of $G_\triangle$ that both contain u and w, to two nodes v_{u,w_1}^*, v_{u,w_2}^*, where $w_1 \in f_1$ and $w_2 \in f_2$, and to $v_{w,u}^*$ and v_u^*. Hence, $|B(v_{u,w}^*)| = 6$ w.r.t. $G_\triangle^*$ for any edge $\{u, w\}$ of $G_\triangle$. Any other virtual node is initially tiled, which implies $|B(v^*)| \leq 6$ for any $v^* \in \mathcal{E}^*$. By construction, each initially tiled node is isolated, i.e., $d(v^*, w^*) \geq 3$ for any $v^*, w^* \in T^{0*}$. Since $G_\triangle^*$ is connected, it follows that $\mathcal{E}^{0*}$ is connected and $B_\mathcal{E}(v^*)$ is connected for any $v^* \in \mathcal{E}^{0*}$, i.e., $\text{LI}^{0*} = \emptyset$. Hence, each property of Definition 3 is satisfied.   □

Lemma 7. *A finite-state agent can emulate Algorithm 1 on $G_\triangle^*$ in $\mathcal{O}(\Delta^2 n^2)$ steps while moving and placing tiles of at most $2^{2\Delta}$ types on $G_\triangle$.*

Proof. Let $F^* \subset L^*$ be the set of virtual nodes v_f^* that correspond to some face f of $G_\triangle$ in the construction of $G_\triangle^*$. Since $G_\triangle^*(L^* \backslash F^*)$ is a subdivision of $G_\triangle$, it can be embedded in the same 3D surface as $G_\triangle$ using vectors that are collinear to vectors in the embedding of $G_\triangle$. It follows that we can use the same fixed order π from the construction of $G_\triangle^*$.

In the following, we define for each node $u \in L$ a bit-sequence $x(u) = (x_1, ..., x_{2\Delta})$ that encodes the occupation of all nodes $v^* \in L^*$ with $\mathcal{R}(v^*) = u$, where a 0 encodes an empty, and a 1 encodes an occupied virtual node. By the construction of $G_\triangle^*$, there are at most 2Δ nodes v^* with $\mathcal{R}(v^*) = u$ such that 2Δ bits suffice. The order of bits in $x(u)$ is uniquely given by π where the first Δ bits encode virtual nodes that correspond to edges of $G_\triangle$, and the following bits encode virtual nodes that correspond to faces of $G_\triangle$. There is no bit for the virtual node $v_u^* \in L^*$ since it is initially occupied and remains occupied until termination by following Algorithm 1. In fact, $\mathcal{R}$ is undefined for $v_u^* \in L^*$.

Consider an agent r on $G_\triangle$ that utilizes $k = 2^{2\Delta}$ types of passive tiles. Each tile type uniquely describes a bit-sequence of length $\log(k) = 2\Delta$ such that r emulates an agent r^* on $G_\triangle^*$ with initial configuration C^{0*} as follows: If r^* moves from v^* to w^*, then r moves from $\mathcal{R}(v^*)$ to $\mathcal{R}(w^*)$ (if $\mathcal{R}(v^*) \neq \mathcal{R}(w^*)$). If r^* places a tile at v^* and $\mathcal{R}(v^*)$ is empty, then r places a tile at $\mathcal{R}(v^*)$ that corresponds to the bit-sequence x in which only v^* is encoded as occupied, otherwise r incorporates the occupation of v^* by changing the tile type. If r^* gathers material and r carries no material, then r also gathers material.

By Theorem 1 and Lemma 6, r^* solves the coating problem on $G_\triangle^*$. Since $\mathcal{R}$ is surjective and any node $\mathcal{R}(v^*) \in L$ is occupied, if $v^* \in L^*$ is occupied, the emulation solves the coating problem on $G_\triangle$ in $\mathcal{O}(|L^*|^2) = \mathcal{O}(\Delta n)$ steps.   □

Our final theorem follows from the virtual graph construction on top of our triangulation $G_\triangle$ of $G_\Diamond$ (with $\Delta \leq 8$) and the previous lemma:

Theorem 3. *A finite-state agent utilizing constantly many tile types can solve the coating problem on arbitrary objects in worst-case optimal $\mathcal{O}(n^2)$ steps.*

5   Future Work

We provided an algorithm that solves the coating problem in the 3D hybrid model in worst-case optimal $\mathcal{O}(n^2)$ steps given that the initial configuration w.r.t. the surface graph fulfills the property of coatability as specified in Definition 3. While the algorithm solves the problem directly in the class of smooth objects, there are certainly surface graphs that violate coatability. We bypassed this problem by emulating our algorithm on a subdivision of these surface graphs using $2^{2\Delta}$ types of tiles. A natural question for future work is whether solving the problem with a single tile type is in fact impossible, and if so, then what is the lowest number of tile types required to solve it. Another open question is how far our worst-case optimal solution is off from the best case solution.

References

1. Akter, M., et al.: Cooperative cargo transportation by a swarm of molecular machines. Sci. Robot. **7**(65), eabm0677 (2022)
2. Blum, M., Kozen, D.: On the power of the compass (or, why mazes are easier to search than graphs). In: 19th Annual Symposium on Foundations of Computer Science (sfcs 1978), pp. 132–142 (1978). https://doi.org/10.1109/SFCS.1978.30
3. Chao, J., et al.: Solving mazes with single-molecule DNA navigators. Nat. Mater. **18**, 273–279 (2019). https://doi.org/10.1038/s41563-018-0205-3
4. Czyzowicz, J., Dereniowski, D., Pelc, A.: Building a nest by an automaton. Algorithmica **83** (2021). https://doi.org/10.1007/s00453-020-00752-0
5. Daymude, J.J., Gmyr, R., Hinnenthal, K., Kostitsyna, I., Scheideler, C., Richa, A.W.: Convex hull formation for programmable matter. In: Proceedings of the 21st International Conference on Distributed Computing and Networking. ICDCN 2020 (2020). https://doi.org/10.1145/3369740.3372916
6. Daymude, J., Gmyr, R., Richa, A., Scheideler, C., Strothmann, T.: Improved leader election for self-organizing programmable matter. In: Algorithms for Sensor Systems - 13th International Symposium on Algorithms and Experiments for Wireless Sensor Networks, ALGOSENSORS 2017, Revised Selected Papers, pp. 127–140 (2017). https://doi.org/10.1007/978-3-319-72751-6_10
7. Derakhshandeh, Z., Dolev, S., Gmyr, R., Richa, A.W., Scheideler, C., Strothmann, T.: Amoebot - a new model for programmable matter. In: Proceedings of the 26th ACM Symposium on Parallelism in Algorithms and Architectures, pp. 220–222. SPAA 2014 (2014). https://doi.org/10.1145/2612669.2612712
8. Derakhshandeh, Z., Gmyr, R., Porter, A.M., Richa, A.W., Scheideler, C., Strothmann, T.: On the runtime of universal coating for programmable matter. Nat. Comput. **17**, 81–96 (2016). https://doi.org/10.1007/s11047-017-9658-6
9. Derakhshandeh, Z., Gmyr, R., Richa, A.W., Scheideler, C., Strothmann, T.: Universal shape formation for programmable matter. In: Proceedings of the 28th ACM Symposium on Parallelism in Algorithms and Architectures, pp. 289–299. SPAA 2016 (2016). https://doi.org/10.1145/2935764.2935784
10. Derakhshandeh, Z., Gmyr, R., Richa, A.W., Scheideler, C., Strothmann, T.: Universal coating for programmable matter. Theoret. Comput. Sci. **671**, 56–68 (2017). https://doi.org/10.1016/j.tcs.2016.02.039
11. Di Luna, G.A., Flocchini, P., Santoro, N., Viglietta, G., Yamauchi, Y.: Shape formation by programmable particles. Distrib. Comput. **33**, 69–101 (2019)
12. Fekete, S., Gmyr, R., Hugo, S., Keldenich, P., Scheffer, C., Schmidt, A.: Cadbots: algorithmic aspects of manipulating programmable matter with finite automata. Algorithmica **83**, 1–26 (2021). https://doi.org/10.1007/s00453-020-00761-z
13. Feldmann, M., Padalkin, A., Scheideler, C., Dolev, S.: Coordinating amoebots via reconfigurable circuits. J. Comput. Biol. **29** (2022). https://doi.org/10.1089/cmb.2021.0363
14. Fraigniaud, P., Ilcinkas, D., Peer, G., Pelc, A., Peleg, D.: Graph exploration by a finite automaton. In: Mathematical Foundations of Computer Science 2004, pp. 451–462 (2004)
15. Gastineau, N., Abdou, W., Mbarek, N., Togni, O.: Leader election and local identifiers for 3d programmable matter. Concurrency Computation: Practice and Experience **34** (2020). https://doi.org/10.1002/cpe.6067

16. Gmyr, R., Hinnenthal, K., Kostitsyna, I., Kuhn, F., Rudolph, D., Scheideler, C.: Shape recognition by a finite automaton robot. In: 43rd International Symposium on Mathematical Foundations of Computer Science (MFCS 2018). LIPIcs, vol. 117, pp. 52:1–52:15 (2018). https://doi.org/10.4230/LIPIcs.MFCS.2018.52

17. Gmyr, R., et al.: Forming tile shapes with simple robots. Nat. Comput. **19** (2020). https://doi.org/10.1007/s11047-019-09774-2

18. Heuer-Jungemann, A., Liedl, T.: From DNA tiles to functional DNA materials. Trends Chem. **1**(9), 799–814 (2019). https://doi.org/10.1016/j.trechm.2019.07.006

19. Hinnenthal, K., Rudolph, D., Scheideler, C.: Shape formation in a three-dimensional model for hybrid programmable matter. In: Proceedings of the 36th European Workshop on Computational Geometry (EuroCG 2020) (2020)

20. Hoffmann, F.: One pebble does not suffice to search plane labyrinths. In: International Symposium on Fundamentals of Computation Theory (1981)

21. Kant, K., Pattanayak, D., Mandal, P.S.: Fort formation by an automaton. In: 2021 International Conference on COMmunication Systems and NETworkS, pp. 540–547 (2021). https://doi.org/10.1109/COMSNETS51098.2021.9352839

22. Kostitsyna, I., Liedtke, D., Scheideler, C.: Universal coating by 3d hybrid programmable matter (2024). https://doi.org/10.48550/arXiv.2303.16180

23. Li, H., Gao, J., Cao, L., Xie, X., Fan, J., Wang, H., Wang, H., Nie, Z.: A DNA molecular robot autonomously walking on the cell membrane to drive the cell motility. Angewandte Chemie International Edition **60** (2021).https://doi.org/10.1002/anie.202108210

24. Nokhanji, N., Flocchini, P., Santoro, N.: Dynamic line maintenance by hybrid programmable matter. Int. J. Netw. Comput. **13**(1), 18–47 (2023). https://doi.org/10.15803/ijnc.13.1_18

25. Patitz, M.: An introduction to tile-based self-assembly and a survey of recent results. Nat. Comput. **13** (2013). https://doi.org/10.1007/s11047-013-9379-4

26. Tan, N., Hayat, A.A., Elara, M.R., Wood, K.L.: A framework for taxonomy and evaluation of self-reconfigurable robotic systems. IEEE Access **8**, 13969–13986 (2020). https://doi.org/10.1109/ACCESS.2020.2965327

27. Thalamy, P., Piranda, B., Bourgeois, J.: 3d coating self-assembly for modular robotic scaffolds. In: 2020 IEEE/RSJ International Conference on Intelligent Robots and Systems (IROS), pp. 11688–11695 (2020).https://doi.org/10.1109/IROS45743.2020.9341324

28. Thalamy, P., Piranda, B., Lassabe, F., Bourgeois, J.: Scaffold-based asynchronous distributed self-reconfiguration by continuous module flow. In: 2019 IEEE/RSJ International Conference on Intelligent Robots and Systems (IROS), pp. 4840–4846 (2019). https://doi.org/10.1109/IROS40897.2019.8967775

29. Thubagere, A.J., et al.: A cargo-sorting DNA robot. Science **357**(6356), eaan6558 (2017)

30. Traversat, W.: Universal Coating by Programmable Matter in 3D. Master's thesis, Eindhoven University of Technology (2020). https://pure.tue.nl/ws/portalfiles/portal/168210057/Traversat_W..pdf

31. Tucci, T., Piranda, B., Bourgeois, J.: A distributed self-assembly planning algorithm for modular robots. In: Proceedings of the 17th International Conference on Autonomous Agents and MultiAgent Systems, pp. 550–558. AAMAS 2018 (2018)

32. Werfel, J., Petersen, K., Nagpal, R.: Designing collective behavior in a termite-inspired robot construction team. Science **343**(6172), 754–758 (2014)

Distributed Binary Labeling Problems in High-Degree Graphs

Henrik Lievonen[1]([✉]), Timothé Picavet[2]([✉]), and Jukka Suomela[1]([✉])

[1] Aalto University, Espo, Finland
{henrik.lievonen,jukka.suomela}@aalto.fi
[2] LaBRI, Université de Bordeaux, Talence, France
timpicavet@gmail.com

Abstract. Balliu et al. (DISC 2020) classified the hardness of solving *binary labeling problems* with distributed graph algorithms; in these problems the task is to select a subset of edges in a 2-colored tree in which white nodes of degree d and black nodes of degree δ have constraints on the number of selected incident edges. They showed that the deterministic round complexity of any such problem is $\mathcal{O}_{d,\delta}(1)$, $\Theta_{d,\delta}(\log n)$, or $\Theta_{d,\delta}(n)$, or the problem is unsolvable. However, their classification only addresses complexity as a function of n; here $\mathcal{O}_{d,\delta}$ hides constants that may depend on parameters d and δ. In this work we study the complexity of binary labeling problems as a function of all three parameters: n, d, and δ. To this end, we introduce the family of *structurally simple* problems, which includes, among others, all binary labeling problems in which cardinality constraints can be represented with a context-free grammar. We classify possible complexities of structurally simple problems. As our main result, we show that if the complexity of a problem falls in the broad class of $\Theta_{d,\delta}(\log n)$, then the complexity for each d and δ is always either $\Theta(\log_d n)$, $\Theta(\log_\delta n)$, or $\Theta(\log n)$. To prove our upper bounds, we introduce a new, more aggressive version of the *rake-and-compress technique* that benefits from high-degree nodes.

1 Introduction

In this work we take the first steps towards characterizing possible distributed computational complexities of graph problems as a function of two parameters: the number of nodes and the maximum degree of the graph. We study so-called *binary labeling problems* [1], previously only studied for constant-degree graphs, and we extend their classification so that it is parameterized also by the degrees of the nodes. To do that, we also introduce a new version of the *rake-and-compress technique* [16] that benefits from high-degree nodes.

1.1 Broader Context

The key goal in the field of distributed graph algorithms is understanding how fast a given graph problem can be solved in a distributed setting. In this work

Y. Emek (Ed.): SIROCCO 2024, LNCS 14662, pp. 402–419, 2024.
https://doi.org/10.1007/978-3-031-60603-8_22

we focus on the LOCAL model of distributed computing (see Sect. 1.8): in each round all nodes can exchange messages with each of their neighbors, and the running time of the algorithm is the number of communication rounds until all nodes stop and announce their own part of the solution (say, their own color).

Landscape of Distributed Complexity. While a lot of early work in the field focused on individual problems, modern theory of distributed graph algorithms makes a heavy use of *structural* results that apply to a broad family of graph problems. The best-known example is *locally checkable labeling problems* (LCLs), first introduced by Naor and Stockmeyer [17]. These are problems in which the task is to label nodes or edges with labels from some finite set, subject to some local constraints—examples of such problems include vertex coloring, edge coloring, maximal independent set, and maximal matching.

By now, we have a very good understanding of the landscape of possible round complexities that *any* LCL problem may have, in settings like cycles, grids, trees, and general graphs [3–5,7–10,12,13,20]. For example, there is no LCL problem with a time complexity between $\omega(\log^* n)$ and $o(\log n)$ in the deterministic LOCAL model. Gap result like this are powerful tools in proving lower bounds and upper bounds. For example, as soon as we have an algorithm that solves some LCL problem in $o(\log n)$ rounds, we can immediately speed it up to $\mathcal{O}(\log^* n)$ rounds for free.

However, there is one key limitation: the landscape of complexities is currently understood well only as a function of the number of nodes n, for a constant maximum degree $\Delta = \mathcal{O}(1)$. For example, there are problems with complexities of the form $\Theta(\log_\Delta n)$, and problems with complexities of the form $\Theta(\log n)$, but we do not have a more fine-grained understanding of all possible complexity classes as a function of both n and Δ.

Beyond Constant Degrees: Fundamental Challenges. LCLs as they were originally introduced by Naor and Stockmeyer [17] only pertain to graphs of some constant maximum degree. To generalize beyond constant Δ, we need a meaningful definition of the problem family.

Unfortunately, many natural generalizations lead to uninteresting results. If the local constraints may depend on the degrees in arbitrarily complicated ways, we can construct artificial problems with complexities of the form $\Theta(\log_{f(\Delta)} n)$ for virtually any function $f(\Delta) = \mathcal{O}(\Delta)$. For example, we could start with a problem that has complexity $\Theta(\log_\Delta n)$, and modify the problem definition so that nodes of degree d will ignore up to $d - f(d)$ adjacent leaf nodes; in essence, we turn nodes of degree d into nodes of degree $f(d)$, and adjust the complexity accordingly. However, this way we will learn nothing about the complexities that *natural* graph problems might have.

Beyond Constant Degrees: Our Take. In this work we study the family of *binary labeling problems*, as defined by Balliu et al. [1] (see Sect. 1.2). These are a special case of LCLs; what is particularly attractive is that the complexity of any given binary labeling problem (in the deterministic LOCAL model) can be automatically deduced, and in the constant-degree case there are only four complexity classes in trees.

Our plan is to see exactly how the characterization of binary labeling problems can be generalized in a meaningful manner beyond constant degrees; the hope here is that this will also guide us when we seek to find the right definitions for generalizing results on all LCLs.

1.2   Binary Labeling Problems

Let us recall the key definitions from [1]. In a binary labeling problem Π, the task is to choose a subset of edges $X \subseteq E$ in a tree $G = (V, E)$, subject to local constraints. The problem is defined as a tuple $\Pi = (d, \delta, W, B)$, where $d \in \{2, 3, \dots\}$ is the *white degree*, $\delta \in \{2, 3, \dots\}$ is the *black degree*, $W \subseteq \{0, 1, \dots, d\}$ is the *white constraint* and $B \subseteq \{0, 1, \dots, \delta\}$ is the *black constraint*. We assume that the input graph is properly 2-colored with colors white and black. We define that $X \subseteq E$ is a solution to problem Π if the following holds:

- If v is a white node of degree d, and v is incident to k edges in X, then $k \in W$.
- If v is a black node of degree δ, and v is incident to k edges in X, then $k \in B$.

That is, we only care about the labeling incident to white nodes of degree d and black nodes of degree δ; these nodes are called *relevant nodes*. It is often useful to imagine that white nodes represent "nodes" and black nodes represent "edges" (if $\delta = 2$) or "hyperedges" (if $\delta > 2$). We will usually assume that $\delta \leq d$ (otherwise we can exchange the roles of black and white nodes).

Examples. Here are a couple of examples of binary labeling problems (adapted from [1]):

1. *Bipartite splitting:* $W = \{1, 2, \dots, d-1\}$ and $B = \{1, 2, \dots, \delta - 1\}$. Here the task is to split the set of edges in two classes: "red edges" X and "blue edges" $E \setminus X$, and all relevant nodes must be incident to at least one red and at least one blue edge.
2. *Bipartite matching:* $W = B = \{1\}$. If we interpret that each edge $\{u, v\} \in X$ indicates that u is matched with v, in this problem each relevant node must be matched with exactly one other node (that may or may not be relevant).

While the problems are well-defined in any 2-colored tree, it is often easiest to consider the case that the tree consists of only leaf nodes and relevant nodes. For example, then bipartite matching is the task of finding a matching in which all internal nodes are matched.

Prior Work. Given a tuple $\Pi = (d, \delta, W, B)$, we can directly look up the round complexity of Π in a table given by Balliu et al. [1]. For example, we immediately obtain:

1. *Bipartite splitting:* For $d = \delta = 2$ this problem requires $\Theta(n)$ rounds, but for any fixed constants $d > 2$, $\delta \geq 2$ the complexity is $\Theta_{d,\delta}(\log n)$.
2. *Bipartite matching:* For $d \geq \delta = 2$ this problem requires $\Theta_d(n)$ rounds, but for any fixed constants $d > 2$, $\delta > 2$ the complexity is $\Theta_{d,\delta}(\log n)$.

However, what their classification does not capture is the complexity as a function of d or δ; we have made this explicit above by writing $\Theta_{d,\delta}$, to emphasize that the hidden constants may depend on d and δ. For example, while there are numerous problems (such as the above two examples) with a complexity $\Theta_{d,\delta}(\log n)$, it is not at all clear what is the base of logarithm in each case. In particular, which binary labeling problems get easier when d and/or δ grows?

1.3 Overview of Contributions and New Ideas

Main Question. In this work we study parameterized families of binary labeling problems

$$\Pi(d, \delta) = \big(d, \delta, W(d), B(\delta)\big),$$

and our main question is this: what can we say about the complexity of any such problem family $\Pi(d, \delta)$, as a function of n, d, and δ?

Key Technical Challenge. We need to be careful not to make the family too broad—otherwise we will end up with an uninteresting result stating that there are artificial problems with virtually any complexity as a function of d and δ, without learning anything about natural graph problems.

Key Results and New Ideas. The main new insight is that we define the family of *structurally simple* problems. This definition captures how $W(d)$ and $B(\delta)$ can depend on d and δ. The aim is to exclude artificial pathological problems.

We then show that this definition is *useful* in the sense that we can prove strong statements about the complexity of structurally simple problems. For example, if the complexity as a function of n is $\Theta_{d,\delta}(\log n)$, then for each d and δ the complexity falls in one of these fine-grained classes: $\Theta(\log_d n)$, $\Theta(\log_\delta n)$, or $\Theta(\log n)$.

Finally, we show that the definition captures a *broad family* of problems: if the constraints $W(d)$ and $B(\delta)$ can be represented as binary strings in a context-free language, then $\Pi(d, \delta)$ is structurally simple. We will discuss this in more detail in Sect. 1.4.

To prove our main result, we also needed to develop a new, more aggressive version of the rake-and-compress technique. We discuss this in more detail in Sect. 1.5.

1.4 Contributions in More Detail

Key Definitions. In a family of binary labeling problems, both W and B are set families of the form $X(k) \subseteq \{0, 1, \ldots, k\}$. If we look at these set families in the examples given by Balliu et al. [1, Table 3], we observe that they all fall in one of two classes, which we call *center-good* and *edge-good*:

Definition 1 (center-good). *A set $X \subseteq \{0, 1, \ldots, k\}$ is ε-center-good if there exists an $x \in X$ such that $k^\varepsilon \leq x \leq k^{1-\varepsilon}$.*

Definition 2 (edge-good). *A set $X \subseteq \{0, 1, \ldots, k\}$ is C-edge-good if for all $x \in X$ we have $x \leq C$ or $x \geq k - C$.*

Now we are ready to give the main definition:

Definition 3 (structurally simple). *A set family $X(k) \subseteq \{0, 1, \ldots, k\}$ is structurally simple if there are some constants $0 < \varepsilon < 1$ and C such that $X(k)$ is either ε-center-good or C-edge-good for each k. A family of binary labeling problems $\Pi(d, \delta) = (d, \delta, W(d), B(\delta))$ is structurally simple if both $W(d)$ and $B(\delta)$ are structurally simple.*

For example, both the bipartite splitting problem and the bipartite matching problem are structurally simple, and so are all problems in [1, Table 3]; in the bipartite splitting problem, $W(d)$ and $B(\delta)$ are center-good, while in the bipartite matching problem, $W(d)$ and $B(\delta)$ are edge-good. To give a pathological example of a set family that is not structurally simple, consider, for example, $X(k) = \{\lfloor \log k \rfloor\}$.

Main Result. Our main result is a complete classification of structurally simple problems in the logarithmic region:

Theorem 1. *Let $\Pi(d, \delta)$ be a structurally simple family of binary labeling problems and assume that the complexity of $\Pi(d, \delta)$ in trees is $\Theta_{d,\delta}(\log n)$. Further assume that $\delta \leq d = \mathcal{O}(n^{1-\alpha})$ for some $\alpha > 0$. Then we can partition the space of parameter values n, d, and δ in three (possibly empty) classes such that the complexity of $\Pi(d, \delta)$ in trees is:*

1. *$\Theta(\log n)$ in the first class,*
2. *$\Theta(\log_\delta n)$ in the second class,*
3. *$\Theta(\log_d n)$ in the third class.*

Remark 1. A few clarifying remarks are in order that help one to interpret the result:

1. Our result is constructive in the sense that we also show how to classify $\Pi(d, \delta)$ for any given d and δ.
2. The complexity class may depend on d and δ. For example, we might have a problem in which for even values of d the complexity is $\Theta(\log_d n)$ and for odd values of d it is $\Theta(\log n)$.
3. Throughout this work, Θ-notation only hides constants that only depend on problem family Π, and not on d and δ; we write $\Theta_{d,\delta}$ explicitly if we are hiding constants that may depend on d and δ.

Language-Theoretic Justification. A set $X \subseteq \{0, 1, \ldots, k\}$ can be also represented as a binary string $\hat{X}$ of length $k+1$: we index the bits with $i = 0, 1, \ldots, k$ and for each $i \in X$, the bit at index i in $\hat{X}$ is 1, and otherwise 0. Given a set family $X(k)$, we can then define a language of binary strings

$$\hat{X} = \{\hat{X}(2), \hat{X}(3), \hat{X}(4), \ldots\}.$$

Note that $\hat{X}$ is by construction *thin*: it contains at most one word of any given length [18].

Example 1. If $X(k) = \{1, 2, \ldots, k-1\}$, then $\hat{X}(k) = 01^{k-1}0$ and

$$\hat{X} = \{010, 0110, 01110, \ldots\} = 01^{+}0.$$

If $X(k) = \{1\}$, then $\hat{X}(k) = 010^{k-1}$ and

$$\hat{X} = \{010, 0100, 01000, \ldots\} = 010^{+}.$$

Here we use 1^{ℓ} to denote a sequence of ℓ 1s, and 1^{+} to denote the sequence of one or more 1s.

Equipped with this notation, we can study families of binary labeling problems from a language-theoretic perspective: any given problem family $\Pi(d, \delta)$ can be specified by giving a pair of languages $(\hat{W}, \hat{B})$. As long as these are languages over the binary alphabet, and there is exactly one string of each length $2, 3, \ldots$, such a pair of languages can be interpreted as a parameterized problem family $\Pi(d, \delta)$.

Many example problems in prior work [1] correspond to *regular languages*. For example, the bipartite splitting problem is $(01^{+}0, 01^{+}0)$, and the bipartite matching problem is $(010^{+}, 010^{+})$, while the *sinkless orientation problem* [7] is $(11^{+}0, 011^{+})$. We take one step beyond regular languages and consider the case of *context-free languages*. The key observation is summarized in the following lemma:

Lemma 1. *Let $X(k) \subseteq \{0, 1, \ldots, k\}$ be a family of sets. If $\hat{X}$ is a context-free language, then $X(k)$ is structurally simple.*

Corollary 1. *Let $\Pi(d, \delta) = (d, \delta, W(d), B(\delta))$ be a family of binary labeling problems. If $\hat{W}$ and $\hat{B}$ are context-free languages, then $\Pi(d, \delta)$ is structurally simple.*

This is the main justification for focusing on structurally simple problems: we have only excluded some pathological cases that are so complicated that they cannot be expressed with context-free grammars (this is also the reason why we call them structurally simple).

1.5   Key Building Block: A New Rake-and-Compress Variant

A key tool for designing $\mathcal{O}(\log n)$-time algorithms for binary labeling problems in trees has been the *rake-and-compress technique* [16]. A typical application of this technique proceeds as follows; we alternate between two steps:

1. Eliminate leaf nodes and isolated nodes.
2. Eliminate sufficiently long paths.

If we apply these steps for $\mathcal{O}(\log n)$ times, in a tree of any shape, we will eliminate all nodes. Then we can construct a solution by working backwards: repeatedly put back one layer of nodes and construct a solution that makes these nodes happy; see [1] for more detailed examples.

From our perspective, the main drawback of the rake-and-compress technique is that it does not benefit from high-degree nodes. For example, if we apply it in a tree in which all internal nodes have degree s, the worst-case complexity is still $\Theta(\log n)$, not $\Theta(\log_s n)$.

We introduce a new version of the technique that strictly benefits from high-degree nodes. Let us first rephrase the usual rake-and-compress procedure as follows (note that nodes in the middle of long paths are also nodes that are not close to higher-degree nodes):

1. Eliminate leaf nodes and isolated nodes.
2. Eliminate nodes that are not close to any node of degree 3 or more.

We generalize this as follows, so that we are much more aggressive with the compression step:

1. Eliminate leaf nodes and isolated nodes.
2. Eliminate nodes that are not close to any node of degree s or more.

We show that this leads to a procedure that completes in $\mathcal{O}(\log_s n)$ rounds, and we show that we can use this more aggressive version of rake-and-compress to solve many binary labeling problems. This is the key building block that leads to the upper bounds $\mathcal{O}(\log_d n)$ and $\mathcal{O}(\log_\delta n)$ in Theorem 1.

We believe that our new rake-and-compress technique will find applications also beyond binary labeling problems.

1.6   Questions for Future Work

While our classification is constructive, it also gives rise to a number of new open questions related to the *decidability* of complexities for entire problem families; we present here one example:

Question 1. Given context-free grammars for thin languages $\hat{W}$ and $\hat{B}$ that define a problem family $\Pi(d, \delta)$, how hard is it to decide if the complexity of $\Pi(d, \delta)$ is $\Theta(\log_d n)$ for all but finitely many values of d and δ?

1.7   Roadmap

We classify structurally simple problems in the logarithmic region as follows:

- In Sect. 2 we prove lower bounds of the forms $\Omega(\log n)$, $\Omega(\log_d n)$, and $\Omega(\log_\delta n)$.
- In Sect. 3 we prove upper bounds of the forms $\mathcal{O}(\log n)$, $\mathcal{O}(\log_d n)$, and $\mathcal{O}(\log_\delta n)$. Here we also introduce and use our new rake-and-compress technique.
- In Sect. 4 we put together the lower and upper bounds and establish a full classification in the logarithmic region. This proves the main result, Theorem 1.

In the full version of this work [14] we show that all problems that can be defined with context-free grammars are indeed structurally simple; this establishes Lemma 1 and Corollary 1. In the full version we also classify problems in the constant and linear complexity classes.

1.8   Model

Let $G = (V, E)$ be a graph with n nodes. We work in the *deterministic LOCAL model* [15,19]: Each node $v \in V$ is assigned a *unique identifier* $\mathrm{id}(v) \in \{1, 2, \ldots, n^c\}$ for some constant c. Initially each node knows its own identifier, degree, the total number of nodes n, and its input label (here: its color, black or white). All nodes execute the same algorithm, and computation proceeds in synchronized rounds. In each round, nodes transmit messages of arbitrary size to their neighbors, receive messages, and perform local deterministic computations of arbitrary complexity. Eventually each node must stop and produce its local output (here: which of its incident edges are in the solution $X \subseteq E$). The running time, or complexity, of the algorithm is defined as the number of rounds required by all nodes to make local output decisions. Note that an algorithm that runs in T rounds can also be interpreted as a mapping from the radius-T neighborhood of each node to its local output.

2   Logarithmic Lower Bounds

We start by proving the main lower bound results. We establish the lower bounds of $\Omega(\log_d n)$, $\Omega(\log_\delta n)$, and $\Omega(\log n)$, depending on the structure of the problem.

2.1   A General Lower Bound

We start by showing a general lower bound that holds for any problem:

Lemma 2. *Let $\Pi(d, \delta)$ be a family of binary labeling problems. For each d and δ the complexity of $\Pi(d, \delta)$ is either $\mathcal{O}(1)$ or $\Omega(\log_d n)$, assuming $\delta \leq d$.*

Proof. As proved by Balliu et al. [1], all binary labeling problems that are not solvable with locality $\mathcal{O}(1)$ have complexity $\Omega_{d,\delta}(\log n)$. They prove this by reducing all such problems to the *forbidden degree or sinkless orientation*, which is proved to be a fixed point of round elimination and not 0-round solvable.

It is known that there exists bipartite graphs with high girth. In particular, for any $a, b \in \mathbb{N}$, there exists (a, b)-biregular graphs with girth $\Theta(\log_{ab} n)$ [11]. By using techniques similar to Balliu et al. [6] on these graphs, we get an $\Omega(\log_d n)$ lower bound for Π.

2.2   Definitions and Problem Transformations

To get more fine-grained lower bounds, we first define a few transformations for problems. With the help of these transformations, we can apply Lemma 2 to prove stronger lower bounds.

We start by introducing the switch and the reverse of a problem and observe that they do not affect the complexity of the problem:

Definition 4 (switch of a problem). *Let* $\Pi(d, \delta) = (d, \delta, W(d), B(\delta))$ *be a family of problems. The* switch *of* $\Pi(d, \delta)$ *is* $\Pi^s(\delta, d) = (\delta, d, B(\delta), W(d))$.

Definition 5 (reverse of a set family). *Let* $X(k) \subseteq \{0, \ldots, k\}$. *The* reverse *of* $X(k)$ *is* $X^r(k) = \{k - x \mid x \in X\}$.

Definition 6 (reverse of a problem). *Let* $\Pi(d, \delta) = (d, \delta, W(d), B(\delta))$ *be a family of problems. The* reverse *of* $\Pi(d, \delta)$ *is* $\Pi^r(d, \delta) = (d, \delta, W^r(d), B^r(\delta))$.

Lemma 3. *Problem* $\Pi^s(\delta, d)$ *has the same complexity as problem* $\Pi(d, \delta)$.

Proof. Given an algorithm solving Π, we get an algorithm solving Π^s by reversing the role of white and black nodes.

Lemma 4. *Problem* $\Pi^r(d, \delta)$ *has the same complexity as problem* $\Pi(d, \delta)$.

Proof. Given an algorithm solving Π, we get an algorithm solving Π^r by replacing the solution output with its complement.

We now define a more complicated transformation called *shift* of a problem:

Definition 7 (shift of a set family). *Let* $X(d) \subseteq \{0, 1, \ldots, d\}$. *A* shift *by* k *of* $X(d)$ *is*

$$X^{\leftarrow k}(d - k) = \{0, 1, \ldots, d - k\} \cap \{x - i \mid x \in X(d), i \in \{0, 1, \ldots, k\}\}.$$

Definition 8 (shift of a problem). *Let* $\Pi(d, \delta) = (d, \delta, W(d), B(\delta))$ *be a family of problems. The* white shift *of* Π *is*

$$\Pi^{\leftarrow wk}(d, \delta) = (d, \delta, W^{\leftarrow k}(d), B(\delta)),$$

and the black shift *of* Π *is*

$$\Pi^{\leftarrow Bk}(d, \delta) = (d, \delta, W(d), B^{\leftarrow k}(\delta)).$$

Lemma 5. *Let $\Pi(d, \delta)$ be a family of problems with complexity $f(n, d, \delta)$. Then the white shift $\Pi^{\leftarrow w k}$ has complexity $\mathcal{O}(f((k+1)n, d+k, \delta))$ and the black shift $\Pi^{\leftarrow B k}$ has complexity $\mathcal{O}(f((k+1)n, d, \delta+k))$*

Proof. Let $\Pi(d, \delta) = (d, \delta, W(d), B(\delta))$ be family of binary labeling problems. We show that the white shift $\Pi^{\leftarrow w k}$ has complexity $\mathcal{O}(f((k+1)n, d+k, \delta))$; the proof for the black shift $\Pi^{\leftarrow B k}$ follows by Lemma 3.

Fix parameters d, δ and k. Let tree G be an instance for problem $\Pi^{\leftarrow w k}$ with n nodes. We are only interested in black nodes of degree δ and white nodes of degree d as the rest of the nodes are unrestricted. Let G' be a copy of G with k additional black leaves attached to each white node of degree d. Note that G' has at most $n' = (k+1)n$ nodes. Let v be a white node of G, and let v' be the corresponding white node in G'. Denote the set of edges incident to v by $E_G(v)$, and edges incident to v' by $E_{G'}(v')$.

Suppose that X is a solution for $\Pi(d+k, \delta)$ on G'; such a solution can be computed from input graph G with locality $\mathcal{O}(f((k+1)n, d+k, \delta))$ by assumption. Then X induces a solution $Y = X \cap E(G)$ for $\Pi^{\leftarrow w k}$ on G. We know that $\deg_X(v') = |E_{G'}(v') \cap X| \in W(d+k)$. We now show that $\deg_Y(v) = |E_G(v) \cap Y| \in W^{\leftarrow k}(d)$.

It is clear that $0 \leq \deg_Y(v) \leq (d+k) - k = d$. Moreover,

$$\deg_Y(v) = \deg_X(v') - |(E_{G'}(v') \setminus E_G(v)) \cap X|, \quad \text{and}$$

$$0 \leq |(E_{G'}(v') \setminus E_G(v)) \cap X| \leq |E_{G'}(v') \setminus E_G(v)| = k.$$

Therefore, $\deg_X(v') - k \leq \deg_Y(v) \leq \deg_X(v')$. Combining these gives

$$\deg_Y(v) \in \{0, 1, \ldots, d\} \cap \{\deg_X(v') - k, \ldots, \deg_X(v')\} \subseteq W^{\leftarrow k}(d),$$

completing the proof.

With the help of these transformations, we are now ready to prove lower bounds for problems. Indeed, if we can show that some problem $\Pi^{\leftarrow w k}(d, \delta)$ has complexity $\Omega(f(n, d, \delta))$, then we also know that Π has complexity

$$\Omega(f(n/(k+1), d-k, \delta)).$$

2.3 Problem Reductions

By putting together Lemmas 2, 3, 4 and 5, we obtain the following results; see the full version [14] for the proofs:

Proposition 1. *Let k and l be integers. The problem*

$$\Pi(d, \delta) = (d, \delta, 0^{d-k+1}1^k, 1^l 0^{\delta-l+1})$$

for $d \geq k, \delta \geq l$ is either unsolvable or has complexity $\Omega(\log n)$ when $d, \delta = \mathcal{O}(n^{1-\varepsilon})$ for some $\varepsilon > 0$.

Proposition 2. *Let k and l be integers. The problem*

$$\Pi(d, \delta) = (d, \delta, 1^k 0^{d-k-l+1} 1^l, 0 1^{\delta-1} 0)$$

has deterministic complexity $\Omega(\log_\delta n)$ when $d = \mathcal{O}(n^{1-\varepsilon})$ for some $\varepsilon > 0$.

3 Logarithmic Upper Bounds

Now we proceed to prove upper bounds; later in Sect. 4 we will see that our lower and upper bounds indeed provide tight bounds for all structurally simple problems in the logarithmic region.

3.1 Center-Good Problems

Recall the concept of center-good constraints that we introduced in Definition 1. We start by showing that problems whose white or black constraint is center-good can be solved efficiently:

Lemma 6. *Assume that $\Pi(d, \delta) = (d, \delta, W(d), B(\delta))$ is a solvable structurally simple problem. If $W(d)$ is center-good, then $\Pi(d, \delta)$ can be solved with locality $\mathcal{O}(\log_d n)$. Similarly, if $B(\delta)$ is center-good, then $\Pi(d, \delta)$ can be solved with locality $\mathcal{O}(\log_\delta n)$.*

To prove the lemma, we will first show that problems of this type are (s, t)-*resilient* [1] for some $s, t \in \mathbb{N}$:

Definition 9 (resilient problem [1]). *A binary labeling problem $\Pi(d, \delta) = (d, \delta, W(d), B(\delta))$ is (s, t)-resilient if*

- *string $\hat{W}(d)$ does not contain a substring of the form 0^{d+1-s} and*
- *string $\hat{B}(\delta)$ does not contain a substring of the form $0^{\delta+1-t}$.*

Resilient problems are useful because they allow partial labelings to be completed:

Definition 10 (partial labeling). *Let $G = (V, E)$ be a tree. We call $\ell \colon E' \to \{0, 1\}$ for some subset of edges $E' \subseteq E$ a partial labeling of G.*

Given two partial labelings $\ell_1 : E_1 \to \{0, 1\}$ and $\ell_2 : E_2 \to \{0, 1\}$ on G, we say that ℓ_2 is a completion of ℓ_1 if $E_1 \subseteq E_2$, that is, it is defined on a larger set of edges.

Given a problem Π and a partial labeling ℓ, we say a node $v \in V$ is labelled in a valid manner if the set of edges incident to v is in the domain of ℓ and if ℓ satisfies the constraints of Π on v.

Lemma 7 (Balliu et al. [2]). *Let $\Pi(d, \delta)$ be a (t, s)-resilient problem, let $G = (V, E)$ be a tree, and let ℓ be a partial labeling on G. There for every node $v \in V$, there exists a completion of ℓ that labels v in a valid manner if one of the following conditions holds:*

- *v is a white node incident to at most t edges with non-empty labels, or*
- *v is a black node incident to at most s edges with non-empty labels.*

Our plan is to recursively remove layers of nodes from the input tree. Then we can put the layers back and complete the layering by exploiting Lemma 7. We first define a procedure DEG:

Definition 11 (DEG). *Let* $G = (V_W \sqcup V_B, E)$ *be a tree with bipartition* (V_W, V_B). *Let*

$$\text{degen}_{s,t}(G) = \{v \in V_W \mid \deg_G(v) \leq s\} \cup \{v \in V_B \mid \deg_G(v) \leq t\}.$$

Procedure $\text{DEG}(s,t)$ *partitions nodes of* G *into non-empty sets* $L_1, L_2, \ldots, L_k = \emptyset$ *for some* k *as follows:*

$$G_0 = G,$$
$$L_{i+1} = \text{degen}_{s,t}(G_i) \quad \text{if } G_i \text{ is not empty,}$$
$$G_{i+1} = G_i \setminus L_{i+1}.$$

We now show that computing decomposition $\text{DEG}(s,t)$ can be computed with locality $\mathcal{O}(\log_{\max\{s,t\}} n)$ in the LOCAL model. We start with the following lemma showing that the number of high-degree nodes in a tree is bounded:

Lemma 8. *Let* $d \geq 1$ *and let* $G(V_W \sqcup V_B, E)$ *be a tree with bipartition* (V_W, V_B), *where* V_W *is the set of white nodes and* V_B *is the set of black nodes. There are at most* $(|V_B| - 1)/(d - 1)$ *white nodes with degree at least* d, *and at most* $(|V_W| - 1)/(d - 1)$ *black nodes with degree at least* d.

Proof. Let d and G be as in the statement. We show the statement for white nodes; the proof for black nodes follows by symmetry.

Denote by $n_{\geq d}^W$ the number of white nodes with degree at least d, and by $n_{<d}^W$ the number of the rest of white nodes. By the handshake lemma for bipartite graphs, we have

$$|E| \geq dn_{\geq d}^W + n_{<d}^W = (d - 1)n_{\geq d}^W + |V_W|. \tag{1}$$

As G is a tree, we have $|E| = |V_W| + |V_B| - 1$. Combining this with (1) gives the result

$$n_{\geq d}^W \leq (|V_B| - 1)/(d - 1).$$

We can now prove the following lemma showing that each round of DEG reduces the number of nodes by a constant factor:

Lemma 9. *Let* $s, t \geq 1$, *and let* $d \geq 1$ *and let* $G(V_W \sqcup V_B, E)$ *be a tree with bipartition* (V_W, V_B). *Running two iterations of* $\text{DEG}(s,t)$ *on* G *reduces the number of nodes in* G *by a factor of* $\Omega(st)$.

Proof. Let s, t and G be as in the statement. We call nodes of V_W white nodes and nodes of V_B black nodes. Let $L_1 = \text{degen}_{s,t}(G)$, $G_1 = G \setminus L_1$, $L_2 = \text{degen}_{s,t}(G_1)$, and $G_2 = G_1 \setminus L_2$. Let n_1^W and n_1^B be the number of white and black nodes in L_1, respectively, and let n_2^W and n_2^B be the number of white and black nodes in L_2.

By Lemma 8, we have

$$n_1^W \leq (|V_B| - 1)/(t - 1) \text{ and } n_1^B \leq (|V_W| - 1)/(s - 1).$$

Applying Lemma 8 again on G_1, we have

$$n_2^W \leq (n_1^B - 1)/(t - 1) \text{ and } n_2^B \leq (n_1^W - 1)/(s - 1).$$

Expanding the definitions of n_1^W and n_1^B gives the result:

$$n_2^W \leq |V_W|/\Omega(st) \text{ and } n_2^B \leq |V_B|/\Omega(st).$$

Combining this lemma with the simple fact that all nodes can locally compute $\mathrm{degen}_{s,t}$ for each round, we immediately get the locality of procedure $\mathrm{DEG}(s,t)$:

Corollary 2. *The locality of* $\mathrm{DEG}(s,t)$ *is* $\mathcal{O}(\log_{st} n) = \mathcal{O}(\log_{\max\{s,t\}} n)$.

We can now show that resilient problems can be solved efficiently in the LOCAL model:

Lemma 10. *Any* (s,t)*-resilient problem has locality* $\mathcal{O}(\log_{\max\{s,t\}} n)$.

Proof. Let G be the input tree. Start by computing the layer decomposition $L_1, \ldots, L_k$ for G; this can be done with locality $\mathcal{O}(\log_{\max\{s,t\}} n)$ by Corollary 2. Label each edge that is between two nodes in the same layer with 0.

We can now proceed to label the rest of the edges layer-by-layer, from $k - 1$ to 1. The nodes on layer i label all their adjacent edges in a valid manner. This is possible by Lemma 7 as white (respectively black) nodes in layer i have at most s (respectively t) edges to layer i or higher layers, and only those edges have a label. Note that this procedure can be done in parallel by all nodes of layer i as all edges with both endpoints in the same layer have their output fixed at 0. Finally, by definition of L_1, white nodes on the layer 1 have at most $s < d$ neighbors, and black nodes on the layer 1 have at most $t < \delta$ neighbors, hence they do not impose any restrictions on their adjacent edges.

Now we are finally ready to prove Lemma 6:

Proof. Let us focus on the case where $W(d)$ is center-good; the case where $B(\delta)$ is center-good is symmetric. By definition, any sufficiently large set $w \in W(d)$ contains an element $p \in [d^\varepsilon, d^{1-\varepsilon}]$. Therefore, the problem is $(d^{\min\{\varepsilon, 1-\varepsilon\}}, 1)$-resilient, and the rest follows by Lemma 10.

3.2   Aggressive Rake-and-Compress

In this section, we introduce the aggressive rake-and-compress algorithm, compute its complexity, and give some of its key properties.

Definition 12. *For a graph* G *and some positive integer* r, *we define the following two sets:*

$$\mathrm{leaves}(G) = \{v \in V(G) \mid \deg_G(v) = 1\},$$
$$\mathrm{ext}_{r,\Delta}(G) = \{v \in V(G) \mid \forall u \in N^r[v], \deg_G(u) < \Delta\}$$
$$= V(G) \setminus N^r[\{v \in V(G) \mid \deg_G(v) \geq \Delta\}],$$

where $N^r[v]$ in the distance-r closed neighborhood of v. Removing the first set from the graph is a rake operation, removing the second set from the graph is a compress operation.

By computing these sets recursively and removing them from the graph, we can partition the graph into layers. This is a very helpful tool to create efficient $\mathcal{O}(\log_\Delta n)$ local algorithms.

Definition 13. *Procedure* $\mathrm{ARC}(r, \Delta)$ *partitions the set of nodes* V *into non-empty sets* $L_1, L_2, \ldots, L_k$ *for some* L *as follows:*

$$G_0 = G,$$
$$L_{i+1} = \mathrm{leaves}(G_i) \cup \mathrm{ext}_{r,\Delta}(G_i) \quad \textit{if } G_i \textit{ is not empty,}$$
$$G_{i+1} = G_i \setminus L_{i+1}.$$

In the full version of this work [14], we prove the following lemma that shows that ARC makes progress fast enough:

Lemma 11. *Let* $G = (V, E)$ *be a tree and* $r \in \{1, 2, \ldots\}$. *Apply successively the distance-r rake operation, compress operation and then rake operation again, r times successively. If* V' *is the remaining vertex set then* $|V'| \leq \mathcal{O}(r)|V|/\Omega(\Delta)$.

Corollary 3. $\mathrm{ARC}(r, \Delta)$ *can be computed in* $\mathcal{O}(r^2 \cdot \log_{\Delta/r} n)$ *rounds.*

Proof. When we apply ARC $r+1$ times, we remove more vertices than by applying the distance-r rake operation, compress operation and then rake operation again, r times successively. Therefore, by Lemma 11, computing $k+1$ ARC layers divides the number of vertices not assigned to a layer by $\Omega(\Delta/r)$. Therefore, after computing $(r+1)k$ layers, the number of vertices not assigned to any layer is divided by $\Omega((\Delta/r)^k)$. We need to compute at most $\mathcal{O}(r \cdot \log_{\Delta/r} n)$ layers to have all vertices assigned to some layer. In total, $\mathcal{O}(r^2 \cdot \log_{\Delta/r} n)$ LOCAL rounds are needed, because every layer takes $\mathcal{O}(r)$ LOCAL rounds to compute.

Here is a key property that is used to build algorithms optimal in n, d and δ:

Lemma 12. *After the* $\mathrm{ARC}(r, \Delta - k + 1)$ *procedure, every relevant degree-Δ vertex in G is adjacent to k raked vertices (i.e. neighbors which were taken due to their degree being 1). Moreover, we can assume those raked vertices are the k lowest-layer neighbors of v.*

Proof. We apply the procedure $\mathrm{ARC}(r, d-k)$ to the input G. Denote by $L_1, \ldots, L_m$ the resulting layer partition. Let v be a degree-Δ node on layer i. Consider the set R of the k lowest-layer neighbors of v. When $u \in R$ is added to its layer, it cannot be due to compress operation, because when at this time, v had at least $\Delta - k$ other neighbors still not assigned to a layer yet, i.e. v had residual degree $\geq \Delta - k + 1$. Therefore, one of its neighbors cannot be added a layer at that time, and we get a contradiction. Hence, u gets assigned to a layer because of a rake operation.

For $r \geq 2$, we have the more general statement:

Lemma 13. *Given a decomposition given by the* ARC$(r, \Delta - k + 1)$ *procedure with* $r \geq 2$, *we have the following. For every relevant degree-Δ vertex v, and every vertex u in the set of k lowest-layer neighbors of v, every neighbor of u different from v is a raked vertex. Moreover, u is also raked.*

Proof. We apply the procedure ARC$(r, d - k)$ to the input G, with $r \geq 2$. Denote by $L_1, \ldots, L_m$ the resulting layer partition. Let v be a relevant degree-Δ vertex, and let u be a layer-j vertex in the set of k lowest-layer neighbors of v. Let w be a neighbor of u different from v, in layer i. Vertex u must be a raked one, because it is one of the k lowest-layer neighbors of v. Therefore, w is at a lower layer than u, i.e. $i < j$. We also get that v is of degree $\geq d - k$ in G_j, as u, one of the k lowest-layer neighbors of w, is still in G_j. Therefore, w is at distance 2 of v in G_i, so w must be a raked vertex because v is of degree $\geq d - k$ in G_i.

We will now use these tools to prove upper-bound results for two problems that play a key role in our classification.

Bipartite Factor Problem. Bipartite (k, l)-factor is the problem $\Pi(d, \delta) = (d, \delta, \{k\}, \{l\})$. Note that bipartite $(1, 1)$-factor is the same problem as bipartite perfect matching.

Lemma 14. *When k and l are non-zero constants, bipartite (k, l)-factor can be solved in $\mathcal{O}(\log_\delta n)$ rounds.*

Proof. We apply the procedure ARC$(1, \min\{d - k, \delta - l\})$ on the input graph G. See the full version [14] for the details.

Quasi-orientation Problem. Quasi-(k, l)-orientation is the problem $\Pi(d, \delta) = (d, \delta, \{k\}, \{0, \delta - l\})$. This problem is easiest to understand if we interpret it so that the edges in X are oriented towards black nodes and edges in $E \setminus X$ are oriented towards white nodes. Then in a quasi-(k, l)-orientation, all relevant white nodes have outdegree k while all relevant black nodes have outdegree l or δ.

Lemma 15. *When k is a constant, quasi-(k, l)-orientation can be solved in $\mathcal{O}(\log_d n)$ rounds.*

Proof. We apply the procedure ARC$(2, d - k)$ to the input graph G. See the full version [14] for the details.

4   Classification of Logarithmic Problems

We now have all the tools to classify logarithmic problems and prove our main theorem.

Theorem 1. *Let $\Pi(d, \delta)$ be a structurally simple family of binary labeling problems and assume that the complexity of $\Pi(d, \delta)$ in trees is $\Theta_{d,\delta}(\log n)$. Further assume that $\delta \leq d = \mathcal{O}(n^{1-\alpha})$ for some $\alpha > 0$. Then we can partition the space of parameter values n, d, and δ in three (possibly empty) classes such that the complexity of $\Pi(d, \delta)$ in trees is:*

1. $\Theta(\log n)$ *in the first class,*
2. $\Theta(\log_\delta n)$ *in the second class,*
3. $\Theta(\log_d n)$ *in the third class.*

Proof. Fix $\delta \leq d$, let $w = \hat{W}(d)$ and $b = \hat{B}(\delta)$. Let Π solvable with $\Theta_{d,\delta}(\log n)$ complexity. We know by Lemma 2 that the problem has complexity $\Omega(\log_d n)$. However, this is not tight: there are some problems that cannot be solved in $\mathcal{O}(\log_d n)$ time. We make a case distinction to classify the problem complexity into multiple classes. First, if w is center-good, then it can be solved in $\mathcal{O}(\log_d n)$ rounds using Lemma 10. Now suppose for the rest of the proof that w is edge-good. Let $w = s_w w_1 0^k w_2 t_w$ with $|w_1|, |w_2| \leq C$, and $b = s_b \ldots t_b$. The complexities split into different classes depending on w and b.

- If $s_w = s_b = 1$ or $t_w = t_b = 1$, then problem is of constant complexity.
- If $s_b = t_b = 1$, then $\Pi(d, \delta)$ can be solved with the quasi-$(k, 0)$-orientation algorithm of Lemma 15 (and possibly Lemma 4) in time $\mathcal{O}(\log_d n)$, because it cannot be that w_1 and w_2 only contains 0's, otherwise we would have $w = 0^{d+1}$ (as $s_w = t_w = 0$).
- Else, either $s_b = 0$ or $t_b = 0$. We will handle this case for the rest of the proof.

Without loss of generality, assume in the following that $t_b = 0$ (otherwise, one can use Lemma 4 and get back to the previous case). If $s_b = 1$, it cannot be that $b = 10^*$, as $s_w = 0$, as the problem would be unsolvable. Therefore, b contains at least 2 1's and $\Pi(d, \delta)$ can be solved in $\mathcal{O}(\log_d n)$ rounds with the quasi-(k, l)-orientation algorithm.

We are now left with the case $s_b = t_b = 0$ for the rest of the proof. Then the problem has complexity $\Omega(\log_\delta n)$ by Proposition 2 (assuming $d = \mathcal{O}(n^{1-\alpha})$ for some $\alpha > 0$). Moreover, by assumption, the complexity is $\mathcal{O}(\log n)$. There are still two possible complexity classes: $\log_\delta n$ and $\log n$. To finish our classification, we need to prove that in all cases, the complexity is either $\Omega(\log n)$ or $\mathcal{O}(\log_\delta n)$. If b is center-good, then the complexity is $\mathcal{O}(\log_\delta n)$ by Lemma 10. For the rest of the proof, suppose b is edge-good. Let $b = s_b b_1 0^l b_2 t_b$ with $|b_1|, |b_2| \leq C$.

- If $s_w = t_w = 1$, then $\Pi(d, \delta)$ can be solved in time $\mathcal{O}(\log_\delta n)$ with Lemma 3, possibly Lemma 4, and the quasi-$(k, 0)$-orientation algorithm of Lemma 15. The algorithm is applicable because b_1 and b_2 do not only contain 0's, otherwise we would have $b = 0^{\delta+1}$.
- Else, either $s_w = 0$ or $t_w = 0$.

For the rest of the proof, we have $s_w = 0$ or $t_w = 0$, and $s_b = t_b = 0$. There are multiple cases:

- Type A: there exists $i \in \{1, 2\}$ such that both b_i and w_i contain a 1. The complexity of a type A problem is $\mathcal{O}(\log_\delta n)$ by Lemma 14 and using Lemma 4 (if necessary).
- Type B: w_1 and b_2 only contain 0's, and w_2 and b_1 contains a 1. If $s_w = 0$, the problem complexity is $\Omega(\log n)$ by Proposition 1. Else, $s_w = 1$ and $t_w = 0$, and the complexity is $\mathcal{O}(\log_\delta n)$ using Lemma 15 and Lemma 3.
- Type C: w_2 and b_1 only contain 0's, and w_1 and b_2 contains a 1. Using Lemma 4, we can get from type C back to type B without loss of generality.
- Type D_c: there exists $c \in \{w, b\}$ such that both c_1 and c_2 only contain 0's. A type D_b or D_w problem is unsolvable. Consider a type D_b problem. Then the problem is unsolvable because $b = 0^{\delta+1}$. Now consider a type D_w problem. If $s_w = t_w = 0$, the problem is unsolvable because $w = 0^{d+1}$. Else, one of s_w and t_w is equal to 0, but not both. So $w = 0^+01$ or $w = 100^+$ and as $s_b = t_b = 0$, the problem is unsolvable.

Acknowledgments. This work was supported in part by the Research Council of Finland, Grant 333837, and it was partly done while TP was doing an internship at Aalto University. We would like to thank the anonymous reviewers for their helpful comments on this work. The full version is available in [14].

References

1. Balliu, A., et al.: Classification of distributed binary labeling problems. In: Attiya, H. (ed.) 34th International Symposium on Distributed Computing, DISC 2020, 12–16 October 2020, Virtual Conference. LIPIcs, vol. 179, pp. 17:1–17:17. Schloss Dagstuhl - Leibniz-Zentrum für Informatik (2020). https://doi.org/10.4230/LIPICS.DISC.2020.17
2. Balliu, A., et al.: Classification of distributed binary labeling problems (2020). https://doi.org/10.48550/arXiv.1911.13294. https://arxiv.org/abs/1911.13294
3. Balliu, A., Brandt, S., Olivetti, D., Suomela, J.: How much does randomness help with locally checkable problems? In: Emek, Y., Cachin, C. (eds.) PODC 2020: ACM Symposium on Principles of Distributed Computing, Virtual Event, Italy, 3–7 August 2020, pp. 299–308. ACM (2020). https://doi.org/10.1145/3382734.3405715
4. Balliu, A., Brandt, S., Olivetti, D., Suomela, J.: Almost global problems in the LOCAL model. Distrib. Comput. **34**(4), 259–281 (2021). https://doi.org/10.1007/S00446-020-00375-2
5. Balliu, A., Hirvonen, J., Korhonen, J.H., Lempiäinen, T., Olivetti, D., Suomela, J.: New classes of distributed time complexity. In: Diakonikolas, I., Kempe, D., Henzinger, M. (eds.) Proceedings of the 50th Annual ACM SIGACT Symposium on Theory of Computing, STOC 2018, Los Angeles, CA, USA, 25–29 June 2018, pp. 1307–1318. ACM (2018). https://doi.org/10.1145/3188745.3188860
6. Balliu, A., et al.: Sinkless orientation made simple. In: Kavitha, T., Mehlhorn, K. (eds.) 2023 Symposium on Simplicity in Algorithms, SOSA 2023, Florence, Italy, 23–25 January 2023, pp. 175–191. SIAM (2023). https://doi.org/10.1137/1.9781611977585.CH17
7. Brandt, S., et al.: A lower bound for the distributed lovász local lemma. In: Wichs, D., Mansour, Y. (eds.) Proceedings of the 48th Annual ACM SIGACT Symposium on Theory of Computing, STOC 2016, Cambridge, MA, USA, 18–21 June 2016, pp. 479–488. ACM (2016). https://doi.org/10.1145/2897518.2897570

8. Chang, Y.J., Kopelowitz, T., Pettie, S.: An exponential separation between randomized and deterministic complexity in the LOCAL model. SIAM J. Comput. **48**(1), 122–143 (2019). https://doi.org/10.1137/17M1117537

9. Chang, Y.J., Pettie, S.: A time hierarchy theorem for the LOCAL model. SIAM J. Comput. **48**(1), 33–69 (2019). https://doi.org/10.1137/17M1157957

10. Fischer, M., Ghaffari, M.: Sublogarithmic distributed algorithms for lovász local lemma, and the complexity hierarchy. In: Richa, A.W. (ed.) 31st International Symposium on Distributed Computing, DISC 2017, Vienna, Austria, 16–20 October 2017. LIPIcs, vol. 91, pp. 18:1–18:16. Schloss Dagstuhl - Leibniz-Zentrum für Informatik (2017). https://doi.org/10.4230/LIPICS.DISC.2017.18

11. Füredi, Z., Lazebnik, F., Seress, Á., Ustimenko, V.A., Woldar, A.J.: Graphs of prescribed girth and bi-degree. J. Comb. Theory Ser. B **64**(2), 228–239 (1995). https://doi.org/10.1006/JCTB.1995.1033

12. Ghaffari, M., Harris, D.G., Kuhn, F.: On derandomizing local distributed algorithms. In: Thorup, M. (ed.) 59th IEEE Annual Symposium on Foundations of Computer Science, FOCS 2018, Paris, France, 7–9 October 2018, pp. 662–673. IEEE Computer Society (2018). https://doi.org/10.1109/FOCS.2018.00069

13. Ghaffari, M., Su, H.H.: Distributed degree splitting, edge coloring, and orientations. In: Klein, P.N. (ed.) Proceedings of the Twenty-Eighth Annual ACM-SIAM Symposium on Discrete Algorithms, SODA 2017, Barcelona, Spain, Hotel Porta Fira, 16–19 January 2017, pp. 2505–2523. SIAM (2017). https://doi.org/10.1137/1.9781611974782.166

14. Lievonen, H., Picavet, T., Suomela, J.: Distributed binary labeling problems in high-degree graphs (2023). https://doi.org/10.48550/arXiv.2312.12243

15. Linial, N.: Locality in distributed graph algorithms. SIAM J. Comput. **21**(1), 193–201 (1992). https://doi.org/10.1137/0221015

16. Miller, G.L., Reif, J.H.: Parallel tree contraction and its application. In: 26th Annual Symposium on Foundations of Computer Science, Portland, Oregon, USA, 21–23 October 1985, pp. 478–489. IEEE Computer Society (1985). https://doi.org/10.1109/SFCS.1985.43

17. Naor, M., Stockmeyer, L.J.: What can be computed locally? SIAM J. Comput. **24**(6), 1259–1277 (1995). https://doi.org/10.1137/S0097539793254571

18. Paun, G., Salomaa, A.: Thin and slender languages. Disc. Appl. Math. **61**(3), 257–270 (1995). https://doi.org/10.1016/0166-218X(94)00014-5

19. Peleg, D.: Distributed computing: a locality-sensitive approach. Soc. Ind. Appl. Math. (2000). https://doi.org/10.1137/1.9780898719772

20. Rozhon, V., Ghaffari, M.: Polylogarithmic-time deterministic network decomposition and distributed derandomization. In: Makarychev, K., Makarychev, Y., Tulsiani, M., Kamath, G., Chuzhoy, J. (eds.) Proceedings of the 52nd Annual ACM SIGACT Symposium on Theory of Computing, STOC 2020, Chicago, IL, USA, 22–26 June 2020, pp. 350–363. ACM (2020). https://doi.org/10.1145/3357713.3384298

Computing Replacement Paths
in the CONGEST Model

Vignesh Manoharan[✉] and Vijaya Ramachandran

The University of Texas at Austin, Austin, TX, USA
{vigneshm,vlr}@cs.utexas.edu

Abstract. We present several results on the round complexity of Replacement Paths and Second Simple Shortest Path which are basic graph problems that can address fault tolerance in distributed networks. These are well-studied in the sequential setting, and have algorithms [18, 20,30,34] that nearly match their fine-grained complexity [3,33]. But very little is known about either problem in the distributed setting.

We present algorithms and lower bounds for these problems in the CONGEST model, with many of our results being close to optimal.

Keywords: Distributed Algorithms · Graph Algorithms · Shortest Paths

1 Introduction

Consider a communication network $G = (V, E)$, with two special nodes s and t in V, and with communication transmission from s to t along a shortest path P_{st}. In the distributed setting, it can be important to maintain efficient communication from s to t in the event that a link (i.e., edge) on this path P_{st} fails. This is the Replacement Paths (RPaths) problem, where for each edge e on P_{st}, we need to find a shortest path from s to t that avoids e. The RPaths problem has been extensively studied in the sequential setting [18,20,30]. The closely related problem of finding a second simple shortest path (2-SiSP), i.e., a shortest simple path from s to t avoids at least one edge on the original shortest path P_{st}, has also been studied in the sequential setting [34]. Surprisingly, there are virtually no results known in the distributed CONGEST model for RPaths or 2-SiSP. In this paper we address this lacuna by obtaining strong round complexity bounds for these problems in the CONGEST model, which are optimal or near-optimal in many cases.

Let $|V| = n$, $|E| = m$. Directed weighted RPaths and 2-SiSP are important problems in sequential fine-grained complexity, being part of the n^3 time complexity class [33] which contains many graph problems: weighted APSP, Negative Triangle Detection, Minimum Weight Cycle (MWC), Radius, Eccentricities and Betweenness Centrality [2]. RPaths and 2-SiSP are also in the fine-grained mn complexity class [3] which is of more relevance in the CONGEST model where

This work was supported in part by NSF grant CCF-2008241.

Y. Emek (Ed.): SIROCCO 2024, LNCS 14662, pp. 420–437, 2024.
https://doi.org/10.1007/978-3-031-60603-8_23

$O(\log n)$ bits of communication occur per round per edge in the graph, leading to $\tilde{O}(m)^1$ communication per round, and the goal is to minimize the number of rounds in the computation.

Despite RPaths and 2-SiSP being well-motivated problems for distributed networks, there is a lack of results in the CONGEST model (see Sect. 1.3 for prior work). In this paper we present several nontrivial results, in many cases near-optimal, for distributed RPaths and 2-SiSP. In addition to algorithms computing weights, we present algorithms that construct paths by computing routing tables in the full version [22].

Although RPaths and 2-SiSP have not been extensively studied in the CONGEST model, other related and more recently defined problems have been studied: CONGEST algorithms for fault-tolerant distance preservers [8,15], which are sparse subgraphs on the network in which replacement path distances are exactly preserved, and fault-tolerant spanners [13,26], which are sparse subgraphs in which replacement path distances are approximations of the distances in the original network have been reported (see Sect. 1.3 for details). The techniques in these results do not readily give efficient algorithms to explicitly compute replacement path weights or to construct a replacement path when a failed edge in known; further, these results mainly deal with undirected unweighted graphs.

Weighted directed RPaths and 2-SiSP are in the mn and n^3 sequential fine-grained complexity classes as discussed above. But these two problems are unique among the problems in these sequential classes in that they become simpler for undirected graphs [18] and for unweighted directed graphs [30]. The results we present in this paper for RPaths and 2-SiSP for the CONGEST model show a similar trend, with the added contribution of an unconditional near linear lower bound for the weighted directed case along with sublinear round algorithms for unweighted and undirected graphs (as long as the network diameter and length of P_{st} are sublinear). Thus we show a proven separation in complexity for RPaths and 2-SiSP between weighted directed graphs and either unweighted directed graphs or undirected graphs (even if weighted) in the CONGEST model. Our results for undirected graphs are in the full version [22].

1.1   Preliminaries

The CONGEST Model. In the CONGEST model [27], a communication network is represented by a graph $G = (V, E)$ where nodes model processors and edges model bounded-bandwidth communication links between processors. Each node has a unique identifier in $\{0, 1, \ldots n-1\}$ where $n = |V|$, and each node only knows the identifiers of itself and its neighbors in the network. Each node has infinite computational power. The nodes perform computation in synchronous rounds, where each node can send a message of up to $\Theta(\log n)$ bits to each neighbor and can receive the messages sent to it by its neighbors. The complexity of an algorithm is measured by the number of rounds until the algorithm terminates.

---

[1]   We use the notation $\tilde{O}, \tilde{\Omega}, \tilde{\Theta}$ to hide poly-logarithmic factors.

We consider both weighted and unweighted graphs G in this paper, where in the weighted case each edge has an integer weight which is known to the vertices incident to the edge. The graph G can be directed or undirected. Following the convention for CONGEST algorithms [4,5,12,16,24], the communication links are always bi-directional and unweighted.

Notation. Let $G = (V, E)$ be a directed or undirected graph with $|V| = n$ and $|E| = m$. Let edge (u, v) have non-negative integer weight $w(u, v)$ according to a weight assignment function $w : E \rightarrow \{0, 1, \ldots W\}$, where $W = poly(n)$. Let $d(s, t)$ denote the weight of a shortest path P_{st} from s to t and h_{st} denote the number of edges (hop distance) on this shortest path. The undirected diameter D is the maximum shortest path distance between any two vertices in the underlying undirected unweighted graph of G.

Note 1. We use *SSSP* and *APSP* to denote the round complexity in the CONGEST model for weighted single source shortest paths (SSSP) and weighted all pairs shortest paths (APSP) respectively. The current best algorithm for weighted APSP runs in $\tilde{O}(n)$ rounds, randomized [7]. For weighted SSSP, recent results [9,31] provide an $\tilde{O}(\sqrt{n} + n^{2/5+o(1)}D^{2/5} + D)$ round randomized algorithm. The current best lower bounds are $\Omega\left(\frac{\sqrt{n}}{\log n} + D\right)$ for weighted SSSP [28] and $\Omega\left(\frac{n}{\log n}\right)$ for (weighted and unweighted) APSP [24].

We now define the two problems we consider in this paper.

Definition 1. *Replacement Paths (RPaths)* : *Given an n-node graph G, two vertices s, t and a shortest path P_{st} from s to t, for each edge $e \in P_{st}$, compute the weight $d(s, t, e)$ of a shortest simple path P_e from s to t that does not contain e.*

Second Simple Shortest Path (2-SiSP) : *Given an n-node graph G, two vertices s, t and a shortest path P_{st} from s to t, compute the weight $d_2(s, t)$ of a shortest simple path P_2 from s to t that differs from P_{st}.*

Our lower bounds for RPaths and 2-SiSP apply even when only one node in the graph is required to know each distance $d(s, t, e)$ or $d_2(s, t)$. In our algorithms, all vertices can learn the distances $d(s, t, e)$ or $d_2(s, t)$ using a simple broadcast in $O(h_{st} + D)$ rounds, which is within the round complexity bounds.

In our results, we assume that the shortest path P_{st} between the vertices s, t is part of the input and that each vertex in the network knows the identities of s and t, and the identities of vertices on P_{st}. The round bounds of our algorithms are unchanged if we are required to compute P_{st} using known CONGEST algorithms for SSSP and broadcast the necessary information in $O(h_{st} + D)$ rounds. Also, note that once we have the weights of the h_{st} replacement paths for P_{st} we can compute the 2-SiSP weight in additional $O(D)$ rounds with a convergecast.

1.2   Our Results

Table 1 lists our results for directed RPaths and 2-SiSP. Our upper bounds are either exact or $(1+\epsilon)$-approximation results, for arbitrarily small constant $\epsilon > 0$,

Table 1. Our results. *SSSP* and *APSP* refer to round complexity of weighted SSSP and APSP (See Note 1). Approximation results hold for approx. ratio α and $(1 + \epsilon)$, where $\alpha > 1$ is an arbitrarily large constant, and $\epsilon > 0$ is an arbitrarily small constant.

Problem	Lower Bound	Upper Bound
Results are for directed graphs. Entries are (*approx. ratio, round bound*)		
Weighted RPaths, 2-SiSP	$1, \Omega\left(\frac{n}{\log n}\right)$ [Theoream 1.A]	$1, O(APSP) = \tilde{O}(n)$ [Theoream 1.B]
Approximate Weighted RPaths, 2-SiSP	$\alpha, \Omega(SSSP) = \Omega\left(\frac{\sqrt{n}}{\log n} + D\right)$ [Theoream 2.A]	$(1 + \epsilon), \tilde{O}(\sqrt{nh_{st}} + D + \min(n^{2/3}, h_{st}^{2/5} n^{2/5 + o(1)} D^{2/5}))$ [Theoream 1.C]
Unweighted RPaths, 2-SiSP	$\alpha, \Omega(SSSP)$ [Theoream 2.A]	$1, \tilde{O}(\min(n^{2/3} + \sqrt{nh_{st}} + D, SSSP \cdot h_{st}))$ [Theoream 2.B]

and our lower bounds are either exact or α-approximation results (for any constant $\alpha > 1$). Our main algorithms and bounds are for computing the *weights* of paths defined in Definition 1. However, we also have distributed algorithms that use routing tables to find such a path when an edge fails in the full version [22].

Directed Weighted Graphs. For an n-node directed weighted graph, we present an RPaths CONGEST algorithm that runs in near-linear $\tilde{O}(n)$ rounds (Sect. 2.2). The classic sequential $\tilde{O}(mn)$-time algorithm for 2-SiSP and RPaths [34] performs a sequence of h_{st} SSSP computations, and a near-linear bound is not achievable on the CONGEST model through implementing this algorithm. Instead, we formulate RPaths as an APSP computation (on an alternate graph) that can be efficiently computed within the APSP bound $\tilde{O}(n)$ in the CONGEST model. We show that our algorithm is nearly optimal by presenting an $\tilde{\Omega}(n)$ lower bound (even when the undirected diameter D is a constant) for both RPaths and 2-SiSP through a reduction from set disjointness (Sect. 2.1).

Our lower bound proof for RPaths is much more involved than the $\tilde{\Omega}(n)$ APSP lower bound in [24], and we also show that RPaths differs from APSP in efficient approximability: the $\tilde{\Omega}(n)$ APSP lower bound in [24] applies to α-approximation for any constant $\alpha > 1$, but for weighted directed RPaths we give in Sect. 3.3 an asymptotically improved algorithm for $(1 + \epsilon)$-approximation (for any constant $\epsilon > 0$) that runs in sublinear rounds ($\tilde{O}(n^{1-c})$ for constant $c > 0$) whenever both h_{st} and D are sublinear.

Theorem 1. *Given a directed weighted graph G on n vertices with undirected diameter D and a shortest path P_{st} of hop length h_{st},*

A. *Any randomized algorithm that computes RPaths or 2-SiSP for P_{st} requires $\Omega\left(\frac{n}{\log n}\right)$ rounds, even if D is constant.*

B. *RPaths and 2-SiSP for P_{st} can be computed in $O(APSP)$ rounds, and hence by a randomized algorithm in $\tilde{O}(n)$ rounds.*

C. *There is a randomized algorithm that computes a* $(1 + \epsilon)$-*approximation of RPaths and 2-SiSP in* $\tilde{O}(\sqrt{nh_{st}} + D + \min(n^{2/3}, h_{st}^{2/5} n^{2/5+o(1)} D^{2/5}))$ *rounds, for any constant* $\epsilon > 0$.

D. *Computing an* α-*approximation of RPaths or 2-SiSP for any constant* $\alpha > 1$ *requires* $\Omega\left(\frac{\sqrt{n}}{\log n} + D\right)$ *rounds.*

Directed Unweighted Graphs. In the case of directed unweighted graphs, the near linear lower bound for the weighted case no longer applies. We give an algorithm based on sampling and computing detours that takes $\tilde{O}(n^{2/3} + \sqrt{nh_{st}} + D)$ rounds. This gives us an algorithm that runs in sublinear rounds whenever both h_{st} and D are sublinear. We also have a simple algorithm taking $O(h_{st} \cdot SSSP)$ rounds that is more efficient when h_{st} is small (Sect. 3.2). We show a lower bound of $\Omega(SSSP) = \tilde{\Omega}(\sqrt{n} + D)$ for computing RPaths and 2-SiSP on directed unweighted graphs (Sect. 3.1), and our algorithm matches this $O(SSSP)$ bound when h_{st} is $O(1)$. This $\tilde{\Omega}(\sqrt{n} + D)$ lower bound shows that computing RPaths is harder in directed unweighted graphs than in undirected unweighted graphs, where we have an $O(D)$ round algorithm (see below).

Theorem 2. *Given a directed unweighted graph G on n vertices with undirected diameter D and a shortest path P_{st} of hop length h_{st},*

A. *Any randomized algorithm that computes RPaths or 2-SiSP requires* $\Omega\left(\frac{\sqrt{n}}{\log n} + D\right)$ *rounds, even if h_{st} and D are as small as $\Theta(\log n)$. These lower bounds also apply to an α-approximation, for any constant $\alpha > 1$.*

B. *There is a randomized algorithm that computes RPaths and 2-SiSP for P_{st} in* $\tilde{O}(\min(n^{2/3} + \sqrt{nh_{st}} + D, SSSP \cdot h_{st}))$ *rounds.*

We adapt our lower bound proof for directed unweighted RPaths to other basic directed graph problems such as *s-t* reachability and *s-t* shortest path in directed unweighted graphs. A folklore lower bound of $\tilde{\Omega}(\sqrt{n} + D)$ for these problems was attributed by [16] to *undirected* lower bound results in [32], and we give explicit proofs here. A lower bound of $\tilde{\Omega}(\sqrt{n} + D)$ for undirected weighted SSSP was known [14,32] (which also applies to directed weighted SSSP) but this does not apply to unweighted directed graphs. These problems are easier in undirected unweighted graphs since undirected BFS can be performed in $O(D)$ rounds. Thus our results indicate that basic problems in directed graphs are asymptotically harder than their undirected unweighted counterparts. We note that this difference is not very surprising since the underlying communication network is the undirected version of the graph regardless of whether the graph is directed or undirected.

Undirected Graphs. For undirected graphs, our upper and lower bounds for RPaths match the round complexity of SSSP (BFS for unweighted) in the CONGEST model, except for weighted RPaths which requires an additional $O(h_{st})$ rounds. The remaining gap between our upper and lower bounds is inherited from the gap between the current best bounds for SSSP. Due to space limitation, details of our undirected graph results are deferred to the full version [22].

Theorem 3. *Given an undirected weighted (or unweighted) graph G on n vertices with diameter D and a shortest path P_{st} of hop length h_{st},*

A. *Any algorithm that computes RPaths or 2-SiSP requires:*
   i. $\Omega(SSSP) = \Omega\left(\frac{\sqrt{n}}{\log n} + D\right)$ *rounds if G is weighted, even if h_{st} is constant. This lower bound applies to α-approximation, for any $\alpha > 1$.*
   ii. $\Omega(D)$ *rounds if G is unweighted, which is a tight bound.*
B. *We can compute RPaths for P_{st} in $O(SSSP+h_{st}) = \tilde{O}(\sqrt{n}+n^{2/5+o(1)}D^{2/5}+ D+h_{st})$ rounds. For 2-SiSP the bound is $O(SSSP)$. If G is unweighted, the bound is $O(D)$ rounds.*

1.3   Prior Work

RPaths and the closely related 2-SiSP are well studied problems in the sequential setting. For weighted directed graphs the classical algorithm of Yen [34] runs in $\tilde{O}(mn)$ time and has a matching fine-grained lower bound of $\tilde{\Omega}(mn)$ assuming a sequential hardness result for MWC [3]. For unweighted directed graphs, a randomized $\tilde{O}(m\sqrt{n})$ algorithm is given in [30]. For undirected graphs, a near-linear time algorithm is given in [18], matching the running time for sequential SSSP. Our bounds for RPaths for these different graph classes in the CONGEST model follow a similar pattern: close to APSP for weighted directed graphs, close to SSSP for undirected graphs, and intermediate bounds for directed unweighted graphs. The more general problem of single source replacement paths (SSRP) was studied in the sequential setting in [10,11].

In the distributed setting, an $O(D \log n)$ algorithm for single source replacement paths in undirected unweighted graphs was given in [15]. We are not aware of any prior results in the CONGEST model for RPaths or 2-SiSP for directed graphs or for weighted graphs. Distributed constructions of fault-tolerant preservers, which construct a sparse subgraph that exactly preserves replacement path distances, have been studied in [8,25] but their constructions do not give an efficient procedure to compute replacement path distances. Fault-tolerant spanners, which construct a sparse subgraph that approximates replacement path distances, have also been studied in CONGEST [13,26].

There are unconditional $\tilde{\Omega}(n)$ CONGEST lower bounds for several graph problems in the sequential n^3 and mn fine-grained complexity class: APSP [24], diameter, radius [1,5], minimum weight cycle (MWC) [23]. CONGEST algorithms for these problems have also been studied: APSP [4,7,21], diameter and radius [1,5], MWC [23] and betweenness centrality [17].

1.4   Our Techniques

CONGEST Lower Bounds. Our lower bounds use reductions either from Set Disjointness or from other graph problems with known CONGEST lower bounds. Set Disjointness is a two party communication complexity problem, where two players Alice and Bob are each given a k-bit string S_a and S_b respectively. Alice and Bob need to communicate and decide if the sets represented by S_a and S_b are disjoint, i.e., whether there is no bit position i, $1 \le i \le k$ with $S_a[i] = 1$ and $S_b[i] = 1$. A classical result in communication complexity is that Alice and Bob must exchange $\Omega(k)$ bits even if they are allowed shared randomness [6,19,29]. Lower bounds using such a reduction also hold against randomized algorithms.

Our reduction from Set Disjointness for the $\tilde{\Omega}(n)$ CONGEST lower bound for directed weighted RPaths is loosely inspired by a construction in a sparse sequential reduction from MWC to RPaths in [3]. Some of our lower bounds use known unconditional CONGEST lower bounds for problems like s-t Subgraph Connectivity and weighted s-t Shortest Path [32].

CONGEST Algorithms. Adapting the sequential algorithm for directed weighted replacement paths [34] directly to the CONGEST model requires up to n SSSP computations, which is not efficient. Instead, our CONGEST algorithm builds on a sequential sparse reduction from RPaths to Eccentricities in [3] which we tailor to work efficiently in the CONGEST model using weighted APSP [7] as a subroutine. Our algorithm for directed unweighted RPaths is loosely based on the sequential algorithm in [30], but we make significant changes to obtain efficiency in the distributed setting. We use a variety of techniques such as sampling, computing shortest paths in skeleton graphs and pipelined BFS.

Many of our algorithms do not use any randomness apart from that used by the randomized algorithms for SSSP and APSP. If we use deterministic CONGEST algorithms for these problems, such as for unweighted APSP [21] and weighted APSP [4], our algorithms will be deterministic as well. The exceptions to this are our directed unweighted RPaths algorithm and approximate weighted directed RPaths, both of which inherently use random sampling.

2   Directed Weighted Replacement Paths

In this section, we prove near-linear unconditional CONGEST lower bounds for directed weighted RPaths and 2-SiSP (Sect. 2.1). We complement this result with an algorithm that runs in a near-linear number of rounds (Sect. 2.2).

2.1   Directed Weighted RPaths Lower Bound

We prove Theorem 1.A by showing an unconditional $\Omega\left(\frac{n}{\log n}\right)$ lower bound for computing 2-SiSP (which extends to RPaths) in directed weighted graphs. Our proof is based on a new reduction from set disjointness pictured in Fig. 1, partly inspired by a sparse sequential reduction from MWC to RPaths in [3].

Consider an instance of the Set Disjointness problem where the players Alice and Bob are given k^2-bit strings S_a and S_b respectively representing sets of at most k^2 elements (the i'th bit being 1 indicates element i belongs to the set). The problem is to determine whether $S_a \cap S_b = \phi$, i.e., that for all indices $1 \le i \le k^2$ either $S_a[i] = 0$ or $S_b[i] = 0$. A classical result in communication complexity

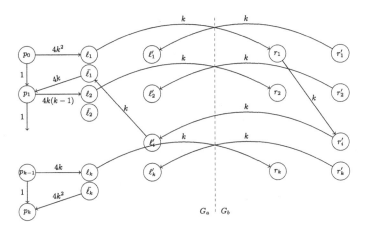

Fig. 1. Directed weighted RPaths, 2-SiSP lower bound construction

is that Alice and Bob must exchange $\Omega(k)$ bits even if they are allowed shared randomness [6,19,29]. Our reduction constructs the graph $G = (V, E)$ described below and we show in Lemma 1 that G has a low-weight 2-SiSP if and only if the sets S_a and S_b are not disjoint.

We will construct $G = (V, E)$ with six sets of vertices (see Fig. 1): $L = \{\ell_i \mid 1 \le i \le k\}, L' = \{\ell'_i \mid 1 \le i \le k\}, R = \{r_i \mid 1 \le i \le k\}, R' = \{r'_i \mid 1 \le i \le k\}, \overline{L} = \{\overline{\ell}_i \mid 1 \le i \le k\}, P = \{p_i \mid 0 \le i \le k\}$. Note that the number of vertices is $n = 6k + 1$. We set $s = p_0$ and $t = p_k$ and for each $1 \le i \le k$, we add the edges (p_{i-1}, p_i) with weight 1; this is the input shortest path $\mathcal{P} = P_{st}$. We add directed edges (ℓ_i, r_i) and (r'_i, ℓ'_i) for each $1 \le i \le k$. Each of these edges has weight k. For each $1 \le i \le k$, we add the edges (p_{i-1}, ℓ_i) with weight $4k(k - i + 1)$ and $(\overline{\ell}_i, p_i)$ with weight $4ki$. This is our base graph.

We now add edges to G based on the set disjointness inputs S_a, S_b. We encode each integer q, $1 \le q \le k^2$, as an ordered pair (i, j) such that $q = (i - 1) \cdot k + j$. If $S_a[q] = 1$, we add the edge $(\ell'_j, \overline{\ell}_i)$ with weight k, if $S_b[q] = 1$, we add the edge (r_i, r'_j) with weight k. For the 2-SiSP problem, the desired output is $d_2(p_0, p_k)$, the weight of a second simple shortest path from p_0 to p_k.

Lemma 1. *If $S_a \cap S_b \ne \emptyset$, then $d_2(p_0, p_k) \le (4k^2 + 9k - 1)$. Otherwise, if $S_a \cap S_b = \emptyset$, then $d_2(p_0, p_k) \ge (4k^2 + 12k)$.*

Proof. If the sets S_a, S_b are not disjoint, then there exists $1 \le i, j \le k$ such that $S_a[(i - 1) \cdot k + j] = S_b[(i - 1) \cdot k + j] = 1$. Then, the path $\langle p_{i-1}, \ell_i, r_i, r'_j, \ell'_i, \overline{\ell}_i, p_i \rangle$ provides a detour for the edge (p_{i-1}, p_i). This can be used along with the shortest paths from p_0 to p_{i-1} and p_i to p_k to obtain a simple path of weight $4k(k+1) + 4k + k - 1$ that does not use edge (p_{i-1}, p_i). So, the second simple shortest path from p_0 to p_k has weight at most $4k^2 + 9k - 1$.

Assume the strings are disjoint. Let $\mathcal{P}_2$ be a second simple shortest path, and let (p_{i-1}, p_i) be the first edge that is not in $\mathcal{P}_2$ but is in the p_0-p_k shortest path $\mathcal{P}$. Since the only other outgoing edge from p_{i-1} is (p_{i-1}, ℓ_i) (with weight

$4k(k - i + 1))$, this edge must be on $\mathcal{P}_2$. Let p_j $(j \geq i)$ be the next vertex from $\mathcal{P}$ that is also on path $\mathcal{P}_2$, such a vertex must exist as p_k is on $\mathcal{P}$ and $\mathcal{P}_2$. By the construction of G, edge $(\bar{\ell}_j, p_j)$ (with weight $4kj$) must be in $\mathcal{P}_2$ which means the path $\mathcal{P}_2$ has weight at least $4k(k - i + 1) + 4kj$ not including edges in the path from ℓ_i to $\bar{\ell}_j$. We also observe that any path from ℓ_i to $\bar{\ell}_j$ requires at least 4 edges, with total weight $4k$. If we have $j > i$, we immediately conclude that $\mathcal{P}_2$ has weight at least $4k(k - i + 1 + j) + 4k \geq 4k(k + 1) + 8k$. If $j = i$, then $\mathcal{P}_2$ contains a path from ℓ_i to $\bar{\ell}_i$. This path can have length 4 if and only if the edges (r_i, r'_j) and $(\ell'_j, \bar{\ell}_i)$ simultaneously exist for some j, which means $S_a[(i - 1) \cdot k + j] = S_b[(i - 1) \cdot k + j] = 1$. This contradicts the assumption that strings S_a and S_b are disjoint. So, this ℓ_i to $\bar{\ell}_i$ path has length at least 8, which means $\mathcal{P}_2$ has weight at least $4k(k + 1) + 8k = 4k^2 + 12k$.   □

To complete the reduction from set disjointness, assume that there is a CONGEST algorithm $\mathcal{A}$ that takes $R(n)$ rounds to compute the weight of a 2-SiSP path in a directed weighted graph on n vertices. Consider the vertex partition V_a, V_b of V with $V_a = L \cup L' \cup \bar{L} \cup P$ and $V_b = R \cup R'$, and let $G_a(V_a, E_a), G_b(V_b, E_b)$ be the subgraphs of G induced by the vertex sets V_a, V_b respectively. Note that G_a is completely determined by S_a and G_b is completely determined by S_b. Alice and Bob will communicate to simulate $\mathcal{A}$ on G. Alice will simulate the computation done in nodes in V_a, and Bob will simulate the computation done in nodes in V_b. If the algorithm communicates from a node in V_a to a node in V_b, Alice sends all the information communicated along this edge to Bob. Since there are $2k$ cut edges, and $\mathcal{A}$ can send $O(\log n)$ bits through each edge per round, Alice and Bob communicate up to $O(2k \cdot \log n)$ bits per round, for a total of $O(2k \cdot \log n \cdot R(n))$ bits. After the simulation, Alice knows $d_2(p_0, p_k)$ and can determine if the sets are disjoint by checking if $d_2(p_0, p_k) > 4k^2 + 9k - 1$ (Lemma 1). Since any communication protocol for set disjointness must use at least $\Omega(k^2)$ bits and $n = \Theta(k)$, $R(n)$ is $\Omega\left(\frac{n}{\log n}\right)$.

Our lower bound also applies to the RPaths problem, since given an algorithm $\mathcal{A}$ that computes replacement path for each edge, Alice can compute the minimum of those to get the second simple shortest path weight and then use Lemma 1 as before. This lower bound applies even for graphs with constant undirected diameter: we can add a 'sink' vertex with incoming edges from all vertices in G, so that Lemma 1 still holds and the undirected diameter is 2.

2.2   Directed Weighted RPaths Algorithm

In this section we present an $\tilde{O}(n)$ round CONGEST algorithm for computing RPaths and 2-SiSP in directed weighted graphs, which is nearly optimal given the near linear lower bound in Sect. 2.1. Our main tool is a reduction from RPaths to weighted APSP that can be simulated efficiently in the CONGEST network. This reduction is inspired by a sequential fine-grained reduction from RPaths to Eccentricities in [3], though some care is needed to ensure that the reduction can be efficiently mapped to the underlying CONGEST network. We present our algorithm to compute replacement path weights, and the second simple shortest

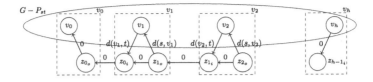

Fig. 2. Directed weighted RPaths reduction to APSP. $v_0 = s$ and $v_h = t$.

path weight $d_2(s,t)$ can be computed by taking the minimum weight replacement path among those computed, with an additional $O(D)$ rounds.

Our algorithm constructs a graph G' pictured in Fig. 2, and runs a weighted APSP algorithm on G'. We show later how communication in the newly constructed G' can be simulated efficiently in the underlying CONGEST network of G, so that the APSP algorithm can be applied to G' in $\tilde{O}(n)$ rounds. The algorithm uses the $\tilde{O}(n)$ round weighted APSP algorithm [7] as a subroutine and has $O(n)$ additive overhead, giving our $\tilde{O}(n)$ round bound.

We construct graph $G'(V', E')$ with $V' = V \cup Z_o \cup Z_i$, where $Z_o = \{z_{j_o} \mid 0 \le j < h\}$, $Z_i = \{z_{j_i} \mid 0 \le j < h\}$. We denote the nodes on the shortest path P_{st} by $s = v_0, v_1, \ldots v_h = t$. E' contains all edges in E with their original weights, except the edges from the given shortest path P_{st} which are removed. Additionally, E' contains directed edges $(z_{j_o}, v_j), (z_{j_i}, z_{j_o}), (v_{j+1}, z_{j_i})$ for $0 \le j < h$. Edge (z_{j_o}, v_j) has weight $d(s, v_j)$, edge (v_{j+1}, z_{j_i}) has weight $d(v_{j+1}, t)$ and edge (z_{j_i}, z_{j_o}) has weight 0 — recall that $d(s, v_j)$ denotes the shortest path distance from s to v_j in G. We use d' to denote shortest path distances in G'. The following lemma shows that we can compute replacement paths in G using distances in G'.

Lemma 2. *The shortest path distance $d'(z_{j_o}, z_{j_i})$ in G' (Figure 2) is equal to the replacement path weight $d(s, t, (v_j, v_{j+1}))$ in the original graph G.*

Proof. Let $\mathcal{P}$ be a replacement path for the edge (v_j, v_{j+1}) with weight $d(s, t, (v_j, v_{j+1}))$. We will construct a path from z_{j_o} to z_{j_i} that has the same weight as $\mathcal{P}$. Let v_a be the first vertex where $\mathcal{P}$ deviates from P_{st}, v_b be the first vertex after v_a where $\mathcal{P}$ rejoins P_{st}. Note that $a \le j$ and $b \ge j + 1$ as it is a replacement path for edge (v_j, v_{j+1}), and $\mathcal{P}$ contains a subpath $\mathcal{P}_{ab}$ from v_a to v_b that does not contain any edge from P_{st}. Construct the path $\langle z_{j_o}, \ldots z_{a_o}, v_a \rangle \circ \mathcal{P}_{ab} \circ \langle v_b, z_{b-1_i}, \ldots z_{j_i} \rangle$, which has weight $w(z_{a_o}, v_a) + w(\mathcal{P}_{ab}) + w(v_b, z_{b-1_i}) = d(s, a) + w(\mathcal{P}_{ab}) + d(b, t) = w(\mathcal{P})$. Thus, we have $d'(z_{j_o}, z_{j_i}) \le d(s, t, (v_j, v_{j+1}))$.

Now, consider any shortest path P from z_{j_o} to z_{j_i}. Observe that any such P must use a unique edge of the form (z_{a_o}, v_a) and a unique edge (z_b, z_{b-1_i}) in order to reach z_{j_i} from z_{j_o}, where $a \le j$ and $b \ge j + 1$. Denote the subpath of P from v_a to v_b as P_{ab}. Now, consider the path $\mathcal{P}$ in G obtained by concatenating s-v_a shortest path, P_{ab} and v_b-t shortest path — this is a replacement path for edge (v_j, v_{j+1}) since $a \le j$, $b \ge j + 1$ and P_{ab} does not contain any edge on P_{st}. The weight of $\mathcal{P}$ is equal to $d(s, v_a) + w(P_{ab}) + d(v_b, t)$ which is equal to $w(P)$

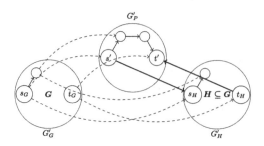

Fig. 3. Directed unweighted RPaths, 2-SiSP lower bound graph G'

since the only nonzero weight edges on P that are outside subpath P_{ab} are the ones with weight $d(s, v_a)$ and $d(v_b, t)$. Hence $d(s, t, (v_j, v_{j+1})) \leq d'(z_{j_o}, z_{j_i})$. $\quad\square$

To simulate an APSP algorithm on G' using the communication network G, we assign vertices v_i, z_{i-1_i}, z_{i_o} of G' to be simulated by CONGEST node v_i of G—this is represented by the dashed boxes in Fig. 2. This ensures that any edge of G' corresponds to either a communication link between nodes in the CONGEST network of G, or the edge is within the same node of G. We can compute the weights required to simulate G' after two SSSP computations with s and t as sources, and use an $\tilde{O}(n)$ algorithm to compute APSP in G' [7]. We show how to augment this algorithm to construct replacement paths using routing tables in the full version [22].

When the hop length of the s-t path h_{st} is small, the simple algorithm of performing h_{st} shortest path computations with each edge on the s-t path removed gives us an $O(h_{st} \cdot SSSP)$ round algorithm. We can obtain an improved round complexity if we only require a $(1 + \epsilon)$-approximation of the replacement path weight. We defer the presentation of this approximation algorithm for directed weighted RPaths to Sect. 3.3 since it uses techniques that build on the directed unweighted RPaths algorithm.

3   Directed Unweighted Replacement Paths

In Sect. 3.1 we show a lower bound of $\tilde{\Omega}(\sqrt{n} + D)$ for computing RPaths and 2-SiSP in directed unweighted graphs, which also proves a folklore lower bound for directed single source reachability. We present an algorithm for directed unweighted RPaths in Sect. 3.2, and extend it to $(1 + \epsilon)$-approximate directed weighted RPaths (Sect. 3.3).

3.1   Directed Unweighted RPaths Lower Bound

Our lower bound method uses a reduction from the undirected *s-t subgraph connectivity* problem defined in [32] as follows: Given an undirected CONGEST network G with n vertices, a subgraph H of G, and two vertices s, t, determine whether s and t are in the same connected component of H. The input subgraph

H is given by letting each vertex know which of its incident edges are in H. It is shown in [32] that this problem has a lower bound of $\Omega\left(\frac{\sqrt{n}}{\log n} + D\right)$ in graphs with D as small as $\Theta(\log n)$. We assume WLOG that network G is connected.

Proof (Proof of Theorem 3.A and Theorem 1.D). Given an instance of s-t subgraph connectivity with undirected network G, vertices s and t and subgraph H, our first attempt is to construct a directed unweighted graph G' with two copies of $V(G)$: G'_H contains only the edges in H, with bidirectional edges, and G'_P contains only a directed shortest path from s' to t' made of edges in G where s', t' are copies of s, t. These copies are connected with directed edges (s', s_H) and (t_H, t') (Fig. 3 without copy G'_G).

This construction has the property that there is a second directed path from s' to t' in G' (apart from the one in G'_P) if and only if there is an s_H-t_H path in G_H. So 2-SiSP weight in G is finite iff s,t are connected in H. But, this construction could have high undirected diameter as we have no control over the diameter of H, and fails to give a meaningful lower bound.

To obtain small undirected diameter, we add a third copy of G, denoted G'_G, which has all edges of G as bidirectional edges, pictured in Fig. 3. This copy is connected to the others with directed edges (v_G, v_H) and (v_G, v') where v_G is the copy in G'_G of $v \in G$. The undirected diameter of G' is now $(D + 2)$ (D is the diameter of G) as we can connect any pair of vertices using a bidirectional path in G'_G along with at most 2 connecting edges. This addition does not add any new directed paths from s' to t'.

Any communication in G' can be simulated in a constant number of rounds in the underlying network of G, as each node v in the network can simulate vertices v_G, v_H, v' of G', and all edges in G' are either within the same node or have an underlying undirected edge of G. Constructing G' requires only an $O(D)$-round computation of undirected shortest path from s to t in G.

This completes the reduction and establishes a lower bound of $\Omega\left(\frac{\sqrt{n}}{\log n} + D\right)$ for 2-SiSP (and RPaths) in unweighted directed graphs by additionally noting that $\Omega(D)$ rounds are necessary, as with other global problems in the distributed model [32], for information to travel to the farthest vertices to determine 2-SiSP. Our lower bound also applies to any α-approximation algorithm ($\alpha > 1$) since we distinguish between 2-SiSP of length $\leq n + 2$ and infinite length.   □

Other Directed Unweighted Graph Problems. Our lower bound for directed unweighted RPaths can be adapted to give the same lower bound for other graph problems on directed unweighted graphs, including the basic problems of s-t directed reachability and s-t directed shortest path. These folklore lower bounds [16] have been attributed to [32] which only deals with undirected graphs, and we make these results explicit.

Lemma 3. *Any algorithm computing s-t directed reachability or s-t directed shortest path in a directed unweighted graph requires $\Omega\left(\frac{\sqrt{n}}{\log n} + D\right)$ rounds, even if the graph has undirected diameter as low as $\Theta(\log n)$.*

Proof. We use a simpler version of the construction in Fig. 3 by removing G'_P from the graph G' to form a graph G''. A directed path from s_H to t_H exists in G'' if and only if s and t are connected in the subgraph H. Using the same arguments as the RPaths lower bound, we note that G'' has undirected diameter $O(D)$ when the network G has undirected diameter D and G'' can be efficiently simulated on the original network G. So we get the desired reduction for both problems from s-t subgraph connectivity [32].   □

3.2   Directed Unweighted RPaths Algorithm

In the sequential setting, there are two approaches to compute replacement paths in directed graphs: (1) remove each edge in the input path P_{st} and compute shortest paths in the resulting graphs separately, using h_{st} shortest path computations [34], (2) compute shortest detour distances in order to compute replacement paths: A *detour* from a to b, where a, b are vertices on P_{st}, is a simple path from a to b with no edge in common with P_{st}. Any replacement path for edge $e \in P_{st}$ can be characterized as the concatenation of an initial s-a subpath of P_{st}, a detour from a to b, and a final b-t subpath of P_{st}, where a, b are vertices in P_{st} such that e is contained in the a-b subpath of P_{st} [30, 34].

Our distributed algorithm uses both these approaches for different ranges of h_{st}, D (as in line 4 of Algorithm 1). In the first method, used in Case 1 of Algorithm 1, we compute replacement paths in $O(h_{st} \cdot SSSP)$ rounds using the obvious algorithm of removing one of the h_{st} edges on the input shortest path P_{st} and computing SSSP from s. We use a directed weighted SSSP algorithm with the weight of the removed edge set to ∞. We do not use unweighted directed BFS since an s-t shortest path could have up to $n - 1$ hops after edge removal.

In the second method, used in Case 2 of Algorithm 1, we present a distributed detour-based algorithm that runs in $\tilde{O}(n^{2/3} + \sqrt{nh_{st}} + D)$ rounds.

To compute short detours (hop length $\leq h$, parameter h determined in line 4), our distributed algorithm exploits pipelining to compute h-hop limited BFS from each vertex on P_{st} in $O(h_{st} + h)$ rounds [17, 21]. We compute these distances in the graph $G - P_{st}$, which is the graph G with edges on P_{st} removed. We denote shortest path distances in graph $G - P_{st}$ by $d^-(u, v)$.

For long detours (hop length $> h$), we sample $\Theta(p)$ vertices (p determined in line 4) in line 6 and compute a 'skeleton graph' on the set of sampled vertices: for $u, v \in S$, we add a directed edge (u, v) to the skeleton graph with weight $d^-(u, v)$ if there is an h-hop directed shortest path from u to v in $G - P_{st}$. The edges of the skeleton graph are computed using an h-hop BFS in line 9. The h-hop distances between all pairs of sampled vertices, and between sampled vertices and vertices on P_{st} are broadcast to all vertices in line 10. Algorithm 2, described below, is run at each vertex $a \in P_{st}$. It uses the h-hop distances (broadcast in line 10 of Algorithm 1) to *locally* compute at vertex a all detours starting from a. It then computes at a the best candidate replacement path among paths first deviating from P_{st} at a for each edge $e \in P_{st}$ that occurs after a on P_{st}, denoted $d^a(s, t, e)$. Finally, Algorithm 1 performs a pipelined minimum operation along P_{st} in line 15 to compute shortest replacement path distances for all h_{st} edges among candidate replacement paths computed by Algorithm 2 at each $a \in P_{st}$.

Algorithm 1 Directed Unweighted RPaths Algorithm

Input: Graph $G = (V, E)$, vertices $s, t \in V$, s-t shortest path P_{st}.

Output: Replacement path distance $d(s, t, e)$ known at s for each $e \in P_{st}$.

1: **Case 1.** $D \leq n^{1/4}, h_{st} \leq n^{1/6}$ or $n^{1/4} < D \leq n^{2/3}, h_{st} \leq n^{1/3}$

2: Perform h_{st} directed weighted SSSP computations in sequence, with each edge $e \in P_{st}$ having its weight set to ∞ to compute $d(s, t, e)$.

3: **Case 2.** $D \leq n^{1/4}, h_{st} > n^{1/6}$ or $n^{1/4} < D \leq n^{2/3}, h_{st} > n^{1/3}$ or $D > n^{2/3}$

4: Fix parameter $p = n^{1/3}$ if $h_{st} < n^{1/3}$ and $p = \sqrt{n/h_{st}}$ if $h_{st} \geq n^{1/3}$, and fix $h = n/p$.

5: Let graph $G - P_{st}$ be G but with all edges in P_{st} removed.

6: Sample each vertex $v \in G$ with probability $\Theta\left(\frac{\log n}{h}\right)$, let the set of sampled vertices be S.

7: **for each** vertex $v \in P_{st} \cup S$ **do**

8: ▷ *Comment: Line 9 computes h-hop $S \times V(P_{st})$ distances and h-hop $S \times S$ distances (edges of the skeleton graph on S)* ◁

9: Perform BFS starting from v on $G - P_{st}$, and the reversed graph, up to h hops to compute unweighted shortest paths: for $u \in P_{st} \cup S$, both v and u know the h-hop limited distance $d^-(v, u)$. Since we have $(|S| + h_{st})$ sources, this takes $O(|S| + h_{st} + h)$ rounds.

10: Broadcast $\{d^-(v, u) \mid u \in S \text{ or } v \in S\}$. At most $(|S|^2 + h_{st}|S|)$ values are broadcast, taking $O(|S|^2 + h_{st}|S| + D)$ rounds.

11: **for each** vertex $a \in P_{st}$ **do**

12: ▷ *Comment: Internally compute replacement paths $d^a(s, t, e)$ that deviate from P_{st} at a, using distances broadcast in line 10* ◁

13: ComputeLocalRPaths(a)   ▷ *Algorithm 2*

14: **for** edge $e \in P_{st}$ **do**

15: Compute $d(s, t, e) \leftarrow \min_a d^a(s, t, e)$ by propagating values from $a \in P_{st}$ up the path P_{st}. The minimum for a single e over all $a \in P_{st}$ takes $O(h_{st})$ rounds, and the computation for all $e \in P_{st}$ can be pipelined in $O(h_{st})$ rounds.

Computation in Algorithm 2 (local computation at each $a \in P_{st}$). Algorithm 2 at vertex $a \in P_{st}$ takes as input the h-hop distances to and from a, computed in line 9 of Algorithm 1, and the h-hop distances broadcast in line 10 of Algorithm 1. In Algorithm 2, all pairs shortest path distances $d^-(u, v)$ for $u, v \in S$ in the skeleton graph are locally computed in line 3 using the h-hop skeleton graph edge distances. These distances, along with h-hop distances $d^-(u, b)$ for $u \in S, b \in P_{st}$, are used to compute long detours in line 5. Short detours are computed at a using the h-hop distance $d^-(a, b)$ to each vertex $b \in P_{st}$. With the best detour distances $\delta(a, b)$ computed in line 5, a locally computes replacement paths using detours starting from a in line 7, which gives the best candidate replacement path distance $d^a(s, t, e)$ among paths that first deviate from P_{st} at a, for each edge e after a on P_{st}.

Lemma 4. *The local computation in Algorithm 2 at $a \in P_{st}$ correctly computes $d^a(s, t, e)$, the minimum weight replacement path for $e \in P_{st}$ among paths that first deviate from P_{st} at a.*

Algorithm 2 Local computation at $a \in P_{st}$ of candidate replacement paths deviating from vertex a

---

Input: Graph $G = (V, E)$, shortest path P_{st}, subset $S \subseteq V$. The following distances in graph $G - P_{st}$ are known to a: h-hop distances $d^-(b, u)$, $d^-(u, b)$ for any $b \in P_{st}$, $u \in S$, $d^-(u, v)$, for $u, v \in S$, and $d^-(a, b)$ for $b \in P_{st}$.

Output: Vertex a computes for each $e \in P_{st}$ after a on P_{st}, the shortest replacement path distance $d^a(s, t, e)$ among paths that first deviate from P_{st} at a.

1: **procedure** COMPUTELOCALRPATHS(a)
2:     ▷ *Comment: All computation is done internally using distances known to a.* ◁
3:     Locally compute all pairs distances in skeleton graph on S: compute $d^-(y, z)$ for each $y, z \in S$ using the skeleton graph h-hop edge distances.
4:     **for each** vertex $b \in P_{st}$ after a along P_{st} **do**
5:         Compute the best (short or long) detour $\delta(a, b)$ from a to b as
            $\delta(a, b) = \min \left(d^-(a, b), \min_{u, v \in S} \left(d^-(a, u) + d^-(u, v) + d^-(v, b) \right) \right)$
6:     **for** edge $e = (x, y) \in P_{st}$ such that a that appears before x on P_{st} or $a = x$ **do**
7:         $d^a(s, t, e) = \min_{b \in P_{st}} \left(d(s, a) + \delta(a, b) + d(b, t) \right)$ (the minimum is over vertices b that appear after y on P_{st} or $b = y$)

Proof. We assume that a knows the correct h-hop distances specified as input to Algorithm 2. Note that the vertices and distances along P_{st} are known to a as part of RPaths input.

Due to our sampling probability, any shortest path between $u, v \in S$ can be decomposed into h-hop subpaths between sampled vertices w.h.p. in n. So in line 3, vertex a correctly computes all pairs shortest path distances between sampled vertices in S using h-hop skeleton graph distances.

Now we show that line 5 computes a shortest detour P_{ab}^d from a to each $b \in P_{st}$ that occurs after a on P_{st}, whose distance is denoted $\delta(a, b)$. If P_{ab}^d is a short detour, with hop length $\leq h$, its distance is equal to the h-hop distance $d^-(a, b)$ which is part of the input to a.

If P_{ab}^d is a long detour, with hop length $> h$, we use the fact that due to our sampling probability, any path of h hops contains a sampled vertex in S w.h.p. in n. We can find a sampled vertex u on the detour within h hops from a and a sampled vertex v on the detour within h hops from b. We will assume WLOG that v occurs after u or $u = v$. Then, the detour distance is $\delta(a, b) = d^-(a, u) + d^-(u, v) + d^-(v, b)$, and line 5 correctly computes this distance.

In any replacement path for edge $e \in P_{st}$ first deviating from P_{st} at a, there is a vertex $b \in P_{st}$ where it rejoins P_{st}. We can characterize such a replacement path as the concatenation of the s-a subpath of P_{st}, a detour P_{ab}^d from a to b, and the b-t subpath of P_{st}. This path has weight $d(s, a) + \delta(a, b) + d(b, t)$. We then compute the minimum over all valid detour endpoints b in line 7. This correctly computes $d^a(s, t, e)$ for edges $e = (x, y)$ that are on the a-b subpath of P_{st} □

Lemma 5. *Algorithm 1 computes replacement path weights in a directed unweighted graph in $\tilde{O}(\min(n^{2/3} + \sqrt{nh_{st}} + D, h_{st} \cdot SSSP))$ rounds.*

Proof. We focus on the analysis of Case 2 since Case 1 is straightforward.

Correctness: The inputs used by Algorithm 2 at vertex $a \in P_{st}$ are correctly computed in Algorithm 1: the h-hop distances from a to other vertices $b \in P_{st}$ and h-hop distances from sampled vertices are computed in line 9. After Algorithm 2 correctly computes $d^a(s,t,e)$, line 15 computes $d(s,t,e) = \min_{a \in P_{st}} d^a(s,t,e)$ for each edge e as the minimum distance among all valid replacement paths which may deviate at any $a \in P_{st}$.

Round Complexity: Recall that local computation does not contribute to the cost of an algorithm in the CONGEST model. So we can ignore Algorithm 2 for the round complexity analysis. In line 9 of Algorithm 1, we use the k-source h-hop BFS algorithm for directed graphs which runs in $O(k + h)$ rounds using pipelining [17,21]. As we have $k = p + h_{st}$ sources and h hops, this takes $O(p + h_{st} + h)$ rounds. We use the standard broadcast operation in line 10 which broadcasts $(p + h_{st}) \cdot p$ values in $O(p^2 + p \cdot h_{st} + D)$ rounds [27]. The global minimum in line 15 involves h_{st} convergecast operations which can be pipelined to take $O(h_{st} + D)$ rounds. The total round complexity is $O(p^2 + p \cdot h_{st} + h + D)$.

Setting parameters $h = n^{2/3}, p = n^{1/3}$ gives us a round complexity of $\tilde{O}(n^{2/3} + n^{1/3} h_{st} + D)$. When $h_{st} \geq n^{1/3}$, the parameters $h = \sqrt{nh_{st}}, p = \sqrt{n/h_{st}}$ are more favorable, giving a round complexity of $\tilde{O}(\sqrt{nh_{st}} + n/h_{st} + D) = \tilde{O}(\sqrt{nh_{st}} + D)$ (since $h_{st} \geq n^{1/3}$). The input parameter h_{st} can be shared to all nodes with a broadcast, so all vertices can choose the setting of h, p appropriately. Thus, Case 2 takes $\tilde{O}(n^{2/3} + \sqrt{nh_{st}} + D)$ rounds.   □

We augment Algorithm 1, which computes only weights, to also construct replacement paths using routing tables in the full version [22].

3.3   Approximate Directed Weighted RPaths Algorithm

We present a $(1+\epsilon)$-approximation algorithm for directed weighted RPaths that runs in $\tilde{O}\left(\sqrt{nh_{st}} + D + \min\left(n^{2/3}, h_{st}^{2/5} n^{2/5+o(1)} D^{2/5}\right)\right)$ rounds.

Proof (Proof of Theorem 1.C). Our algorithm is based on the directed unweighted RPaths algorithm described earlier. The key tool is to replace h-hop BFS computation in line 9 of Algorithm 1 with $(1 + \epsilon)$-approximate h-hop limited shortest path computation, using an algorithm in ([24], Theorem 3.6), which gives us a $\tilde{O}(k + h)$-round algorithm for k sources.

With this change, approximate distances are computed in the skeleton graph in line 3 of Algorithm 2. Thus, the local detour distances (both short and long) are $(1 + \epsilon)$-approximate detour distances in line 5. The final replacement paths add these approximate detours to exact distances (line 15 of Algorithm 1) and are hence $(1 + \epsilon)$-approximate. Using the same analysis as Lemma 5, we get an algorithm with round complexity $\tilde{O}\left(n^{2/3} + \sqrt{nh_{st}} + D\right)$.

When h_{st} is small, we can improve the $h_{st} \cdot SSSP$ round algorithm used in the exact unweighted algorithm. A recent result in [23] shows that k-source approximate directed weighted SSSP can be performed in $\tilde{O}(\sqrt{nk} + D)$ rounds if $k \geq n^{1/3}$ and in $\tilde{O}(\sqrt{nk} + D + k^{2/5} n^{2/5+o(1)} D^{2/5})$ rounds if $k < n^{1/3}$. We compute all detours using an h_{st}-source SSSP computation by treating each $a \in P_{st}$ as a source and computing shortest path distances in $G - P_{st}$. This method is efficient when $h_{st} < n^{1/3}$. Combining the two methods proves our result.   □

4 Further Research

We have presented several nontrivial algorithms and lower bounds for RPaths and 2-SiSP in the CONGEST model, with many of our results being near-optimal. A key avenue for further research is to narrow or close the gap between the $\tilde{O}(n^{2/3} + \sqrt{nh_{st}} + D)$ upper bound and $\tilde{\Omega}(\sqrt{n} + D)$ lower bound for directed unweighted RPaths and approximate weighted directed RPaths. Another question is whether the dependence on h_{st} can be improved in our algorithms.

References

1. Abboud, A., Censor-Hillel, K., Khoury, S.: Near-linear lower bounds for distributed distance computations, even in sparse networks. In: Gavoille, C., Ilcinkas, D. (eds.) Proceedings of DISC 2016, pp. 29–42. Springer, Heidelberg (2016). https://doi.org/10.1007/978-3-662-53426-7_3
2. Abboud, A., Grandoni, F., Williams, V.V.: Subcubic equivalences between graph centrality problems, apsp and diameter. In: Proceedings of SODA 2015, pp. 1681–1697. SIAM (2015)
3. Agarwal, U., Ramachandran, V.: Fine-grained complexity for sparse graphs. In: Proceedings of STOC 2018, pp. 239–252 (2018)
4. Agarwal, U., Ramachandran, V.: Faster deterministic all pairs shortest paths in congest model. In: SPAA 2020, pp. 11–21. ACM (2020)
5. Ancona, B., Censor-Hillel, K., Dalirrooyfard, M., Efron, Y., Williams, V.V.: Distributed distance approximation. In: OPODIS 2020, vol. 184, pp. 30:1–30:17. Schloss Dagstuhl - Leibniz-Zentrum für Informatik (2020)
6. Bar-Yossef, Z., Jayram, T.S., Kumar, R., Sivakumar, D.: An information statistics approach to data stream and communication complexity. J. Comput. Syst. Sci. **68**(4), 702–732 (2004)
7. Bernstein, A., Nanongkai, D.: Distributed exact weighted all-pairs shortest paths in near-linear time. In: Proceedings of STOC 2019, pp. 334–342. ACM (2019)
8. Bodwin, G., Parter, M.: Restorable shortest path tiebreaking for edge-faulty graphs. J. ACM **70**(5), 1–24 (2023)
9. Cao, N., Fineman, J.T.: Parallel exact shortest paths in almost linear work and square root depth. In: Proceedings of SODA 2023, pp. 4354–4372. SIAM (2023)
10. Chechik, S., Cohen, S.: Near optimal algorithms for the single source replacement paths problem. In: Proceedings of SODA 2019, pp. 2090–2109. SIAM (2019)
11. Chechik, S., Magen, O.: Near optimal algorithm for the directed single source replacement paths problem. In: ICALP 2020, vol. 168, pp. 81:1–81:17 (2020)
12. Chechik, S., Mukhtar, D.: Single-source shortest paths in the CONGEST model with improved bounds. Distrib. Comput. **35**(4), 357–374 (2022)
13. Dinitz, M., Robelle, C.: Efficient and simple algorithms for fault-tolerant spanners. In: PODC 2020, pp. 493–500. ACM (2020)

14. Elkin, M.: An unconditional lower bound on the time-approximation trade-off for the distributed minimum spanning tree problem. SIAM J. Comput. **36**(2), 433–456 (2006)
15. Ghaffari, M., Parter, M.: Near-optimal distributed algorithms for fault-tolerant tree structures. In: Proceedings of SPAA 2016, pp. 387–396. ACM (2016)
16. Ghaffari, M., Udwani, R.: Brief announcement: distributed single-source reachability. In: Proceedings of PODC 2015, pp. 163–165. ACM (2015)
17. Hoang, L., et al.: A round-efficient distributed betweenness centrality algorithm. In: Proceedings of PPoPP 2019, pp. 272–286. ACM (2019)
18. Katoh, N., Ibaraki, T., Mine, H.: An efficient algorithm for k shortest simple paths. Networks **12**(4), 411–427 (1982)
19. Kushilevitz, E., Nisan, N.: Communication Complexity. Cambridge University Press, Cambridge (1996)
20. Lawler, E.L.: A procedure for computing the k best solutions to discrete optimization problems and its application to the shortest path problem. Manag. Sci. **18**(7), 401–405 (1972)
21. Lenzen, C., Patt-Shamir, B., Peleg, D.: Distributed distance computation and routing with small messages. Distrib. Comput. **32**(2), 133–157 (2019)
22. Manoharan, V., Ramachandran, V.: Near optimal bounds for replacement paths and related problems in the congest model. arXiv preprint arXiv:2205.14797 (2022)
23. Manoharan, V., Ramachandran, V.: Improved approximation bounds for minimum weight cycle in the congest model. In: Proceedings of PODC 2024, ACM (2024, to appear)
24. Nanongkai, D.: Distributed approximation algorithms for weighted shortest paths. In: Proceedings of STOC 2014, pp. 565–573. ACM (2014)
25. Parter, M.: Distributed constructions of dual-failure fault-tolerant distance preservers. In: DISC 2020. Schloss Dagstuhl-Leibniz-Zentrum für Informatik (2020)
26. Parter, M.: Nearly optimal vertex fault-tolerant spanners in optimal time: sequential, distributed, and parallel. In: Proceedings of STOC 2022, pp. 1080–1092. ACM (2022)
27. Peleg, D.: Distributed Computing: A Locality-Sensitive Approach. SIAM (2000)
28. Peleg, D., Rubinovich, V.: A near-tight lower bound on the time complexity of distributed minimum-weight spanning tree construction. SIAM J. Comput. **30**(5), 1427–1442 (2000)
29. Razborov, A.A.: On the distributional complexity of disjointness. Theor. Comput. Sci. **106**(2), 385–390 (1992)
30. Roditty, L., Zwick, U.: Replacement paths and k simple shortest paths in unweighted directed graphs. ACM Trans. Algor. (TALG) **8**(4), 1–11 (2012)
31. Rozhoň, V., Haeupler, B., Martinsson, A., Grunau, C., Zuzic, G.: Parallel breadth-first search and exact shortest paths and stronger notions for approximate distances. In: Proceedings of STOC 2023, pp. 321-334. ACM (2023)
32. Sarma, A.D., et al.: Distributed verification and hardness of distributed approximation. SIAM J. Comput. **41**(5), 1235–1265 (2012)
33. Williams, V.V., Williams, R.R.: Subcubic equivalences between path, matrix, and triangle problems. J. ACM **65**(5), 27:1–27:38 (2018)
34. Yen, J.Y.: Finding the k shortest loopless paths in a network. Manag. Sci. **17**(11), 712–716 (1971)

Reaching Agreement Among k out of n Processes

Gadi Taubenfeld$^{(\boxtimes)}$ (iD)

Reichman University, 46150 Herzliya, Israel
tgadi@runi.ac.il
https://faculty.runi.ac.il/gadi

Abstract. In agreement problems, each process has an input value and must choose a decision (output) value. Given $n \geq 2$ processes and $m \geq 2$ possible different input values, we want to design an agreement algorithm that enables as many processes as possible to decide on the (same) input value of one of the processes in the presence of t crash failures. Without communication, when each process simply decides on its input value, at least $\lceil (n - t)/m \rceil$ of the processes are guaranteed to always decide on the same value. Can we do better with communication? For some cases, for example, when $m = 2$, even in the presence of a single crash failure, the answer is negative in a deterministic asynchronous system where communication is either by using atomic read/write registers or by sending and receiving messages. The answer is positive in other cases.

Keywords: Partial agreement · Shared Memory · Message passing

1 Introduction

The problem of reaching agreement is a fundamental coordination problem and is at the core of many algorithms for fault-tolerant distributed applications. The problem is to design an algorithm in which *all* the participants reach a common decision based on their initial opinions. This problem is a special case of the (n, k)-partial agreement problem, introduced and defined below, in which it is required that at least k of the n participants reach a common decision. When the exact values of n and k are not important, we will refer to this problem as the *partial agreement* problem.

The relation of the notion of partial agreement with the interesting notions of X agreement [9], almost everywhere agreement (in which all but a small number of correct participants must choose a common decision value) [9], almost-t-resilient agreement [19], and bounded disagreement [5], is discussed in details in the related work section.

1.1 The (n, k)-Partial Agreement Problem

The *t-resilient (n, k)-partial agreement* problem is to design an algorithm for n processes that supports a single operation called propose() and can tolerate

© The Author(s), under exclusive license to Springer Nature Switzerland AG 2024
Y. Emek (Ed.): SIROCCO 2024, LNCS 14662, pp. 438–455, 2024.
https://doi.org/10.1007/978-3-031-60603-8_24

t crash failures. The operation takes an input parameter, called the *proposed* value, and returns a result, called the *decided* value. It is assumed that each of the n processes invokes the propose operation at most once. The problem requirements are that there exists a *decision value v* such that:

- *Agreement*: At most $n - k$ processes may decide on values other than v. Thus, when all the n processes decide, at least k of them decide on (the same value) v.
- *Weak validity*: v is the input (proposed) value of at least one of the processes.
- *t-resiliency*: Each process that does not crash, invokes the propose operation, and eventually decides and terminates, as long as no more than t processes crash.

We notice that the agreement requirement means that, in every execution, there must exist a value v such that the number of processes that have decided on v plus the number of processes that haven't decided (possibly crashed) is at least k. The (n, n)-partial agreement problem is the familiar consensus (i.e., full agreement) problem for n processes [15], in which all the non-faulty processes must eventually decide on the same value, which must be a proposed value. When there are only two (resp. more than two) possible input values, the problem is called the partial binary (resp. partial multi-valued) agreement problem. Weak validity only requires that v be a proposed value. A stronger requirement is,

- *Strong validity*: Every decided value must be a proposed value.

The necessary conditions proved in Sect. 3 hold only for strong validity. All the other results hold for both the weak and strong validity requirements.

1.2 The (n, k, l)-Partial Set Agreement Problem

The *t-resilient (n, k, ℓ)-partial set agreement* problem captures a weaker form of the (n, k)-partial agreement problem in which the agreement property is weakened. The problem is to design an algorithm for n processes that supports a single operation called propose() and can tolerate t crash failures. The operation takes an input parameter, called the *proposed* value, and returns a result, called the *decided* value. It is assumed that each of the n processes invokes the propose operation at most once. The requirements of the problem are that there exists a set of decision values V of size at most ℓ such that:

- *Agreement*: At most $n - k$ processes may decide on values not in V. Thus, when all the n processes decide, at least k of them decide values in V.
- *Weak validity*: Each $v \in V$ is the input (proposed) value of at least one of the processes.
- *t-resiliency*: Each process that does not crash, invokes the propose operation, and eventually decides and terminates, as long as no more than t processes crash.

As before, a stronger validity requirement, called *strong validity*, is that every decided value must be a proposed value. The $(n, n, 1)$-partial set agreement problem is the familiar consensus problem. The $(n, k, 1)$-partial set agreement problem is the (n, k)-partial agreement. The (n, n, ℓ)-partial set agreement problem, with strong validity, is the familiar ℓ-set agreement problem for n processes, which is to find a solution where each process starts with an input value from some domain and must choose some process' input as its output, and all the n processes together may choose no more than ℓ distinct output values [6].

1.3 Motivation

The first and foremost motivation for this study is related to the basics of computing, namely, increasing our knowledge of what can (or cannot) be done in the context of failure-prone distributed systems. Providing necessary and sufficient conditions for the solvability of the partial agreement problem helps us determine the limits of synchronization algorithms and identify to what extent communication is helpful when solving weak variants of the fundamental *full* agreement problem. Furthermore, as was pointed out in [9], in many practical situations, we may be willing to settle for cooperation between the vast majority of the processes, which raises the question of when this is possible.

Another inspiration for this work is related to biology. The partial agreement problem arises in biological systems where there is a predefined threshold, and it is only required that the number of participants that reach agreement exceeds the threshold for a specific action to take place. A well-known and extensively studied example is *quorum sensing*. Many species of bacteria use quorum sensing to coordinate gene expression according to the density of their local population. Quorum sensing is triggered to begin when the number of bacteria, that sense that a sufficient number of bacteria are present, reaches a certain threshold. Quorum sensing allows bacteria to synchronize and, by doing so, enables them to successfully infect and cause disease in plants, animals, and humans [3,11]. Several groups of insects, like ants and honey bees, have been shown to use quorum sensing in a process that resembles collective decision-making [16,18].

1.4 Models of Computation

Our model of computation consists of a collection of n deterministic processes. Each process has a unique identifier. We denote by t the maximum number of processes that may fail. The only type of failure considered in this paper is a process *crash* failure. A crash is a premature halt. Thus, until a process possibly crashes, it behaves correctly by reliably executing its code. We consider the following shared memory (SM) and message passing (MP) models.

1. *The asynchronous RW model.* There is no assumption on the relative speeds of the processes. Processes communicate by atomically reading and writing shared registers. At each atomic step, a process can read or write, but not both. A register that can be written and read by any process is a multi-writer multi-reader (MWMR) register. If a register can be written by a single (predefined) process and read by all, it is a single-writer multi-reader (SWMR) register.
2. *The asynchronous MP model.* There is no assumption on the relative speeds of the processes. Processes communicate by sending and receiving messages. There is no assumption on the speed of the messages.

3. *The synchronous MP model.* It is assumed that the processes communicate in "rounds" of communications. At the beginning of a round, each process may send messages to other processes, and all the messages sent during a round arrive at their destinations by the end of this round. Processes start each round at the same time. That is, a process may start participating in a round only when all other processes have finished participating in the previous round.

4. *The asynchronous SM(g) model.* There is no assumption on the relative speeds of the processes. Processes communicate by (1) atomically reading and writing shared registers (at each atomic step, a process can read or write, but not both), and (2) using full agreement objects for g processes that can tolerate any number of failures – also called wait-free full agreement objects for g processes.

1.5 Known Results

We will use the following known results:

1. There is no solution for the (n, n)-partial agreement problem (i.e., consensus problem) for $n \geq 2$ processes and $m \geq 2$ input (proposed) values that can tolerate a single crash failure in an asynchronous system where communication is done either by sending messages or by reading and writing atomic registers [10, 14].

2. For any $\ell \geq 1$, there is no solution for the (n, n, ℓ)-partial set agreement problem, assuming strong validity (i.e., ℓ-set agreement problem), for $n \geq \ell + 1$ processes and $m \geq \ell + 1$ input values that can tolerate ℓ crash failures in an asynchronous system where communication is done either by sending messages or by reading and writing atomic registers [4, 13, 17].

3. In a synchronous message-passing system in which up to $1 \leq t \leq n - 2$ processes may crash, every full agreement algorithm requires at least $t + 1$ rounds [1, 8], and there exists a full agreement algorithm with $t + 1$ rounds [2]. When $t = n - 1$, t rounds are necessary and sufficient.

4. There is no solution for the (n, n)-partial agreement problem for $n \geq t + 1$ processes that can tolerate t crash failures in an asynchronous system using atomic registers and wait-free full agreement objects for t processes [12].

1.6 Content of the Article

Let n be the number of processes, m the number of possible different input values, t an upper bound on the number of crash failures, and g the size (# of processes) of the full agreement objects when assuming the $SM(g)$ model. In all the results, unless stated otherwise, it is assumed that $n \geq 2$. Given two positive integers a and b, the notation $a \bmod b$ (i.e., a modulo b) is used for the remainder of the division of a by b. Table 1 summarizes the main results presented in this article regarding the solvability and complexity of the (n, k)-partial agreement problem (none is for partial *set* agreement).

Table 1. Summary of the results

Necessary and sufficient conditions for solving the (n, k)-partial agreement problem
The necessary conditions proved in Sect. 3 (i.e., R4 & R5) hold only assuming strong validity.
All the other results hold for both the weak and strong validity conditions.

Result	Model	Values	Necessary condition	Sufficient condition	Comm. helps?	Section
R1	Asynchronous RW + MP	$m = 2$ $t \geq 1$	$k \leq \lceil n/2 \rceil$ Theorem 1	$k \leq \lceil n/2 \rceil$ Theorem 1	No	2
R2	Asynchronous RW + MP	$m \geq 2$ $t = 1$	$k \leq \lceil n/2 \rceil$ Corollary 2	$k \leq \lceil n/2 \rceil$ Corollary 2	Yes, when $m > 2$	2
R3	Asynchronous RW + MP	$m \geq 2$ $t \geq 1$	$k \leq \lceil n/2 \rceil$ Corollary 1			2
R4	Asynchronous RW + MP	$m \geq 2$ $t \geq 1$ $\min(m, t + 1)$ divides n	$k \leq$ $n / \min(m, t + 1)$ Corollary 3	$k \leq$ $n / \min(m, t + 1)$ Corollary 3	No, when $m \leq t + 1$ Yes, if not	3
R5	Asynchronous RW + MP	$m \geq 2$ $t \geq 1$	$k \leq$ $\lfloor n / \min(m, t + 1) \rfloor +$ $n \bmod \min(m, t + 1)$ Theorem 2	$k \leq$ $\lceil n / \min(m, t + 1) \rceil$ Theorem 2	No, when $m \leq t + 1$ Yes, if not	3
R6	Synchronous MP	$m \geq 2$ $t \geq 1$	t rounds are necessary for every $k \geq \lceil (n + t + 1)/2 \rceil$ Theorem 3			4
R7	Synchronous MP	$m \geq 2$ $t \geq 1$ $\ell \geq 1$		$\lfloor t/\ell \rfloor + 1$ rounds are sufficient for $k \leq \lceil n/\ell \rceil$ Theorem 4	Yes	4
R8	Asynchronous SM(g)	$m \geq 2$ $n > t \geq 1$ $g = t$	$k \leq \lceil (n + t - 1)/2 \rceil$ Theorem 5			5
R9	Asynchronous SM(g)	$m \geq 2$ $t \geq 1$ $g \geq 1$		$k \leq \max$ $(\lceil n / \min(m, t + 1) \rceil,$ $g, 3\lfloor \min(\lfloor n/2 \rfloor, g)/2 \rfloor)$ Theorems 6 & 2 (R5)	Yes	5
R10	Asynchronous SM(g)	$m \geq 2$ $g = t = n/2$ 4 divides n	$k \leq 3n/4$ Corollary 5	$k \leq 3n/4$ Corollary 5	Yes	5

A few remarks.

1. It follows from R3 that in the presence of failures, the best we can hope for is to solve (n, k)-partial agreement for $k = \lceil n/2 \rceil$. This can be achieved when either $m = 2$ (R1) or $t = 1$ (R2).

2. It is interesting to note that for proving R1–3, it suffices to use the (above) known impossibility result #1. In contrast, for proving R5, there is a need to use the (much stronger) known impossibility result #2.

3. Strong validity is assumed in proving the necessary condition of R5 (and of R4). Results R1–3 hold for *both* weak and strong validity, and hence do not follow from R5.

4. R4, which follows from R5, provides a tight bound for partial multi-valued agreement in the special case where $\min(m, t+1)$ divides n.

5. When looking at the round (time) complexity of synchronous partial agreement, R6 shows that in many cases (i.e., when $\lceil (n+t+1)/2 \rceil \leq k < n$) we might be able to save only one round, compared to the solvability of full agreement for which $t+1$ rounds are necessary and sufficient.

6. R7 shows that in some cases, it is possible to significantly reduce the number of rounds, compared to the solvability of full agreement for which $t+1$ rounds are necessary and sufficient. R7 follows easily from a known result regarding the number of rounds sufficient for solving the set agreement problem [2,7].

7. It follows from R8 and R9 that for the $SM(n/2)$ model (i.e., with consensus objects for $n/2$ processes), the bound is tight when $n = t/2$ and n is divisible by 4 (R10).

2 Asynchronous Partial Agreement: The Binary Case with Implications

Let $n \geq 2$ be the number of processes, $m \geq 2$ the number of possible different input values, and $t \geq 1$ an upper bound on the number of crash failures. We show that in the presence of failures, the best we can hope for is to solve (n, k)-partial agreement for $k = \lceil n/2 \rceil$ (Corollary) 1). This bound is tight when either $m = 2$ (Theorem 1) or $t = 1$ (Corollary 2). We first show that for binary partial agreement, in the presence of failures, at most $\lceil n/2 \rceil$ processes are guaranteed to decide on the same value and that this can be achieved without communication. All the results in Sect. 2 hold under both the weak and the strong validity requirements, and hence do not follow from the results in Sect. 3 which hold under strong validity only.

Computational Model. The results presented in Sect. 2 and Sect. 3 hold for a shared memory model that supports atomic read/write registers and a message passing model that supports send and receive messages. The necessary conditions (impossibility results) in these sections will be proven for the shared memory model. However, we observe that a shared memory system that supports atomic registers can simulate a message passing system that supports send, receive and even broadcast operations. Hence the necessary conditions (impossibility results) proved for the shared memory model in Sect. 2 and Sect. 3 also hold for such a message passing system. The simulation is as follows. With each process p we associate an unbounded array of shared registers which all processes can read from, but only p can write into. To simulate a broadcast (or sending) of a message, p writes to the next unused register in its associated array. When p has to receive a message, it reads the new messages from each process.

Theorem 1. *For $n \geq 2$, $m = 2$ and $t \geq 1$, there exists an (n, k)-partial agreement algorithm that can tolerate t crash failures if and only if $k \leq \lceil n/2 \rceil$. Furthermore, for every $k \leq \lceil n/2 \rceil$, there exists such an algorithm in which the processes do not need to communicate.*

Informally, the essence of the proof is in showing that an $(n, \lceil n/2 \rceil + 1)$-partial agreement algorithm (object) has the same computational power as an (n, n)-partial agreement algorithm, in the presence of a single failure. Thus, since it is impossible to solve (n, n)-partial agreement, it is also impossible to solve $(n, \lceil n/2 \rceil + 1)$-partial agreement. In the proof, the known result #1, regarding the impossibility of solving (n, n)-partial agreement in the presence of a single faulty process, is used.

Proof. Since, in an asynchronous system, a crashed process cannot be distinguished from a very slow process, the agreement requirement is equivalent to (i.e., can be simplified as follows): "When all the n processes decide, at least k of them decide on the same value." Without communication, when each process simply decides on its input value, at least $\lceil n/2 \rceil$ of the processes are guaranteed to decide on the same value in runs where all the n processes decide. (A run is a sequence of atomic steps by the processes.) Obviously, since there is no communication, this simple algorithm satisfies both the weak and strong validity requirements for any number $t \geq 1$ of failures. This completes the proof of the *if* direction.

To prove the *only if* direction, we assume to the contrary that there exists an (n, k)-partial agreement algorithm where $k = \lceil n/2 \rceil + 1$, called A, that can tolerate 1 crash failure, and shows that this assumption leads to a contradiction. Obviously, proving the result for $t = 1$ and $k = \lceil n/2 \rceil + 1$ implies the same result for $t \geq 1$ and $k = \lceil n/2 \rceil + 1$.

By definition, in any (fault-free) run of A in which all the n processes decide, there must exist a (proposed) value v such that the number of processes that decide on v minus the number of processes that decide on any other possible value is at least two (two when n is even, and three when n is odd). Moreover, in any run in which *exactly* one process fails, and all the other processes decide, there must exist a (proposed) value v such that the number of processes that decide on v minus the number of processes that decide on any other possible value is at least one (one when n is even, and two when n is odd). Thus, in any run of A in which *at most* one process fails, there is a (proposed) value v such that a strict majority (i.e., more than half) of the processes decide on v.

We use A to construct an (n, n)-partial agreement algorithm that can tolerate a single crash failure, called B, as follows: B works in two (asynchronous) phases of computation:[1]

1. Phase one: Each process p participates in A and decides on some value denoted $decision_p(A)$.
2. Phase two: Each process p owns a single-writer register, and initially writes $decision_p(A)$ in a single-writer register. Then p repeatedly reads the single-writer registers of the other processes until it learns the decision values from the first phase of all the other processes except maybe one of them (since one process may fail).

---

[1] There are no synchrony assumptions whatsoever. A process that finishes phase one, immediately starts participating in phase two.

As explained above, in the $n - 1$ decision values from phase one that p knows about (including its own value), there must be one (proposed) value v that was decided upon by more than half of the processes. So, at the end of phase two, p decides on that value v, and terminates. This completes the description of algorithm B. We prove that when $t = 1$, in B all the non-faulty processes decide on the same value v. Consider two possible cases:

1. All the n processes successfully write their decision values from phase one into their single-writer registers. In such a case, as explained above, a value v must exist such that (in phase one) the number of processes that decided on v minus the number of processes that decided on any other possible value is at least two. Thus, in any subset of size $n - 1$ of these n values (that some process may know about at the end of phase one) v is the majority value.
2. Some process failed, and only $n - 1$ processes succeeded in writing their decision values from phase one into their single-writer registers. In such a case, as explained above, a value v s must exist such that (in phase one) the number of processes that decided on v minus the number of processes that decided on any other possible value is at least one. Thus, since all the non-faulty processes see (at the end of phase two) the same subset of size $n - 1$, they will all decide on the same (proposed) value v.

Thus, B is an (n, n)-partial agreement algorithm that can tolerate one faulty process, violating the known result #1 (as stated in the introduction), regarding the impossibility of solving (n, n)-partial agreement in the presence of a single faulty process [10, 14]. □

Corollary 1. *For $n \geq 2$, $m \geq 2$, and $t \geq 1$, there exists an (n, k)-partial agreement algorithm that can tolerate t crash failures only if $k \leq \lceil n/2 \rceil$.*

Proof. By definition, any algorithm that solves the (n, k)-partial agreement problem in the presence of (up to) t failures when the maximum number of possible input values is m, must also solve the (n, k)-partial agreement problem in the presence of t failures when the maximum number of possible input values is strictly less than m. The result follows. □

Corollary 2. *For $n \geq 2$, $m \geq 2$, and $t = 1$, there exists an (n, k)-partial agreement algorithm that can tolerate a single crash failure if and only if $k \leq \lceil n/2 \rceil$.*

Proof. The only if direction follows immediately from Corollary 1. For proving the if direction, consider the following $(n, \lceil n/2 \rceil)$-partial agreement algorithm that can tolerate a single failure. Each process p writes its input value into a single writer register (resp. sends its input to everybody) and continuously reads the single-writer registers of the other processes (resp. waits to receive messages from the other processes) until it knows the inputs of $n - 1$ processes, including itself. Since there is at most one failure, this procedure will always terminate. Then, p decides on the maximum input value it knows about. This reduces the number of decision values to at most 2. Hence, this algorithm solves the $(n, \lceil n/2 \rceil)$-partial agreement problem. □

3   Asynchronous Partial Agreement: The Multi-valued Case

We provide necessary and sufficient conditions for the solvability of multi-valued partial agreement. These conditions also indicate when communication might help. The sufficient conditions hold under both the weak and the strong validity requirements; the necessary conditions hold only under the strong validity requirement. Given two positive integers a and b, the notation $a \bmod b$ (i.e., a modulo b) is used for the remainder of the division of a by b.

Theorem 2. *For $n \geq 2$, $m \geq 2$ and $t \geq 1$, there exists an (n, k)-partial agreement algorithm that can tolerate t crash failures,*

1. ***if** $k \leq \lceil n/\min(m, t+1) \rceil$. Furthermore, when $m \leq t+1$ there exists such an algorithm in which the processes do not need to communicate;*
2. ***only if** $k \leq \lfloor n/\min(m, t+1) \rfloor + (n \bmod \min(m, t+1))$.*

A very interesting special case is when n is divisible by $\min(m, t+1)$.

Corollary 3. *For $n \geq 2$, $m \geq 2$ and $t \geq 1$, when $n \bmod \min(m, t+1) = 0$, there exists an (n, k)-partial agreement algorithm that can tolerate t crash failures if and only if $k \leq n/\min(m, t+1)$.*

The proof of the *if* direction of Theorem 2 follows from Lemma 1. The proof of the *only if* direction follows from Lemma 2. Informally, the essence of the proof of Lemma 2 is in showing that the computational power of an $(n, \lfloor n/\min(m, t+1) \rfloor + (n \bmod \min(m, t+1)) + 1)$-agreement algorithm (object) is at least as strong as the computational power of an $(n, n, \min(m, t+1) - 1)$-partial set agreement algorithm, in the presence of $\min(m, t+1) - 1$ failures. Thus, since it is impossible to solve $(n, n, \min(m, t+1)-1)$-partial set agreement in the presence of $\min(m, t+1) - 1$ failures, it is also impossible to solve $(n, \lfloor n/\min(m, t+1) \rfloor + (n \bmod \min(m, t+1)) + 1)$-partial agreement. In the proof the known result #2, regarding the impossibility of solving $(n, n, t-1)$-partial set agreement in the presence of $t-1$ faulty processes and $m \geq t$ possible input values, is used.

Lemma 1 (if direction). *For $n \geq 2$, $m \geq 2$, and $t \geq 1$, there exists an (n, k)-partial agreement algorithm that can tolerate t crash failures if $k \leq \lceil n/\min(m, t+1) \rceil$. Furthermore, when $m \leq t+1$, an algorithm exists in which the processes do not need to communicate.*

Proof. Since, in an asynchronous system, a crashed process cannot be distinguished from a very slow process, the agreement requirement can be simplified as follows: "When all the n processes decide, at least k of them decide on the same value."

For $m \leq t+1$, consider the following $(n, \lceil n/m \rceil)$-partial agreement algorithm. Without communication, each process simply decides on its own input value. Thus, at least $\lceil n/m \rceil$ of the processes are guaranteed to decide on the same value in runs where all the n processes decide (i.e., in fault-free runs). Since there is no communication, this simple algorithm satisfies both the weak and strong validity requirements, for any number of $t \geq 1$ of failures.

For $m > t + 1$, consider the following $(n, \lceil n/t + 1 \rceil)$-partial agreement algorithm. Each process p writes its input value into a single writer register (resp. sends its input to everybody) and continuously reads the single-writer registers of the other processes (resp. waits to receive messages from the other processes) until it knows the inputs of $n - t$ processes, including itself. Since there are at most t failures, this procedure will always terminate. Then, p decides on the maximum input value it knows about. This reduces the number of decision values to $t + 1$. Hence, this algorithm solves the $(n, \lceil n/t + 1 \rceil)$-partial agreement problem.   □

Lemma 2 (only if). *For $n \geq 2$, $m \geq 2$ and $t \geq 1$, there exists an (n, k)-partial agreement algorithm that tolerates t crash failures **only if** $k \leq \lfloor n/\min(m, t + 1)\rfloor + (n \bmod \min(m, t + 1))$.*

Proof. Recall that strong validity is assumed. We assume to the contrary that for $n \geq 2$, $m \geq 2$ and $t \geq 1$, there exists an (n, k)-partial agreement algorithm where $k = \lfloor n/\min(m, t + 1)\rfloor + (n \bmod \min(m, t + 1)) + 1$, called A, that can tolerate t crash failures, and shows that this assumption leads to a contradiction. Clearly algorithm A is correct with the additional restriction that $m \leq t + 1$. Thus, for the rest the proof of Lemma 2, it is assumed that $m \leq t+1$, and hence, $m = \min(m, t + 1)$.

For a run ρ, we denote by $I(\rho)$ the number of (different) input (proposed) values in ρ; clearly, $1 \leq I(\rho) \leq n$. By the above assumption,

for every run ρ of A in which all the n processes decide,   (1)
there exists a proposed value v such that at least
$\lfloor n/m \rfloor + (n \bmod m) + 1$ processes decide on v in ρ.

It follows from (1) and the fact that $n = \lfloor n/m \rfloor * m + (n \bmod m)$, that,

for every run ρ of A in which all the n processes decide,   (2)
either $I(\rho) < m$ or there exists a proposed value u such that
at most $\lfloor n/m \rfloor - 1$ processes decide on u in ρ.

It also follows from (1), the fact that $n - (\lfloor n/m \rfloor - 1) * m > m - 1$, and the assumption that $m \leq t + 1$, that

for every run ρ of A in which at least $n - (m - 1)$ processes decide   (3)
(i.e., there are at most $m - 1$ failures in ρ), there exist a proposed
value v such that at least $\lfloor n/m \rfloor$ processes decide on v.

It follows from (2) and (3) that,

for every run ρ of A in which at least $n - (m - 1)$ processes decide   (4)
(i.e., there are at most $m - 1$ failures in ρ),
   1. there exist a proposed value v such that in ρ, and in any extension of
      ρ, at least $\lfloor n/m \rfloor$ processes decide on v, and

2. either $I(\rho) < m$ or there exist a proposed value u such that in ρ, and in any extension of ρ, at most $\lfloor n/m \rfloor - 1$ processes decide on u.

We use A to construct an $(n, n, m - 1)$-partial set agreement (i.e., $(m - 1)$-set agreement for n processes) algorithm for m different input values that can tolerate $m - 1$ crash failures, called B, as follows: B works in two (asynchronous) phases of computation:

1. Phase one: Each process p participates in A and decides on some proposed value denoted $decision_p(A)$.
2. Phase two: Each process owns a single-writer register. Each process p writes $decision_p(A)$ into its single-writer register and repeatedly reads all the n single-writer registers until it notices the decision values from the first phase of all the other processes except maybe $m - 1$ of them (since $m - 1$ processes may fail).

Let us denote by $V(p)$ the multi-set (i.e., with possible repetitions) of decision values that p noticed in the second phase. Clearly, $n - (m - 1) \leq |V(p)| \leq n$. Next, p considers only the values in $V(p)$ with the largest number of repetitions, decides on the largest value among these values, and terminates. For example, if $V(p) = \{1, 1, 1, 2, 2, 2, 3, 3\}$ then p decides on the value 2.

By property (4) above, we have that for every run ρ of B in which at least $n - (m - 1)$ processes decide,

1. for every process p, there exists a proposed value v, such that $v \in V(p)$ and appears at least $\lfloor n/m \rfloor$ times in $V(p)$.
2. either $I(\rho) < m$ or there exist a proposed value u such that in ρ, and in any extension of ρ, for every process p, u and appears at most $\lfloor n/m \rfloor - 1$ times in $V(p)$.

Thus, there exists a proposed value u that no process will decide on in B; and for each process p there is a proposed value (which appears at least $\lfloor n/m \rfloor$ times in $V(p)$) that p can decide on. (We notice that, if v is the value that $\lfloor n/m \rfloor + (n \bmod m) + 1$ processes decide on in some execution then in a prefix of that execution in which $m - 1$ processes fail it is *not* required that at least $\lfloor n/m \rfloor$ processes decide on v.) Thus, the processes will decide on at most $m - 1$ different proposed values. This implies that B is an $(n, n, m - 1)$-partial set agreement algorithm that can tolerate $m - 1$ faulty process, violating the known result #2 (stated in the introduction), regarding the impossibility of solving $(n, n, m - 1)$-partial set agreement in the presence of $m - 1$ faulty processes and m possible input values [4, 13, 17]. □

Corollary 4. *For $n \geq 2$, $m \geq 2$ and $t \geq 1$, there exists an (n, k)-partial agreement algorithm that can tolerate t crash failures **only if*** $k \leq \min_{2 \leq \ell \leq \min(m, t+1)} \{ \lfloor n/\ell \rfloor + (n \bmod \ell) \}$.

Proof. By definition, any algorithm that solves the (n, k)-partial agreement problem in the presence of (up to) t failures when the maximum number of possible input values is m, must also solve the (n, k)-partial agreement problem in the presence of t failures when the maximum number of possible input values is strictly less than m. The result follows from the above observation and Lemma 2.   □

4   Synchronous Partial Agreement

In a synchronous message-passing system in which up to t processes may crash, there is a simple full agreement algorithm with $t + 1$ rounds. Furthermore, it is known that every full agreement algorithm requires at least $t + 1$ rounds [1]. Can we do better for partial agreement? We show that, in some cases, we might be able to reduce the number of rounds to t, while in other cases we can do much better. All the results in Sect. 4 hold under both the weak and the strong validity requirements.

Theorem 3. *For $n \geq 2$, $m \geq 2$, $n - 2 \geq t \geq 1$, and $k \geq \lceil (n + t + 1)/2 \rceil$, in a synchronous message-passing system in which up to t processes may crash, every (n, k)-partial agreement algorithm requires at least t rounds.*

Informally, the essence of the proof is in showing that if an $(n, \lceil (n + t + 1)/2 \rceil)$-partial agreement algorithm can be solved in less than t rounds in the presence of t failures, then an (n, n)-partial agreement algorithm can be solved in less than $t + 1$ rounds in the presence of t failures. Since it is impossible to solve (n, n)-partial agreement in less than $t + 1$ rounds in the presence of t failures, it is also impossible to solve $(n, \lceil (n + t + 1)/2 \rceil)$-partial agreement in less than t rounds in the presence of t failures. In the proof, the known result #3 is used.

Proof. By definition, any algorithm that solves the (n, k)-partial agreement problem in the presence of (up to) t failures when the maximum number of possible input values is m, must also solve the (n, k)-partial agreement problem in the presence of t failures when $m = 2$. So, since we are proving a lower bound, for the rest of the proof we assume that $m = 2$.

We assume to the contrary that there exists an (n, k)-partial agreement algorithm where $k = \lceil (n + t + 1)/2 \rceil$, called A, that can be solved in less than t rounds and can tolerate t crash failures, and shows that this assumption leads to a contradiction. Obviously, $k = \lceil (n + t + 1)/2 \rceil$ implies the same result for $k \geq \lceil (n + t + 1)/2 \rceil$.

By definition, in any (fault-free) run of A in which all the n processes decide, there must exist a value v such that the number of processes that decide on v minus the number of processes that decide on any other value is at least $t + 1$. To see this, observe that since the value v appears at least $\lceil (n+t+1)/2 \rceil$ times, the other values appear at most $n - (\lceil n+t+1)/2 \rceil$ times. So, $\lceil (n+t+1)/2 \rceil - (n - \lceil (n+t+1)/2 \rceil) \geq t + 1$.

This implies that in any run in which at most t processes fail, and all the other processes decide, there must exist a value v such that the number of processes that decide on v minus the number of processes that decide on any other possible value is at least one. Thus, in any run of A in which *at most* t processes fail, there is a value v such that a strict majority (i.e., more than half) of the processes decide on v.

We use A to construct an (n, n)-partial agreement algorithm, called B, that can be solved in t rounds and can tolerate t crash failures. B works in two phases, the first takes less than t rounds and the second takes exactly one round.

1. Phase one: Each process p participates in A and decides on some value denoted $decision_p(A)$. This takes at most $t - 1$ rounds. We notice that it is possible that no process fails during the first phase.
2. Phase two: Each process p send the value $decision_p(A)$ (from the first phase) to all the other processes. Then p waits to receive messages from the other processes until it learns the decision values from the first phase of all the other processes except maybe t of them (since t processes may fail).

Recall that we assume $m = 2$. As explained above, in the $n - t$ (or more) decision values from phase one that p knows about (including its own value), there must be exactly one value v that was decided upon by more than half of the processes. So, at the end of phase two, p decides on v, and terminates. This completes the description of algorithm B.

We prove that for $t \geq 1$, in B all the non-faulty processes decide on the same proposed value v. For a given run, assume that f processes, where $1 \leq f \leq t$ have failed. Thus $n - f$ processes succeeded in sending their decision values from phase one to all the other processes. In such a case, as explained above, there must exist a value v such that (in phase one) the number of processes that decided on v minus the number of processes that decided on any other possible value is at least $t - f + 1 \geq 1$. Thus, since each of the non-faulty processes see (at the end of phase two) a subset of size at least $n - t$ values, they will all decide on the same value v.

Thus, B is an (n, n)-partial agreement algorithm that requires less than $t + 1$ rounds in the presence of t failures. However, this violates the known result #3 (as stated in the introduction), regarding the impossibility of solving (n, n)-partial agreement in less than $t + 1$ rounds in the presence of t faulty processes [1]. □

Next, we observe that it is possible to significantly reduce the number of rounds in some cases. This simple observation follows easily from a known result regarding the number of rounds that are sufficient for solving the set agreement problem [2,7].

Theorem 4. *For $n \geq 2$, $m \geq 2$, $t \geq 1$, and $\ell \geq 1$, in a synchronous message-passing system in which up to t processes may crash, there exists an $(n, \lceil n/\ell \rceil)$-partial agreement algorithm with $\lfloor t/\ell \rfloor + 1$ rounds.*

Proof. A simple algorithm was presented in [2,7] that solves ℓ-set agreement and requires only $\lfloor t/\ell \rfloor + 1$ rounds. This algorithm clearly solves also $(n, \lceil n/\ell \rceil)$-partial agreement. For completeness, we give a description of the algorithm and an explanation below.

The algorithm [2,7]: The algorithm consists of exactly $\lfloor t/\ell \rfloor + 1$ rounds. In each round, every process sends a message with its preferred value (initially its input) to all the processes (including herself) and waits until the end of the round to receive all the messages that were sent to it during the round. From the set of all messages that the process has received in a given round, it chooses the minimum value as its new preferred value. Then, it continues to the next round supporting this (minimum) value as its new preferred value. The algorithm terminates after $\lfloor t/\ell \rfloor + 1$ rounds, and each process decides on its preferred value at the end of the last round.

Explanation. If at some round, x processes fail, then in the next round the number of different values will be *at most* $x + 1$, and this number will never be increased in subsequent rounds. Thus, the worst case is when the number of faults is the same in each round. Given t faults and $\lfloor t/\ell \rfloor + 1$ rounds, it is only possible to arrange that in *each* round there will be at least

$$x = \left\lfloor \frac{t}{\lfloor t/\ell \rfloor + 1} \right\rfloor = \ell - 1$$

faults in each round. Thus, the maximum possible number of *different* decision values that is $x + 1 = \ell$. □

5 Partial Agreement Using Strong Shared Objects

The results presented in this section are for the $SM(g)$ computational model. Recall that wait-free full agreement objects for g processes are shared objects that solve the full agreement problem for g processes in the presence of any number of failures [12]. All the results in Sect. 5 hold under both the weak and the strong validity requirements.

Theorem 5. *For $n \geq 2$, $m \geq 2$, and $n > t \geq 1$, there exists an (n, k)-partial agreement algorithm that can tolerate t crash failures using atomic registers and wait-free full agreement objects for t processes, only if $k \leq \lceil (n + t - 1)/2 \rceil$.*

The proof is an adaptation of the one used for proving Theorem 1. Informally, the essence of the proof is in showing that an $(n, \lceil (n + t - 1)/2 \rceil + 1)$-partial agreement algorithm (object) has the same computational power as an (n, n)-partial agreement algorithm, in the presence of t failures, in the $SM(t)$ model. Thus, since it is impossible to solve (n, n)-partial agreement, it is also impossible to solve $(n, \lceil (n + t - 1)/2 \rceil + 1)$-partial agreement. In the proof, the known result #4, is used.

Proof. By definition, any algorithm that solves the (n, k)-partial agreement problem in the presence of (up to) t failures when the maximum number of possible input values is m, must also solve the (n, k)-partial agreement problem in the presence of t failures when $m = 2$. So, for the rest of the proof, we assume that $m = 2$.

We assume to the contrary that there exists an (n, k)-partial agreement algorithm where $k = \lceil(n + t - 1)/2\rceil + 1$, called A, that can tolerate t crash failures, and shows that this assumption leads to a contradiction.

By definition, in any (fault-free) run of A in which all the n processes decide, a proposed value v must exist such that the number of processes that decide on v minus the number of processes that decide on any other possible value is at least $t+1$. To see this, observe that since the value v appears at least $\lceil(n+t-1)/2\rceil+1$ times, the other possible values appear at most $n - (\lceil(n + t - 1)/2\rceil + 1)$ times. So, $(\lceil(n + t - 1)/2\rceil + 1) - (n - (\lceil(n + t - 1)/2\rceil + 1)) \geq t + 1$.

This implies that, in any run in which at most t processes fail, and all the other processes decide, there must exist a proposed value v such that the number of processes that decide on v minus the number of processes that decide on the other possible value is at least one. Thus, in any run of A in which at most t processes fail, there is a proposed value v such that a strict majority of the processes decide on v.

We use A to construct an (n, n)-partial agreement algorithm that can tolerate t crash failures, called B, as follows: B works in two (asynchronous) phases of computation. A process that finishes phase one, immediately starts participating in phase two.

1. Phase one: Each process p participates in A and decides on some proposed value denoted $decision_p(A)$.
2. Phase two: Each process p owns a single-writer register, and initially writes $decision_p(A)$ in a single-writer register. Then p repeatedly reads the single-writer registers of the other processes until it learns the decision values from the first phase of all the other processes except maybe t of them (since t processes may fail).

Recall that $m = 2$. As explained above, in the $n - t$ decision values from phase one that p knows about (including its own value), there must be one proposed value v that was decided upon by more than half of the processes. So, at the end of phase two, p decides on that value v, and terminates. This completes the description of algorithm B. We prove that when $t \geq 1$, in B all the non-faulty processes decide on the same value v. For a given run, assume that f processes, where $1 \leq f \leq t$ have failed. Thus $n - f$ processes succeeded in writing their decision values from phase one into their single-writer registers. In such a case, as explained above, a value v must exist such that (in phase one) the number of processes that decided on v minus the number of processes that decided on the other possible value is at least $t - f + 1 \geq 1$. Thus, since each of the non-faulty processes see (at the end of phase two) a subset of size at least $n - t$ values, they will all decide on the same proposed value v.

Thus, B is an (n, n)-partial agreement algorithm that can tolerate t faulty process, violating the known result #4 (as stated in the introduction), regarding the impossibility of solving (n, n)-partial agreement in the presence of a t faulty process using atomic registers and wait-free full agreement objects for t processes [12]. □

Next, we demonstrate that in the $SM(g)$ model, when $g \geq 2$, we can do better than the sufficient condition presented in Theorem 1.

Theorem 6. *For $n \geq 4$, $m \geq 2$, and $t \geq 1$, there exists an (n,k)-partial agreement algorithm that can tolerate t crash failures using atomic registers and wait-free full agreement objects for g processes if $k \leq \max(g, 3\lfloor\min(\lfloor n/2\rfloor, g)/2\rfloor)$.*

Proof. We consider two cases. The first case is when $g > 3\lfloor n/4\rfloor$. To solve this case, a single wait-free full agreement object for g processes, called A, is used. We let g processes participate in A and decide on the same value. Each one of the other $n - g$ processes simply decides on its input.

The second case is when $g \leq 3\lfloor n/4\rfloor$. Let $\hat{g} = \min(\lfloor n/2\rfloor, g)$. Three wait-free full agreement objects for g processes, called A, B and C, are used for the algorithm. Two disjoint groups of processes, G_1 and G_2 are created each of size $\hat{g}$. Then, we choose $\lfloor\hat{g}/2\rfloor$ processes from G_1 and $\lfloor\hat{g}/2\rfloor$ processes from G_2, and create the groups G_3 and G_4, respectively. Each process that does not belong to G_1 or G_2 simply decides on its input and terminates. The processes in G_1 first participate in A, and then the processes in G_3 use the decision value from A as their new input, participate in C and use the decision value from C as their final decision value. The other processes in G_1 use the decision value from A as their final decision value. Similarly, the processes in G_2 first participate in B. Then the processes in G_4 use the decision value from B as their new input, participate in C and use the decision value from C as their final decision value. The other processes in G_2 use the decision value in A as their final decision value. The result follows.   □

We notice that, by Theorems 5 and 6, for the $SM(n/2)$ model, the bound is tight when $n = t/2$ and n is divisible by 4.

Corollary 5. *Assume that $m \geq 2$ and n is divisible by 4. There exists an (n,k)-partial agreement algorithm that can tolerate $n/2$ crash failures using atomic registers and wait-free full agreement objects for $n/2$ processes, if only if $k \leq 3n/4$.*

6   Related Work

For lack of space, most of the related work section is omitted from this conference version. Please see the full version of the paper for the complete section, which also includes many open problems [20].

Motivated by the need to achieve Byzantine agreement on sparse networks, the notion of *almost everywhere agreement* was introduced in [9], in which all but a small number of the correct processes must choose a common decision value. Intuitively, the correctness condition is relaxed by "giving up for lost" those correct processes whose communication paths to the remainder of the network are excessively corrupted by faulty processes. As an intermediate step in defining almost everywhere agreement, the notion of t-resilient X agreement was introduced, in which (1) when at most t processes fail, all but X of the correct processes must eventually decide on a common value, and (2) "if all correct processes begin with the same value v, then v must be the common decision value" [9]. Condition #1 above is similar to our agreement requirement (in the definition of (n,k)-partial agreement); condition #2 is much weaker than our weak validity requirement when $m > 2$ and is similar when $m = 2$.

Another related definition was introduced in [19], in which the traditional notion of fault tolerance is generalized by allowing a limited number of participating correct processes not to terminate in the presence of faults. Every process that does terminate is required to return a correct result. Various results regarding the solvability of problems like election, consensus and renaming using atomic registers are presented for this generalization,.

The most recent related definition is that of bounded disagreement, which limits the number of processes that decide differently from the plurality [5]. Here the agreement requirement is similar to that of X agreement (and hence also ours); the validity requirement used is strong validity. The main result of [5] is that there are infinitely many instances of the bounded disagreement task that are not equivalent to any consensus task and any set agreement task. None of our results overlap or can be derived from the results in [5,9,19].

7 Discussion

The fundamental full agreement problem has been intensively studied for over 40 years. It is intriguing to revisit some of the numerous questions and results for this problem and study them in the context of the partial agreement problem. For example, can randomized partial agreement algorithms with better time, space, and message complexities be designed than those of known randomized full agreement algorithms? Similarly, what are the time and space complexities of obstruction-free partial agreement algorithms? How many Byzantine failures can partial agreement algorithms tolerate, and what is the complexity of such Byzantine partial agreement algorithms? Studying the partial set agreement problem defined in the introduction would be interesting. Apart from presenting the problem and the new technical results, the significance of the article is in exposing open problems that hopefully will stimulate further research on the partial agreement problem.

References

1. Aguilera, M.K., Toueg, S.: A simple bivalency proof that t-resilient consensus requires $t + 1$ rounds. Inf. Process. Lett. **71**(3), 155–158 (1999)
2. Attiya, H., Welch, J.: Distributed Computing: Fundamentals, Simulations and Advanced Topics, 2nd edn. Wiley, Hoboken (2004)
3. Bassler, B.: Quorum Sensing: How Bacteria Communicate. The Explorer's Guide to Biology, 48 p. (2019). https://explorebiology.org/summary/cell-biology/quorum-sensing:-how-bacteria-communicate

4. Borowsky, E., Gafni, E.: Generalized FLP impossibility result for t-resilient asynchronous computations. In: Proceedings of the 25th ACM Symposium on Theory of Computing, pp. 91–100 (1993)

5. Chan, D., Hadzilacos, V., Toueg, S.: Bounded disagreement. Theor. Comput. Sci. **826–827**, 12–24 (2020). Special issue on OPODIS 2016

6. Chaudhuri, S.: More choices allow more faults: set consensus problems in totally asynchronous systems. Inf. Comput. **105**(1), 132–158 (1993)

7. Chaudhuri, S., Herlihy, M., Lynch, N., Tuttle, M.: Tight bounds for k-set agreement. J. ACM **47**(5), 912–943 (2000)

8. Dolev, D., Strong, H.R.: Authenticated algorithms for byzantine agreement. SIAM J. Comput. **12**(4), 656–666 (1983)

9. Dwork, C., Peleg, D., Pippenger, N., Upfal, E.: Fault tolerance in networks of bounded degree. SIAM J. Comput. **17**(5), 975–988 (1988)

10. Fischer, M.J., Lynch, N.A., Paterson, M.S.: Impossibility of distributed consensus with one faulty process. J. ACM **32**(2), 374–382 (1985)

11. Fuqua, W.C., Winans, S.C., Greenberg, E.P.: Quorum sensing in bacteria: the LuxR-LuxI family of cell density-responsive transcriptional regulators. J. Bacteriol. **176**(2), 269–275 (1994)

12. Herlihy, M.P.: Wait-free synchronization. ACM Trans. Program. Lang. Syst. **13**(1), 124–149 (1991)

13. Herlihy, M.P., Shavit, N.: The topological structure of asynchronous computability. J. ACM **46**(6), 858–923 (1999)

14. Loui, M., Abu-Amara, H.: Memory requirements for agreement among unreliable asynchronous processes. Adv. Comput. Res. **4**, 163–183 (1987)

15. Pease, M., Shostak, R., Lamport, L.: Reaching agreement in the presence of faults. J. ACM **27**(2), 228–234 (1980)

16. Pratt, S.C.: Quorum sensing by encounter rates in the ant Temnothorax albipennis. Behav. Ecol. **16**(2), 488–496 (2005)

17. Saks, M., Zaharoglou, F.: Wait-free k-set agreement is impossible: the topology of public knowledge. SIAM J. Comput. **29**, 1449–1483 (2000)

18. Seeley, T.D., Visscher, P.K.: Group decision making in nest-site selection by honey bees. Apidologie **35**(2), 101–16 (2004)

19. Taubenfeld, G.: A closer look at fault tolerance. Theory Comput. Syst. **62**, 1085–1108 (2018). Conf. version appeared in PODC 2012

20. Taubenfeld, G.: Reaching agreement among k out of n processes (2023). https://arxiv.org/abs/2205.04873. arXiv:2205.04873v3

On the Bit Complexity of Iterated Memory

Guillermo Toyos-Marfurt[✉] and Petr Kuznetsov

LTCI, Institute Polytechnique de Paris, Palaiseau, France
{guillermo.toyos,petr.kuznetsov}@telecom-paris.fr

Abstract. Computability, in the presence of asynchrony and failures, is one of the central questions in distributed computing. The celebrated asynchronous computability theorem (ACT) characterizes the computing power of the read-write shared-memory model through the geometric properties of its *protocol complex*: a combinatorial structure describing the states the model can reach via its finite executions. This characterization assumes that the memory is of unbounded capacity, in particular, it is able to store the exponentially growing states of the *full-information* protocol.

In this paper, we tackle an orthogonal question: what is the *minimal* memory capacity that allows us to simulate a given number of rounds of the full-information protocol? In the *iterated immediate snapshot* model (IIS), we determine necessary and sufficient conditions on the number of bits an IIS element should be able to store so that the resulting protocol is equivalent, up to isomorphism, to the full-information protocol. Our characterization implies that $n \geq 3$ processes can simulate r rounds of the full-information IIS protocol as long as the bit complexity per process is within $\Omega(rn)$ and $O(rn \log n)$. Two processes, however, can simulate *any* number of rounds of the full-information protocol using only 2 bits per process, which implies, in particular, that just 2 bits per process are sufficient to solve ε-agreement for arbitrarily small ε.

Keywords: Theory of computation · Distributed computing models · Iterated Immediate Snapshot · Combinatorial Topology · Communication Complexity · Approximate Agreement

1 Introduction

One of the central questions in the theory of distributed computing is how to characterize the class of problems that can be solved in a given system model. There exists a plethora of models, varying synchrony assumptions, fault types and communication media. Evaluating their computing power and establishing meaningful comparisons between these models poses a formidable challenge.

The combinatorial approach [10,12] has proved to be instrumental in analyzing the computing power of a large range of shared-memory models [7,11,12,16].

© The Author(s), under exclusive license to Springer Nature Switzerland AG 2024
Y. Emek (Ed.): SIROCCO 2024, LNCS 14662, pp. 456–477, 2024.
https://doi.org/10.1007/978-3-031-60603-8_25

A set of executions produced by a model can be represented as a *simplicial complex*, a combinatorial structure, whose geometric properties allow us to reason about the model's computing power.

For example, the *asynchronous computability theorem* (ACT) [12] characterizes wait-free task solvability of the *read-write shared-memory* model. Here a task is defined as a tuple $(\mathcal{I}, \mathcal{O}, \Delta)$, where $\mathcal{I}$ and $\mathcal{O}$ are, resp., the *input complex* and the *output complex*, describing the task's input and output configurations, and $\Delta : \mathcal{I} \to 2^{\mathcal{O}}$ is a map relating each input assignment to all output assignments allowed by the task. A task is solvable, if and only if there exists a continuous map from $\mathcal{I}$ to $\mathcal{O}$ that *respects* Δ.

Notice that the characterization is defined entirely by the task, i.e., regardless of the "operational" behavior of the model. The core observation behind ACT and its followups [7,11,16] is that finite executions of the read-write model can be described via iterations of the *immediate snapshot* [2,3,8]. It turns out that the *iterated immediate snapshot* model (IIS) has a very nice regular structure, precisely captured by recursive applications of the *standard chromatic subdivision* [15] (Fig. 1). This representation assumes, however, the *full-information protocol*: in each iteration, a process writes its current state to the memory and updates its state with a snapshot of the states written by the other processes. As a result, the state grows exponentially with the number of iterations, which requires memory locations of unbounded capacity.

In this paper, we consider a more realistic scenario of *bounded* iterated immediate snapshot memory (B-IIS). More precisely, we determine necessary and sufficient conditions on the capacity of the shared iterated memory so that its protocol complex is isomorphic to r iterations of the standard chromatic subdivision and, thus, maintains the properties of r IIS rounds. Intuitively, this boils down to defining an optimal *encoding* of a process state at the end of round r using the available number of bits, so that, given a snapshot of such encoded values, one can consistently map these states to the vertices of the r iterations of the standard chromatic subdivision. One can then derive the conditions on the *bit complexity*, i.e., the amount of memory required to reproduce the *full-information* protocol in the B-IIS model.

Our contributions are the following. Given an input complex $\mathcal{I}$ of a task, we determine necessary and sufficient conditions on the number of bits that enable an encoding function for simulating r rounds of (full-information) IIS. We show that for 3 or more processes, the per-process bit complexity (i.e., the amount of memory a process must have at its disposal in each iteration) is *at least linear* both in the number of rounds and the number of processes and *at most linearithmic*: within $\Omega(rn)$ and $O(rn \log n)$.[1] To obtain these results, we establish an insightful combinatorial characterization of the iterated standard chromatic subdivision via f-vectors, a common tool in polyhedral combinatorics [22].

---

[1] For the case of two processes, we show (in the accompanying technical report [21]) that the bit complexity is exactly 2 bits.

Roadmap. The paper is organized as follows. We overview the related work in Sect. 2. In Sect. 3, we recall the combinatorial-topology notions used in this work. In Sect. 4, we describe the B-IIS model. Then in Sect. 5, we give the conditions that the B-IIS model has to satisfy in order to replicate the iterated full information protocol. Next, in Sect. 6, we characterize these conditions by studying the f-vector of the iterated chromatic subdivision. Section 7 contains the proofs of the asymptotic bounds derived from the equations in Sect. 6. Finally, in Sect. 8, we give the asymptotic bounds of bit complexity of the B-IIS model and a discussion of the obtained results. The special case of a 2-process system and the approximate agreement task is relegated to the extended online version of this work [21].

2   Related Work

The asynchronous computability theorem (ACT) was formulated by Herlihy and Sha-vit [12], using elements of simplicial topology, which helped resolving the long-standing conjecture of the impossibility of set agreement in the wait-free read-write model (concurrently resolved by Saks and Zacharoglou [19], and Borowsky and Gafni [2]). The characterization [12] was later extended to more general classes of models [7,11,16]. Herlihy, Kozlov and Rajsbaum [10] later gave a thorough introduction to the use of combinatorial topology in distributed computing.

Borowsky and Gafni [3] introduced the simplex approximation task and presented the first wait-free read-write protocol for simplex approximation on the iterated standard chromatic subdivision. Nishimura [17] considered variants of the IIS model in order to study the time complexity of simplex agreement.

Communication complexity is a well-studied sub-field of complexity theory, which considers the amount of information required for multiple distributed agents to acquire new knowledge [18]. Yet, the problem of bit complexity of full-information protocols has only been addressed recently. Delporte-Gallet et al. [6] show that for the particular case of two processes, it is possible to solve any wait-free solvable task using mostly 1-bit messages in wait-free dynamic networks. Notably, the algorithm for achieving a 2-process ϵ-agreement has the same characteristics as the one presented in this work. It can be said that our contribution lies in delivering a communication-efficient ϵ-agreement within the IIS protocols. Another version of 1-bit approximate agreement on dynamic networks is presented in [1].

Recently, Delporte-Gallet et al. [5] gave a comprehensive study of the computational power of memory bounded registers. Using combinatorial tools, they determine bit complexity required for the existence of wait-free and t-resilient read-write solutions for a given task. However, their approach solely focuses on task solvability, relying on simulations and other techniques that result in protocols requiring an exponential number of rounds for simulating a single round of IIS. In this paper, we aim to simulate IIS using *the same* number of rounds, thus, ensuring the same time complexity.

3   Preliminaries

Simplicial Complex. We represent the IIS and B-IIS model as topological spaces, defined as *simplicial complexes* [10]: a set of *vertices* and an inclusion-closed set of vertex subsets, called *simplices*. We call the simplicial complex of $n+1$ vertices with its power set the n-dimensional simplex Δ^n. The *dimension* of a simplex Δ, denoted as $dim(\Delta)$, is its number of vertices $V(\Delta)-1$. The dimension of a simplicial complex is equal to the dimension of the largest simplex it contains. We call *faces* the simplices contained in a simplicial complex, where a 0-face is a vertex. We call *facets* the simplices which are not contained in any other simplex. We denote the set of vertices of a simplicial complex $\mathcal{A}$ as $V(\mathcal{A})$. We call a simplicial complex *chromatic* when it comes with a *coloring function* that assigns every vertex to a unique process identifier. When considering vertices, we use the notation $|_p$ when we consider only the vertices of color p and we denote v_p a vertex with color p.

Standard Constructions. Given a simplicial complex $\mathcal{A}$, the *star* of $\mathcal{S} \subseteq \mathcal{A}$, $\mathrm{St}(\mathcal{A}, \mathcal{S})$, is a subcomplex made of all simplices in $\mathcal{A}$ containing a simplex of $\mathcal{S}$ as face. The *link*, $\mathrm{Lk}(\mathcal{A}, \mathcal{S})$ is the subcomplex consisting of all simplices in $\mathrm{St}(\mathcal{A}, \mathcal{S})$ which do not have a common vertex with $\mathcal{S}$.

Standard Chromatic Subdivision. The *standard chromatic subdivision* of a simplicial complex $\mathcal{A}$, denoted as $\mathrm{Ch}\,\mathcal{A}$, is a complex whose vertices are tuples (c, σ) where c is a color and σ is a face of $\mathcal{A}$ containing a vertex of color c. A set of vertices of $\mathrm{Ch}\,\mathcal{A}$ defines a simplex if for each pair (c, σ) and (c', σ'), $c \neq c'$ and either $\sigma \subseteq \sigma'$ or $\sigma' \subseteq \sigma$. For Δ^n, the vertices (c, Δ^n) define a *central simplex* in $\mathrm{Ch}\,\mathcal{A}$. The standard chromatic subdivision can be seen as the "colored" analog of the *standard barycentric subdivision* [10]. Figure 1 illustrates the application of Ch to the 2-dimensional simplex. In general, a subdivision operator $\tau : \mathcal{A} \to \mathcal{B}$ is a map that "divides" the simplices in $\mathcal{A}$ into smaller simplices. For a rigorous definition see [10].

Tasks. A *distributed task* (or simply a *task*) is defined as a tuple $(\mathcal{I}, \mathcal{O}, \Delta)$, where $\mathcal{I}$ and $\mathcal{O}$ are, resp., the *input complex* and the *output complex*, describing the task's input and output configurations, and $\Delta : \mathcal{I} \to 2^{\mathcal{O}}$ is a map relating each input assignment to all output assignments allowed by the task. In this work we consider general colored tasks, where $\mathcal{I}$ and $\mathcal{O}$ are chromatic simplicial complexes, allowing processes to have different sets of inputs and outputs. To solve a task, a distributed system uses a (communication) *protocol* to share information between processes - *IIS model* is an example of such a protocol. The different possible configurations reached after communication can be character-ized by a protocol map Ξ. From the protocol complex $\Xi(\mathcal{I})$, a decision function $\delta : \Xi(\mathcal{I}) \to \mathcal{O}$ maps the simplices from the protocol complex to outputs, respect-ing the specification of the problem given by Δ: $\delta \circ \Xi(\mathcal{I}) \subseteq \Delta(\mathcal{I})$. For a formal treatment of task solvability using these concepts, please refer to [10].

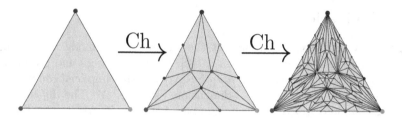

Fig. 1. Iterated application of the standard chromatic subdivision to Δ^2. Leftmost diagram illustrate the Δ^2 simplex. The diagram in the middle shows Ch Δ^2, which is subdivided again resulting in the rightmost image Ch^2 Δ^2.

4   System Model

Computational Model. The IIS protocol is *wait-free*, i.e., each process is expected to reach an output in a finite number of its own steps, regardless of the behavior of other processes. Computation is structured in a sequence of *immediate-snapshot layers* $M[1], \ldots, M[r]$, where each layer is represented as an array of shared variables (one per process), initialized to $\bot$. Algorithm 1 shows the pseudo-code of the IIS protocol.

Algorithm 1: Iterated Immediate Snapshot (IIS) protocol. Code for process $p_i \in \Pi$ and $r > 0$ rounds.

Shared: $M[r]$ array of r memory-bounded snapshots with $|\Pi|$ entries.
Initial: $v = input(i)$ ▷ What the process sees. At first, its input.
for $k := 1$ *to* r **do**
    | $M[k, i] \leftarrow v$;
    | $v \leftarrow snapshot(M[k])$;
end
return $\delta_i(v)$

Note that the protocol is *full-information* and *generic*: in each layer, a process writes its *complete view* and then updates its view as a snapshot of views written by (a subset of) other processes. After a certain number of rounds, the process applies the task-specific decision function $\delta_i(v)$ based on the process's view v, to compute the task's output.

Algorithm 2 shows the bounded IIS communication model, the pseudo-code is written in the same style as the full-information protocol. In B-IIS, instead of writing their entire state, processes write a representation of their states using an *encoding function*. Then, depending on what the process reads on the next snapshot, the internal state changes according to the $next\_state_i$ function. Notice that each process keeps track of their state by using a local variable s. Finally, the decision function δ_i now takes the internal state of the process.

Algorithm 2: Bounded Iterated Immediate Snapshot (B-IIS) protocol.
Code for process $p_i \in \Pi$ and $r > 0$ rounds.

Shared: $M[r]$ array of r memory-bounded snapshots with $|\Pi|$ entries.
Initial: $v = input(i)$ ▷ What the process sees. At first, its input.
Initial: $s = next\_state_i(v, \bot, 0)$ ▷ Process state. At first, determined by the process input.
for $k := 1$ *to* r **do**
  $\quad M[k, i] \leftarrow encode_i(s, k);$
  $\quad v \leftarrow snapshot(M[k]);$
  $\quad s \leftarrow next\_state_i(v, s, k);$
end
return $\delta_i(s)$

Consequently, depending on the $encode_i$ and $next\_state_i$ functions we have different communication protocols. Note that removing the round parameter k on the $encode$ and $next\_state_i$ functions does not restrict the expressiveness of the algorithm, as the round number can be inferred from the internal state s. A trivial observation is that if both the encoding and the state function are the identity (that is, $encode_i : state, round \mapsto state$ and $next\_state_i : view, state, round \mapsto view$) the protocol is equal to the original (full-information) IIS and the bit complexity is $|\Pi|^r$ in the r-th round. (Here Π is the set of processes.) Thus, the bounds on bit complexity can be imposed by restricting the output space of the $encode_i$ functions: $|\operatorname{Im} encode_i| < B$, where $\operatorname{Im} encode_i$ is the image of $encode_i$. We also call $\operatorname{Im} encode_i$ the *encoding set*, denoted as E. Therefore, we need to find an "intelligent" $encode$ function that uses as few bits as possible to encode a state s, while still being able to be decoded by the $next\_state_i$ functions of the other processes. In Sect. 5 we will show necessary and sufficient conditions which allow us to determine exactly how many bits are needed.

Topological Model. We consider a system with $n + 1$ asynchronous processes $\Pi = \{p_0, \ldots, p_n\}$, where any processes can fail at any time. Processes communicate using the IIS and B-IIS protocol as described in Sect. 4. We denote Ξ and Ξ_b the protocol maps of IIS and B-IIS respectively. As both protocols are iterated we can denote Ξ^r and Ξ_b^r for the protocol maps of the $r - th$ iteration of Algorithm 1 and 2 respectively. Note that Ξ_b depends on the $encode$ and $next\_state$ functions. When not stated explicitly, we assume these functions are well defined and have the same behavior as the full-information protocol. That is, the protocol complex $\Xi_b(\mathcal{I})$ is isomorphic to $\Xi(\mathcal{I})$.

5   Equivalence Between Bounded and Full-Information IIS

The goal of this work is to explore how, by leveraging the B-IIS model, we can attain a protocol complex equivalent to the (full-information) IIS. In other words, we aim to identify the conditions that *encode* and *next_state* functions must satisfy for B-IIS to possess the same computational power as IIS, and to determine the memory requirements for achieving this equivalence.

The central idea for achieving this is *process distinguishability*. Our objective is for a process to be capable of deducing the state of another process by reading the values stored in shared memory. The underlying intuition is that when a process knows the potential configurations within the protocol complex, it can infer the possible states of the other processes. Consequently, if the other processes can effectively communicate their respective states, there is no need to write their entire state in shared memory. Instead, a signal suffices for identification within the protocol complex. The following definitions formalize such notions:

Definition 1 (Distinguishability of a vertex). *Let $\mathcal{A}$ a chromatic complex, $encode : V(\mathcal{A}) \rightarrow E$, $p, q \in \Pi$ and s_p, t_q two adjacent vertices in $\mathcal{A}$. We say that s_p **distinguishes** t_q under encode if there is no other vertex $w_q \in \mathrm{Lk}(\mathcal{A}, s_p)|_q$ sharing the same encoding: $encode(t_q) = encode(w_q)$.*

Definition 2 (Distinguishability of a chromatic simplicial complex). *Let $\mathcal{A}$ a chromatic complex and encode : $V(\mathcal{A}) \rightarrow E$. We say that $\mathcal{A}$ is **distinguishable** under encode if for every pair of adjacent vertices $s, t \in V(\mathcal{A})$, s distinguishes t and vice versa.*

In other words, Definition 1 stipulates that a vertex in the protocol complex is distinguishable by another vertex if the former can tell apart the latter just by reading the written value in the snapshot memory. Then, if this relation is true for all vertices on the simplicial complex, we say that the whole complex is distinguishable.

Theorem 1 establishes that when the input complex is distinguishable, the protocol complex of B-IIS exhibits an identical structure to the full-information protocol, namely the standard chromatic subdivision.

Theorem 1 (Characterization of Ξ_b). *Let $\mathcal{I}$ a chromatic input complex, encode : $V(\mathcal{A}) \rightarrow E$ the encoding function of the B-IIS protocol, and Ξ_b its associated protocol map. The following equivalence holds:*

$$\Xi_b(\mathcal{I}) \cong \mathrm{Ch}\,\mathcal{I} \iff \mathcal{I} \text{ is distinguishable under encode}$$

Proof. For the sufficient condition, we will show that $\Xi_b(\mathcal{I}) \cong \Xi(\mathcal{I})$, as we have that the protocol map of the unbounded IIS protocol is equivalent to the standard chromatic subdivision: $\Xi \cong \mathrm{Ch}$. Take any facet σ of $\mathcal{I}$, let $v_p, w_q \in V(\sigma)$. Because v_p distinguishes w_q, there exists a locally invertible function *decode* for each state of p such that $decode_p(encode_q(w_q), v_p) = w_q$. Thus if $next\_state_p$ has the form $view, state \mapsto decode_p(view, state)$, then $next\_state_p$ is injective in $view$ and returns the state of each vertex in $\mathrm{Lk}(\mathcal{I}, v_p)$. Next, if the values written by *encode* makes all pair of vertices distinguishable, then, from all states in $\mathcal{I}$, the behavior of Ξ_b is the same: from a snapshot, it decodes the views of the other processes, obtaining their respective local states. Thus, $\Xi_b(\mathcal{I})$ has the same behavior as the *full information* protocol Ξ: from reading the snapshot we get the views of the other processes. Thus $\Xi_b(\mathcal{I}) \cong \Xi(\mathcal{I}) \implies \Xi_b(\mathcal{I}) \cong \text{Ch } \mathcal{I}$.

Now for the necessary condition we will prove it by contradiction. Take an input complex composed of two processes p, q where the first process has one state which is adjacent to two different states of the second process. That is $\mathcal{I} = \{v_p, w_q, t_q, \{v_p, w_q\}, \{v_p, t_q\}, \{\}\}$. And suppose that $\Xi_b(\mathcal{I}) \cong \text{Ch } \mathcal{I}$ but v_p does not distinguish w_q nor t_q as $encode_q(w_q) = encode_q(t_q) = e$, making $\mathcal{I}$ not distinguishable under $encode$.

There are only two possibilities for process p: It reads the value e encoded by both states of q or it reads $\perp$. For q, there are two possibilities from each state: q reads the value written by p or not. Therefore, the states where p reads e with all four states of q are compatible configurations. Then $\Xi_b(\mathcal{I})$ will have a vertex of degree 4, which implies that $\Xi_b(\mathcal{I}) \not\cong \text{Ch } \mathcal{I}$ - a contradiction. $\square$

Theorem 1 gives the requirements that the B-IIS protocol must satisfy, in terms of distinguishability, for the protocol complex to be equivalent to the full information one. Note that Theorem 1 takes an arbitrary input complex. Thus, we can iteratively apply the theorem to obtain the protocol complex after an arbitrary number of rounds.

Corollary 1. *Let $\mathcal{I}$ be a chromatic input complex, $r \geq 1$, and $encode : V(\mathcal{A}) \to E$ the encoding function of the B-IIS protocol. The following equivalence holds:*

$$\Xi_b^r(\mathcal{I}) \cong \text{Ch}^r \mathcal{I} \iff \forall r' \in [0, r-1], \text{Ch}^{r'} \mathcal{I} \text{ is distinguishable under } encode$$

Figure 2 shows a commutative diagram illustrating the process. Starting from the initial input complex $\mathcal{I}$, we can equivalently apply r times either Ξ_b or Ch to yield the r-th round protocol complex. Subsequently, the decision function is applied yielding the output complex.

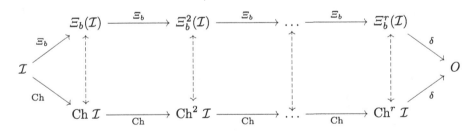

Fig. 2. Commutative diagram illustrating the equivalence between Ξ_b and Ch over an input complex $\mathcal{I}$ after r iterations. Dashed arrows indicate the existence of an isomorphism.

We will further develop the requirements for the encoding function to ensure distinguishability of a simplicial complex. To accomplish this, we define the indistinguishability graph. The idea is that for each process we get a graph that describes the vertices that need to have a different encoding for the simplicial complex to be distinguishable under an arbitrary encoding function. Figure 3 presents an illustrative instance of a simplicial complex and its associated indistinguishability graph for one of the processes.

Definition 3 (Indistinguishability graph of a chromatic complex). *Let $\mathcal{A}$ chromatic simplicial complex, $p \in \Pi$ and $v, w \in V(\mathcal{A})|_p : v \neq w$. We define the **indistinguishability graph** $G_p : \mathcal{A} \times \Pi \to (V(\mathcal{A}), V(\mathcal{A}) \times V(\mathcal{A}))$ of $\mathcal{A}$ with respect to process p as follows:*

$$(v, w) \in G_p(\mathcal{A}) \iff \exists t \in V(\mathcal{A}) : v, w \in V(\mathrm{Lk}(\mathcal{A}, t))|_p$$

Fig. 3. Example of a simplicial complex along with its corresponding indistinguishability graph of the green-labeled process. The relative positions of the green vertices are preserved in the drawing. (Color figure online)

The following theorem characterizes the problem of determining whether an encoding renders an input complex distinguishable. Additionally, it addresses the challenge of finding an encoding function for an arbitrary simplicial complex. This is achieved by establishing the equivalence between these problems and the task of finding a vertex coloring for the indistinguishability graphs associated with a simplicial complex.

Theorem 2. *Let $\mathcal{A}$ be a chromatic input complex with process labelling in Π and $\mathrm{encode} : V(\mathcal{A}) \to E$ an encoding function,*

$\mathcal{A}$ is distinguishable under $\mathrm{encode} \iff \forall p \in \Pi$, encode is a proper vertex coloring of $G_p(\mathcal{A})$

Proof. We will prove the necessary and sufficient conditions by contradiction. For the former, suppose that there is an edge $(v_p, w_p) \in G_p(\mathcal{A})$ such that $\mathrm{encode}(v_p) = \mathrm{encode}(w_p)$. Because such edge is in $G_p(\mathcal{A})$, by definition of indistinguishability graph, it implies that there exists $t_q \in V(\mathcal{A})$ such that t_q is adjacent to v_p and w_p in the protocol complex. However, as both vertices have the same value under encode, t_q cannot distinguish v_p and w_p. A contradiction, as we assumed that $\mathcal{A}$ is distinguishable under encode. Now for the sufficient condition, suppose that $\mathcal{A}$ is not distinguishable but encode is a proper vertex coloring of $G_p(\mathcal{A})$. That implies that there is a vertex t_q that is adjacent to w_p and v_p such that $\mathrm{encode}(v_p) = \mathrm{encode}(w_p)$. As $w_p, v_p \in \mathrm{Lk}(\mathcal{A}, t_q)$ we have that (w_p, v_p) in $G_p(\mathcal{A})$. A contradiction, because both vertices have the same color in encode. $\square$

As an outcome of Theorem 2, we can now provide insights into the amount of information that the encoding function needs to write by leveraging well-known results from graph theory. Hence, the bit complexity required to satisfy Theorem 1 corresponds to the maximum chromatic number among all graphs in $\{G_p(\mathcal{A})\}_{p\in\Pi}$. The encoding function does not have to encode any information about the process state but a distinct value that renders the process state distinguishable within the protocol complex. Then, the *next_state* function can decode the message using the coloring of the input complex. Thus, the amount of information to write can be expressed in terms of the cardinality of the encoding set: $|\operatorname{Im} encode|$. It follows that to implement the encoding function we will need at least $\log_2(|\operatorname{Im} encode|)$ bits, giving the bit complexity of the B-IIS protocol.

Corollary 2. *Let $\mathcal{A}$ a chromatic simplicial complex, $encode : V(\mathcal{A}) \to E$ encoding function and $\omega(G)$ the size of the largest clique in G. The following condition gives a lower bound on the size of the encoding set:*

$$\mathcal{A} \text{ is distinguishable under } encode \implies |\operatorname{Im} encode| \geq \max_{p\in\Pi} \omega(G_p(\mathcal{A}))$$

Proof. If $\mathcal{A}$ is distinguishable under *encode*, then by Theorem 2, *encode* is a proper vertex coloring of the family of graphs $G_\Pi(\mathcal{A}) = \{G_p(\mathcal{A})\}_{p\in\Pi}$. Let ω be the size of the biggest clique of the graphs in $G_\Pi(\mathcal{A})$. In order to color a clique of size ω, we need at least ω different colors. As a result, the cardinality of the encoding set - $|\operatorname{Im} encode|$ - has to be at least as big as ω. □

Corollary 3. *Let $\mathcal{A}$ chromatic simplicial complex and $\Delta(G)$ the biggest vertex degree in G. There exists an encoding function $encode : V(\mathcal{A}) \to E$ such that $\mathcal{A}$ is distinguishable under $encode$ and $|\operatorname{Im} encode| \leq \max_{p\in\Pi} \Delta(G_p(\mathcal{A})) + 1$.*

Proof. Let *encode* be an arbitrary encoding function, in order for $\mathcal{A}$ to be distinguishable under *encode*, we require *encode* to be a proper vertex coloring of all graphs in $\{G_p(\mathcal{A})\}_{p\in\Pi}$. By Brook's theorem [4], the chromatic number of a graph G is less or equal to $\Delta(G)+1$. That means that there exists a proper vertex coloring function *encode* whose size of its encoding set $|\operatorname{Im} encode| \leq \Delta(G) + 1$. Let δ_p be a proper coloring of $G_p(\mathcal{A})$, because each G_p takes the vertices of different processes, we can define the *encode* by using each δ_p for the vertices of process p. Then, we have that *encode* is a proper coloring of all the graphs in $\{G_p(\mathcal{A})\}_{p\in\Pi}$ and by Theorem 2, $\mathcal{A}$ is distinguishable under *encode*. Moreover, by Brook's theorem $|\operatorname{Im} \delta_p| \leq \Delta(G_p(\mathcal{A})) + 1$. Thus, the cardinality of the encoding set will be as small as the biggest $\Delta(G_p(\mathcal{A})) + 1$. □

It follows directly from Theorem 2 that finding the smallest encoding set is computationally intractable. As it is equivalent to determining the chromatic number of a graph. Note that the hardness arises from the input complex, which can be arbitrarily large.

Corollary 4. *Given a chromatic simplicial complex $\mathcal{A}$, determining a function $encode$ such that (i) $\mathcal{A}$ is distinguishable under $encode$ and (ii) $|\operatorname{Im} encode|$ is minimal, is NP-Hard.*

Proof. We will show that given any graph H, we can build a chromatic input complex $\mathcal{H}$, such that $G_{p_0}(\mathcal{H}) = H$. Thus, the set of indistinguishability graphs is the set of all graphs.

For each edge $(v_i, v_j) \in H$, we build the simplicial complex $\mathcal{H}_{ij} = \{\{(p_0, v_i), (p_1, v_{ij})\},$

$\{(p_1, v_{ij}), (p_0, v_j)\}, \{(p_0, v_i)\}, \{(p_1, v_{ij})\}, \{(p_0, v_j)\}, \emptyset\}$. Then we build $\mathcal{H}$ by doing the union of all these simplices: $\mathcal{H} = \bigcup_{(v_i, v_j) \in H} \mathcal{H}_{ij}$. That is, we build a 2-process simplicial complex by adding a pair of edges for each edge in H. Then, we have that $(p_0, v_i), (p_0, v_j) \in V(\mathrm{Lk}(\mathcal{H}, (p_1, v_{ij}))) \ \forall i, j$. As these are the only vertices in the link, we have that $H = G_{p_0}(\mathcal{H})$.

As a result, the optimization problem requires determining the chromatic number of arbitrary big graphs, which is NP-Hard. □

6   f-Vector Analysis for Distinguishability and Iterated Chromatic Subdivision

Corollaries 2 and 3 establish lower and upper bounds on the cardinality of the encoding set $|\mathrm{Im}\ encode|$ w.r.t the clique size and degree of the indistinguishability graphs. We can now forget about the task to solve, concurrency and the B-IIS protocol and define the information needed to be written as a counting problem over a topological space: the iterated standard chromatic subdivision of a chromatic simplicial complex.

First, in Sect. 6.1 we give the necessary definitions used throughout this section. Then in Sect. 6.2, we establish the relationship between the clique size and vertex degree of indistinguishability graphs and their associated simplicial complex. Then in Sect. 6.3 we investigate the behavior of the f-vector under chromatic subdivision. In Sect. 7, we conduct an asymptotic analysis of the equations obtained to derive asymptotic lower and upper bounds on the size of the encoding set.

6.1   Definitions

Additional Standard Constructions on Simplicial Complexes. Given a simplicial complex $\mathcal{A}$, the *open star* of $\mathcal{S} \subseteq \mathcal{A}$, $\mathrm{St}^\circ(\mathcal{A}, \mathcal{S})$ is defined as: $\mathrm{St}^\circ(\mathcal{A}, \mathcal{S}) = \{\sigma \in \mathcal{A} : \mathcal{S} \subseteq \sigma\}$. Note that the open star does not yield a simplicial complex but just a topological space. The boundary of $\mathcal{A}$ is defined as $\partial(\mathcal{A}) = \{\sigma \in \mathcal{A} : \sigma \subset G \text{ for a unique facet } G \in \mathcal{A}\} \cup \{\emptyset\}$. Then we can define the interior of $\mathcal{A}$ as $Int\ \mathcal{A} = \mathcal{A} \setminus \partial(\mathcal{A})$. Notice that the boundary is a simplicial complex but the interior is not. The simplicial join of two simplicial complexes $\mathcal{A}$ and $\mathcal{B}$, denoted as $\mathcal{A} * \mathcal{B}$ is the simplicial complex with set of vertices $V(\mathcal{A}) \cup V(\mathcal{B})$ whose faces are all the unions $\alpha \cup \beta, \alpha \in \mathcal{A}$ and $\beta \in \mathcal{B}$. Given a non-negative integer l, the l-skeleton of a simplicial complex $\mathcal{A}$, denoted as $skel^l(\mathcal{A})$, is the set of simplices of $\mathcal{A}$ with dimension at most l. In particular, $skel^0(\mathcal{A}) = V(\mathcal{A})$.

f-Vector. The f-*vector* of a simplicial complex $\mathrm{f}(\mathcal{A}) = (\mathrm{f}_{-1}(\mathcal{A}), \mathrm{f}_0(\mathcal{A}), \dots, \mathrm{f}_n(\mathcal{A}))$ has the number of k-dimensional faces in its k-th coordinate: $\mathrm{f}_k(\mathcal{A}) = |\{\sigma \in \mathcal{A} :$

$dim(\sigma) = k\}|$ for $-1 \leq k \leq n$. In particular, if $\mathcal{A} \neq \emptyset$ we have that $\emptyset \in \mathcal{A}$ and $f_{-1}(\mathcal{A}) = 1$. Enumerating the faces of a simplicial complex is a recurrent problem in algebraic combinatorics and discrete geometry. An overview of these kind of problems is presented in [13].

6.2   f-Vector and the Indistinguishability Graphs

Lemma 1. *Let $\mathcal{A}$ be a chromatic simplicial complex, the following is a bound on the clique number of the indistinguishability graphs:*

$$\max_{p \in \Pi} \omega(G_p(\mathcal{A})) \geq \max_{p \in \Pi} \max_{v \in V(\mathcal{A})} f_1(\mathrm{St}^\circ(\mathcal{A}, v)|_p)$$

Proof. Let $v_p \in V(\mathcal{A})$ such that v_p has the biggest number of adjacent vertices in $\mathcal{A}$ of a single process $q \in \Pi$. This can be enumerated by taking the f-vector of the open star of v_p, and taking only the 1-faces, which by definition is the number of edges that v_p has. Note that in $\mathcal{A}$, all vertices adjacent to v_p of process q are contained in $\mathrm{Lk}(\mathcal{A}, v_p)|_q$, then by definition of indistinguishability graph, all vertices of $\mathrm{Lk}(\mathcal{A}, v_p)|_q$ will form a clique in $G_p(\mathcal{A})$ with the cardinality of the set. Thus, the biggest clique has to be greater or equal to the size of such clique.   □

Lemma 2. *Let $\mathcal{A}$ be a chromatic simplicial complex, the following equality holds:*

$$\max_{p \in \Pi} \Delta(G_p(\mathcal{A})) = \max_{p \in \Pi} \max_{v \in V(\mathcal{A})} f_0(\mathrm{Lk}(\mathcal{A}, \mathrm{St}(\mathcal{A}, v))|_p)$$

Proof. Let $v_p \in V(\mathcal{A})$ be the vertex that gives the maximum argument on the right side of the equality, we have that $\mathrm{St}(\mathcal{A}, v_p)$ gives the subcomplex of facets containing v_p. Recall that all vertices that are adjacent to v_p are contained in the star. Then, we take the f_0 of the link of the star, enumerating the vertices that are adjacent to the star (and then we restrict to a particular process label). Note that if we take a vertex $t \in \mathrm{St}(\mathcal{A}, v_p)$, then if $w_p \in \mathrm{Lk}(\mathcal{A}, \mathrm{St}(\mathcal{A}, v_p)) \implies (v_p, w_p) \in G_p(\mathcal{A})$. But also if we have an edge $(v'_p, w'_p) \in G_p(\mathcal{A})$ it implies that it is in the link of the star. So finding the maximum link of the star of a vertex v restricting to a process p gives the maximum vertex degree of $G_p(\mathcal{A})$.   □

6.3   f-Vector of the Standard Chromatic Subdivision

The objective of this section is to derive equations enabling the computation of the f-vector of the iterated chromatic subdivision using the f-vector of the original simplicial complex. Initially, we present a general theorem that characterizes the f-vector of any subdivision map. Subsequently, we demonstrate its instantiation for both the standard chromatic subdivision and open stars. The motivation behind these equations is to establish a method for computing the f-vectors in Lemmas 1 and 2 based on the f-vector of the initial simplicial complex.

Theorem 3 (f-vector of a subdivision). *Let τ be a subdivision operator, Δ^i the i-dimensional simplex, and $\mathcal{A}$ an n-dimensional simplicial complex. The following identity holds for the f-vector of $\tau(\mathcal{A})$:*

$$f_k(\tau(\mathcal{A})) = \sum_{i=k}^{n} f_i(\mathcal{A})\, f_k(Int\ \tau(\Delta^i))$$

Proof. Let k be a proper coordinate of the f-vector, $-1 \leq k \leq n$, we will prove the identity by induction over the skeleton of $\mathcal{A}$. Let $S(i) = f_k(\tau(Skel^i \mathcal{A}))$, we will prove that $S(i) = S(i-1) + f_i(\mathcal{A}) \cdot f_k(Int\ \tau(\Delta^i))$. Note that $Skel^n \mathcal{A} = \mathcal{A}$. For the base case, notice that if $i < k$, then $f_k(\tau(Skel^i \mathcal{A})) = 0$. Because by definition of skeleton, $dim(Skel^i \mathcal{A}) \leq i$ and a subdivision operator cannot yield a simplicial complex of higher dimension than its input. For the inductive step, we define $K := \tau(Skel^i \mathcal{A}) - \tau(Skel^{i-1} \mathcal{A})$, which is the set of $(i+1)$-tuples of $\tau(Skel^i \mathcal{A})$. Note that we use the term tuples because these are not proper simplices but subsets of $V(\mathcal{A})$ of size $i+1$. Then we have that $K \cup \tau(Skel^{i-1} \mathcal{A}) = \tau(Skel^i \mathcal{A})$ and $K \cap \tau(Skel^{i-1} \mathcal{A}) = \emptyset$. Thus, $f_k(K) + f_k(\tau(Skel^{i-1} \mathcal{A})) = f_k(\tau(Skel^i \mathcal{A}))$.

From the inductive hypothesis we have that $f_k(\tau(Skel^{i-1} \mathcal{A})) = S(i-1)$. It remains only to enumerate $f_k(K)$. First, note that all the $(i+1)$-tuples of K come from the subdivision of an i-face of $\mathcal{A}$ and we have in total $f_i(\mathcal{A})$ i-faces in $\mathcal{A}$. Thus, we have to enumerate how many k-faces are generated from subdividing the i-faces of $\mathcal{A}$. To know how many i-faces are yielded we apply τ to each i-face of $\mathcal{A}$. Let $\sigma \in Skel^i \mathcal{A}$ an i-face, note that σ itself is a simplex of dimension i - that is, Δ^i. It remains to recall that $f_k(\tau(\sigma)) = f_k(\tau(\partial(\sigma))) + f_k(\tau(Int\ \sigma))$. Where $f_k(\tau(\partial(\sigma)))$ is already counted on $S(i-1)$, as $\tau(\partial(\sigma)) \subseteq \tau(Skel^{i-1}\mathcal{A})$. This leaves that the new faces yielded are in $Int\ \tau(\sigma^i)$. However, by definition of interior if we take $\sigma' \in Skel^i \mathcal{A} : \sigma' \neq \sigma$, we get $Int\ \tau(\sigma) \cap \tau(\sigma') = \emptyset$. Thus for the purposes of enumerating faces, we can replace σ with Δ^i.

As we have $f_i(\mathcal{A})$ i-faces, the new i-faces generated by the subdivision are $f_i(\mathcal{A}) \cdot f_k(Int\ \tau(\Delta^i))$. This concludes the inductive step. By solving the now proven recurrence relation, we get the final summation expression. $\qquad\square$

Theorem 3 makes it possible to express the f-vector of the subdivision of any simplicial complex in terms of its given f-vector and the application of the subdivision operator on Δ^n. Thus, if we can provide an expression for $f_k(Int\ \tau(\Delta^n))$, we can compute the subdivision for any simplicial complex. Note that we cannot use Theorem 3 to compute the f-vector of open stars since they do not form a simplicial complex. Thus, we present a reformulation for the specific case of open stars.

Lemma 3. *Let τ be a subdivision operator, $\mathcal{A}$ a n-dimensional simplicial complex, $v \in V(\mathcal{A})$ and $r \in V(\Delta^i)$. The following identity holds for the f-vector of $St^\circ(\tau(\mathcal{A}), v)$:*

$$f_k(St^\circ(\tau(\mathcal{A}), v)) = \sum_{i=k}^{n} f_i(St^\circ(\mathcal{A}, v))\, f_k(Int\ St^\circ(\tau(\Delta^i), v))$$

Proof. We apply can apply the same reasoning as in Theorem 3. However, as we are now enumerating the k-faces in an open star, some of the i-faces yielded in $Int\ \tau(\Delta^i)$ are not included in $St°(\tau(\Delta^i)$. Indeed, the ones that are in the open star are $Int\ St°(\tau(\Delta^i), r)$. Substituting the new internal k-faces using this term in Theorem 3 yields the desired expression. □

Lemma 4. *Let $v \in \Delta^n$ and $k > 0$, the following identity holds:*

$$\mathrm{f}_k(Int\ \mathrm{St}°(\mathrm{Ch}\ \Delta^n, v)) = \sum_{i=1}^{k} \binom{n}{i} \mathrm{f}_{k-i}(\mathrm{St}°(\mathrm{Ch}\ \Delta^{n-i}, v))$$

Proof. We aim to enumerate the k-faces in $Int\ \mathrm{Ch}\ \Delta^n$ that include the vertex v. Moreover, as the faces are in the interior, they must necessarily include at least one vertex from the corresponding facet in the central simplex which has nodes with colors different from v. We note this facet as σ.

Hence, the number of faces will be all possible simplicial joins of v with this facet and then with faces on the boundary of $St°(\mathrm{Ch}\ \Delta^n, v)$.

Let $k > 1$ (by definition of interior, if k $= 0$, we have $\mathrm{f}_0(Int\ St°(\mathrm{Ch}\ \Delta^n, v)) = 0$). Suppose we want to construct a k-face from an $(i - 1)$-face of σ. To do this, we take a face $\delta \in \sigma : dim(\delta) = i - 1$ and perform the simplicial join with v: $\delta * v$. Note that we can choose $\binom{n}{i}$ different $i - 1$ faces. We perform the simplicial join $\delta * v$ and obtain an i-face in the interior. Now, if we want a k-face, we have no other option but to perform a second simplicial join with another face of dimension $(k - i)$ on $\partial\ St°(\mathrm{Ch}\ \Delta^n, v)$.

The faces we can join with δ are those that do not contain any of the colors of δ; these are found in the subdivision of one of the $(k - i)$-faces of Δ^n. Note that, as it is a subdivision, if we restrict ourselves to one of these faces, it is equal to the subdivision of a simplex of the dimension of such face. Therefore, the faces we can join are $\mathrm{f}_{k-i}(\Delta^{n-i})$. To obtain all faces, we have to enumerate from taking a single vertex from σ $(dim(\delta * v) = 1)$ to taking a $k - 1$ face $(dim(\delta * v) = k)$, thus obtaining the formulated equation. □

Note that Lemma 3 applies to a generic simplicial complex. Consequently, we can use an already subdivided complex as input to obtain an expression for the iterated chromatic subdivision. Corollary 5 directly follows from Lemma 3 and Lemma 4.

Corollary 5. *Let $r > 0$, $\mathcal{A}$ a simplicial complex, $v \in V(\mathcal{A})$, and $v' \in V(\Delta^i)$. The f-vector of the open star in v of the r-th chromatic subdivision of $\mathcal{A}$ is:*

$$\mathrm{f}_k(\mathrm{St}°(\mathrm{Ch}^r\ \mathcal{A}, v)) = \sum_{i=k}^{n} \mathrm{f}_i(\mathrm{St}°(\mathrm{Ch}^{r-1}\ \mathcal{A}, v)) \sum_{j=1}^{k} \binom{i}{j} \mathrm{f}_{k-j}(\mathrm{St}°(\mathrm{Ch}\ \Delta^{i-j}, v'))$$

Note that $\mathrm{f}_i(\mathrm{St}°(\mathcal{A}, v))$ represents an input value for the problem. Specifically, it denotes the i-faces containing the vertex v. A straightforward yet crucial observation is that we now have an expression to compute the f-vector of the open star of $\mathrm{Ch}\ \Delta^n$:

Corollary 6. *Let* $v \in V(\Delta^n)$. *The* f-*vector of the open star in* v *of* Ch Δ^n *is:*

$$f_k(\mathrm{St}^\circ(\mathrm{Ch}\ \Delta^n, v)) = \sum_{i=k}^{n} \binom{n}{i} \sum_{j=1}^{k} \binom{i}{j} f_{k-j}(\mathrm{St}^\circ(\mathrm{Ch}\ \Delta^{i-j}, v))$$

Proof. The result follows directly from Corollary 5. Notice that to calculate $f_k(Int\ \mathrm{St}^\circ(\mathrm{Ch}\ \Delta^n, v))$ we need to know $f_{k'}(\mathrm{St}^\circ(\mathrm{Ch}\ \Delta^{n'}, v))$ where $k' \in [0, k-1]$ and $n' \in [0, n-1]$, thus the recursion is well defined. □

An interesting observation from Corollary 6 is that $\{f_n(\mathrm{St}^\circ(\mathrm{Ch}\ (\Delta^n, v))\}_{n \geq 0}$ is a well-known sequence known as ordered Bell numbers or Fubini numbers, which count the number of weak orderings on a set of n elements [9, 20].

We derive an equation to compute the f-vector in Lemma 2, which is used to determine the vertex degree on the indistinguishability graphs.

Lemma 5. *Let* $p \in \Pi$, $\mathcal{A}$ *a simplicial complex,* $v_p \in V(\mathcal{A})|_p$. *The following identity holds:*

$$f_0(\mathrm{Ch}\ \mathcal{A}, \mathrm{Lk}(\mathrm{St}(\mathrm{Ch}\ \mathcal{A}, v_p))|_p) = \sum_{i=1}^{n} f_i(\mathrm{St}^\circ(\mathcal{A}, v_p)))$$

Proof. We want to count all vertices with label p that are at distance 2 from v_p in Ch $\mathcal{A}$. St(Ch $\mathcal{A}, v_p$) gives the faces that contain v_p, and by counting the vertices in the link of the latter we are counting all vertices that are at distance two of v_p. We have that for each face of $\mathrm{St}^\circ(\mathcal{A}, v_p)$, when subdivided by Ch, will yield a single new vertex with label p. Moreover, such vertex will be at distance 2 from v_p, as v_p will be adjacent to the interior vertices with different label than v_p while these vertices will be adjacent to the interior vertex with label p. Moreover, all vertices of $\mathrm{Lk}(\mathcal{A}, v_p)$ are at distance 2 from v_p in Ch $\mathcal{A}$, so the new vertices yielded are only the ones at distance 2 in Ch $\mathcal{A}$. □

Finally, we need to consider that, in Lemmas 2 and 1, we search for the vertex with the biggest f-vector at coordinates 0 and 1 in Ch^r $\mathcal{A}$. Lemma 6 shows that, for iterated subdivisions, the biggest f-vector of an open star comes from the vertices that are at the first subdivision. This is a crucial result, as it allows us to fix the vertex with the biggest f-vector and perform asymptotic analysis of the f-vector on this vertex under iterated subdivisions.

Lemma 6. *Let* τ *be a subdivision operator and* $\mathcal{A}$ *a* n-*dimensional simplicial complex.*

$$\underset{v \in V(\tau^r(\mathcal{A}))}{\arg\max}\ f_k(\mathrm{St}^\circ(\tau^r(\mathcal{A}), v)) \in V(\tau(\mathcal{A})), \qquad \forall k \leq n, \forall r \geq 1$$

Proof. We will proceed by induction on r for an arbitrary k. The base step $r = 1$ is trivial. For the inductive step, let $v = \arg\max_{v \in V(\tau^r(\mathcal{A}))} f_k(\mathrm{St}^\circ(\tau^r(\mathcal{A}), v))$. By inductive hypothesis, $v \in V(\tau(\mathcal{A}))$. We have that $\forall \tilde{v} \in$

$V(\tau^r(\mathcal{A})), \mathrm{f}_k(\mathrm{St}^\circ(\tau^r(\mathcal{A}), \tilde{v})) \leq \mathrm{f}_k(\mathrm{St}^\circ(\tau^r(\mathcal{A}), v))$. From Lemma 3, we have that $\mathrm{f}_k(\mathrm{St}^\circ(\tau^{r+1}(\mathcal{A}), v))$ is computed as a sum of the faces in $\mathrm{St}^\circ(\tau^r(\mathcal{A}), v)$ multiplied by a constant that come from the simplex subdivision $\mathrm{f}_k(Int\ \mathrm{St}^\circ(\tau(\Delta^i), v))$. Thus, $\forall \tilde{v} \in V(\tau^r(\mathcal{A})), \mathrm{f}_k(\mathrm{St}^\circ(\tau^{r+1}(\mathcal{A}), \tilde{v})) \leq \mathrm{f}_k(\mathrm{St}^\circ(\tau^{r+1}(\mathcal{A}), v))$. Next, the star of the new vertices $V(\tau^{r+1}(\mathcal{A})) \setminus V(\tau^r(\mathcal{A}))$ will have a constant f-vector, irregardless of r, based on the dimension of the face they are a central simplex of. Hence, the vertex with the biggest f-vector (in all its coordinates) of its open star will still be v.   □

Remark 1. It follows directly from Lemma 6 that $\max_{v \in V(\mathrm{Ch}^r\ \mathcal{A})} \mathrm{f}_0(\mathrm{Lk}(\mathrm{Ch}^r\ \mathcal{A}, \mathrm{St}(\mathrm{Ch}^r\ \mathcal{A}, v)))|_p \in V(\mathrm{Ch}\,(\mathcal{A}))$ as Lemma 5 shows that it is computed as the sum of open stars.

Putting everything together, we can use the derived equations from Lemmas 2 and 1, along with Lemma 6, to establish upper and lower bounds on the existence of an encoding set with size within these bounds:

Theorem 4. *Let $\mathcal{I}$ a chromatic input complex and $r > 0$. There exists a function encode : $V(\mathcal{I}) \to E$ such that when used as encoding function in the B-IIS protocol, $\Xi_b^r \cong \Xi^r$ and $|E|$ is bounded by:*

$$\max_{v \in V(\mathrm{Ch}\ \mathcal{I})} \mathrm{f}_1(\mathrm{St}^\circ(\mathrm{Ch}^r\ \mathcal{I}, v)) \leq |\mathrm{Im}\ encode| \leq \max_{v \in V(\mathrm{Ch}\ \mathcal{I})} \mathrm{f}_0(\mathrm{Lk}(\mathrm{Ch}^r\ \mathcal{I}, \mathrm{St}(\mathrm{Ch}^r\ \mathcal{I}, v))) + 1$$

7 Asymptotic Analysis of f-Vector of the Iterated Chromatic Subdivision

Recalling that our objective is to obtain asymptotic bounds on the size of the encoding set. Now, using the inequality of Theorem 4 and the f-vector equations derived in Sect. 6, we can perform an asymptotic analysis w.r.t the number of processes in the system (directly corresponding to the dimension of the input complex) and the number of rounds r. Thus, our objective will be to find tight asymptotic bounds for $\mathrm{f}_1(\mathrm{St}^\circ(\mathrm{Ch}^r\ \mathcal{A}, v))$ and $\mathrm{f}_0(\mathrm{Lk}(\mathrm{Ch}^r\ \mathcal{A}, \mathrm{St}(\mathrm{Ch}^r\ \mathcal{A}, v)))$.

We will use the conventional definitions of asymptotic analysis in computer science. If the reader is not familiar with this, we recommend checking [14] for reference.

The following combinatorial identity will be used in the proof of Lemma 8:

Lemma 7. *Let $n, k, r \in \mathbb{N}$ such that $r \leq r \leq n$, the following identity holds:*

$$\sum_{i=k}^{n} \binom{n}{i}\binom{i}{r} b^{i-\alpha}(i-r)_{k-r} = \frac{b^{k-\alpha}}{r!}(b+1)^{n-k}(n)_k$$

where $(n)_k$ is the falling factorial.

Proof. The result follows from the binomial theorem and the factorial formulation of the binomial coefficient:

$$\sum_{i=k}^{n} \binom{n}{i}\binom{i}{r} b^{i-\alpha}(i-r)_{k-r} = \sum_{i=k}^{n} \frac{n!}{(n-i)!r!(i-k)!} b^{i-\alpha} = \frac{n!}{r!}\frac{(n-k)!}{(n-k)!}\sum_{i=0}^{n-k}\frac{b^{i+k-\alpha}}{(n-k-i)!i!}$$

$$= \frac{(n)_k}{r!} b^{k-\alpha}\sum_{i=0}^{n-k}\binom{n-k}{i}b^{i} = \frac{b^{k-\alpha}}{r!}(b+1)^{n-k}(n)_k$$

$\square$

Lemma 8. *Let Δ^n be the n-dimensional simplex, $v \in V(\Delta^n)$, $(n)_k$ the falling factorial, and $k \leq n$, the following is a tight asymptotic bound on the f-vector of the open star of v:*

$$f_k(\mathrm{St}^{\circ}(\mathrm{Ch}\ \Delta^n, v)) \in \Theta\left(\frac{(k+1)^{n-k}(n)_k}{\ln(2)^{k-1}}\right)$$

Proof. First, to ease notation, let $T(k,n) := f_k(\mathrm{St}^{\circ}(\mathrm{Ch}\ \Delta^n, v))$, and define the bounding function $f(k,n) := \frac{(k+1)^{n-k}(n)_k}{\ln(2)^{k+1}}$. It is important to notice that in order to compute $T(k,n)$, we require the values $T(k', n') : k' \in [0, k-1] \wedge n' \in [0, n-1]$. Consequently, we can prove the result through a general induction over n for all k.

The base case is trivial: $T(0,n) \stackrel{\text{def}}{=} 1 \in \Theta(1) = \Theta(f(0,n)), \forall n \in \mathbb{N}$. For the induction step, we will prove it by computing the limit of $\frac{T(k,n)}{f(k,n)}$ as n goes to infinity for all $k \leq n$ and showing this limit converges to a non-zero constant.

$$f_k(\mathrm{St}^{\circ}(\mathrm{Ch}\ \Delta^n, v)) \in \Theta(f(k,n)) \stackrel{\mathrm{IH}}{\Longleftrightarrow} \sum_{i=k}^{n}\binom{n}{i}\sum_{j=1}^{k}\binom{i}{j}\frac{(k-j+1)^{i-k}(i-j)_{k-j}}{\ln(2)^{k-j-1}}$$

$$\stackrel{\text{Lemma 7}}{=} \sum_{j=1}^{k}\frac{\ln(2)^j}{\ln(2)^{k-1}j!}(k-j+2)^{n-k}(n)_k = \frac{(n)_k}{\ln(2)^{k-1}}\sum_{j=1}^{k}\frac{(k-j+2)^{n-k}\ln(2)^j}{j!} \in \Theta(f(k,n))$$

$$\stackrel{\forall k \leq n}{\Longleftrightarrow} \lim_{n\to\infty}\frac{\frac{(n)_k}{\ln(2)^{k-1}}\sum_{j=1}^{k}\frac{(k-j+2)^{n-k}\ln(2)^j}{j!}}{(n)_k(k+1)^{n-k}\ln(2)^{-k+1}} = C > 0 \stackrel{\forall k \leq n}{\Longleftrightarrow} \lim_{n\to\infty}\sum_{j=1}^{k}(\frac{k-j+2}{k+1})^{n-k}\frac{\ln(2)^j}{j!} = C > 0$$

We will now show that the limit of the last equation is in the interval $(\ln 2, 1)$ for all values of $k \leq n$. For the lower bound, noting that the term inside the sum is strictly positive, we take the first term of the sum:

$$\lim_{n\to\infty}\sum_{j=1}^{k}(\frac{k-j+2}{k+1})^{n-k}\frac{\ln(2)^j}{j!} > \lim_{n\to\infty}(\frac{k+1}{k+1})^{n-k}\frac{\ln 2}{1!} = \ln 2 > 0 \ \ \forall k \leq n$$

For the upper bound we use the fact that $\sum_{j=1}^{\infty}\frac{\ln(2)^j}{j!} = 1$ and $\frac{k-j+2}{k+1} \leq 1 \forall k \leq n$. Thus the following series is a multiplication of a convergent sequence by another one which their values are in $(0, 1]$:

$$\lim_{n\to\infty}\sum_{j=1}^{k}(\frac{k-j+2}{k+1})^{n-k}\frac{\ln(2)^j}{j!} < \lim_{n\to\infty}\sum_{j=1}^{\infty}(\frac{k-j+2}{k+1})^{n-k}\frac{\ln(2)^j}{j!} < 1 \ \ \forall k \leq n$$

Consequently, the asymptotic ratio between $\frac{T(k,n)}{f(k,n)}$ is in the interval $(\ln 2, 1)$ $\forall k \leq n$. It follows that $f(k,n)$ is a tight asymptotic bound of $T(k,n)$ □

Now that we have a tight asymptotic bound on $f_k(\mathrm{St}^\circ(\mathrm{Ch}\ (\Delta^n), v))$, we can now give a bound for the equation in Lemma 4. Which will be useful for computing the final result regarding the f-vector of the iterated chromatic subdivision:

Lemma 9. *Let Δ^n be the n-dimensional simplex, and $1 \leq k \leq n$,*

$$\sum_{j=1}^{k} \binom{n}{j} f_{k-j}(\mathrm{St}^\circ(\mathrm{Ch}\ \Delta^{n-j}, v)) \in \Theta\left(\frac{(k+1)^{n-k}(n)_k}{\ln(2)^{k-1}}\right)$$

Proof. Defining $T(k,n)$ as in Lemma 8, let $R(k,n) := \sum_{j=1}^{k} \binom{n}{j} T(k-j, n-j)$ and $f(k,n) := \frac{(k+1)^{n-k}(n)_k}{\ln(2)^{k+1}}$.

$$R(k,n) \in \Theta(f(k,n)) \overset{\text{Lemma } 8}{\Longleftrightarrow} \sum_{j=1}^{k} \binom{n}{j} \frac{(k-j+1)^{i-k}(i-j)_{k-j}}{\ln(2)^{k-j-1}}$$

$$= \frac{(n)_k}{\ln(2)^{k-1}} \sum_{j=1}^{k} \frac{(k-j+1)^{n-k}\ln(2)^j}{j!} \in \Theta(f(k,n)) \overset{\forall k \leq n}{\Longleftrightarrow} \lim_{n \to \infty} \sum_{j=1}^{k} (\frac{k-j+1}{k+1})^{n-k} \frac{\ln(2)^j}{j!} = C > 0$$

where the previous limit is the same as in Lemma 8 but with $k - j + 1$ instead of $k - j + 2$ on the numerator and $\frac{k-j+1}{k+1} < 1$. Thus the limit is non-zero convergent.   □

We are now ready to give a tight asymptotic bound on the iterated chromatic subdivision of a simplicial complex.

Theorem 5 (Asymptotic bound on the iterated chromatic subdivision). *Let $r > 0$, $\mathcal{A}$ a n-dimensional simplicial complex and $v \in \mathcal{A}$. The following is a tight asymptotic bound on the f-vector of the star of v in $\mathrm{Ch}^r\ \mathcal{A}$:*

$$f_k(\mathrm{St}^\circ(\mathrm{Ch}^r\ \mathcal{A}, v)) \in \Theta\left(\left(\frac{(k+1)^{n-k}(n)_k}{\ln(2)^{k-1}}\right)^r\right)$$

Proof. Again, to relief notation we write $P(k,r) := f_k(\mathrm{St}^\circ(\mathrm{Ch}^r\ \mathcal{A}, v))$, T and R defined as in the proof of Lemma 9. Then we denote the bounding function $f(k,r) := \left(\frac{(k+1)^{n-k}(n)_k}{\ln(2)^{k-1}}\right)^r$. Note that n is implicitly defined as the dimension of $\mathcal{A}$. Now instead of doing induction over n, we will prove it by induction over r for all $k \leq n$.

The base step:

$$P(k,1) \in \Theta(f(k,r)) \overset{\text{Lemma } 9}{\Longleftrightarrow} \sum_{i=k}^{n} P(i,0) \frac{(k+1)^{i-k}(i)_k}{\ln(2)^{k-1}} \in \Theta(f(k,r))$$

$$\Longleftrightarrow \frac{1}{\ln(2)^{k-1}} \sum_{i=k}^{n} P(i,0)(k+1)^{i-k}(i)_k \in \Theta(f(k,r))$$

$P(k,0)$ are constants of the input complex: the f-vector of the open star of a vertex in the simplicial complex. Now we prove that the latter is bounded by showing it is a high and lower bound:

$$\frac{1}{\ln(2)^{k-1}} \sum_{i=k}^{n} P(i,0)(k+1)^{i-k}(i)_k < \frac{1}{\ln(2)^{k-1}} \sum_{i=k}^{n} P(i,0)(k+1)^{n-k}(n)_k$$

$$= \frac{(k+1)^{n-k}(n)_k}{\ln(2)^{k-1}} \sum_{i=k}^{n} P(i,0) \implies P(k,1) \in O(f(k,1))$$

$$\frac{1}{\ln(2)^{k-1}} \sum_{i=k}^{n} P(i,0)(k+1)^{i-k}(i)_k > \frac{(k+1)^{n-k}(n)_k}{\ln(2)^{k-1}} P(n,0) \implies P(k,1) \in \Omega(f(k,1))$$

Thus $P(k,1) \in \Theta(f(k,r))$. Now for the inductive step:

$$P(k,r) \in \Theta(f(k,r)) \overset{\text{HI}}{\Longleftrightarrow} \sum_{i=k}^{n} f(k,r-1) \sum_{j=1}^{k} \binom{i}{j} T(k-j, i-k) \in \Theta(f(k,r))$$

$$\overset{\text{Lemma 9}}{=} f(k,r-1) \sum_{i=k}^{n} \frac{(k+1)^{i-k}(i)_k}{\ln(2)^{k-1}} = f(k,r-1)\frac{1}{\ln(2)^{k-1}} \sum_{i=k}^{n}(k+1)^{i-k}(i)_k \in \Theta(f(k,r))$$

$$\Longleftrightarrow f(k,r-1) \cdot f(k,1) \in \Theta(f(k,r))$$

Notice that $\sum_{i=k}^{n}(k+1)^{i-k}(i)_k \in \Theta((k+1)^{n-k}(n)_k)$ as the expression is the sum of two polynomials over k and i respectively multiplied. The highest degree of both polynomials is obtained when $i = n$. Finally, $f(k,r-1)\cdot f(k,1) = f(k,r)$ completing the proof. □

Note that an intuition for the asymptotic growth of the iterated subdivision is that it is invariant to the round number. As such, we could expect the asymptotic bound to have a functional form $h(k,n)^r$.

An interesting fact is that for $f_n(\mathrm{St}^\circ(\mathrm{Ch}\,\Delta^n, v))$, the following is an approximation of the Fubini numbers [9]: $f_n(\mathrm{St}^\circ(\mathrm{Ch}\,\Delta^n, v)) \sim \frac{n!}{2\ln(2)^{n+1}}$, which corresponds with the tight asymptotic bound of Theorem 5.

The final bound needed to compute the upper limit of the size of the encoding set follows directly from Theorem 5.

Corollary 7. *Let $\mathcal{A}$ be a n-dimensional simplicial complex and $v \in V(\mathcal{A})$:*

$$f_0(\mathrm{Lk}(\mathrm{Ch}^r\,\mathcal{A}, \mathrm{St}(\mathrm{Ch}^r\,\mathcal{A}, v))) \in \Theta\left(\left(\frac{n!n^n}{\ln(2)^{n-1}}\right)^r\right)$$

Proof. As before, to ease notation let $f_n(k,r) := \left(\left(\frac{n!n^n}{\ln(2)^{n-1}}\right)^r\right)$ be the bounding function.

$$f_0(\mathrm{Lk}(\mathrm{Ch}^r\,\mathcal{A}, \mathrm{St}(\mathrm{Ch}^r\,\mathcal{A}, v))) \in \Theta(f_n(k,r)) \overset{\text{Lemma 5}}{\Longleftrightarrow} \sum_{i=1}^{n} f_i(\mathrm{St}^\circ(\mathrm{Ch}^r\mathcal{A}, v)) \in \Theta(f_n(k,r))$$

$$\overset{\text{Theorem 5}}{\Longleftrightarrow} \sum_{i=1}^{n}\left(\frac{(i+1)^{n-i}(n)_i}{\ln(2)^{i-1}}\right)^r \in \Theta(f(k,r)) \Longleftrightarrow \lim_{n\to\infty} \frac{\sum_{i=1}^{n}\left(\frac{(i+1)^{n-i}(n)_i}{\ln(2)^{i+1}}\right)^r}{\left(\frac{n!n^n}{\ln(2)^{n-1}}\right)^r}$$

$$= \lim_{n\to\infty} \sum_{i=1}^{n}\left(\frac{\ln(2)^{n-i}}{(n-i)!} \cdot \left(\frac{i+1}{n}\right)^n \cdot \frac{1}{(i+1)^i}\right)^r = C > 0$$

Notice that in the final equation, the sum can be separated by 3 independent sequences, where each one of them is positive and convergent. Moreover, these sequences are multiplied to the power of r. As all terms are strictly positive, the sum will be grater than 0. And it will converge as the series of multiplied positive sequences are convergent. $\square$.

8   Main Results and Ramifications

Using the asymptotic bounds calculated in Sect. 7, we can derive upper and lower bounds on the (asymptotic) bit complexity of the B-IIS protocol.

Corollary 8. *Let Ξ^r and Ξ_b^r be the protocol maps of the full information and bounded IIS protocols for r rounds, respectively. Let $n > 2$ be the number of processes in the system, and $\mathcal{I}$ be a chromatic input complex. If $\Xi^r(\mathcal{I}) \cong \Xi_b^r(\mathcal{I})$, then $\Omega(rn)$ is a lower bound on the bit complexity of the r-th round of the B-IIS protocol.*

Proof. If $\Xi^r(\mathcal{I}) \cong \Xi_b^r(\mathcal{I})$, by Corollary 1, we need Ch $\mathcal{I}$ to be distinguishable under the encode function associated to Ξ_b. Then by Theorem 4 we need $|\operatorname{Im} encode| \geq \max_{v \in V(\text{Ch } \mathcal{A})} f_1(\text{St}^\circ(\text{Ch}^r \mathcal{A}, v))$. From Theorem 5, we have that $f_1(\text{St}^\circ(\text{Ch}^r \mathcal{A}, v)) \in \Theta((2^{n-1}n)^r)$. We require the function $encode$ to be capable of encoding each one of these values on an entry of $M[k]$. Therefore, we need at least $\Theta(\log_2((2^{n-1}n)^r))$ bits to encode all those values, hence the bit complexity is at least $\Omega(rn)$. $\square$

Corollary 9. *Let Ξ^r and Ξ_b^r be the protocol maps of the full information and bounded IIS protocols for r rounds, respectively. Let $n > 2$ be the number of processes in the system, and $\mathcal{I}$ be a chromatic input complex. There exists an encoding function for Ξ_b^r with bit complexity $O(rn \log n)$ such that $\Xi^r(\mathcal{I}) \cong \Xi_b^r(\mathcal{I})$.*

Proof. By Theorem 1, we know that there exists a function $encode$ such that $\Xi_b^r \cong \Xi^r$ and $|\operatorname{Im} encode| \leq \max_{v \in V(\text{Ch } \mathcal{A})} f_0(\text{Lk}(\text{Ch}^r \mathcal{A}, \text{St}(\text{Ch}^r \mathcal{A}, v))) + 1$. From Lemma 7, we have that $encode$ will require to write at most $\Theta(\log_2((\frac{n!n^n}{\ln(2)^{n-1}})^r))$ bits on an entry of $M[k]$. Which gives $O(rn \log n)$ as an upper bound on the bit complexity. $\square$

Notice that in the two-process case ($n = 1$), Theorem 5 and Corollary 7 give a tight bound $\Theta(1)$ on the bit complexity. In the full version of this work [21], we give an algorithm that solves ϵ-agreement for arbitrarily small ϵ with the bit complexity of 2.

It is noteworthy that the asymptotic complexity does only depend on the number of processes and iterations of immediate snapshot. The input size (the number of simplices in the input complex) contributes only a constant factor when calculating the f-vector over iterated subdivisions. Indeed, in each subdivision, the number of indistinguishable states only increases *locally* within each simplex.

Concluding Remarks and Open Questions. This paper appears to be the first to address simulations of *full-information* protocols using bounded iterated memory. By using tools of combinatorial topology, we derive lower and upper bounds on the bit complexity required to simulate IIS. Our results extend and complement recent findings on two-process systems [5,6], shedding light on the feasibility of simulating IIS with a constant bit complexity. Our results underscore the practical application of combinatorial topology for distributed systems by identifying necessary and sufficient properties that the protocol complex must possess in order to yield a desired result.

Future lines of work involve using the same algebraic tools to characterize different protocol maps and complexes. This could lead to optimal bit complexity protocols to solve different kind of tasks. For instance, we have that for colorless tasks, the chromatic subdivision can be replaced by the simpler and more studied barycentric subdivision, which has a different f-vector.

In this paper, we focused on simulating iterated protocols using the same number of iterations. What if we allow for using multiple iterations in simulating a single IIS round? Recently, it has bee shown that one can simulate a round of IIS with 2-bit entries using an exponential number of iterations [5]. Determining the precise trade-off between the number of iterations and the amount of memory remains an open question.

Acknowledgments. Guillermo Toyos-Marfurt was supported by Uruguay's National Agency for Research and Innovation (ANII) under the code "POS_EXT_2021_1_171849". The authors also acknowledge the support of ANR project DUCAT (ANR-20-CE48-0006) and TrustShare Innovation Chair (financed by Mazars Group).

References

1. Armenta-Segura, J., Ledent, J., Rajsbaum, S.: Two-agent approximate agreement from an epistemic logic perspective. Computación y Sistemas **26**(2), 769–785 (2022)
2. Borowsky, E., Gafni, E.: Immediate atomic snapshots and fast renaming. In: PODC, pp. 41–51. ACM Press, New York (1993). https://doi.org/10.1145/164051.164056
3. Borowsky, E., Gafni, E.: A simple algorithmically reasoned characterization of wait-free computation (extended abstract). In: PODC 1997: Proceedings of the Sixteenth Annual ACM Symposium on Principles of Distributed Computing, pp. 189–198. ACM Press, New York (1997). https://doi.org/10.1145/259380.259439
4. Brooks, R.L.: On colouring the nodes of a network. Math. Proc. Cambridge Philos. Soc. **37**(2), 194–197 (1941). https://doi.org/10.1017/S030500410002168X
5. Delporte, C., Fauconnier, H., Fraigniaud, P., Rajsbaum, S., Travers, C.: The computational power of distributed shared-memory models with bounded-size registers (2023)
6. Delporte-Gallet, C., Fauconnier, H., Rajsbaum, S.: Communication complexity of wait-free computability in dynamic networks. In: Richa, A.W., Scheideler, C. (eds.) SIROCCO 2020. LNCS, vol. 12156, pp. 291–309. Springer, Cham (2020). https://doi.org/10.1007/978-3-030-54921-3_17
7. Gafni, E., Kuznetsov, P., Manolescu, C.: A generalized asynchronous computability theorem. In: PODC (2014)

8. Gafni, E., Rajsbaum, S.: Distributed programming with tasks. In: Lu, C., Masuzawa, T., Mosbah, M. (eds.) OPODIS 2010. LNCS, vol. 6490, pp. 205–218. Springer, Heidelberg (2010). https://doi.org/10.1007/978-3-642-17653-1_17

9. Gross, O.A.: Preferential arrangements. Am. Math. Mon. **69**(1), 4–8 (1962)

10. Herlihy, M., Kozlov, D.N., Rajsbaum, S.: Distributed Computing Through Combinatorial Topology. Morgan Kaufmann (2014)

11. Herlihy, M., Rajsbaum, S.: The topology of shared-memory adversaries. In: PODC, pp. 105–113 (2010)

12. Herlihy, M., Shavit, N.: The topological structure of asynchronous computability. J. ACM **46**(2), 858–923 (1999)

13. Klee, S., Novik, I.: Face enumeration on simplicial complexes. In: Beveridge, A., Griggs, J.R., Hogben, L., Musiker, G., Tetali, P. (eds.) Recent Trends in Combinatorics. TIVMA, vol. 159, pp. 653–686. Springer, Cham (2016). https://doi.org/10.1007/978-3-319-24298-9_26

14. Kleinberg, J., Tardos, E.: Algorithm Design. Addison-Wesley Longman Publishing Co. (2005)

15. Kozlov, D.N.: Chromatic subdivision of a simplicial complex. Homology Homotopy Appl. **14**(1), 1–13 (2012)

16. Kuznetsov, P., Rieutord, T., He, Y.: An asynchronous computability theorem for fair adversaries. In: Proceedings of the ACM Symposium on Principles of Distributed Computing, PODC 2018, pp. 1–10 (2018)

17. Nishimura, S.: Schlegel diagram and optimizable immediate snapshot protocol. In: 21st International Conference on Principles of Distributed Systems, OPODIS 2017, pp. 22:1–22:16 (2017)

18. Razborov, A.A.: Communication complexity. In: Schleicher, D., Lackmann, M. (eds.) An Invitation to Mathematics, pp. 97–117. Springer, Heidelberg (2011). https://doi.org/10.1007/978-3-642-19533-4_8

19. Saks, M., Zaharoglou, F.: Wait-free k-set agreement is impossible: the topology of public knowledge. In: Proceedings of the 25th ACM Symposium on Theory of Computing, pp. 101–110 (1993)

20. Sloane, N.J.A.: The on-line encyclopedia of integer sequences. Fubini numbers, Entry A000670 (2023)

21. Toyos-Marfurt, G., Kuznetsov, P.: On the bit complexity of iterated memory (extended version). CoRR abs/2402.12484 (2024). https://arxiv.org/abs/2402.12484

22. Ziegler, G.M.: Shellability and the upper bound theorem. In: Ziegler, G.M. (ed.) Lectures on Polytopes. GTM, vol. 152, pp. 231–290. Springer, New York (1995). https://doi.org/10.1007/978-1-4613-8431-1_8

Brief Announcements

Sharding in Permissionless Systems in Presence of an Adaptive Adversary

Emmanuelle Anceaume, Davide Frey, and Arthur Rauch[✉]

IRISA/Université de Rennes/CNRS/Inria, Rennes, France
arthur.rauch@inria.fr

Abstract. We present SplitChain, a protocol intended to support the creation of scalable proof of stake and account-based blockchains without undermining decentralization and security. This is achieved by using sharding. SplitChain is designed to dynamically adapt the number of shards to the system load to avoid over-dimensioning issues encountered in static sharding-based solutions.

Keywords: Blockchain · Sharding · Distributed ledger · Scalability

1 Introduction

Blockchain technology is well known to provide a tamper-proof "append-only" distributed-ledger abstraction. The immutable nature of this ledger implies a constant growth of its storage requirements, with every block being retained for eternity. Not only does this impact storage, but it also increases the communication requirements for a new honest party to join the system as it needs to download the entire blockchain from the network in order to have a consistent view of the transaction history. Sharding is a scaling solution involving splitting the blockchain into several smaller blockchains, called "shards", each processed and stored by its respective set of validators. Allocating validators to separate shards better balances communication and storage loads, accommodating more efficient network growth to achieve near-linear throughput scalability as the number of shards increases. However, sharding poses several challenges. Firstly, since the system is partitioned, compromising a shard demands a much smaller portion of dishonest validators compared to what is required for classic blockchains. In order to ensure shard safety, it is necessary to implement an unbiased and verifiable random allocation of validators to prevent the adversary from targeting a particular shard. It is also necessary to periodically relocate validators to prevent the adversary from amassing corrupted validators inside a shard. Secondly, it is necessary to ensure both the verification and atomicity of cross-shard transactions. A cross-shard transaction refers to a transaction made between users managed by different shards. As each shard only knows the state of the users stored in its blockchain, it is necessary to ensure that *(i)* an invalid

© The Author(s), under exclusive license to Springer Nature Switzerland AG 2024
Y. Emek (Ed.): SIROCCO 2024, LNCS 14662, pp. 481–487, 2024.
https://doi.org/10.1007/978-3-031-60603-8_26

transaction will not be accepted by any shard, and *(ii)* a valid transaction partially accepted by the involved shards can be aborted so that funds cannot be locked up indefinitely or duplicated. To solve both problems there must exist some minimal amount of synchronization among shards. Most sharding solutions, including Elastico, Omniledger, RapidChain or Ethereum 2.0, achieve this using a global synchronization blockchain maintained by every validator in addition to their shard's blockchain. In addition, sharding solutions based on a fixed number of shards (static sharding) are penalized by an uneven distribution of transactions, as revealed by our experiments on Ethereum.

Contributions. We propose SplitChain, a fully decentralized state sharding solution such that *(i)* shards progress at their own pace without requiring the maintenance of a synchronization blockchain or any heavy synchronization mechanisms, *(ii)* shards keep a loosely synchronized view of each other's state to guarantee the processing of any cross-shard state updates in a bounded number of consensus executions, *(iii)* shards are tolerant to a fast adaptive adversary controlling less than a third of all validators by relying on a novel distributed attribution mechanism, *(iv)* shards self-adapt to the current payload of the system by self merging and splitting the set of accounts and *(v)* shards efficiently forward transactions by leveraging the properties of hypercubic routing protocols.

2   Model of the System

Splitchain uses the account model to provide durable identities to its users and to enforce single-input single-output transactions for easier transaction management. Accounts are persistent balances identified by a public key hash. They are divided into user accounts and validator accounts. Users can send and receive transactions, while validators participate in SplitChain's protocol. All validator accounts possess the same constant amount of currency called stake. We consider an asynchronous communication model, implemented by a peer-to-peer network of validators that self-divide into several chains. Both the number of chains and validators vary over time to withstand a dynamic and open environment. We consider a Byzantine adversary that never controls more than a fraction $0 \leq \mu < 1/3$ of validators.[1] Furthermore, the adversary is adaptive over a period $\delta \geq 3$, in the sense that if the adversary chooses to corrupt some newly attributed validator, this corruption will be effective after δ successive consensus executions. We call such an adversary a (μ,δ)-adaptive adversary. Beyond common cryptographic functions, i.e. a hash function and asymmetric signatures, and Merkle trees, validators use verifiable random functions(VRF) [5].

---

[1] Given that validators all hold the same amount of stake, this equates to controlling less than μ of the total validation stake.

3   Handling Multiple Chains

There are two types of blocks in SplitChain: *chaining blocks* and *transaction blocks*, each of them being constructed during the same consensus execution. Each transaction block contains all the transactions accepted by validators during the consensus execution. It is paired with a chaining block, that stores the hash of the previous chaining block of their chain and the Merkle root of the transaction block. The index of a block refers to its position in the chain.

Definition 1 (Cue block). *Let N_{cue} be some positive integer. $\forall k \geq 0$, every chaining block at position $k \times N_{cue}$ of a chain is called a* cue block. *A cue block points to both the previous chaining block and the previous cue block.*

Chaining blocks are akin to traditional block headers. A chaining block of a chain C contains a list of hashes. In particular, it provides the Merkle roots of the transaction block produced during their consensus execution, and of the accounts managed by the chain as well as the fingerprints needed for cross-chain transaction verification. It also contains the list of the latest known chaining block indices of the chains other than C, which we refer to as the fingerprint of the block. For synchronization purposes, chaining blocks are disseminated to all chains using SplitChain's dedicated routing protocol described in Sect. 4. In contrast, transaction blocks of chain C are only stored by the validators participating in C's consensus executions. To limit the amount of data stored by validators, the full version of this paper includes a pruning method that leverages cue blocks to guarantee near-constant bootstrapping costs.

3.1   Assigning Validators to Chains

Each block of each chain is created via the local execution of a consensus algorithm. Consensus committees are periodically renewed (at each consensus execution) and their members are selected via two stages. The rationale of both stages is to prevent the (μ, δ)-adversary from predicting, and thus manipulating, members of consensus committees. Briefly, starting from their *initialization chains*, validators are uniformly distributed over their *reference chains* (first stage), and then uniformly distributed over their *consensus chains* (second stage). Validators stay forever in their initialization chain (that is as long as then want to actively participate in the creation of SplitChain's blocks), they stay for N_{cue} blocks in their reference chain, and finally they stay for no more than the duration of three consensus executions in their consensus chains. Initialization chains provide a stable anchor point for validators and allow them to prove the existence of their account, while the combination of both reference and consensus chains allow SplitChain to face a (μ, δ)-adaptive adversary by moving validators to randomly chosen consensus chains every δ consensus executions. Hence, for any chain C_i, and for any validators v, v' and v'' of Splitchain, at time t, C_i can be the initialization chain of v, while it is the reference chain of v' and finally the consensus chain of v''.

3.2   Creation of the Blocks of a Chain

As described above, each chain C_i of Splitchain plays the role of a consensus chain. By construction the consensus committee of C_i contains V/N validators. (Note that Sect. 3.1 presents a partial attribution policy that can be used to replace the one previously presented, when SplitChain is made of sufficiently many chains). For performance reasons, V/N validators cannot altogether be involved in the execution of a consensus algorithm. The solution we propose is to rely on a particular asynchronous consensus called the merge consensus algorithm [6]. This algorithm leverages the cryptographic sortition lottery introduced by Algorand [2] to elect, in a non-interactive and private way, a subset of the committee members. This subset has a bounded expected size that is large enough to handle adversarial behaviors, but independent from the size of the consensus committee. The consensus algorithm runs a series of asynchronous rounds, such that at each round, a new subset of validators is elected via cryptographic sortition. The algorithm ensures, with any high probability (whp) that all the transactions proposed by honest validators at the beginning of a consensus execution are included in a block after a finite number of rounds.

3.3   Cross Chain Transactions

Splitchain's transactions take place between any two user accounts. When both accounts of a transaction are not managed by the same chain, a transaction is said to be cross-chain. Cross-chain transactions are handled in two steps: the withdrawal operation and then the deposit operation.

The withdrawal operation occurs when the transaction is inserted in a transaction block of chain C_{send}, whose label prefixes the emitter account. During the creation of the new transaction and chaining blocks, the consensus committee members of C_{send} generate corresponding *relay transactions*, which they organize as a Merkle tree T. A relay transaction contains the user's initial transaction and the Merkle path of the relay transaction inside T. The Merkle root "relay TX" of T is included into the new chaining block under construction. Note that transactions are always redirected to the chain whose label prefixes the identifier of the emitter account, whereas relay transactions are redirected to the chain of the receiver account.

4   Dedicated Routing Protocol

SplitChain initially starts with a single chain C whose label is the empty chain of bits ε. If during N_{cue} consecutive blocks, the average number of transactions per block exceeds some threshold T_{split}, then C triggers a split into two sibling chains, each taking over half of the accounts of C. Conversely, if during N_{cue} consecutive blocks the average number of transactions per block is lower than a threshold T_{merge}, both sibling chains merge in a single one whose label corresponds to the maximal prefix of their previous labels. By design, each chain label is unique.

Our topology is inspired by hypercubic networks [1]. We use the following approach to match each chain to one or more vertices of a hypercube: Let k be the number of bits of the longest chain label. Splitchain conforms to a subgraph of a hypercube of dimension k, where each chain whose label has exactly k bits is mapped to the vertex of the same label, and each chain whose label contains strictly less than k bits is mapped to all vertices whose labels are prefixed by the chain's label. As a result, when all chain labels are k bits long, the network corresponds to a hypercube of dimension k. A hypercube of dimension k can be constructed recursively by connecting two hypercubes of dimension $k - 1$. This allows the dimension of the hypercube to be changed dynamically with k: if k increases, each vertex of label l of size $k - 1$ is divided into two vertices of labels $l|0$ and $l|1$. Similarly, if k decreases, vertices are merged in the same way as for chains. This ensures that no vertex represents two chains simultaneously.

To limit the number of messages transmitted by validators and hence to improve the usage of network resources, we introduce the notion of core validators. Core validators maintain two routing tables: A core routing table containing the list of the core validators of neighboring chains for cross-chain communication, and a consensus table containing the list of their consensus committee members. By using their core routing table, core validators forward transactions to the core validators of the neighboring chain whose label is closest to the transaction's destination. On the other hand, the consensus table is used to broadcast messages within their consensus committee. Only core validators keep track of the list of committee members. The number of core validators is chosen to ensure that at least one honest validator belongs to the core. Election of core validators is done using the cryptographic sortition lottery at consensus round 0. If successful, an elected validator v is part of the core of its chain until its consensus credentials expire. Once elected, core validators send their proof of election along with their consensus messages, guaranteeing with any high probability that all committee members of $C^{\mathrm{cons}(v)}$ will agree on the list of core validators at the end of the merge consensus execution. The hash of the list of new core validators will be included in the chaining block decided as the outcome of the merge consensus.

5   Related Work

Several sharded blockchains have been proposed in recent years: here, we focus on those whose concepts have influenced the design of SplitChain. Elastico [4], partitions validators into shards, each creating a block of transactions, which are then aggregated into a "global-block" in the system's unique blockchain. The shards are renewed after each global-block, ensuring strong safety against adaptive adversaries at the expense of synchronization and storage costs, i.e., validators store the entire system. Elastico cannot ensure atomicity in cross-shard transactions [3]. In comparison, SplitChain splits the management and storage of transactions and user accounts between the different chains at the cost of a slower adaptive adversary. Omniledger [3] features shards with separate ledgers, but it still requires a global blockchain for validator allocation. In contrast, SplitChain shuffles a few validators of a chain after each block to withstand an adversary adaptive over a small, configurable number of consensus instances and does not require any global blockchain to manage validator identities. RapidChain [8] reduces the size of its shards thanks to a synchronous consensus algorithm tolerating a proportion of 1/2 Byzantine nodes. But it inherits problems of Elastico and Omniledger, namely the use of PoW for validator enrollment (Elastico) and the use of the UTXO model (Omniledger). On the other hand, Splitchain is asynchronous, does not require PoW and routes transactions through a small subset of validators (the core). Finally, Monoxide [7] uses PoW for shard consensus and allows each miner to finalize and propose new transaction blocks to multiple shards simultaneously. But to guarantee safety, the majority of miners are required to work for most if not all shards. This causes Monoxide to resemble more to a parallelization solution than to an actual sharded system.

6   Conclusion

We have presented SplitChain, a protocol supporting the creation of scalable account-based blockchains without undermining decentralization and security. SplitChain distinguishes itself from other sharded blockchains by minimizing the synchronization constraints among shards while maintaining security guarantees. Specifically, SplitChain is the first permissionless sharded blockchain that does not require a dedicated shard or a global blockchain to attribute validators to their consensus chain. This avoids the need for a global reconfiguration of the shards each time a new batch of validators is added to the system. A dedicated routing protocol enables transactions to be redirected between shards with a low number of hops and messages. Finally, SplitChain dynamically adapts the number of shards to the system load to avoid over-dimensioning issues encountered in static sharding-based solutions. Further research will investigate the practical performance of SplitChain through its implementation and the potential use of sharding to enhance user privacy.

References

1. Anceaume, E., Ludinard, R., Sericola, B.: Performance evaluation of large-scale dynamic systems. SIGMETRICS Perform. Eval. Rev. **39**, 108–117 (2012)
2. Gilad, Y., Hemo, R., Micali, S., Vlachos, G., Zeldovich, N.: Algorand: scaling byzantine agreements for cryptocurrencies. In: ACM SOSP (2017)
3. Kokoris-Kogias, E., Jovanovic, P., Gasser, L., Gailly, N., Syta, E., Ford, B.: OmniLedger: a secure, scale-out, decentralized ledger via sharding. In: IEEE S&P (2018)
4. Luu, L., Narayanan, V., Zheng, C., Baweja, K., Gilbert, S., Saxena, P.: A secure sharding protocol for open blockchains. In: ACM SIGSAC CCS (2016)
5. Micali, S., Rabin, M., Vadhan, S.: Verifiable random functions. In: Symposium on Foundations of Computer Science (1999)

6. Saunois, G., Robin, F., Anceaume, E., Sericola, B.: Permissionless consensus based on proof-of-eligibility. In: IEEE NCA (2020)
7. Wang, J., Wang, H.: Monoxide: scale out blockchains with asynchronous consensus zones. In: USENIX NSDI (2019)
8. Zamani, M., Movahedi, M., Raykova, M.: RapidChain: scaling blockchain via full sharding. In: ACM SIGSAC CCS (2018)

Online Drone Scheduling for Last-Mile Delivery

Saswata Jana[1] , Giuseppe F. Italiano[2] , Manas Jyoti Kashyop[2],
Athanasios L. Konstantinidis[2], Evangelos Kosinas[3],
and Partha Sarathi Mandal[1]([✉])

[1] Indian Institute of Technology Guwahati, Guwahati, India
psm@iitg.ac.in
[2] Luiss University, Rome, Italy
[3] University of Ioannina, Ioannina, Greece

Abstract. Delivering a parcel from the distribution hub to the customer's doorstep is called the *last-mile delivery* step in delivery logistics. In this paper, we study a hybrid *truck-drones* model for the last-mile delivery step, in which a truck moves on a predefined path carrying parcels and drones deliver the parcels. We define the ONLINE DRONE SCHEDULING problem, where the customer's requests for the parcels appear online during the truck's movement. The objective is to schedule a drone for every request, aiming to minimize the number of drones used subject to the battery budget of the drones and compatibility of the schedules. We propose a 3-competitive deterministic algorithm with $O(\log n)$ worst-case time per request, where n is the total number of requests being served at any instance. We further improve the competitive ratio to 2.7 with the same time complexity. We also introduce ONLINE VARIABLE-SIZE DRONE SCHEDULING problem (OVDS). Here, all customer requests are available in advance, but the drones with different battery capacities appear online, and the objective is the same as the online drone scheduling problem. We propose a $(2\alpha+1)$-competitive algorithm for the OVDS problem with running time $O(n \log n)$, where n is the total customer requests and α is the ratio of maximum to minimum drone battery capacities.

Keywords: Online Algorithm · Optimization · Drone-Delivery Scheduling · Last-mile Delivery

1 Introduction

Rapid developments in drone technology motivate scientists and researchers to use drones in various operations. Delivery giant companies, e.g., Amazon, DHL, and FedEx, have started using drones for the *last-mile delivery*. Last-mile delivery,

S. Jana—Acknowledge financial support from the Prime Minister's Research Fellowship (PMRF) scheme of the Govt. of India (PMRF-ID: 1902165)

P. S. Mandal—This work was done while Partha Sarathi Mandal was in the position of Visiting Professor at Luiss University, Rome, Italy.

Y. Emek (Ed.): SIROCCO 2024, LNCS 14662, pp. 488–493, 2024.
https://doi.org/10.1007/978-3-031-60603-8_27

the final and costliest step in delivery logistics, involves collecting packages from the distribution hub and delivering them to the customer using a vehicle (delivery truck). The problem of delivering goods using vehicles is studied as the *routing problem*, where the aim is to minimize the total make-span(e.g., [8,9]). Murray and Chu [10] proposed a model called the *flying sidekicks traveling salesman problem* (FSTSP), where a delivery truck and a single drone cooperate to deliver the packages. The authors in [3] and [11] modified the FSTSP model and introduced multiple drones (extended FSTSP model). Daknama and Kraus [4] proposed a variation of the extended FSTSP model where every drone has a limited battery capacity (budgeted extended FSTSP model). Recent studies [2] suggest that the use of drones in the delivery logistic system reduces CO_2 emissions compared to the use of only delivery trucks or human-operated ground vehicles. Further, use of drones facilitates contactless delivery, avoiding traffic congestion by flying and following a shorter route toward the customer's doorstep. However, using only drones has certain limitations, such as the size of the packages and the distance between the customer and the warehouse. Therefore, a hybrid model with a delivery truck and several drones (truck-drones model) is more desirable [12]. In the truck-drones model, the truck containing all the drones and the packages follows conventional ground routes. A customer will be delivered either by truck or by drone. In [1], authors considered that the truck has some specific discrete stopping points, which are also the places for launching and landing points for drones. Based on the customer's position, launching and landing points are computed. Several works (e.g., [6]) considered that customer requests will appear online, and the goal is to minimize the total make-span.

In this work, we use the hybrid approach and consider the truck-drones model. For simplicity, we do not consider customer requests that are close to the truck path and are served by the truck. Further, we assume that the truck has some specific discrete stopping points where the drone can take off from the truck and land on the truck. We compute the launch and rendezvous point of each customer request for a drone, and these requests appear online. Based on the launch and rendezvous point, we develop an algorithm (interval generator) that generates an interval for every customer request. The interval generator assigns the amount of battery consumption of a drone used to serve a specific request as the cost of the interval corresponding to that request. Due to brevity we differ the discussion on the interval generator to the full version [7] of this paper. For the rest of the paper, we assume that a black-box interval generator provides an interval for every customer request with appropriate cost. Our objective is to use the minimum number of drones to serve all the requests. Note that our objective differs from the previous works that considered online customer requests intending to minimize the total make-span.

2   Drone Scheduling Through Bin Packing with Conflicts

In this section, we formally define the ONLINE DRONE SCHEDULING problem and discuss our approach to solve it. We differ the details about data structures, correctness, time-complexity analysis, and competitive analysis to the full version [7]. The input for this problem consists of intervals corresponding to each customer's requests, along with the associated costs. A new request from a customer corresponds to insertion of a new interval. A request served or withdrawn from a customer corresponds to deletion of an existing interval. Let G denote the underlying interval graph. Let $I_t = [l_t, r_t]$ be the interval inserted at time t. Let v_t be the vertex in G that corresponds to I_t. Therefore, insertion or deletion of an interval is insertion or deletion of the corresponding vertex in the graph G. Let $\mathsf{cost}(I_t)$ be the cost associated with interval I_t. We consider the ONLINE DRONE SCHEDULING problem as proper coloring of intervals with a fixed budget B for every color, where B is a positive integer and is part of the input. A color assigned to an interval corresponds to a drone assigned to serve the request represented by the interval and budget B of the color corresponds to the battery capacity of the drone. We define the following problem.

Definition 1 (ONLINE DRONE SCHEDULING). *Intervals are appearing online. Suppose at time step t, interval $I_t = [l_t, r_t]$ appears with cost $\mathsf{cost}(I_t)$ and $l_t \geq t$. The goal is to assign a color to the interval with the following constraints:*
1. *Any two intervals with the same color are disjoint (non-overlapping).*
2. *For a particular color, the summation of the costs of all intervals with that color is at most B.*

To solve the problem we will use algorithms for online bin packing with conflicts [5]. For interval I_t at time t, our goal is to compute a color $\mathsf{Color}(I_t)$. A color $\mathsf{Color}(I_t)$ is a 2-tuple, $\mathsf{Color}(I_t) = (\mathsf{idNumber}(I_t), \mathsf{binNumber}(I_t))$, with the following properties.

1. (**conflict free**) No two overlapping intervals have the same idNumber. Thus, the idNumber ensures a conflict-free property.
2. (**budget constraint**) For all the intervals with the same idNumber and binNumber (in other words, intervals with the same color), the summation of their costs must be at most B. Therefore, the budget constraint associated with every color is satisfied.
3. (**proper coloring**) The above two properties imply that for any two intervals $I_t \neq I_{t'}$ and $I_t \cap I_{t'} \neq \emptyset$, $(\mathsf{idNumber}(I_t), \mathsf{binNumber}(I_t)) \neq (\mathsf{idNumber}(I_{t'}), \mathsf{binNumber}(I_{t'}))$.

We use the following approach to compute the color.

1. idNumber is computed using an online interval coloring algorithm where the colors are represented using positive integer numbers. Note that the coloring algorithm used at this step does not have any budget associated with the colors and thus can be assigned to any number of non-overlapping intervals. For an interval I_t, the color (the non-negative integer) allotted to I_t at this step using the online coloring algorithms is assigned as the idNumber(I_t).
2. Once the idNumber(I_t) is computed for an interval I_t in the previous step, next step is to compute binNumber(I_t) ensuring that all the intervals with the

same color as $(\mathsf{idNumber}(I_t), \mathsf{binNumber}(I_t))$ satisfies the budget constraint B. Therefore, we use an online bin packing algorithm to compute the $\mathsf{binNumber}$ for I_t considering all the previously inserted intervals with the $\mathsf{idNumber}$ equal to $\mathsf{idNumber}(I_t)$.

For an interval $I_{t'} = [l_{t'}, r_{t'}]$, color of the interval is computed at time $t = l_{t'}$. Further, we assume that intervals have distinct endpoints. Therefore, at the time step t, the following scenarios occur during handling the update.

1. If a new interval I_t appears, then I_t is inserted. Further, at time step t at most one interval appears.
2. If there is an existing interval $I_{t'}$ with $t' < t$ and $r_{t'} = t$, then $I_{t'}$ is deleted.
3. If there is an interval $I_{t'}$ with $t' \leq t$ and $l_{t'} = t$, then color for $I_{t'}$ is computed.
4. As intervals have distinct endpoints, scenario 2 and 3 are mutually exclusive.

Because of the properties of our original ONLINE DRONE SCHEDULING problem, the intervals appearing online satisfy the following properties.

- Interval appearing at time step t has left endpoint at least t.
- Our deletion scheme is that at time step t, if there is an interval $I_{t''}$ with $t'' < t$ and $r_{t''} = t$, then we delete $I_{t''}$. Otherwise, if there is an interval $I_{t'}$ with $t' \leq t$ and $l_{t'} = t$, the color for $I_{t'}$ is computed. Note that because of our deletion scheme, while computing the color for $I_{t'}$, all the intervals overlapping with $I_{t'}$ must contain the point t forming a clique. Further, because of our coloring scheme, colors are assigned to the intervals in the increasing order of their left endpoints.

2.1 Computing idNumber

At time step t, we consider the interval $I_{t'} = [l_{t'}, r_{t'}]$ with $l_{t'} = t$ and $t' \leq t$ to compute $\mathsf{idNumber}(I_{t'})$. Let $\mathcal{I}_t$ be the set of intervals overlapping with $I_{t'}$ and contain the point t. Since all the intervals have distinct endpoints, the left endpoint of every interval in $\mathcal{I}_t$ is strictly smaller than t. We observe the following properties about the set of intervals $\mathcal{I}_t$.

1. $\mathcal{I}_t$ forms a clique because all the intervals in $\mathcal{I}_t$ contains the point t.
2. Because of our deletion scheme, $\mathcal{I}_t$ is the only set of intervals overlapping with the interval $I_{t'}$ and with left endpoints strictly less than t.
3. As of our coloring scheme, color for each interval in $\mathcal{I}_t$ is already computed.
4. Any interval not in $\mathcal{I}_t$ and overlapping with $I_{t'}$ must have left endpoint strictly greater than t. Because of our coloring scheme, the color for such an interval is not yet computed and we do not consider such an interval while computing the color for $I_{t'}$ in the current step.

Our goal is to find the smallest $\mathsf{idNumber}$ not assigned to any interval in $\mathcal{I}_t$ and assign it as $\mathsf{idNumber}(I_{t'})$. We obtain the following result.

Lemma 1. *For each interval $I_{t'}$, there exists an algorithm to compute $\mathsf{idNumber}(I_{t'})$ with $O(\log n)$ worst-case time complexity.*

2.2   Computing binNumber

At time step t, after computing idNumber($I_{t'}$) for the interval $I_{t'}$, the next task is to compute binNumber for $I_{t'}$. First we use the next-fit strategy from online bin packing to compute the binNumber and obtain the following result.

Lemma 2. *For each interval $I_{t'}$, there exists an algorithm to compute* binNumber($I_{t'}$) *with $O(\log n)$ worst-case time complexity.*

We design a deterministic algorithm using the algorithms developed in Lemma 1 and Lemma 2, and obtain the following result.

Theorem 1. *There exists a deterministic 3-competitive algorithm for* ONLINE DRONE SCHEDULING *problem that assigns every drone in $O(\log n)$ worst-case time, where n is the maximum number of requests being served by some drone at any instance (active requests) through out handling of the request sequence.*

We improve the the result in Theorem 1 using first-fit strategy to compute binNumber. Use of first-fit strategy leads to the use of slightly complicated data structures resulting in a different time-complexity analysis and competitive ratio analysis. We obtain the following result.

Theorem 2. *There exists a deterministic 2.7-competitive algorithm for* ONLINE DRONE SCHEDULING *problem that performs every drone allotment in $O(\log n)$ worst-case time, where n is the total number of requests received through out handling of the request sequence.*

3   Online Variable-Size Drone Scheduling

In this section, we formally define the ONLINE VARIABLE-SIZE DRONE SCHEDULING (OVDS) problem and discuss our approach to solve it. In the OVDS problem, we know all the customer requests in advance. So, we compute the underlying interval graph and partition the intervals into independent sets before the drones appear online. Each interval is associated with a cost reflecting the drone's battery consumption for the corresponding customer request. Drones with different battery capacities appear online. We allocate as many requests as possible to the current drone from one of the pre-computed independent sets. The battery capacity of every drone is at least the maximum cost of an interval. The goal is to use a minimum number of drones. Our motivation for the OVDS problem comes from the following scenario: there is an allocation unit for drones that allocates a drone as and when requested by a delivery unit. The delivery unit is not aware of the battery capacities of the drones in advance. Therefore, upon request, when a drone is allotted, depending upon the drone's battery capacity, the delivery unit aims to schedule as many requests as possible for that drone. The delivery unit keeps raising requests to the allocation unit until all customer requests are allocated to some drones. The goal is to minimize the number of requests raised by the delivery unit. We assign a color to the intervals to allocate drones to the intervals, similar to Sect. 2. The color of an interval I denoted by $\mathsf{Color}(I) = (\mathsf{idNumber}(I), \mathsf{binNumber}(I))$ corresponds to the drone allotted to serve the corresponding request. There is a list of intervals corresponding to each of the $\mathsf{idNumber}$. For each such list, we use the *any-fit* strategy from the bin-packing problem and compute the $\mathsf{binNumber}$ for the intervals in that list. We state the following result and differ the details to the full version of the paper.

Theorem 3. *There exists a deterministic $(2\alpha + 1)$-competitive algorithm for the* OVDS *problem that takes total $O(n \log n)$ time, where α is the ratio between the maximum and minimum battery capacity and n is the total number of requests.*

4   Conclusion

This paper studies the theoretical aspects of the ONLINE DRONE SCHEDULING problem and the ONLINE VARIABLE-SIZE DRONE SCHEDULING problem and presents deterministic algorithms providing upper bounds on the competitive ratio. In future, we aim to explore the lower bound direction for both problems and implement a simulation for possible practical applications.

References

1. Boysen, N., Briskorn, D., Fedtke, S., Schwerdfeger, S.: Drone delivery from trucks: drone scheduling for given truck routes. Networks **72**, 506–527 (2018)
2. Chiang, W.C., Li, Y., Shang, J., Urban, T.L.: Impact of drone delivery on sustainability and cost: Realizing the UAV potential through vehicle routing optimization. Appl. Energy **242**, 1164–1175 (2019)
3. Crişan, G.C., Nechita, E.: On a cooperative truck-and-drone delivery system. In: Procedia Computer Science, vol. 159, pp. 38–47 (2019). knowledge-Based and Intelligent Information & Engineering Systems: Proceedings of the 23rd International Conference KES2019
4. Daknama, R., Kraus, E.: Vehicle routing with drones. ArXiv abs/1705.06431 (2017)
5. Epstein, L., Levin, A.: On bin packing with conflicts. SIAM J. Optim. **19**(3), 1270–1298 (2008). https://doi.org/10.1137/060666329
6. Jaillet, P., Wagner, M.R.: Online vehicle routing problems: A survey. The Vehicle Routing Problem: Latest Advances and New Challenges, pp. 221–237 (2008)
7. Jana, S., Italiano, G.F., Kashyop, M.J., Konstantinidis, A.L., Kosinas, E., Mandal, P.S.: Online drone scheduling for last-mile delivery. arXiv preprint arXiv:2402.16085 (2024), https://arxiv.org/abs/2402.16085
8. Laporte, G.: The traveling salesman problem: an overview of exact and approximate algorithms. Eur. J. Oper. Res. **59**(2), 231–247 (1992)
9. Laporte, G.: The vehicle routing problem: an overview of exact and approximate algorithms. Eur. J. Oper. Res. **59**(3), 345–358 (1992)
10. Murray, C.C., Chu, A.G.: The flying sidekick traveling salesman problem: optimization of drone-assisted parcel delivery. Transp. Res. Part C: Emerg. Technol. **54**, 86–109 (2015)
11. Murray, C.C., Raj, R.: The multiple flying sidekicks traveling salesman problem: Parcel delivery with multiple drones. Transp. Res. Part C: Emerg. Technol. **110**, 368–398 (2020)
12. Wang, X., Poikonen, S., Golden, B.: The vehicle routing problem with drones: several worst-case results. Optim. Lett. **11**, 679–697 (2017)

Efficient Self-stabilizing Simulations of Energy-Restricted Mobile Robots by Asynchronous Luminous Mobile Robots

Keita Nakajima[1]($\boxtimes$) iD, Kaito Takase[2] iD, and Koichi Wada[2] iD

[1] Tokyo Institute of Technology, Tokyo, Japan
nakajima.k.au@m.titech.ac.jp
[2] Hosei University, Tokyo, Japan
kaito.takase.6z@stu.hosei.ac.jp, wada@hosei.ac.jp

Abstract. In this study, we explore efficient simulation implementations to demonstrate computational equivalence across various models of autonomous mobile robot swarms. Our focus is on RSYNCH, a scheduler designed for energy-restricted robots, which falls between FSYNCH and SSYNCH. We propose efficient protocols for simulating $n(\geq 2)$ luminous ($\mathcal{LUMI}$) robots operating in RSYNCH using $\mathcal{LUMI}$ robots in SSYNCH or ASYNCH. Additionally, we confirm that all our proposed simulation protocols are self-stabilizing, ensuring functionality from any initial configuration.

1 Introduction

1.1 Background and Motivation

The computational issues of autonomous mobile entities operating in a Euclidean space in *Look-Compute-Move* (*LCM*) cycles have been the subject of extensive research in distributed computing. In the *Look* phase, an entity, viewed as a point and usually called *robot*, obtains a snapshot of the space; in the *Compute* phase it executes its algorithm (the same for all robots) using the snapshot as input; it then moves toward the computed destination in the *Move* phase. Repeating these cycles, the robots can collectively perform some tasks and solve some problems. The research interest has been on determining the impact that *internal* capabilities (e.g., memory, communication) and *external* conditions (e.g. synchrony, activation scheduler) have on the solvability of a problem.

In the most common model, $\mathcal{OBLOT}$, in addition to the standard assumptions of *anonymity* and *uniformity* (robots have no IDs and run identical algorithms), the robots are *oblivious* (no persistent memory to record information

---

This work was supported in part by JSPS KAKENHI Grant Number 20K11685, and 21K11748.

of previous cycles) and *silent* (without explicit means of communication). Computability in this model has been the object of intensive research since its introduction in [12]. Extensive investigations have been carried out to clarify the computational limitations and powers of these robots for basic coordination tasks such as Gathering, Pattern Formation, Flocking.

A model which provides robots with persistent memory, albeit limited, and communication means is the $\mathcal{LUMI}$ model, formally defined and analyzed in [3], following a suggestion in [11]. In this model, each robot is equipped with a constant-sized memory (called *light*), whose value (called *color*) can be set during the *Compute* phase. The light is visible to all the robots and is persistent in the sense that it is not automatically reset at the end of a cycle. Hence, these luminous robots are capable of both remembering and communicating a constant number of bits in each cycle.

An important result is that, despite these limitations, the simultaneous presence of both persistent memory and communication renders luminous robots strictly more powerful than oblivious robots [3]. This, in turns, has opened the question about the individual computational power of the two internal capabilities, memory and communication, and motivated the investigations on two sub-models of $\mathcal{LUMI}$: $\mathcal{FST\!A}$ where the robots have a constant-size persistent memory but are silent, and $\mathcal{FCOM}$, where robots can communicate a constant number of bits but are oblivious (e.g., see [1,2,5,6,9,10]).

All these studies across various models have highlighted the crucial role played by two interrelated *external* factors: the level of synchronization and the activation schedule provided by the system. As in other types of distributed computing systems, there are two different settings; the synchronous and the asynchronous settings. In the *synchronous* (also called *semi-synchronous*) (SSYNCH) setting, introduced in [12], time is divided into discrete intervals, called *rounds*. In each round, an arbitrary but nonempty subset of the robots is activated, and they simultaneously perform exactly one atomic *Look-Comp-Move* cycle. The selection of which robots are activated at a given round is made by an adversarial scheduler, constrained only to be fair, i.e., every robot is activated infinitely often. Weaker form of synchronous adversaries have also been introduced and investigated. The most important and extensively studied is the *fully-synchronous* (FSYNCH) scheduler, which activates all the robots in every round. Other interesting synchronous schedulers are RSYNCH, studied for its use to model energy-restricted robots [2], as well as the family of *Sequential* schedulers (e.g., *Round Robin*), where in each round only one robot is activated.

In the *asynchronous* setting (ASYNCH), introduced in [4], there is no common notion of time and each robot is activated independently of the others. it allows for arbitrary but finite delays between the *Look*, *Comp* and *Move* phases, and each movement may take an arbitrary but finite amount of time. The duration of each robot's cycle, as well as the timing of robot's activation, are controlled by an adversarial scheduler, constrained only to be fair, i.e., every robot must be activated infinitely often.

1.2 Contributions

Like in other types of distributed systems, understanding the computational difference between various levels of synchrony and asynchrony has been a primary research focus. In the robot model, to "separate" between the computational power of robots in two settings, we demonstrate that certain problems are solvable in one model but unsolvable in another. For example, in $\mathcal{OBLOT}$, *Rendezvous* is unsolvable in SSYNCH, but solvable in FSYNCH[12], indicating a separation between FSYNCH and SSYNCH in $\mathcal{OBLOT}$. Conversely, to show that a weaker model is equivalent to a stronger model, we devise a *simulation protocol* that allows the correct execution of any protocol from the stronger model in the weaker model. The first attempt in the robot model involves constructing a simulation protocol for any $\mathcal{LUMI}$ protocol in SSYNCH using $\mathcal{LUMI}$ robots in ASYNCH [3]. This protocol employs $5k$ colored lights in $\mathcal{LUMI}$ robots to simulate SSYNCH protocols using k colors[1].

In this paper, we focus on making such simulations more efficient, specifically, considering simulations that involve $\mathcal{LUMI}$ robots operating in RSYNCH and $\mathcal{LUMI}$ robots operating in ASYNCH under the most unrestricted adversary. Though RSYNCH is introduced for modeling energy-restricted robots [2], it is interesting in its own right because $\mathcal{LUMI}$ robots in ASYNCH have the same power as those in RSYNCH [2,3], and $\mathcal{FCOM}$ robots in RSYNCH have the same power as $\mathcal{LUMI}$ robots in RSYNCH [2].

The simulator of $\mathcal{LUMI}$ robots with k colors in SSYNCH uses $\mathcal{LUMI}$ robots in ASYNCH and utilizes $5k$ colors [3], including 5 colors to control the simulation. Therefore, we aim to reduce the number of colors used to control the simulation.

Table 1 presents both the previous simulation results and our new results. Here for any model, $M \in \{\mathcal{FCOM}, \mathcal{LUMI}\}$ and any adversarial scheduler $A \in \{$ RSYNCH, SSYNCH, ASYNCH$\}$, M^A denotes the robot model M working in A.

In this paper, we first establish that by utilizing the property of the simulator detailed in [3],

(1) k-color $\mathcal{LUMI}$ robots in RSYNCH can be simulated by $4k$-color $\mathcal{LUMI}$ robots in SSYNCH.

(2) k-color $\mathcal{LUMI}$ robots in RSYNCH can be simulated by $5k$-color $\mathcal{LUMI}$ robots in ASYNCH.

Previously, case (1) required $36k$ colors [2]. In contrast, our simulator for case (2) also uses $5k$ colors, effectively simulating $\mathcal{LUMI}$ robots in RSYNCH using $\mathcal{LUMI}$ robots in ASYNCH. Additionally, we demonstrate that when the number of robots is limited to 2 ($n = 2$), the simulator in the case (1) can be implemented more efficiently, Specifically, we show that:

(3) k-color $\mathcal{LUMI}$ robots in RSYNCH can be simulated by $3k$-color $\mathcal{LUMI}$ robots in ASYNCH.

---

[1] In the case of $k = 1$, this protocol simulates $\mathcal{OBLOT}$ robots working in SSYNCH with $\mathcal{LUMI}$ robots with 5 colors in ASYNCH.

(4) In the case of $k = 1$, $\mathcal{OBLOT}$ robots in RSYNCH cannot be simulated by 2-color $\mathcal{LUMI}$ robots in ASYNCH. This demonstrates that the number of colors used in the simulator shown in (3) is optimal.

We also confirm that all our proposed simulation protocols are self-stabilizing, ensuring functionality from any initial configuration. These self-stabilization can be done without increasing the number of colors.

Table 1. The previous results and this paper's results.

$n \geq 2$

simulating model	simulated model	# colors	Ref.
$\mathcal{LUMI}^A$	$\mathcal{LUMI}^S$	$5k$	[3]
$\mathcal{LUMI}^S$	$\mathcal{LUMI}^{RS}$	$36k$	[2]
$\mathcal{LUMI}^A$	$\mathcal{LUMI}^S$ *	$3k$	[7]
$\mathcal{LUMI}^S$	$\mathcal{LUMI}^{RS}$	$4k$	This paper
$\mathcal{LUMI}^A$	$\mathcal{LUMI}^{RS}$	$5k$	This paper

* unfair SSYNCH

$n = 2$

simulating model	simulated model	# colors	Ref.
$\mathcal{LUMI}^A$	$\mathcal{LUMI}^{RS}$	$3k$	This paper

2   Simulation of RSYNCH on $\mathcal{LUMI}$

2.1   Simulation of RSYNCH with Any Number of Robots

The details of the model of autonomous mobile robots, schedulers, events, computational Relationships, and all proofs are included in [8].

4-Color Simulation of RSYNCH *by* SSYNCH. We show that semi-synchronous systems equipped with a light with 4 colors are at least as powerful as *restricted-repetition* system (RSYNCH) without lights. More precisely, we have:

Theorem 1. $\forall R \in \mathcal{R}, \mathcal{OBLOT}^{RS}(R) \subseteq \mathcal{LUMI}_4^S(R)$.[2]

Figure 1(a) shows the transition diagram representation of protocol SIM_S^{RS}, which uses four colors: T, M, S, S'. Initially, all lights are set to T.

The protocol simulates a sequence of mega-cycles, each of which starts with some robots trying to execute protocol $\mathcal{P}$ and ends with all robots finishing the mega-cycle having executed $\mathcal{P}$ once. After this end configuration, it transits to start one, and a new mega-cycle starts. This protocol can correctly simulate SSYNCH by controlling a robot that is activated within a mega-cycle.

Theorem 2. *Protocol* SIM_S^{RS} *is correct, i.e. any execution of protocol* SIM_S^{RS} *in* SSYNCH *corresponds to a possible execution of* $\mathcal{P}$ *in* RSYNCH.

---

[2] $\mathcal{LUMI}_4^S(R)$ means a set of problems 4-color $\mathcal{LUMI}$ robots R can solve in SSYNCH.

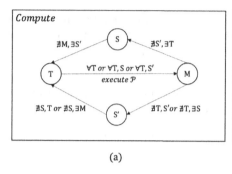 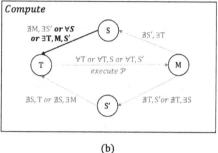

(a)   (b)

Fig. 1. Transition diagram representations of protocol SIM_S^{RS}((a)), and self-stabilizing protocol ss-SIM_S^{RS}((b)). The label in the nodes represents the color of the light of the executing robot. The label of an edge expresses the condition on the lights of all the other robots that must be satisfied for the transition to occur. The notation "$\forall A, B$" means: "$\{\mathrm{Light}[\mathrm{r}] \mid \forall r \in R\} = \{A, B\}$", "$\exists A$" means: "$\exists r \in R, \mathrm{Light}[\mathrm{r}] \in \{A\}$", "$\nexists A$" means: "$\{\mathrm{Light}[\mathrm{r}] \mid \forall r \in R\} \cap \{A\} = \emptyset$". Conditions, written in bold face in (b) are newly added.

Making SIM_S^{RS} *Self-stabilizing.* We can make the protocol SIM_S^{RS} self-stabilizing (denoted as ss-SIM_S^{RS} and shown in Fig. 1(b)) and show that ss-SIM_S^{RS} works correctly from any global configuration.

Theorem 3. *Protocol ss-SIM_S^{RS} is correct and self-stabilizing, i.e. from any initial global configuration, any execution of protocol SIM_S^{RS} in* SSYNCH *corresponds to a possible execution of* $\mathcal{P}$ *in* RSYNCH.

5-Color Self-stabilizing Simulation of RSYNCH *by* ASYNCH. We can show if we use one more color (that is, use 5 colors), protocol SIM_S^{RS} (resp. ss-SIM_S^{RS}) can be extended so that it works in ASYNCH to simulate RSYNCH from an initial configuration (resp. any initial configuration). They are called SIM_A^{RS} and ss-SIM_A^{RS}.

Theorem 4. *Protocols SIM_A^{RS} and ss-SIM_A^{RS} are correct, i.e. any execution of protocol SIM_A^{RS} in* ASYNCH[5] *corresponds to a possible execution of* $\mathcal{P}$ *in* RSYNCH. *Also ss-SIM_A^{RS} is self-stabilizing.*

Thus, we have the following theorem.

Theorem 5. $\forall R \in \mathcal{R}, \mathcal{OBLOT}^{RS}(R) \subseteq \mathcal{LUMI}_5^A(R)$.

2.2   Optimal Simulation of RSYNCH by ASYNCH with Two Robots

In the case of two robots, we show that asynchronous systems equipped with a light with 4 colors are at least as powerful as *restricted-repetition* system (RSYNCH) without lights. More precisely, we have:

Theorem 6. $\forall R \in \mathcal{R}_2, \mathcal{OBLOT}^{RS}(R) \subseteq \mathcal{LUMI}_3^A(R)$.

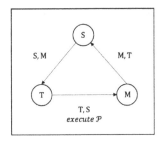

Fig. 2. Transition diagram of protocol SIM-2_A^{RS}.

Figure 2 shows SIM-2_A^{RS} protocol that produces RSYNCH execution of any $\mathcal{OBLOT}^{RS}$ protocol $\mathcal{P}$. Also, we can show that any simulation of two $\mathcal{OBLOT}^{RS}$ robots by two $\mathcal{LUMI}^S$ robots with two colors is impossible.

Theorem 7. *Protocol SIM-2_A^{RS} is correct and self-stabilizing. SIM-2_A^{RS} is an optimal simulating protocol with respect to the number of colors.*

References

1. Buchin, K., Flocchini, P., Kostitsyna, I., Peters, T., Santoro, N., Wada, K.: Autonomous mobile robots: refining the computational landscape. In: APDCM 2021, pp. 576–585 (2021)
2. Buchin, K., Flocchini, P., Kostitsyna, I., Peters, T., Santoro, N., Wada, K.: On the computational power of energy-constrained mobile robots: algorithms and cross-model analysis. In: Proceedings of the 29th International Colloquium on Structural Information and Communication Complexity (SIROCCO), pp. 42–61 (2022)
3. Das, S., Flocchini, P., Prencipe, G., Santoro, N., Yamashita, M.: Autonomous mobile robots with lights. Theor. Comput. Sci. **609**, 171–184 (2016)
4. Flocchini, P., Prencipe, G., Santoro, N., Widmayer, P.: Hard tasks for weak robots: the role of common knowledge in pattern formation by autonomous mobile robots. In: 10th International Symposium on Algorithms and Computation (ISAAC), pp. 93–102 (1999)
5. Flocchini, P., Santoro, N., Viglietta, G., Yamashita, M.: Rendezvous with constant memory. Theor. Comput. Sci. **621**, 57–72 (2016)
6. Flocchini, P., Santoro, N., Wada, K.: On memory, communication, and synchronous schedulers when moving and computing. In: Proceedings of the 23rd International Conference on Principles of Distributed Systems (OPODIS), pp. 25:1–25:17 (2019)
7. Nakai, R., Sudo, Y., Wada, K.: Asynchronous gathering algorithms for autonomous mobile robots with lights. In: Proceedings of the 23rd International Symposium (SSS), pp. 410–424 (2021)
8. Nakajima, K., Takase, K., Wada, K.: Efficient self-stabilizing simulations of energyrestricted mobile robots by asynchronous luminous mobile robots. arXiv, cs(ArXivArXiv:2403.05542) (2024)
9. Okumura, T., Wada, K., Défago, X.: Optimal rendezvous $\mathcal{L}$-algorithms for asynchronous mobile robots with external-lights. In: Proceedings of the 22nd International Conference on Principles of Distributed Systems (OPODIS), pp. 24:1–24:16 (2018)
10. Okumura, T., Wada, K., Katayama, Y.: Brief announcement: optimal asynchronous rendezvous for mobile robots with lights. In: Proceedings of the 19th International Symposium on Stabilization, Safety, and Security of Distributed Systems (SSS), pp. 484–488 (2017)

11. Peleg, D.: Distributed coordination algorithms for mobile robot swarms: new directions and challenges. In: Distributed Computing-IWDC 2005, pp. 1–12 (2005)
12. Suzuki, I., Yamashita, M.: Distributed anonymous mobile robots: formation of geometric patterns. SIAM J. Comput. **28**, 1347–1363 (1999)

Network Abstractions for Characterizing Communication Requirements in Asynchronous Distributed Systems

Hugo Rincon Galeana[(✉)] and Ulrich Schmid

TU Wien, Vienna, Austria
hugorincongaleana@gmail.com, s@ecs.tuwien.ac.at

Abstract. Whereas distributed computing research has been very successful in exploring the solvability/impossibility border of distributed computing problems like consensus in representative classes of computing models with respect to model parameters like failure bounds, this is not the case for characterizing necessary and sufficient communication requirements. In this paper, we introduce network abstractions as a novel approach for modeling communication requirements in asynchronous distributed systems. A network abstraction of a run is a sequence of directed graphs on the set of processes, where the i-th graph defines the potential message chains that may arise in the i-th portion of the run. Formally, it is defined via associating (potential) message sending times with the corresponding message receiving times in a message schedule. Network abstractions allow to reason about the future causal cones that might arise in a run, hence also facilitate reasoning about liveness properties, and are inherently compatible with temporal epistemic reasoning frameworks. We demonstrate the utility of our approach by providing necessary and sufficient network abstractions for solving the canonical firing rebels with relay (FRR) problem, and variants thereof, in asynchronous systems with up to f byzantine processes. FRR is not only a basic primitive in clock synchronization and consensus algorithms, but also integrates several distributed computing problems, namely triggering events, agreement and even stabilizing agreement, in a single problem instance.

Keywords: Dynamic Networks · Byzantine Fault Tolerance · Asynchronous Systems · Graph Sequences · Causal Cones

A substantial part of the existing distributed computing research is devoted to the question of when some distributed computing problem, say, consensus [20], is solvable in a given model of computation. Impossibility results, like the celebrated FLP consensus impossibility in asynchronous systems where just one

---

H. R. Galeana—This work was supported by the FWF projects ByzDEL (P33600) and DMAC (P32431), and the Doctoral College *Resilient Embedded Systems*, which is run jointly by the TU Wien's Faculty of Informatics and the UAS Technikum Wien.

Y. Emek (Ed.): SIROCCO 2024, LNCS 14662, pp. 501–506, 2024.
https://doi.org/10.1007/978-3-031-60603-8_29

process may crash [11], tell when this is not possible. Otherwise, solution algorithms, along with their correctness proofs, are usually provided.

Proving impossibility results is easier in weaker models of computation, where the adversary has more power. In addition, impossibility results a fortiori carry over to weaker computing models. For example, the FLP impossibility implies that consensus is also impossible in asynchronous systems where $f \geq 1$ processes may crash. Consequently, one usually tries to establish an impossibility result for a computing model that is as strong as possible. By contrast, designing and proving correct algorithms is easier in strong computing models, with the solutions remaining valid in stronger models only. Consequently, it is desirable to find algorithms for computing models that are as weak as possible. Sometimes, the respective attempts can be made matching, which allows one to precisely identify the possibility/impossibility border. For example, in synchronous systems with up to f byzantine faulty processes, it is known that $n = 3f + 1$ processes are necessary and sufficient for solving consensus [20].

Unfortunately, exploring this possibility/impossibility border with respect to communication requirements is much harder. On the one hand, there is a substantial body of distributed computing research devoted to information-theoretic communication complexity [18], which has been sparked by Yao's seminal work [31] on distributed function computation for two processes. A few classic examples, among many possible others, are symmetry breaking in chains and rings [8] and lower bounds for all pair shortest paths [16].

Consequently, most traditional research on complex distributed computing problems considered networks of processes that are fully connected by means of bidirectional links that are reliable [11] or fair-lossy [2]. In not fully-connected but still static network topologies, solving consensus with up to f byzantine processes requires a $2f + 1$-connected undirected communication graph [10], and some more involved connectivity constraints for directed communication graphs [27]. Time-varying communication graphs are considered in research on dynamic networks [17], where it is usually assumed that a message adversary [1] supplies the directed communication graph for every particular round. Both impossibility results and solution algorithms have been provided for a wide variety of different message adversaries, ranging from oblivious ones [7,29] to ones that eventually ensure a short period of stability only [30]. Albeit most of this research focuses on consensus, other distributed computing problems like k-set agreement [14,15] and stabilizing consensus [5,25] have been studied in such models as well.

Unfortunately, most solution algorithms depend heavily on the underlying system assumptions, which are often incomparable, in particular, in the case of message adversaries. Determining necessary and sufficient communication requirements for solving a given problem, which are reasonably independent of the underlying model, and thus facilitate a rating of the overall communication efficiency of a particular solution, appears rather hopeless. Indeed, besides the already mentioned connectivity results [10,27] for distributed systems with static communication topologies, we are only aware of very few attempts on characterizing the problem solvability/impossibility border for dynamic networks under message adversaries in general, which all rely on very abstract properties: In the case of consensus, Nowak, Schmid and Winkler [22] showed that the non-connectedness/connectedness of the topological space of the infinite admissible executions generated by a solution algorithm defines the border. In [24], Schmid, Schwarz and Winkler introduced message adversary simulations, which define a message adversary hierarchy and the notion of a strongest message adversary for solving a given distributed computing problem. In any case, such characterizations are way too abstract for distilling necessary and sufficient communication requirements out of those.

By contrast, exactly this is possible in the two-step approach advocated by Ben-Zvi and Moses in [3]. The authors considered the simple *ordered response* (OR) problem in a distributed computing model with time-bounded end-to-end communication here. Based on the epistemic analysis of the necessary knowledge for solving OR, the occurrence of a a communication pattern called a *centipede*, which generalizes the usual message chains in purely asynchronous systems, was shown to be necessary in every run.

In this paper,[1] we propose a novel approach for characterizing necessary and sufficient communication requirements for asynchronous message-passing systems: *network abstractions*. In a nutshell, a network abstraction of a run is a finite or infinite sequence $G_1, G_2, \ldots$ of directed graphs on the set of processes, which defines the *potential* causality possible in the run. The index in the graph sequence represents the progress of time, and G_i specifies the *potential* message chains, i.e., ones that *might* occur in the period of time represented by index i. Note carefully, though, that the i-th period of time at different processes need *not* be synchronized (which is in stark contrast to the traditional modeling of dynamic networks, see [13] for details). Technically, we accomplish this by associating (potential) message sending times with the corresponding message receiving times in a *message schedule*. Overall, network abstractions also allow one to reason about the *future* causal cones [19] that might arise in a run, hence also facilitate reasoning about liveness properties. Moreover, they are inherently compatible with temporal epistemic reasoning frameworks, see [13].

To showcase the utility of our approach, we identify necessary and sufficient network abstractions for a (weak) agreement problem called *firing rebels with relay* (FRR), and its simpler variant *firing rebels* (FR), in asynchronous systems with up to f byzantine processes, which has been introduced and epistemically analyzed by Fruzsa, Kuznets and Schmid in [12]. FRR requires all correct agents to execute an action FIRE in an all-or-nothing fashion when sufficiently many agents learned about the occurrence of sufficiently many START events at correct processes. Informally, its specification comprises the following three properties:

(C) *Correctness*: If sufficiently many processes observed START, then all correct processes must execute FIRE eventually.

(U) *Unforgeability*: If a correct process executes FIRE, then START was observed by a correct process.

---

[1] An Arxiv version of this paper may be consulted at [13].

(R) *Relay*: If a correct process executes FIRE, then eventually all correct processes execute FIRE.

For FR, only (C) and (U) need to hold. The FRR problem is particularly suitable
for our purpose, since it is a canonical problem that combines several interesting
aspects of distributed computing problems in a single instance: It is not only
closely related to the firing squad problem [4,6], which is a weak terminating
agreement problem, but is also the core of an information-less version of the
pivotal *consistent broadcasting* (CB) primitive introduced by Srikanth and Toueg
in [26]. CB has been used in fault-tolerant clock synchronization [23,26,28], in
byzantine synchronous consensus [9,26], and (in a slightly extended form) in the
simulation of crash-prone protocols in byzantine settings proposed in [21]. FRR
itself is relevant in practice for (almost) simultaneously triggering an event at
different processes, and is interesting theoretically also because of its relation
to stabilizing consensus [5], namely, in the case where no correct process ever
executes FIRE.

Summary of our Contributions:

- A novel framework that enables us to represent the potential communication
  in an asynchronous message-passing system as sequences of directed graphs
  called network abstractions.
- A characterization of the necessary and sufficient network abstractions for
  solving FR and FRR.
- Novel asynchronous protocols for FR and FRR that, unlike the existing solutions, also work correctly under the necessary and sufficient network abstractions identified above.

More concretely, our main results are stated as follows:

Theorem 1 (FR solvability). *FR is solvable in a byzantine asynchronous
system with $n \geq 2f + 1$ processes controlled by an adaptive or non-adaptive
adversary, iff every run is governed by a communication abstraction $ST \cdot G$,
where ST is a START event graph with participating set S_{ST} of processes in ST
such that either (i) $|S_{ST}| < 2f + 1$ and G is arbitrary or (ii) there is a subset
$S \subseteq S_{ST}$ that is a $2f + 1$ strong root of G.*

Theorem 2 (FRR solvability). *FRR is solvable in a byzantine asynchronous
distributed system with $n \geq 3f + 1$ processes controlled by an adaptive or non-
adaptive adversary, iff any run r is governed by a communication abstraction
based on a dynamic network abstraction $\mathcal{G} = G^{\omega}$, where G is a $2f + 1$-strongly
vertex-connected graph that is also $(2f + 1, f + 1, f)$-co-rooted.*

Overall, our findings demonstrate that our novel approach indeed allows one to
identify necessary and sufficient *time-evolving* network connectivity conditions
for solving a given distributed computing problem in a byzantine asynchronous
system, and to develop optimal solution algorithms.

References

1. Afek, Y., Gafni, E.: Asynchrony from synchrony. In: Frey, D., Raynal, M., Sarkar, S., Shyamasundar, R.K., Sinha, P. (eds.) ICDCN 2013. LNCS, vol. 7730, pp. 225–239. Springer, Heidelberg (2013). https://doi.org/10.1007/978-3-642-35668-1_16
2. Basu, A., Charron-Bost, B., Toueg, S.: Crash failures vs. crash + link failures. In: PODC 1996, p. 246. ACM Press, Philadelphia (1996). https://doi.org/10.1145/248052.248102
3. Ben-Zvi, I., Moses, Y.: Beyond Lamport's happened-before: on time bounds and the ordering of events in distributed systems. J. ACM **61**(2), 13:1–13:26 (2014). https://doi.org/10.1145/2542181, http://doi.acm.org/10.1145/2542181
4. Charron-Bost, B., Moran, S.: The firing squad problem revisited. Theoret. Comput. Sci. **793**, 100–112 (2019). https://doi.org/10.1016/j.tcs.2019.07.023
5. Charron-Bost, B., Moran, S.: Minmax algorithms for stabilizing consensus. Distrib. Comput. **34**(3), 195–206 (2021). https://doi.org/10.1007/s00446-021-00392-9
6. Coan, B.A., Dolev, D., Dwork, C., Stockmeyer, L.: The distributed firing squad problem. In: STOC 1985, pp. 335–345. ACM, New York (1985). https://doi.org/10.1145/22145.22182
7. Étienne Coulouma, Godard, E., Peters, J.: A characterization of oblivious message adversaries for which consensus is solvable. Theor. Comput. Sci. **584**, 80–90 (2015). https://doi.org/10.1016/j.tcs.2015.01.024, https://www.sciencedirect.com/science/article/pii/S0304397515000602
8. Dinitz, Y., Moran, S., Rajsbaum, S.: Bit complexity of breaking and achieving symmetry in chains and rings. J. ACM **55**(1), 3:1–3:28 (2008). https://doi.org/10.1145/1326554.1326557
9. Dwork, C., Lynch, N., Stockmeyer, L.: Consensus in the presence of partial synchrony. J. ACM **35**(2), 288–323 (1988)
10. Fischer, M.J., Lynch, N., Merritt, M.: Easy impossibility proofs for distributed consensus problems. Distrib. Comput. **1**(1), 26–39 (1986)
11. Fischer, M.J., Lynch, N.A., Paterson, M.S.: Impossibility of distributed consensus with one faulty process. J. ACM **32**(2), 374–382 (1985)
12. Fruzsa, K., Kuznets, R., Schmid, U.: Fire! In: TARK 2021. EPTCS, vol. 335, pp. 139–153 (2021). https://doi.org/10.4204/EPTCS.335.13
13. Galeana, H.R., Schmid, U.: Network abstractions for characterizing communication requirements in asynchronous distributed systems (2023). https://doi.org/10.48550/arXiv.2310.12615
14. Rincon Galeana, H., Winkler, K., Schmid, U., Rajsbaum, S.: A topological view of partitioning arguments: reducing k-set agreement to consensus. In: Ghaffari, M., Nesterenko, M., Tixeuil, S., Tucci, S., Yamauchi, Y. (eds.) SSS 2019. LNCS, vol. 11914, pp. 307–322. Springer, Cham (2019). https://doi.org/10.1007/978-3-030-34992-9_25
15. Godard, E., Perdereau, E.: k-set agreement in communication networks with omission faults. In: OPODIS 2016. LIPICS, vol. 70, pp. 8:1–8:17. Schloss Dagstuhl - Leibniz-Zentrum für Informatik (2016). https://doi.org/10.4230/LIPIcs.OPODIS.2016.8
16. Holzer, S., Wattenhofer, R.: Optimal distributed all pairs shortest paths and applications. In: PODC 2012, pp. 355–364. ACM, New York (2012). https://doi.org/10.1145/2332432.2332504
17. Kuhn, F., Oshman, R.: Dynamic networks: models and algorithms. SIGACT News **42**(1), 82–96 (2011)

18. Kushilevitz, E., Nisan, N.: Communication Complexity. Cambridge University Press (1997). https://books.google.at/books?id=yiV6pwAACAAJ

19. Kuznets, R., Prosperi, L., Schmid, U., Fruzsa, K.: Causality and epistemic reasoning in byzantine multi-agent systems. In: TARK 2019. EPTCS, vol. 297, pp. 293–312. Open Publishing Association (2019). https://doi.org/10.4204/EPTCS.297.19

20. Lamport, L., Shostak, R., Pease, M.: The byzantine generals problem. ACM Trans. Program. Lang. Syst. **4**(3), 382–401 (1982). https://doi.org/10.1145/357172.357176

21. Mendes, H., Tasson, C., Herlihy, M.: Distributed computability in byzantine asynchronous systems. In: STOC 2014, pp. 704–713. ACM, New York (2014). https://doi.org/10.1145/2591796.2591853

22. Nowak, T., Schmid, U., Winkler, K.: Topological characterization of consensus under general message adversaries. In: PODC 2019, pp. 218–227. ACM, New York (2019).https://doi.org/10.1145/3293611.3331624

23. Robinson, P., Schmid, U.: The asynchronous bounded-cycle model. Theoret. Comput. Sci. **412**(40), 5580–5601 (2011). https://doi.org/10.1016/j.tcs.2010.08.001

24. Schmid, U., Schwarz, M., Winkler, K.: On the strongest message adversary for consensus in directed dynamic networks. In: SIROCCO 2018, pp. 102–120 (2018).https://doi.org/10.1007/978-3-030-01325-7_13

25. Schwarz, M., Schmid, U.: Round-oblivious stabilizing consensus in dynamic networks. In: Johnen, C., Schiller, E.M., Schmid, S. (eds.) SSS 2021. LNCS, vol. 13046, pp. 154–172. Springer, Cham (2021). https://doi.org/10.1007/978-3-030-91081-5_11

26. Srikanth, T.K., Toueg, S.: Simulating authenticated broadcasts to derive simple fault-tolerant algorithms. Distrib. Comput. **2**(2), 80–94 (1987). https://doi.org/10.1007/BF01667080

27. Tseng, L., Vaidya, N.H.: Fault-tolerant consensus in directed graphs. In: PODC 2015, pp. 451–460. ACM, New York (2015). https://doi.org/10.1145/2767386.2767399

28. Widder, J., Schmid, U.: The Theta-model: achieving synchrony without clocks. Distrib. Comput. **22**(1), 29–47 (2009). https://doi.org/10.1007/s00446-009-0080-x, http://www.vmars.tuwien.ac.at/documents/extern/1724/paper.pdf

29. Winkler, K., Paz, A., Rincon Galeana, H., Schmid, S., Schmid, U.: The time complexity of consensus under oblivious message adversaries. In: ITCS 2023, vol. 251, pp. 100:1–100:28. Schloss Dagstuhl – Leibniz-Zentrum für Informatik, Dagstuhl, Germany (2023). https://doi.org/10.4230/LIPIcs.ITCS.2023.100, https://drops.dagstuhl.de/opus/volltexte/2023/17603

30. Winkler, K., Schwarz, M., Schmid, U.: Consensus in directed dynamic networks with short-lived stability. Distrib. Comput. **32**(5), 443–458 (2019). https://doi.org/10.1007/s00446-019-00348-0

31. Yao, A.C.: Some complexity questions related to distributive computing (preliminary report). In: STOC 79, 30 April–2 May 1979, Atlanta, Georgia, USA, pp. 209–213 (1979). https://doi.org/10.1145/800135.804414, http://doi.acm.org/10.1145/800135.804414

On the Existence of Consensus Converging Organized Groups in Large Social Networks

Vasiliki Liagkou[1,4], Panagiotis E. Nastou[5,6], Paul Spirakis[2],
and Yannis C. Stamatiou[3(✉)]

[1] Computer Technology Institute and Press - "Diophantus",
University of Patras Campus, 26504 Patras, Greece
liagkou@cti.gr
[2] Department of Computer Science, University of Liverpool, Liverpool, UK
P.Spirakis@liverpool.ac.uk
[3] Business Administration Department, University of Patras, 26504 Patras, Greece
stamatiu@upatras.gr
[4] Department of Informatics and Telecommunications, University of Ioannina,
Kostakioi Arta, 47100 Ioannina, Greece
[5] Department of Mathematics, University of the Aegean, Applied Mathematics and
Mathematical Modeling Laboratory, Samos, Mytilene, Greece
pnastou@aegean.gr
[6] Center for Applied Optimization, University of Florida, Gainesville, USA

Abstract. In this paper we investigate the emergence of highly organized communities in graphs modeling social networks and interactions among their members. We show that the formation of large organized communities requires exponentially large social networks. Our approach is based on Kolmogorov complexity of graphs represented as finite size strings of bits. We apply this approach to the problem of existence of organized communities which have the structure of the Zig-Zag class of graphs, with information diffusion properties. We provide a lower bound for the number of nodes a social network must have to allow the formation of a large Zig-Zag like community and conditions upon which a large Zig-Zag graph can be located in sufficiently large social networks by providing a randomized linear algorithm which locates this structure.

Keywords: Kolmogorov Complexity · Organized Structures · Social Networks

1 Introduction

In this paper, we address the problem of the existence and efficient identification of certain configurations, termed *highly organized structures* in our context, in networks of agents based on *Kolmogorov Complexity* (see [6]) and *Random Graphs*. We apply our approach to the class of *Zig-Zag* outerplanar graphs which

© The Author(s), under exclusive license to Springer Nature Switzerland AG 2024
Y. Emek (Ed.): SIROCCO 2024, LNCS 14662, pp. 507–512, 2024.
https://doi.org/10.1007/978-3-031-60603-8_30

have the important property of being *close-knit* (see in [10]). We also show how to locate a large Zig-Zag structure, which is also a close-knit graph, in a random graph giving a linear time probabilistic algorithm. Thus, we provide, to the best of our knowledge, the first (randomized) algorithm for locating a large close-knit structure in a random graph. In our recent work in [7], we applied this approach to Sierpinski Graphs and focused on the study of the Ramsey Numbers of incompressible graphs containing regular substructures.

2   Graph Connectivity and Fast Diffusion of Information

Close-knit agent structures, as showed in [10], play an important role in *diffusing* throughout society ideas and beliefs that are, initially, held by only a relatively small group of interacting agents. A group of nodes S in a given graph G is *r-close-knit* if the following condition is true for every non empty $S' \subseteq S$,

$$\min_{S' \subseteq S} \frac{d(S', S)}{\sum_{i \in S'} d_i} \geq r \tag{1}$$

where $d(S', S)$ is the internal connections of S' i.e. the number of links $\{i, j\}$ where $i, j \in S'$ while d_i is the degree of i. Given a positive integer k and a real number $0 < r \leq 1/2$, we call a graph G (r, k)-close-knit if every agent belongs to a group S of cardinality at most k which is r-close-knit as defined by Inequality (1). Finally, a *class* of graphs is *close-knit* if for every $0 < r \leq 1/2$ there exists an integer k such that every graph in the class is (r, k)-close-knit. Young proved that for classes of close-knit agent graphs *all* community members will eventually adopt the innovation in a number of interactions which is bounded independently from the size of community. Now, we describe the class of Zig-Zag outerplanar graphs, a *close-knit graph family* with good information diffusion properties (see [8]). First, we define k-outerplanar graphs and some basic concepts.

Definition 1 (k-Outerplanar Graph). *For $k \geq 2$, an embedding of a graph G is k-outerplanar, if it is planar and when all vertices on the outer face are deleted, then one obtains a $(k-1)$-outerplanar embedding of the resulting graph. If G has a k-outerplanar embedding, then it is called a k-outerplanar graph.*

A maximal outerplanar graph (*MOP graph*) is a graph such that the addition of an edge results in a *non-outerplanar* graph ([4]). In MOPs the maximum degree of the vertices equals 4, regardless of the network's size. A vertex x_i of a graph G is called an ear if the link $\{x_{(i-1)}, x_{(i+1)}\}$ that bridges x_i lies entirely in the graph.

Definition 2 (Θ-primitive graph). *The Θ-primitive graph is a graph of four vertices formed by a cycle graph of four vertices in which one of the two non-adjacent pairs of vertices is connected with an edge.*

Definition 3 (Zig-Zag, $k > 4$). Z_k *is obtained from* Z_{k-1} *by connecting a new vertex to the vertices of degree 2 and 3 of one of the two ears of* Z_{k-1}. *The new vertex along with the vertex of degree 2 of the selected ear of* Z_{k-1} *is one of the ears of* Z_k. *The other is the same ear of* Z_{k-1}.

In [8], the following was proved:

Theorem 1. *The family of Zig-Zag Graphs is close-knit.*

3 Emergence of Organized Structures in Social Networks

In this section, we investigate the conditions on which highly organized Zig-Zag outerplanar subgraphs appear in sufficiently large networks of interacting entities. The *Kolmogorov Complexity* of a (binary) string x, denoted by $C(x)$, is the *length* of the *shortest algorithmic description* of x (see [3,5,9]). Similarly, the *conditional* Kolmogorov Complexity of x given y, denoted by $C(x|y)$, is the length of the shortest program which produces x as output given y as input. Our focus is on finite strings encoding graphs (see [6]):

Definition 4. *Each labeled graph G on n nodes can be represented by a binary string $E(G)$ of length $\binom{n}{2}$. We assume a fixed ordering of the $\binom{n}{2}$ possible edges in the graph, e.g. lexicographically, and let the ith bit in the string indicate presence (1) or absence (0) of the ith edge.*

Definition 5. *Let a labeled graph G on n nodes and a labeled graph H on k nodes. Each subset of k nodes of G induces a subgraph G_k of G which is an ordered labeled* occurrence *of H if it is isomorphic to H.*

Definition 6. *A labeled graph G on n nodes has* randomness deficiency *at most $\delta(n)$, and is called $\delta(n)$-*random *or $\delta(n)$-*incompressible, *if it satisfies*

$$C(E(G)|n) \geq \binom{n}{2} - \delta(n). \tag{2}$$

If $\delta(n) = 0$, we will call G random or incompressible. In [6], the following is proved (Theorem 2.6.2):

Theorem 2. *Let x of length n satisfy $C(x) \geq n - \delta(n)$. Then for sufficiently large n, each block y of length $l = \log n - \log \log n - \log(\delta(n) + \log n) - O(1)$ occurs at least once in x.*

We prove the following along the ideas leading to a similar incompressibilty result for cliques stated in [6]:

Theorem 3. *Let Z_k be a labeled Zig-Zag outerplanar graph on k vertices and G be a labeled incompressible graph on n vertices that contains Z_k as an induced subgraph. Then $n \geq 2^{\frac{k-1}{2}}$.*

Proof. Let G to be a labeled *incompressible* graph with n vertices whose encoding $E(G)$ has length $l(E(G))$. Since G is incompressible, it holds

$$C(E(G)|n, P) \geq \frac{n(n-1)}{2} \tag{3}$$

where P is a program that can reconstruct $E(G)$ from the value n and an alternative encoding $E'(G)$ we will describe below. From Theorem 2, $E(G)$ contains all possible strings of the size l and, thus, a substring for Z_k as an *induced* subgraph of G for appropriate values of k that satisfy, in conjunction with n, the expression for l. Then, we can describe G in the following alternative encoding $E'(G)$. We add to $E(G)$ the description of the Zig-Zag outerplanar subgraph of G. For that description, we need $\log \binom{n}{k}$ bits to denote the subset of the k vertices and $\log k!$ bits to denote the specific ordering $i_1 i_2 \ldots i_{k-1} i_k$ where i_1, i_k are the "ear" vertices and i_2, i_{k-1} are the vertices of degree 3 adjacent to vertices i_1 and i_k respectively while the rest of the vertices follow the "ziz-zag" ordering. So, we need at most $\log \binom{n}{k} + \log k! \leq \log \frac{n^k}{k!} + \log k! = k \log n$ bits. We delete, from $E(G)$, *all* the bits that encode the edges of the Zig-Zag subgraph, saving $\frac{k(k-1)}{2}$ bits. The length of the new encoding is

$$l(E'(G)) \leq l(E(G)) + k \log n - \frac{k(k-1)}{2}. \tag{4}$$

Given the value of n, the program P can reconstruct $E(G)$ from $E'(G)$,

$$C(E(T)|n, P) \leq l(E'(T)). \tag{5}$$

Since G is incompressible $l(E'(G)) \geq l(E(G))$, which from (4), holds if

$$k \log n \geq \frac{k(k-1)}{2} \Leftrightarrow n \geq 2^{\frac{k-1}{2}}. \tag{6}$$

Now we will examine the problem of locating a large Zig-Zag outerplanar subgraph in *randomly* evolving graph structures. For $n \in \mathbb{N}$ and $0 \leq p \leq 1$ we denote by $\mathcal{G}_{n,p}$ the random graph model for the generation of a binomial random graph on n vertices where every edge appears with probability p independently of all other edges (see [1]). One of the most central questions in random graph theory is to find a *threshold function* for a graph property $\mathcal{P}$. A threshold function is a function $p^* : \mathbb{N} \to [0, 1]$ ensuring that $G(n, p)$ satisfies $\mathcal{P}$ when $p = \omega(p^*)$ and does not satisfy $\mathcal{P}$ when $p = o(p^*)$ with probability tending to 1 as $n \to \infty$. Based on the approach in [1], we show that in an n vertex random graph G, we can define conditions according to which there exists in G a Zig-Zag outerplanar subgraph of linear, in n, size. We provide a randomized algorithm that succeeds in locating a linear size subgraph of a random graph with high probability. For balanced graphs, the following result holds (see [2]):

Definition 7 (Balanced graphs). *Let H be a graph with v vertices and e edges. We call $\rho(H) = e/v$ the density of H. We call H balanced if every subgraphs H' has $\rho(H') \leq \rho(H)$. We call H strictly balanced if every proper subgraph H' has $\rho(H') < \rho(H)$.*

Theorem 4 (Threshold functions for balanced graphs). *Let H be a balanced graph with v vertices and e edges and $A(G)$ be the event that H is a subgraph of a random graph G in the $\mathcal{G}_{n,p}$ random graph model. Then $p = n^{-v/e}$ is the threshold function for $A(G)$.*

It is not hard to check that the Θ-primitive graph is balanced. Therefore the following holds, according to Theorem 4:

Corollary 1. *The threshold function for the appearance of a Θ-primitive subgraph in a random graph of the $\mathcal{G}_{n,p}$ random graph model is $p = n^{-4/5}$.*

This means that a Θ-primitive graph appears almost certainly in a random graph in the $\mathcal{G}_{n,p}$ random graph model if $p = \omega(n^{-4/5})$. We now prove the following with respect to the appearance of a large Zig-Zag outerplanar subgraph in a random graph in the $\mathcal{G}_{n,p}$ random graph model:

Theorem 5. *For the $\mathcal{G}_{n,p}$ random graph with for $p = \frac{c}{n^\alpha}$, $c > 0$, for any positive real number $\alpha < 1/2$ there exists a Zig-Zag outerplanar subgraph in $\mathcal{G}_{n,p}$ of size αn, with high probability.*

Proof. We provide an algorithm that succeeds, with high probability, in constructing a large (i.e. of linear size) Zig-Zag outerplanar subgraph of size αn of a graph G in $\mathcal{G}_{n,p}$ for $p = \frac{c}{n^\alpha}$, with $c > 0$ and $\alpha < 1/2$. The variables x_2 and x_3 contain the vertices of degree 2 and 3 respectively of the ear to be extended with the adition of a new vertex.

Step 1: Construct a Θ-primitive subgraph of G involving four vertices of the graph, say $v_{i_1}, v_{i_2}, v_{i_3}$, and v_{i_4} - wlog, assume that v_{i_1} and v_{i_4} are the two vertices of degree 2.

Step 2: Set $x_2 = v_{i_4}$ and $x_3 = v_{i_3}$, and fix an ordering $v_{i_5}, v_{i_5}, \ldots, v_{i_n}$ of the rest of the vertices. Also, set $k \leftarrow 5$.

Step 3: repeat:

If vertex v_{i_k} is adjacent to *both* x_2 and x_3 then include v_{i_k} in the current subgraph and set $x_3 \leftarrow x_2$, $x_2 \leftarrow x_{i_k}$, $k \leftarrow k + 1$.

else set $k \leftarrow k + 1$.

until $k > n$.

4   Conclusions

We considered *Kolmogorov Complexity* for studying the concept of complexity of *societies and their interacting agents*. This theory leads to conditions upon the existence of certain substructures in large structures such as networks of interacting agents. We study the appearance of regular structures in societies with certain desirable properties such as *close-knittedness* and we prove that the Zig-Zag graphs cannot emerge in graphs unless they reach a sufficient size, exponential in the size of the structure.

References

1. Bollobás, B.: Random Graphs, 2nd edn. Cambridge University Press, Cambridge (2001)
2. Bollobás, B., Frieze, A.M.: Spanning maximal planar subgraphs of random graphs. Random Struct. Algor. **2**(2), 225–231 (1991)
3. Chaitin, G.H.: On the length of programs for computing finite binary sequences. J. ACM **1k**, 547–570 (1966)
4. Claverol, M., et al.: Metric dimension of maximal outerplanar graphs. Bull. Malay. Math. Sci. Soc. **44**, 2603–2630 (2021)
5. Kolmogorov, A.N.: Three approaches to the quantitative definition of information. Prob. Inf. Trans. **1**(1), 1–7 (1965)
6. Li, M., Vitányi, P.M.B.: An Introduction to Kolmogorov Complexity and its Applications, 4th edn. Springer, New York (2019). https://doi.org/10.1007/978-0-387-49820-1
7. Liagkou, V., Nastou, P.E., Spirakis, P., Stamatiou, Y.C.: On the existence of highly organized communities in networks of locally interacting agents. In: Proceedings 7th International Symposium on Cyber Security, Cryptology and Machine Learning (CSCML 2023) (2023)
8. Manolopoulos, C., Stamatiou, Y.C., Efstathiadou, R.: Globally emergent behavioral patterns as a result of local interactions in strongly interrelated individuals. In: Proceedings 29th Eurasian Business and Economics Society (EBES) Conference, Lisbon (2019)
9. Solomonoff, R.J.: A formal theory of inductive inference. Inf. Control **7**(1), 1–22 (1964)
10. Young, H.P.: The diffusion of innovations in social networks. In: Blume, L.E., Durlauf, S.N. (eds.) Proceedings of the Economy as a Complex Evolving System, vol. III. Oxford University Press, Oxford (2003)

Multi-agent Online Graph Exploration on Cycles and Tadpole Graphs

Erik van den Akker$^{(\boxtimes)}$ [ID], Kevin Buchin [ID], and Klaus-Tycho Foerster [ID]

TU Dortmund, Dortmund, Germany
erik.vandenakker@tu-dortmund.de

Abstract. We study the problem of multi-agent online graph exploration, in which a team of k agents has to explore a given graph, starting and ending on the same node. The graph is initially unknown. Whenever a node is visited by an agent, its neighborhood and adjacent edges are revealed. The agents share a global view of the explored parts of the graph. The cost of the exploration has to be minimized, where cost either describes the time needed for the entire exploration (time model), or the length of the longest path traversed by any agent (energy model). We investigate graph exploration on cycles and tadpole graphs for 2–4 agents, providing optimal results on the competitive ratio in the energy model (1-competitive with two agents on cycles and three agents on tadpole graphs), and for tadpole graphs in the time model (1.5-competitive with four agents). We also show competitive upper bounds of 2 for the exploration of tadpole graphs with three agents, and 2.5 for the exploration of tadpole graphs with two agents in the time model.

Keywords: Graph Exploration · Cycles · Tadpole Graphs · Agents

1 Introduction

In the Online Graph Exploration problem, all nodes of a given weighted graph $G = (V, E, l)$ have to be visited by an agent a, starting and ending on a starting node $s \in V$. The graph is initially unknown, except for s and its neighborhood $N(s)$. Whenever the agent reaches a yet unvisited node v, all adjacent edges and $N(v)$ are revealed. Nodes can be distinguished, but the labels do not provide any information about the graph to the agent. We consider graph exploration with $k \in \mathbb{N}$ agents $a_1, ..., a_k$, for which we assume unlimited computational power and shared knowledge, meaning that as soon as any agent learns something about the neighborhood of a node, all other agents instantly receive the same information. A graph is considered explored, when each node in V has been visited by any agent, and all agents have returned to s. All agents move at identical speed, taking time $l(e)$ for traversing some edge e. An agent can decide to wait at a node, while other agents are traversing edges. The goal of a graph exploration strategy is to minimize the cost of the exploration.

© The Author(s), under exclusive license to Springer Nature Switzerland AG 2024
Y. Emek (Ed.): SIROCCO 2024, LNCS 14662, pp. 513–519, 2024.
https://doi.org/10.1007/978-3-031-60603-8_31

Table 1. The new bounds for the exploration of cycles and tadpole graphs.

Graph Class (# of Agents)	Time Model		Energy Model	
	Lower Bound	Upper Bound	Lower Bound	Upper Bound
Cycles (2 Agents)	1.5 [10]	1.5 [10]	1	1 (Thm. 2)
Tadpole Graphs (2 Agents)	1.5 (Thm. 3)	2.5 (Thm. 5)	1.5 (Thm. 4)	2.5 (Thm. 5)
Tadpole Graphs (3 Agents)	1.5 (Thm. 3)	2 (Thm. 6)	1	1 (Thm. 6)
Tadpole Graphs (4+ Agents)	1.5 (Thm. 3)	1.5 (Thm. 7)	1	1 (Thm. 6)

This problem is widely used to describe the exploration of unknown terrain by multiple autonomous robots. Some variations focus on the general exploration time [3,8,10] (here called the **Time Model**), while others focus on the maximum distance traversed by any individual agent, modeling the maximum energy consumption of a single robot [5,6] (here called the **Energy Model**). We consider both models for the cost of an exploration.

In the single-agent case, many graph classes like directed graphs, tadpole graphs or cactus graphs have been investigated [2,7,9], while the main focus of the literature in the case of multi-agent exploration lies on general graphs and trees [3–5,8], with the exception of cycles (in the time model) and $n \times n$-grid graphs [10,11]. We consider cycles (in the energy model) and tadpole graphs, which consist of a cycle and a path attached to one of its nodes, called the tail. An overview of our results is given in Table 1.

Online algorithms are analyzed in terms of their competitive ratios, which describe the relationship between the costs of online and optimal solutions. Note that a c-competitive exploration strategy for the time model is also at most c-competitive for the energy model, since in an exploration that takes time $\mathcal{T}$, no agent can traverse a distance greater than $\mathcal{T}$. In the optimal case the maximum distance traversed by any agent matches the general exploration time.

In the following Sections we provide an overview of our results on Cycles in Sect. 2 and Tadpole Graphs in Sect. 3. A detailed version of this BA is provided on arXiv [1].

2   Cycles

Preliminaries. We call the point on a cycle with exactly distance $L_c/2$ in both directions to the starting point s, the midpoint m of the cycle. If m falls onto an edge, this edge is called e_{mid}. If m falls onto a node, the node is called v_{mid}. In this case we consider $l(e_{mid}) = 0$. We consider $e_{mid} = (v_l, v_s)$. If $e_{mid} = (v_l, v_s)$ does not exist, we consider $v_l = v_s = v_{mid}$. We denote p_l and p_s with length d_l and d_s as edge-disjoint paths from s to v_l and v_s respectively, where p_l and p_s combined contain all edges of the cycle except (v_l, v_s). We always assume w.l.o.g that $d_s \leq d_l$. An example is given in Fig. 1a.

Higashikawa et al. presented the *ALE (avoid longest edge) Algorithm*, which is a greedy 1.5-competitive online algorithm for the exploration of cycles in the

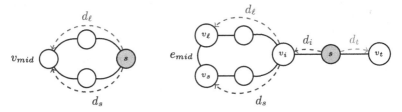

(a) Cycle example. p_ℓ and p_s with (b) Tadpole Graph example. lengths d_ℓ, d_s and v_{mid} are marked. d_ℓ, d_s, d_i, d_t and e_{mid} are marked.

Fig. 1. Examples of a cycle (a) and a tadpole graph (b).

time model [10], that sends two agents in different directions and always chooses the shorter edge. We first show that *ALE* is 1.5-competitive in the energy model and then present a modification we call the *AMP (avoid midpoint) Algorithm,* reaching a competitive ratio of 1 in the energy model, while still being 1.5-competitive in the time model. For proving this we make use of an observation made by Higashikawa et al. [10]: an optimal strategy for exploring cycles with 2 Agents takes time $2d_\ell$.

Theorem 1 (The *ALE* algorithm and the energy model). *The ALE algorithm has a competitive ratio of* 1.5 *in the energy model on cycles.*

Proof. Consider the graph shown in Fig. 2a. Let $0.5 > \varepsilon > 0$. Since (s, x_2) is the longest edge in the graph, one agent explores the graph clockwise from s to x_2. After the agent reaches x_2 the shortest path to s is the edge (x_2, s), leading to a cost of 3. An optimal strategy would have sent both agents from s to x_1 and x_2 respectively, leading to an exploration cost of $2(1 + \varepsilon)$. The overhead of ALE in this case is $\frac{3}{2(1+\varepsilon)}$, leading to a lower bound of 1.5. Since ALE is 1.5-competitive in the time model, the upper bound of 1.5 for the energy model follows.

Theorem 2 (AMP on Cycles). *For the energy model, the AMP (Avoid Midpoint) Algorithm 1 explores a cycle with a competitive ratio of* 1. *For the time model the algorithm explores a cycle with a competitive ratio of* 1.5.

Proof. We can prove the 1-competitiveness of the *AMP algorithm* for the energy model, by showing that the agents only traverse the cycle to v_s and v_ℓ (or v_{mid}) before returning, and never traverse the edge e_{mid}. We then use this result to prove the 1.5-competitive ratio for the time model.

Energy Model. If both agents reach v_{mid} at the same time, all nodes have been visited and both agents have traversed a distance of $L/2$ before and L after backtracking, which matches the result of the optimal offline strategy.

If an agent a_1 reaches v_{mid} first, having traversed a distance $d_1 = L/2$, the other agent a_2 must have traversed a distance $d_2 < d_1$. Thus, a_2 moves to v_{mid}

Algorithm 1. AMP (Avoid Midpoint)

Require: A unknown cycle graph $G = (V, E)$, Two agents a_1, a_2, a starting node $s \in V$

    $d(a_1) \leftarrow 0; d(a_2) \leftarrow 0$ ▷ Agents track their already traversed distance

    $Exp \leftarrow \{s\}$ ▷ Keep track of already explored vertices

    Let $n(a_1)$ and $n(a_2)$ describe the next nodes seen by the agents

    Assign the two neighbors of s randomly to $n(a_1)$ and $n(a_2)$

    while $n(a_1) \notin Exp \lor n(a_2) \notin Exp$ **do**   ▷ While graph is not explored

        if $d(a_1) + l(n(a_1)) < d(a_2) + l(n(a_2))$ **then**

            a_1 traverses edge to $n(a_1)$

            $Exp \leftarrow Exp \cup \{n(a_1)\}$

            $d(a_1) \leftarrow d(a_1) + l(n(a_1))$

            $n(a_1) \leftarrow$ next revealed node

        else

            a_2 traverses edge to $n(a_2)$

            $Exp \leftarrow Exp \cup \{n(a_2)\}$

            $d(a_2) \leftarrow d(a_2) + l(n(a_2))$

            $n(a_2) \leftarrow$ next revealed node

        end if

    end while

    a_1 and a_2 return to s using shortest paths.

---

without stopping. Then both agents have traversed a distance of $L/2$ when all nodes have been visited and L after backtracking.

Assume w.l.o.g an agent a_1 is currently located on v_s after having traversed distance d_s and has not yet traversed e_{mid}, while the other agent a_2 has not reached v_ℓ yet. Since $d_\ell < L/2 < d_s + l(e_{mid})$ the agent a_2 traverses to v_ℓ without stopping. At this point all nodes have been visited and the agents backtrack, having taken the same paths as they would have in an optimal offline exploration.

Time Model. Recall that an optimal strategy takes time $2d_\ell$. In the AMP algorithm, the agents traverse the distances d_s and d_ℓ. Since only one agent is traversing an edge at a time until all nodes have been visited, the maximum time needed to visit all nodes is $d_s + d_\ell \leq 2d_\ell$. After backtracking to s the full exploration time is at most $3d_\ell$, leading to a competitive ratio of 1.5.

3   Tadpole Graphs

Preliminaries. A tadpole graph contains exactly one node with degree 1, which we call the end of the tail v_t and exactly one node with degree 3, we call the intersection v_i. All other nodes of the graph have degree 2. We define v_s, v_ℓ, v_{mid} similar to cycles, but starting at v_i instead of s, when s is located on the tail of the tadpole graph. An Example is given in Fig. 1b. Note that any optimal exploration strategy on tadpole graphs takes at least time $2 \max(d(s, v_\ell), d(s, v_t))$.

We first prove that the lower bound of 1.5 holds on tadpole graphs for at least two agents in the time model, and for two agents in the energy model in

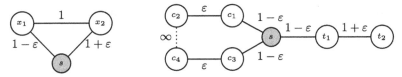

(a) Lower bound construction for ALE (energy model).

(b) Lower bound construction for Tadpole Graphs (energy model).

Fig. 2. Lower Bound constructions for: ALE and Tadpole Graphs (energy model).

Theorem 3 and Theorem 4. For the following Theorem 5–7, we roughly sketch the exploration strategies on tadpole graphs, and how they achieve the stated upper bounds for the given number of agents. A detailed descripton and analysis is provided on arXiv [1].

Theorem 3 (Time Model: Lower bound for exploring tadpole graphs). *For any number of agents $k \geq 2$, any online exploration strategy for tadpole graphs has a competitive ratio of at least 1.5 in the time model.*

Proof. We can adapt the cycle from the proof of Thm.2.2 by Higashikawa et al. [10], by adding a tail of length ε to the starting node. This does not influence the offline exploration time, since the path can always be explored by some agent in parallel and thus keeps the lower bound of 1.5 on tadpole graphs with any number of agents ≥ 2.

Theorem 4 (Energy Model: Lower bound for exploring tadpole graphs). *For $k = 2$ agents, no online exploration strategy on tadpole graphs can have a competitive ratio better than 1.5 in the energy model.*

Proof. We can adapt the proof idea from Lemma 7 by Dynia et al. [5], using a tree with three branches and connecting two branches with an arbitrarily long edge. The resulting tadpole graph is shown in Fig. 2b. Let $\varepsilon > 0$ be an arbitrary small and ∞ be an arbitrary large number. The maximum distance traversed by any agent online here is at least $6 - 2\varepsilon$, while an optimal exploration has length 4. This leads to a lower bound of 1.5.

Theorem 5 (Two-agent randomized tadpole graph exploration). *Using the AMP strategy with $k = 2$ agents for the exploration of tadpole graphs, choosing random paths as soon as the intersection is found leads to a competitive upper bound of 2.5.*

Using two agents, we randomly choose a direction as soon as the intersection is found and apply the AMP strategy on the chosen paths. As soon as the tail or the cycle are explored, one agent returns to s and explores the remaining path of the graph. This leads to an exploration time of at most $5 \max(d(s, v_\ell), d(s, v_t))$ and a competitive ratio of 2.5 in the time model, which also applies to the energy model.

Theorem 6 (Three agent tadpole graph exploration). *For $k = 3$ agents there exists an exploration strategy for tadpole graphs with a competitive ratio of 2 for the time-, and 1 for the energy model.*

Using three agents, when starting at the intersection one agent is sent in each direction. When starting on a node with degree 2, one agent waits at s until the intersection is found. Then each direction is assigned to an agent. At each step, only the agent with the shortest distance to s after the next edge, traverses the next edge. This leads to an exploration time of at most $4\max(d(s, v_\ell), d(s, v_t))$ and a competitive ratio of 2 in the time model. Since any agent only traverses one path and returns, this leads to a maximum distance of $2\max(d(s, v_\ell), d(s, v_t))$ and a competitive ratio of 1 in the energy model.

Theorem 7 (Four agent tadpole graph exploration). *Using $k = 4$ agents, a competitive ratio of 1.5 can be achieved on tadpole graphs in the time model.*

Using four agents, we send three teams containing one agent in each direction when starting on the intersection, and two teams of two agents in each direction when starting on a different node. When the intersection is first discovered, the team on the intersection splits into two single-agent teams. Only one team (the one that sees the largest distance to s) is waiting at any point in time. This leads to an exploration time of at most $3\max(d(s, v_\ell), d(s, v_t))$ and a competitive ratio of 1.5 in the time model.

References

1. van den Akker, E., Buchin, K., Foerster, K.: Multi-agent online graph exploration on cycles and tadpole graphs. CoRR **abs/2402.13845** (2024)
2. Brandt, S., Foerster, K., Maurer, J., Wattenhofer, R.: Online graph exploration on a restricted graph class: optimal solutions for tadpole graphs. Theor. Comput. Sci. **839**, 176–185 (2020)
3. Dereniowski, D., Disser, Y., Kosowski, A., Pajak, D., Uznanski, P.: Fast collaborative graph exploration. Inf. Comput. **243**, 37–49 (2015)
4. Disser, Y., Mousset, F., Noever, A., Skoric, N., Steger, A.: A general lower bound for collaborative tree exploration. Theor. Comput. Sci. **811**, 70–78 (2020)
5. Dynia, M., Korzeniowski, M., Schindelhauer, C.: Power-aware collective tree exploration. In: Grass, W., Sick, B., Waldschmidt, K. (eds.) ARCS 2006. LNCS, vol. 3894, pp. 341–351. Springer, Heidelberg (2006). https://doi.org/10.1007/11682127_24
6. Dynia, M., Łopuszański, J., Schindelhauer, C.: Why robots need maps. In: Prencipe, G., Zaks, S. (eds.) SIROCCO 2007. LNCS, vol. 4474, pp. 41–50. Springer, Heidelberg (2007). https://doi.org/10.1007/978-3-540-72951-8_5
7. Foerster, K., Wattenhofer, R.: Lower and upper competitive bounds for online directed graph exploration. Theor. Comput. Sci. **655**, 15–29 (2016)
8. Fraigniaud, P., Gasieniec, L., Kowalski, D.R., Pelc, A.: Collective tree exploration. Networks **48**(3), 166–177 (2006)

9. Fritsch, R.: Online graph exploration on trees, unicyclic graphs and cactus graphs. Inf. Process. Lett. **168**, 106096 (2021)

10. Higashikawa, Y., Katoh, N., Langerman, S., Tanigawa, S.: Online graph exploration algorithms for cycles and trees by multiple searchers. J. Comb. Optim. **28**(2), 480–495 (2014)

11. Ortolf, C., Schindelhauer, C.: Online multi-robot exploration of grid graphs with rectangular obstacles. In: SPAA 2012, pp. 27–36. ACM (2012)

Author Index

© The Editor(s) (if applicable) and The Author(s), under exclusive license
to Springer Nature Switzerland AG 2024
Y. Emek (Ed.): SIROCCO 2024, LNCS 14662, pp. 521–522, 2024.
https://doi.org/10.1007/978-3-031-60603-8